Springer Series in the Data Sciences

Springer Series in the Data Sciences focuses primarily on monographs and graduate level textbooks. The target audience includes students and researchers working in and across the fields of mathematics, theoretical computer science, and statistics. Data Analysis and Interpretation is a broad field encompassing some of the fastest-growing subjects in interdisciplinary statistics, mathematics and computer science. It encompasses a process of inspecting, cleaning, transforming, and modeling data with the goal of discovering useful information, suggesting conclusions, and supporting decision making. Data analysis has multiple facets and approaches, including diverse techniques under a variety of names, in different business, science, and social science domains. Springer Series in the Data Sciences addresses the needs of a broad spectrum of scientists and students who are utilizing quantitative methods in their daily research. The series is broad but structured, including topics within all core areas of the data sciences. The breadth of the series reflects the variation of scholarly projects currently underway in the field of machine learning.

Mayer Alvo

Statistical Inference and Machine Learning for Big Data

Springer

Mayer Alvo
Department of Mathematics and Statistics
University of Ottawa
Ottawa, ON, Canada

ISSN 2365-5674 ISSN 2365-5682 (electronic)
Springer Series in the Data Sciences
ISBN 978-3-031-06783-9 ISBN 978-3-031-06784-6 (eBook)
https://doi.org/10.1007/978-3-031-06784-6

Mathematics Subject Classification (2022): 62-00, 62Fxx, 62Gxx, 62Hxx, 62N99, 62M99

This Springer imprint is published by the registered company Springer Nature Switzerland AG
The registered company address is: Gewerbestrasse 11, 6330 Cham, Switzerland

Preface

This book grew from a desire to provide both students and practitioners a reference text, which describes several modern and useful methods in statistics. In recent years, the subject of statistics has expanded at an exponential rate. The volume of data presently available has imposed new demands on both theory and implementation, stretching the capacity of sophisticated high-speed computers. This book represents an attempt to exhibit some of the mainstream topics in the literature. In Part I, we display various types of data most often encountered in applications and cross-reference some of the tools most often used for the analysis. Readers may use this part to direct their attention to the corresponding tools later described in this book. In Part II, we provide a brief summary of the basics in probability and statistical inference. Readers may wish to refresh their knowledge by quickly perusing the relevant chapters. The important Pearson system of distributions is discussed in preparation for its application to microarray data and to approximations to optimal nonparametric rank tests. Estimation and hypothesis testing are described followed by an exposition of classical Bayesian methods. The subject of multivariate analysis appears in Chap. 4. It includes references and applications to principal component analysis, factor analysis, canonical correlation, linear discriminant analysis, and multidimensional scaling. Nonparametric statistics forms the basis of Chap. 5. Much is included in addition to the usual one and two sample tests of location and scale. Specifically, the unified theory of hypothesis testing presented is then applied to the problems of testing for location and the detection of umbrella alternatives. Also included in this chapter are methods for spatial analysis which are then applied on COVID-19 data. In Chap. 6, we introduce the topic of exponential tilting and its implications to Tweedie's formula. Here, we see the importance of the Pearson system of distributions. In Chap. 7, we discuss the subject of contingency tables. Classical time series methods form the subject of Chap. 8 where we also include methods of state space modeling. Estimating equations are briefly described in Chap. 9 followed by a detailed focus on symbolic data analysis in Chap. 10. The latter topic is a modern addition to the toolbox for statisticians who deal with *big data*.

In our contemporary digitized society, *big data* is highly prized for its commercial and scientific value. One can argue that there are three general stages for processing *big data*: the preparation stage, the analysis stage, and the presentation stage. Statisticians are especially interested in the analysis stage, which has proven to be very challenging to handle using traditional methods that require human involvement. Consider for example the online resource ranking system. With millions of new entries appearing every day, no research group would be able to analyze all these entries, assign ranks, and maintain the online search and recommended system in a timely manner. Heterogeneous *big data* sets collected from multiple sources also present a challenge given the need to determine correlation between the elements. The addition of interaction terms to statistical models can improve performance and prediction accuracy. It is however difficult to choose appropriate interaction terms in a multidimensional data set.

The human brain is an excellent data processor. However, it can be enormously challenged for processing *big data*. Artificial neural networks constitute a mathematical model inspired by the general structure of biological neural networks. They have proven to be useful for modeling complex relationships between inputs and outputs or to find patterns in data. They far surpass human or traditional model capabilities in applications such as audio *big data* analysis, visual *big data* analysis, and natural language processing.

Big data and neural networks are natural companions. Neural networks are very good at uncovering complex relations and abstract features from large data sets by combining multiple information sources, processing heterogeneous data, and effectively capturing dynamic change in data. In turn, the availability of extremely large volumes of data provides the necessary training examples to enable an effective training of deep neural networks. Part III focuses on the tools for machine learning. Chapter 11 deals with regression methods at length seen through the lens of machine learning. Neural networks are presented in Chap. 12. We conclude in Part IV with a look at Bayesian computational methods, which emphasize Bayesian Markov Chain Monte Carlo, and at Bayesian nonparametric statistics. The latter has been a popular topic since the 1970s when it was introduced by Ferguson.

This book provides both graduate and advanced undergraduate students with a reasonably detailed presentation of a variety of current subjects in statistics. It may also serve as a reference text for practitioners. Recommended prerequisites are two one-semester courses, one in probability and one in statistical inference at the third-year undergraduate level. In writing this book, I have benefited substantially from the assistance provided by my present postdoctoral fellow and former graduate student, Dr. Oleksii Volkov. He was the driving force behind the presentation on machine learning, and I am indebted to him for allowing me to share his knowledge of the subject and for his attention to detail in the presentation. I am grateful as well to my graduate MSc students Xiuwen Duan and Jingrui Mu. Xiuwen exploited the use of the Pearson system in connection with Tweedie's formula whereas Jingrui modeled COVID-19 data using Bayesian spatial-temporal methods. My undergraduate students Keyi Liu applied symbolic analysis to analyze data on Parkinson's and heart diseases whereas Josh Larock provided the computing analyses for the applications of multivariate analyses and nonparametric tests. To all four students many thanks for your insights and very hard work. It was a real pleasure sharing these times together.

Ottawa, ON, Canada Mayer Alvo

Acknowledgments

This book was funded by the Natural Sciences and Engineering Research Council of Canada (grant OGP0009068), for which I am grateful.

I would also like to thank my wife Helen for her patience and support throughout the writing of this book. My children Anita and Daniel (Jennifer) as well as my grandchildren, Alyssa, Ryan, Maya, and Jacob often inquired about my progress and provided the additional motivation to complete the work. All in all, it was a labor of love rooted in a continuing admiration for the subjects of mathematics and statistics.

Contents

List of Acronyms

ACF	Autocorrelation function
ANOVA	Analysis of variance
ARMA	Autoregressive moving average
ARIMA	Autoregressive integrated moving average
CART	Classification and regression trees
CDF	Cumulative distribution function
CLT	Central limit theorem
GIS	Geographic information systems
i.i.d.	Independent and identically distributed
LDA	Linear discriminant analysis
LSTM	Long short-term memory
MISE	Mean integrated square error
MLE	Maximum likelihood estimator
MLP	Multi-layer perceptron
PACF	Partial autocorrelation function
PCA	Principal components analysis
RNN	Recurrent neural network
SVM	Support vector machines

List of Nomenclatures

$\boldsymbol{X}=(X_1,\ldots,X_n)$	A random vector
$X_{(1:n)}$	The vector $(X_1,\ldots,X_n)$
$\boldsymbol{X}_{-i}$	$(X_1,\ldots X_{i-1},X_{i+1},X_n)$
θ,β,ω,w	A parameter, possibly a vector
H_0	Null hypothesis
H_1	Alternative hypothesis
$f_X(x;\theta)$	The density of X
$F_X(x;\theta)$	The cumulative distribution function of X
$\Phi(x)$	The cumulative distribution of a standard normal
$\mathcal{N}_k(\mu,\Sigma)$	The k dimensional multivariate normal distribution with mean μ and variance-covariance Σ
$\pi(x;\theta)$	A smooth alternative density of X
$R_1,\ldots,R_n$	The ranks of $X_1,\ldots,X_n$ from smallest to largest
$a_n(R)$	Score function evaluated at rank R
$L(\theta)$	The likelihood function of $X_1,\ldots,X_n$
$l(\theta)$	log of the likelihood function of $X_1,\ldots,X_n$
$I_n(\theta)$	The Fisher information function for θ based on $X_1,\ldots,X_n$
$\bar{X}_n$	The sample mean of $X_1,\ldots,X_n$
S_n^2	The sample variance of $X_1,\ldots,X_n$
$X_n \xrightarrow{P} X$	The sequence of random variables X_n converges in probability to X
$X_n \xrightarrow{\mathcal{L}} X$	The sequence of random variables X_n converges in distribution to X
$X_n \xrightarrow{a.s.} X$	The sequence of random variables X_n converges almost surely to X
$X \xrightarrow{d} Y$	X and Y have the same distribution
$E[X]$	mean of X
$\mathrm{Var}[X]$	variance of X
$A^\top$	Transpose of the matrix A

List of Figures

List of Tables

Part I.

Introduction to Big Data

The term "*big data*" refers to very large data sets which may not be easily analyzed by the usual methods of classical statistics or by standard software. *Big data* has been characterized by the three dimensions of volume, variety, and velocity. Volume refers to both the quantity of data and the large number of separate variables on each object or individual. Variety refers to the several "flavors" of data which arise in a multitude of contexts in our world today including all fields of science, social networks, the environment, and medicine. Velocity refers to the challenges posed by the analysis and processing speed of data. Examples of data occur in medical studies wherein multitudes of variables are recorded for each patient. In genetic studies, gene expression data may be collected on thousands of genes for each patient. The internet presently provides a rich source of data which are enticing for market research and for social studies. Image data obtained on facial recognition would task the largest computers. The mathematical analyses of *big data* go beyond the calculation of simple descriptive statistics, simple uses of linear regression for trend detection, and analysis of variance for treatment comparisons. The analyses of *big data* require sophisticated optimization and classification methods to uncover nonlinear relationships and to make relevant predictions.

In this book, we present a variety of advanced statistical methods at a level suitable for advanced undergraduate and graduate students as well as for individuals wishing to familiarize themselves with these important subjects. Real-life applications exploiting these methods are presented in areas such as citation analysis, genetics, and in environmental problems which enable the reader to gain an appreciation of the challenges and the utility of statistics. As a result, the practitioner may exploit a variety of the methods described in order to analyze a particular data set.

The book begins in Part I by outlining various data types and by indicating how these are normally represented graphically and subsequently analyzed. In Part II, Chaps. 2 and 3, we introduce the basic tools in probability and statistics. Here, we have retained the most useful and relevant results pertinent to this book. In Chap. 4, we proceed with an introduction to multivariate methods and to copula methods. We illustrate a number of applications by presenting real-life examples. In Chap. 5, we introduce nonparametric methods which are particularly useful in the analysis of *big data* when the underlying distributions are often unknown. Some emphases are placed on the use of ranking methods. We continue with a discussion of exponential tilting and its applications in Chap. 6. There we discuss the subject of empirical Bayes and its application to microarray data. In Chap. 7, we touch on counting data analysis and survival analysis. In Chap. 8, time series methods are briefly described both from the usual classical and from the state space modeling approaches. Estimating equations and empirical likelihood are discussed in Chap. 9. We present their application in nonparametric testing. Symbolic data analysis is a relatively new field which aims to reduce the dimension of the data through a process of aggregation. It forms the subject of Chap. 10 wherein traditional statistical Methods are applied to aggregated medical data. In Part III, we focus first on the subject of regression through the lens of machine learning. In Chap. 11, we describe regression methods from the machine learning point of view along with support vector machines often used to study interactions and classification. We then continue in Chap. 12 with the important topics of neural networks and text analytics. We conclude with Part IV by presenting the computational aspects of *big data* with special attention devoted to Markov chain Monte Carlo methods and to Bayesian nonparametric statistics.

1. Examples of *Big Data*

Big data comes in many flavors and dimensions. It may take the form of medical data on a patient where several characteristics are measured, or it may involve several years of economic Dow Jones closing daily averages. Environmental data involves the measurement of several variables in both time and space. Genetic data may involve the measurement of gene expression level on thousands of genes. Ranking data is multifaceted and arises whenever individuals are asked to order a set of candidates or objects. Categorical data are qualitative in nature and involve principally counting data. Symbolic data consists of data that can be aggregated by a variable.

The methods of displaying data are equally diverse and revealing as are the methods by which data are analyzed. In this chapter, we describe various types of *big data* and suggest some of the common methods by which they are analyzed. We begin by tabulating various types of data along with their corresponding graphical representations and common methods of analysis. We then proceed with more detailed descriptions and illustrations (Table 1.1).

1.1. Multivariate Data

Formally, multivariate data arise as several realizations of a p dimensional vector X usually modeled by a multidimensional distribution indexed by parameters such as the mean and the variance-covariance matrix. The distribution of choice is the multivariate normal. The sample data are then used to draw inference in the nature of estimation and hypothesis testing on the parameters. Some examples of multivariate data are:

1. In medical studies, p measurements are taken on each of n patients. Body characteristics such as weight, height, cholesterol level, systolic, and diastolic blood pressures are measured on each patient along with indicator data on various symptoms.

2. In agriculture, crop yields are monitored through measurements of soil qualities, choices of fertilizer, insecticides, planting schedules, and environmental factors.

3. Data from cruise lines may have different fare prices, advertising budgets, amenities offered, scheduling choices, fuel prices, and administrative salaries to determine their impact on revenues from various routes.

M. Alvo, *Statistical Inference and Machine Learning for Big Data*, Springer Series in the Data Sciences, https://doi.org/10.1007/978-3-031-06784-6_1

Table 1.1.: Examples of *big data*

Data type	Graphical representations	Common methods of analyses
Multivariate	Correlation matrix, multidimensional scaling	ANOVA, regression, PCA, factor analysis, classification, discriminant, clustering
Categorical: nominal	Bar graphs, pie charts	frequency comparison methods
Categorical: ordinal	Bar graphs, pie charts	contingency table analyses, survey methods
Environmental	line plots, spatial maps	regression, Kriging
Genetic		Empirical Bayes, false discovery rate, local false discovery Rate
Time series: frequency domain	Line plots	Spectral analysis, wavelet analysis
Time series: time domain	Autocorrelation, cross correlation	Spearman test of trend, Durbin-Watson serial correlation test, ARIMA models
Ranking		sign test, Wilcoxon test, Kruskal-Wallis, Friedman test
Social network		Random graph theory
Symbolic	Histograms, intervals	adaptation of multivariate methods
Images	Digital pictures	

4. In psychological studies, groups of subjects may be compared with respect to their ability to solve a set of problems.

5. In comparative studies, different methods of teaching mathematics to school children may be evaluated with respect to their effectiveness.

6. In studies on intelligence, the interest may be to determine the influence of creativity on achievement. In another example, we may have a set of data on personality variables and another set on aggressive behavior.

7. Daily reports of the weather include measurements of wind speed and direction, air pressure, humidity level, and temperature.

In examples 1, 2, and 3, the dependent variables are the symptoms, the crop yields, and the route revenue, respectively. The other variables are labeled independent variables which may be correlated among themselves. When auxiliary data are available, regression analysis is used to assess the relationship of the dependent or response variables on the independent or auxiliary variables. In examples 4 and 5, the interest is in comparing groups of subjects. In example 6, we are interested in correlating linear combinations of variables related to creativity with linear combinations of variables related to achievement. In such cases, multivariate analysis of variance methods are utilized. In example 7, the interest may be in a time series analysis. In all cases, multivariate analysis aims to take into account those correlations among the variables when drawing inferences.

At times, the dimension p of the vector variable X may be large. It is then popular to make of use of principal component analysis in order to extract uncorrelated linear combinations of the variables that capture most of the variance. Factor analysis is often used as a follow-up to uncover the underlying unobserved or latent factors behind the data itself.

Multivariate data arise in problems of classification, discrimination, and clustering when it is desired to group together data that exhibit common characteristics. Which of these methods is used depends on what is known about the structure of the groups themselves. In this connection, logistic regression is used to predict a categorical (usually dichotomous) variable from a set of predictor variables. In medical research, the dependent variable is whether or not a patient has a disease. For a logistic regression, the predicted dependent variable is the estimated probability that a particular subject will be in one of the categories.

In problems of classification resulting from application of logistic regression, it is common to compute several measures of how effective the classification procedure is. The sensitivity is the percentage of occurrences correctly predicted (e.g., of all who actually have the disease, what percentage were correctly predicted to have the disease). The specificity is the percentage of nonoccurrences correctly predicted (of all who actually are free of the disease, what percentage were correctly predicted not to have the disease). Focusing on error rates, the false positive rate is the percentage of predicted occurrences which are incorrect (i.e., of all who were predicted to have the disease, what percentage actually did not have the disease), and the false negative rate is the percentage of predicted nonoccurrences which are incorrect (i.e., of all who were predicted not to have the disease, what percentage actually did have the disease).

In discriminant analysis, one wishes to predict group membership from a set of two or more continuous variables. The analysis creates a set of discriminant functions (that is, weighted combinations of the predictors) that enable the prediction into which group a case falls, based on scores on the predictor variables (usually continuous, but could include dichotomous variables and dummy coded categorical predictors).

In cluster analysis, the goal is to place into groups, cases (research units) that share similar characteristics. An example relates to the sushi preferences of the Japanese referenced later in this book.

1.2. Categorical Data

Categorical data are qualitative in nature and consist of counting data that can be grouped into categories. In general, the categories have no particular ordering. Nominal data is a type of data used to name variables without providing any numerical value. This data type is a subcategory of categorical data. Nominal data is sometimes called "labeled" or "named" data. Examples of nominal data include name, hair color, and sex.

On the other hand, an ordinal variable is one that incorporates an obvious ordering. Hence, classifying rents as high, medium, or low implies a clear ordering in accordance with level of rent. Some ordinal data examples include measurements on a Likert scale or interval scale, disease severity, and customer satisfaction from survey data. Categorical data are often represented as a two-way contingency table which illustrate how two categorical variables are related. The table may represent the counts or percentages of persons who belong to a group for two or more quantitative variables. This makes it easier to uncover different relationships among the variables. Some everyday examples of categorical data may be found in E-commerce, online dating, customer service organizations, and personality tests. Such data are frequently collected using surveys or questionnaires where respondents often list their gender, opinions, and preferences. The simplest example consists of the number of individuals *for* or *against* a specific candidate or motion. For multiple categories, such data can be displayed in a bar graph where the length of the bar represents the size of the counts. Some examples of categorical data along with the appropriate possible levels are:

1. Ethnicity (Hispanic, African American, Native American, Asian, Other)
2. Eye color (green, blue, brown, black)
3. Household composition (single, married, widowed, divorced)
4. Employment status (employed for wages, self-employed, a homemaker, a student, retired, etc.)
5. Film genres (action, adventure, comedy, crime, mystery, drama, historical, etc.)

6. Home country (Canada, the USA, Australia, India, Germany)
7. Car color (red, green, grey, black, white, etc.)
8. Religion (Muslin, Buddhist, Christian)
9. Seasons (winter, spring, summer, autumn)
10. Holidays (Thanksgiving, Halloween)
11. Types of pet (dog, cat, hamster)
12. Blood groups (Group A, Group B, Group AB, Group O)

A bar chart is mostly used to analyze frequency, while a pie chart analyzes percentage. This is done after grouping data into a table. In the case of ordinal data, which has a given order or scale, the scale does not have a standardized interval. This is not applicable for nominal data. The distinction between the different types of data is important because it influences the choice of statistical analyses conducted.

(a) Nominal Data Variable: This type of categorical data variable has no intrinsic ordering with respect to its categories. For example, marital status is a categorical variable having two categories (single and married) with no intrinsic ordering. There are two main categories of nominal data variables, namely, matched and unmatched category. Below are the tests carried out on each category:

- Matched Category: The McNemar Test is a distribution-free test for paired nominal data (two groups), whereas the Cochran's Q test is for three or more groups.
- Unmatched Category: The Fisher's exact test is used when the expected frequency is less than 5. The chi-square Test is used when the expected frequency is 5 or more.

(b) Ordinal Data Variable: This type of categorical variable has an intrinsic ordering. For example, when studying the severity of a disease, severity is a categorical variable with ordered categories which may be critical, medium, and low. There are two main categories of ordinal data variables, namely, matched and unmatched category. Below are the tests usually carried out on each category:

- Matched Category: The Wilcoxon signed-rank test is used to assess the differences between two groups of matched samples. The Friedman two-way layout is used to find differences in matched sets of three or more groups.
- Unmatched Category: The Wilcoxon rank-sum test is used to investigate two groups of independent samples. The Kruskal-Wallis test is used to investigate three or more groups.

Table 1.2.: Sources of environmental data

Agency	Website
EPA	https://edg.epa.gov/metadata/catalog/main/home.page
OECD	https://data.oecd.org/environment.htm
UK	https://www.ecn.ac.uk/data
UN	https://geodata.grid.unep.ch/
UN	https://unstats.un.org/unsd/envstats/qindicators.cshtml
Wikipedia	https://en.wikipedia.org/wiki/List_of_biological_databases
World Bank	http://data.worldbank.org/topic/environment

1.3. Environmental Data

Environmental data is typically collected by government. This involves the measurement of variables in part on their impact on ecosystems. For example, data collected to assess the impact of acid rain involves the measurement of sulfates deposited over a large region at specific periods of time. Agencies collecting environmental data include those listed in Table 1.2.

Environmental data as well as health-related data (such as COVID-19) are analyzed using the tools in multivariate and time series analysis. Spatial modeling which exploits the use of Bayesian spatial-temporal models and Kriging is often used to create spatial maps that can be used to monitor the effectiveness of control measures. In another example, the monitoring of trend in pH or other variables may be done using rank correlation methods in nonparametric statistics.

1.4. Genetic Data

Genetics dates back to the Augustinian friar and scientist Gregor Mendel (1822–1884), whose studies of pea plants in the mid-1800s established many of the rules of heredity for living organisms. In the majority of cases, with the exception of identical twins, an individual's genetic sequence is considered unique to that individual and is regularly used in the identification of individuals to determine innocence or guilt in legal proceedings. In biomedical research, scientists try to understand the influence of genes in the cause of disease. Gene expression levels presented in microarrays can be used to discover genes that require further investigation.

Genomics is the study of the entirety of an organism's genes. Genomics researchers involved in bioinformatics make use of high-performance computing and mathematical techniques to analyze enormous amounts of DNA sequence data to find variations that affect health, disease, or drug response. There are about 3 billion units of DNA across 23,000 genes for humans. Genomics is a much newer field than genetics which became possible only in the last couple of decades due to technical advances in DNA sequencing and computational biology. Some links to genetic data are:

1. A good source of information on available genomic data is Kovalevskaya et al. [2016].
2. See also the Stanford Genomic resources website http://genome-www.stanford.edu/.
3. See https://en.wikipedia.org/wiki/List_of_genetics_research_organizations for a list of genetics research organizations.

1.5. Time Series Data

Data indexed by time is referred to as a time series. Typically, it consists of a sequence taken at successive equally spaced points in time. Examples include the daily closing value of the Dow Jones industrial average and the heights of ocean tides. They are used in any domain of applied science and engineering which involves temporal measurements such as econometrics, finance, signal processing, pattern recognition, weather forecasting, and earthquake prediction.

In analyzing time series data, one wishes to extract meaningful characteristics of the data such as trends. Time series forecasting is the use of a model to predict future values based on previously observed values. The parametric approach to analyzing a time series focuses on decomposing the series in terms of trend, autoregressive, and moving average components, all involving a small number of parameters. Intervention analysis deals with the analysis of interrupted time series due to a change in the nature of the data. Nonparametric methods explicitly estimate the covariance or the spectrum of the process without assuming that the process has any particular structure. More recently Bayesian and state space modeling have been used to analyze time series data. Methods of time series analysis may also be divided into linear and nonlinear as well as univariate and multivariate. Further examples of time series data may be found in the following website.

https://machinelearningmastery.com/time-series-datasets-for-machine-learning/

1.6. Ranking Data

Ranking data commonly arise from situations where it is desired to rank a set of individuals or objects in accordance with some criterion. Such data may be observed directly or it may come from a ranking of a set of assigned scores. Alternatively, ranking data may arise when transforming continuous or discrete data in a nonparametric analysis. Examples of ranking data may be found in politics [Barnes and Kaase, 1979; Croon, 1989; Inglehart, 1977; Moors and Vermunt, 2007; Vermunt, 2004], voting and elections [Gormley and Murphy, 2008; Murphy and Martin, 2003; Skrondal and Rabe-Hesketh, 2003; Stern, 1993], market research [Beggs et al., 1981; Chapman and Staelin, 1982; Dittrich et al., 2000], food preference [Kamishima and Akaho, 2006; Nombekela et al., 1993;

Vigneau et al., 1999], psychology [Bockenholt, 2001; Decarlo and Luthar, 2000; Maydeu-Olivares and Bockenholt, 2005; Regenwetter et al., 2007; Riketta and Vonjahr, 1999], health economics [Craig et al., 2009; Krabbe et al., 2007; McCabe et al., 2006; Ratcliffe et al., 2006, 2009; Salomon, 2003], medical treatments [Plumb et al., 2009], candidates in an election [Diaconis, 1988; Koop and Poirier, 1994], types of sushi [Kamishima and Akaho, 2006], place of living [Duncan and Brody, 1982], choice of occupations [Goldberg, 1975; Yu and Chan, 2001], and even horse racing [Benter, 1994; Henery, 1981; Stern, 1990].

For the analysis of ranking data as well as for a discussion of ranking models, the reader is referred to Alvo and Yu [2014]. A further approach for the analysis of ranking data in a parametric setting is given in Alvo and Yu [2018].

Some software packages for analyzing ranking data are

1. PerMallows: An R Package for Mallows and Generalized Mallows Models

2. An R package for analyzing and modeling ranking data Paul H Lee and Philip LH Yu

https://bmcmedresmethodol.biomedcentral.com/articles/10.1186/1471--2288--13-6

1.7. Social Network Data

Social network data sets are widely used in data mining to uncover relationships and preferences among subjects. They can be found in many forms and obtained from observations, surveys, archives, databases, etc. Network data can also be generated from other types of data and semantic web or even be randomly generated. Activities on the Internet, such as e-mail, chat, and forums, leave traces that can be used as sources for network data.

For network analysis, there are archives formed by different social networking services such as Facebook, Twitter, LinkedIn, and Google+. These organizations collect data about users, their posts, or tweets, all of which regrettably are not publicly available. A large amount of data is stored in Internet Movie Database (IMDb) and services such as Amazon, lastFM, Pandora, or Netflix. Converting data into multiple two-mode networks and combining them in network analysis allows us to obtain information about collaboration between actors, producers and composers, similarity of the movies according to different measures, etc. Other repositories of networks and data sets of other types are available as, for example, repositories of data sets, KDnuggets data sets for data mining, data surfing on the World Wide Web, public data sets on Amazon Web Services, TunedIT, the Internet2 Observatory Data Collections, Infochimps, the Cooperative Association for Internet Data Analysis (CAIDA data), network data sources on Pajek's web page, KONECT, SNAP, GDELT, NYC Taxi and Uber Trips, bike sharing, and MAG.

Universities have archives of past students' achievements and research. Network data can be derived from such archived data. As well, archived data about the inhabitants of a city or an area can be used for genealogical analysis. For "scientific" genealogies used in anthropological research, see the website KinSource. A good source of social network data may be found in the following websites:

1. https://snap.stanford.edu/data/
2. https://link.springer.com/referenceworkentry/10.1007%2F978-1-4939-7131-2_313

Social network data is often analyzed by using graph theory. Paul Erdős and Alfréd Rényi proposed an approach to formalize the notion of a random graph. The Erdős-Rényi model, denoted by G (n,m), where n is the number of nodes and m is the number of edges, generates a random graph on n nodes and m edges (uniformly) randomly selected among the $n(n-1)/2$ potential edges. Closely related to the Erdős-Rényi model is the Gilbert model G (n,p) where n is the number of nodes and p is the probability that an edge is included in the random graph. In this model the $n(n-1)/2$ potential edges of a simple undirected graph G (n,p) $\in$ G (n,p) are included independently with probability p. We do not pursue the study of graph theory in this book.

1.8. Symbolic Data

Symbolic data is an extension of standard data and includes as input, tables, histograms, and intervals. The data units are called symbolic. They not only contain values or categories but also may include internal variation and structure. Symbolic data analysis (SDA) is based on four spaces: the space of individuals, the space of concepts, the space of descriptions, and the space of symbolic objects. The space of descriptions models individuals, while the space of symbolic objects models concepts. As an example, we may have film data listed by director, years in which films were made, and locations for filming. References to a description of symbolic data and to accompanying software are the following:

1. https://chance.amstat.org/2013/09/big_picture_26--3/
2. http://citeseerx.ist.psu.edu/viewdoc/download?doi=10.1.1.95.4237&rep=rep1&type=pdf
3. https://www.ceremade.dauphine.fr/SODAS/Introdiday.htm

1.9. Image Data

The term image inherently reflects a graphic representation and, in the GIS world, differs significantly from raster data which consists of a matrix of cells (or pixels) where each cell contains a value representing information. Most often, image data is used to store remotely sensed imagery, e.g., satellite scenes or orthophotos or ancillary graphics such as photographs, scanned plan documents, etc. Image data is typically used in GIS systems as background display data or as a graphic attribute. Remote sensing software makes use of image data for image classification and processing. Typically, this data must be converted into a raster format (and perhaps vector) to be used analytically with the GIS. Image data is typically stored in a variety of industry standard proprietary formats. A reference to image data is the following:

https://desktop.arcgis.com/en/arcmap/10.3/manage-data/raster-and-images/what-is-raster-data.htm.

Part II.

Statistical Inference for Big Data

In Chap. 2, we describe the language of probability and statistics. We introduce the important tools of moment and cumulant generating functions. The various notions of convergence play an important role in statistical estimation and hypothesis testing. For reference, we list some common distributions along with their mean, variance, skewness, and kurtosis. Karl Pearson's system of distributions which plays an important role both in empirical Bayes and in the optimality of nonparametric tests is described. We conclude with a description of both univariate and multivariate forms of the central limit theorem.

2. Basic Concepts in Probability

The study of probability theory is based on the concept of random variables. The *sample space* is the set of all possible outcomes of an experiment. A real valued random variable X is defined as a function from the sample space to the set of real numbers. As an example, suppose in an experiment of rolling two independent dice, X represents the sum of the face values of the two dice. Then the sample space consists of the ordered pairs $\{(i,j) : i,j = 1,2,\ldots,6\}$ and the range of X is $\{2,3,4,\ldots,12\}$. We assume that the reader is familiar with the basic notions of the algebra of sets or events and the elementary notions of a probability function, denoted by P.

The *cumulative distribution function* of a random variable X is the probability that X takes on a value less than or equal to x, denoted by $F_X(x) = P(X \le x)$. Suppose 30% of students have heights less than or equal to 150 cm. Then the probability that the height X of a randomly chosen student is less than or equal to 150 cm is 0.3 can be expressed as $F_X(150) = P(X \le 150) = 0.3$. The cumulative distribution function $F_X(x)$ in general is a non-decreasing right continuous function of x which satisfies

$$\lim_{x\to-\infty} F_X(x) = 0, \lim_{x\to\infty} F_X(x) = 1$$

Most random variables are either discrete or continuous. A random variable is called *discrete* if it takes a finite or a countably infinite number of values. Examples are the results of a hockey game (win, lose, or draw), the number of sunny days in a week, or the number of patients infected with a disease. The probability mass function (or probability density function) of a discrete random variable is defined as $f_X(x) = P(X = x)$. The probability density function for a discrete random variable X, $f(x)$ satisfies the following conditions:

$$f_X(x) \ge 0 \quad \text{and} \quad \sum_{x\in\Omega} f_X(x) = 1$$

where Ω is the range of the discrete random variable X.

We say that a random variable is *continuous* if its cumulative distribution function is a continuous function. A continuous random variable such as weight, length, or lifetime takes any numerical value in an interval or on the real line. Typically, a continuous cumulative distribution function has a derivative except at some points. This derivative denoted by

$$f_X(x) = \frac{d}{dx} F_X(x) = F'_X(x),$$

M. Alvo, *Statistical Inference and Machine Learning for Big Data*, Springer Series in the Data Sciences, https://doi.org/10.1007/978-3-031-06784-6_2

is called the *probability density function of* X . The cumulative distribution function of a continuous random variable X on the entire real line satisfies

$$F_X(x) = \int_{-\infty}^{x} f_X(t)\, dt, \qquad f_X(x) \geq 0 \quad \text{and} \quad \int_{-\infty}^{\infty} f_X(x) = 1.$$

The joint cumulative distribution function for a vector of $p\,(\geq 1)$ random variables, $\boldsymbol{X} = (X_1, X_2, \ldots, X_p)$, is defined as

$$F(x_1, x_2, \ldots, x_p) = P\left(X_1 \leq x_1, \ldots, X_p \leq x_p\right).$$

The joint density can be defined as

$$f_{\boldsymbol{X}}(x_1, x_2, \ldots, x_p) = \frac{\partial^p F(x_1, x_2, \ldots, x_p)}{\partial x_1 \ldots \partial x_p}$$

provided the multiple derivatives exist. The p random variables $X_1, X_2, \ldots, X_p$ are said to be *independent* if their joint probability density function is the product of the p individual marginal densities, denoted $\{f_{X_i}(x)\}$, i.e.,

$$f_{\boldsymbol{X}}(x_1, x_2, \ldots, x_p) = f_{X_1}(x_1) f_{X_2}(x_2) \ldots f_{X_p}(x_p).$$

A *random sample* of the random variable X is a set of random variables $X_1, X_2, \ldots,$ X_p, which are independent and identically distributed (i.i.d.) with the same probability density function as X.

In probability and statistics, we are interested in properties of the distribution of a random variable. The expected value of a function $g(X)$ of a real valued random variable X, denoted by $E[g(X)]$, is defined as

$$E[g(X)] = \begin{cases} \int_{-\infty}^{\infty} g(x) f(x)\, dx & \text{if } X \text{ is continuous} \\ \\ \sum_{x \in \Omega} g(x) f(x) & \text{if } X \text{ is discrete.} \end{cases}$$

The ith *moment* of a random variable X is $E\left[X^i\right]$ and the ith *central moment* about its mean is $\mu_i = E[(X - \mu)^i]$ where $\mu = E\left[X\right]$ is the mean of X. Note that $\mu_1 = 0$ and $\mu_2 = Var(X)$ is the variance of X, usually denoted by σ^2.

The third and fourth central moments are used to define other global properties of a distribution function such as the skewness and kurtosis. Skewness measures the lack of symmetry in a distribution and is defined to be equal to $\frac{\mu_3}{\mu_2^{3/2}}$.

For a normal distribution, the skewness is 0. Positive values of skewness imply the tail on the right is longer or fatter, whereas negative values indicate the tail on the left side is longer or fatter. The mode and median are on the left of the mean for distributions with positive skewness, whereas they are on the right when the distribution is negatively skewed. We denote the square of the skewness by

$$\beta_1 = \frac{\mu_3^2}{\mu_2^3}$$

Kurtosis is used to describe the extreme values and measures the "fatness" of the tails of the distribution. It is defined to be

$$\beta_2 = \frac{\mu_4}{\mu_2^2}.$$

For a normal distribution, the kurtosis is 3.

A distribution with heavy tails or outliers will have a large kurtosis, whereas one with light tails will have a small kurtosis. A leptokurtic distribution refers to one with a kurtosis greater than 3, while a platykurtic has a kurtosis less than 3. A distribution is called mesokurtic if it resembles the normal.

We note that for a random sample $X_1, \ldots, X_n$, from some population, the sample estimates for the mean μ_1 and variance μ_2 are respectively

$$\begin{aligned} \bar{X}_n &= \sum X_i/n \\ S_n^2 &= \sum \left(X_i - \bar{X}_n\right)^2 / (n-1), n > 1 \end{aligned}$$

The sample estimate of β_1 is

$$b_1 = \left[\frac{\Sigma \left(X_i - \bar{X}\right)^3}{n}\right]^2 \left[\frac{\Sigma \left(X_i - \bar{X}\right)^2}{n}\right]^{-3}$$

whereas the sample estimate of the kurtosis β_2 is

$$b_2 = \left[\frac{\Sigma \left(X_i - \bar{X}\right)^4}{n}\right] \left[\frac{\Sigma \left(X_i - \bar{X}\right)^2}{n}\right]^{-2}$$

The moment generating function of a distribution, if it exists, is defined to be

$$M(t) = E\left[e^{tX}\right]$$

for t in an open interval around 0. The moment generating function yields the moments of a distribution by setting the derivatives with respect to t equal to 0. The moment generating function does not always exist as in the case of the Cauchy distribution. On the other hand, the characteristic function of a random variable, defined as

$$\phi(t) = E\left[e^{itX}\right], t\ real$$

where i is the complex operator for which $i^2 = -1$ always exists and uniquely characterizes the distribution of a random variable [Van der Vaart, 2000, p.13–15]. In fact,

$$|\phi(t)| \leq \int \left|e^{itx}\right| dF(x) = 1$$

In connection with the moments of a distribution, we cite the following result from Van der Vaart [2000, p.18].

Theorem 2.1. *Let X_n and X be random variables such that $E[X_n^p] \to E[X^p] < \infty$ for every positive integer p. If the distribution of X is uniquely determined by its moments, then $X_n \xrightarrow{d} X$.*

Another important set of constants of a distribution are the cumulants denoted $\{\kappa_r\}$. These are defined as the coefficients of $\log \phi(t)$, also known as the cumulant generating function,

$$\phi(t) = \exp\left(\sum_{r=1}^{\infty} \kappa_r \frac{(it)^r}{r!}\right).$$

In terms of the cumulants, the square of the skewness is given by

$$\beta_1 = \frac{\kappa_3^2}{\kappa_2^3}$$

and the excess kurtosis by

$$\beta_2 - 3 = \frac{\kappa_4}{\kappa_2^2}$$

The cumulant generating function uniquely determines a distribution and is statistically more appealing as it generates the mean and the variance directly unlike the moment generating function which only generates the moments.

Example 2.1. To illustrate, the moment generating function of a normal variable with mean μ and variance σ^2 is

$$M(t) = e^{\mu t + \frac{\sigma^2 t^2}{2}}$$

Differentiating repeatedly with respect to t and setting $t = 0$, we obtain the first and second moments, respectively:

$$M'(0) = \mu, M''(0) = \mu^2 + \sigma^2$$

On the other hand, the characteristic function is given by

$$\phi(t) = e^{i\mu t - \frac{\sigma^2 t^2}{2}}$$

and the cumulant generating function is

$$\log \phi(t) = i\mu t - \frac{\sigma^2 t^2}{2}.$$

Differentiating the latter repeatedly with respect to t and setting $t = 0$, we see that the first and second cumulant are the mean and variance, respectively:

$$\kappa_1 = \mu, \kappa_2 = \sigma^2, \kappa_r = 0, r > 2$$

Cumulants have the important property that for two independent random variables, the cumulant generating function of their sum is the sum of their cumulant generating functions. As well, a distribution with given cumulants can be approximated through an Edgeworth series [Field and Ronchetti, 1990]. We do not pursue this topic further but instead refer the reader to Kendall and Stuart [1979, p.223 and p.373].

Using the linear properties of the expectation operator, we can determine the expected value and moments of a linear combination of a set of random variables. For instance, for p random variables $X_1, \ldots, X_p$ and real valued constants $a_i, i = 1, \ldots, p$, we have

$$\begin{aligned} E\left[\sum_{i=1}^{p} a_i X_i\right] &= \sum_{i=1}^{p} a_i E\left[X_i\right] \\ Var(\sum_{i=1}^{p} a_i X_i) &= \sum_{i=1}^{p} a_i^2 Var(X_i) + 2\sum_{i=1}^{p}\sum_{j=i+1}^{p} a_i a_j Cov(X_i, X_j) \end{aligned}$$

where

$$Cov(X_i, X_j) = E\left[X_i X_j\right] - E\left[X_i\right] E\left[X_j\right]$$

We note that for a random sample $X_1, \ldots, X_n$ from some population

$$\begin{aligned} E\left[\bar{X}_n\right] &= \mu, Var\left(\bar{X}_n\right) = \sigma^2/n \\ E\left[S_n^2\right] &= \sigma^2 \\ Var\left(S_n^2\right) &= \sigma^4\left(\frac{2}{n-1} + \left(\frac{\kappa_4}{\kappa_2^2}\right)\frac{1}{n}\right) \end{aligned} \tag{2.1}$$

2.1. Pearson System of Distributions

In later chapters, we shall make use of an important result due to Karl Pearson who in a series of papers [Pearson, 1895, 1901, 1916] devised a system of distributions to determine a density $f(x)$ given the first four moments if they exist

$$\frac{d \log f(x)}{dx} = \frac{1}{f(x)}\frac{df(x)}{dx} = \frac{x-a}{b_0 + b_1 x + b_2 x^2} \tag{2.2}$$

This system includes many, though not all, distribution functions. The constants in (2.2) can be shown to be functions of the mean, variance, skewness, and kurtosis (see

Kendall and Stuart 1979), namely,

$$\begin{aligned} b_1 &= a = -\frac{\mu_3\left(\mu_4 + 3\mu_2^2\right)}{10\mu_4\mu_2 - 12\mu_3^2 - 18\mu_2^3} \\ &= -\frac{\sqrt{\mu_2}\sqrt{\beta_1}\left(\beta_2 + 3\right)}{A} \\ b_0 &= -\frac{\mu_2\left(4\beta_2 - 3\beta_1\right)}{A} \\ b_2 &= -\frac{\left(2\beta_2 - 3\beta_1 - 6\right)}{A} \\ A &= 10\beta_2 - 12\beta_1 - 18 \end{aligned}$$

where β_1 *and* β_2 refer to the square of the skewness and to the kurtosis, respectively. Here, b_0 cannot be equal to zero. Moreover, $\sqrt{\beta_1}$ is interpreted to be equal to $\frac{\mu_3}{\mu_2^{3/2}}$.

Chabert and Tourneret [2011] introduced a multivariate extension of the Pearson system and provided the convenient Table 2.1. We note, however, that multivariate extensions are best addressed through the use of copulas [Nelsen, 2006], a subject to be addressed in Chap. 4 of this book.

We shall have occasion to make use of the Pearson representation in connection with applications of Tweedie's formula as well as in the discussion of optimal rank tests for location and dispersion. For now, we mention a simple application to Stein's lemma.

Example 2.2. Stein's lemma. An immediate application of (2.2) is to the case of a normal density with mean 0 and variance σ^2. Consequently, $b_1 = a = b_2 = 0, b_0 = -\sigma^2$ leading to the Pearson system:

$$\frac{f'(x)}{f(x)} = -\frac{x}{\sigma^2}$$

Table 2.1.: Pearson system of distributions

Type	Distribution	β_1	β_2
0	Gaussian	0	3
I	Beta	$\neq 0$	
II	Beta	0	< 3
III	Gamma		
IV	Nonstandard		
V	Inverse gamma		
VI	Fisher		
VII	Student	0	> 3

from which we get

$$xf(x) = -\sigma^2 f'(x)$$

Let $g(x)$ be a function for which the expectations below are finite. Then integrating by parts

$$\begin{aligned}
\int xg(x) f(x) \, dx &= -\int \sigma^2 g(x) f'(x) \, dx \\
&= -\sigma^2 \left[g(x) f(x)\right]_{-\infty}^{\infty} + \sigma^2 \int g'(x) f(x) \, dx \\
&= \sigma^2 E\left[g'(X)\right] \qquad (2.3)
\end{aligned}$$

The latter expression is Stein's lemma, and it may be used to find the first four moments $E\left[X^k\right]$ of the normal by choosing $g(x) = x^{k-1}, k = 1, \ldots, 4$. Stein's lemma has found numerous applications in statistics including proving the Central limit theorem [Ross, 2011]. Quoting from Adcock et al. [2019] "Stein's Lemma provides a useful tool for deriving covariances between functions of component random variables. This lemma has applications in finance, notably for portfolio selection and hence for the capital asset pricing model."

Table 2.2 exhibits the densities of some important discrete and continuous random variables as well as their mean, variance, skewness, and excess kurtosis γ_2. An important result in connection with the skewness and the excess kurtosis is the Cornish-Fisher asymptotic expansion for quantiles of the distribution of a random variable. Setting $x = \Phi^{-1}(\alpha)$, the $100(1-\alpha)$ upper percentage point z_α can be approximated by

$$z_\alpha \simeq \mu + \sigma g(x)$$

where

$$\begin{aligned}
g(x) &= x + \left[\gamma_1 h_1(x)\right] + \left[\gamma_2 h_2(x) + \gamma_1^2 h_{11}(x)\right] \qquad (2.4) \\
&\quad + \left[\gamma_3 h_3(x) + \gamma_1 \gamma_2 h_{12}(x) + \gamma_1^3 h_{111}(x)\right] + \ldots \qquad (2.5)
\end{aligned}$$

with

$$\begin{aligned}
h_1(x) &= \frac{1}{6} H_2(x), h_2(x) = \frac{1}{24} H_3(x) \\
h_{11}(x) &= -\frac{1}{36}\left(2H_3(x) + H_1(x)\right) \\
h_3(x) &= \frac{1}{120} H_4(x)
\end{aligned}$$

$$h_{12}(x) = -\frac{1}{24}\left(H_4(x) + H_2(x)\right)$$

$$h_{111}(x) = \frac{1}{324}\left(12H_4(x) + 19H_2(x)\right)$$

The coefficients γ are given in terms of the cumulants by

$$\gamma_{r-2} = \frac{\kappa_r}{\kappa_2^{r/2}}, r \geq 2.$$

whereas the Hermite polynomials are

$$H_n(x) = n! \sum_{i=0}^{[n/2]} \frac{(-1)^i x^{2-2i}}{2^i i! (n-2i)}$$

If the distribution being approximated is that of the sum of n independent identically distributed random variables, the bracketed terms in (2.4) are of the same order with respect to n. Next, we record in the following examples other results of interest which will have applications for simulation of the Dirichlet distribution.

Example 2.3. (Breiman [1992], Section 13.6). Let $(U_i)_{i=1}^n$ be independent identically distributed uniform variables on the interval $(0,1)$, and let

$$U_{(1)} \leq \cdots \leq U_{(n)}$$

be the order statistics. Then

$$\left(U_{(1)}, \ldots, U_{(n)}\right) \overset{d}{=} \left(\frac{\Gamma_1}{\Gamma_{n+1}}, \ldots, \frac{\Gamma_n}{\Gamma_{n+1}}\right)$$

where

$$\Gamma_i = \sum_{k=1}^{i} X_k,$$

and $\{X_k\}_{k=1}^n$ are a random sample from $Exp(1)$.

Example 2.4. Let $Z_i \sim G(\alpha_i, 1), i = 1, \ldots, n$ be independent. Then, the distribution of $\left(\frac{Z_1}{S}, \ldots, \frac{Z_n}{S}\right)$ where $S = \sum_{i=1}^n Z_i$ is given by the Dirichlet $D(\alpha)$. This result enables us to generate the Dirichlet distribution.

Table 2.2.: Some important discrete and continuous random variables

Name		Mean	Variance	Skewness	Ex. kurtosis
Discrete					
	Probability mass function				
Uniform on $1, 2, \ldots, m$	$\frac{1}{m}, x = 1, \ldots, m$	$\frac{m+1}{2}$	$\frac{m^2-1}{12}$	0	$-\frac{6(m^2+1)}{5(m^2-1)}$
Binomial	$\binom{n}{x} p^x (1-p)^{n-x}$ $x = 0, 1, \ldots, n,\ 0 \leq p \leq 1$	np	$np(1-p)$	$\frac{1-2p}{\sqrt{np(1-p)}}$	$\frac{1-6p(1-p)}{np(1-p)}$
Multinomial	$\frac{n!}{x_1! x_2! \ldots x_k!} p_1^{x_1} \ldots p_k^{x_k}, \sum x_i = n, \sum p_i = 1$	$E[X_i] =$ np_i	$np_i(1-p_i)$		
Poisson	$\frac{\lambda^x e^{-\lambda}}{x!}, \lambda > 0, x = 0, 1, \ldots$	λ	λ	$\lambda^{-1/2}$	λ^{-1}
Negative binomial	$\frac{\Gamma(x+k)}{\Gamma(k)\Gamma(x+1)} \left(1 - \frac{k}{\mu+k}\right)^x \left(\frac{k}{\mu+k}\right)^k, x =$ $0, 1, \ldots; \mu, k > 0$	μ	$\mu + \frac{\mu^2}{k}$		

(continued)

Table 2.2.: (continued)

Name		Mean	Variance	Skewness	Ex. kurtosis
Continuous					
	Probability density function				
Uniform on (a, b)	$\frac{1}{b-a}$, $a < x < b$	$\frac{a+b}{2}$	$\frac{(b-a)^2}{12}$	0	$-\frac{6}{5}$
Beta $Be\,(\alpha, \beta)$	$\frac{\Gamma(\alpha+\beta)}{\Gamma(\alpha)\Gamma(\beta)} p^{\alpha-1}(1-p)^{\beta-1}, 0 < p < 1$	$\frac{\alpha}{\alpha+\beta}$	$\frac{\alpha\beta}{(\alpha+\beta)^2(\alpha+\beta+1)}$		
Dirichlet $D\,(\alpha)$	$\frac{\Gamma(\sum \alpha_i)}{\prod \Gamma(\alpha_i)} \prod p_i^{\alpha_i - 1}, \sum p_i = 1, 0 < p_i < 1$				
Normal $N\,(\mu, \sigma^2)$	$\frac{1}{\sqrt{2\pi\sigma^2}} \exp\left(-\frac{(x-\mu)^2}{2\sigma^2}\right) - \infty < x < \infty$	μ	σ^2	0	0
Gamma $G\,(\alpha, \beta)$	$\frac{\beta^\alpha}{\Gamma(\alpha)} x^{\alpha-1} e^{-\beta x}\ x \geq 0, \alpha, \beta > 0$	$\frac{\alpha}{\beta}$	$\frac{\alpha}{\beta^2}$	$\frac{2}{\sqrt{\alpha}}$	$\frac{6}{\alpha}00$
Exponential $G\,(1, \lambda)$	$\lambda e^{-\lambda x}\ x \geq 0$	$\frac{1}{\lambda}$	$\frac{1}{\lambda^2}$	2	6
Chi-square χ^2_ν	$\frac{1}{2^{\frac{\nu}{2}}\Gamma(\frac{\nu}{2})} x^{\frac{\nu}{2}-1} e^{-\frac{x}{2}}, x \geq 0$	ν	2ν	$\sqrt{\frac{8}{\nu}}$	$\frac{12}{\nu}$
Laplace	$\frac{1}{2\sigma} \exp\left(-\frac{\lvert x-\mu \rvert}{\sigma}\right), -\infty < x < \infty$	μ	$2\sigma^2$	0	3
Logistic	$\frac{\exp(-\frac{x-\mu}{\sigma})}{\sigma(1+\exp(-\frac{x-\mu}{\sigma}))^2}, -\infty < x < \infty$	μ	$\frac{\pi^2\sigma^2}{3}$	0	1.2
Student's t_ν	$\frac{\Gamma(\frac{\nu+1}{2})}{\sqrt{\pi\nu}\,\Gamma(\frac{\nu}{2})} \left(1 + \frac{x^2}{\nu}\right)^{-\frac{\nu+1}{2}}, -\infty < x < \infty$	0	$\frac{\nu}{\nu-2}$	0	$\frac{6}{\nu-4}$

2.2. Modes of Convergence

Many of the important results in probability and statistics are asymptotic in nature. Knowing the limiting distribution of functions of random variables has important consequences in statistics. For example, knowing the asymptotic distribution of the sample mean enables us to construct confidence intervals and to perform tests of hypotheses on the population mean. We shall be concerned with three basic modes of convergence: weak convergence (equivalently, convergence in distribution), convergence in probability, and convergence almost surely. An important tool in proving limit results is the Borel-Cantelli lemma which provides conditions under which a limiting event will have probability zero.

Lemma 2.1. *(Borel-Cantelli) (i) Let $A_1, \ldots, A_n$ be a sequence of events in some probability space such that*

$$\sum P(A_i) < \infty.$$

Then the probability that the set of outcomes which occur infinitely often in the sequence of events is 0, that is,

$$P\left(\limsup_{n\to\infty} A_n\right) = P\left(\bigcap_{n=1}^{\infty} \bigcup_{k\geq n}^{\infty} A_k\right) = 0$$

(ii) If the events $A_1, \ldots, A_n$ are independent and

$$\sum P(A_i) = \infty$$

then the probability that the set of outcomes which occur infinitely often in the sequence of events is 1, that is,

$$P\left(\limsup_{n\to\infty} A_n\right) = 1$$

As an example, suppose that in coin tossing, $(X_n = 1)$ is the event that we observe heads on the nth toss and that we have $P(X_n = 1) = \frac{1}{n^2}$. Since, $\sum P(X_n = 1) = \frac{\pi^2}{6} < \infty$, the probability that $(X_n = 1)$ occurs infinitely often is 0. This result still holds even if $P(X_n = 1) = (0.9)^n$. On the other hand, if the tosses are independent and $P(X_n = 1) = \frac{1}{n}$, then the lemma states we will still observe infinitely many heads. The Borel-Cantelli lemma is used to demonstrate various theorems in probability including Kolmogorov's $0-1$ law [Breiman, 1992].

The distribution of various functions of the data is important in statistics for the purpose of drawing inference. Hence, in estimation theory, we are interested in the distribution of the sample mean which serves as an estimator of the population mean. In hypothesis testing, we may be interested in calculating the probability that a certain interval, centered at the sample mean, contains a hypothesized population mean. Knowledge of the underlying distribution of the estimator is thus critical. Unfortunately, the exact distribution of a statistic is often intractable, and we then need to rely on asymptotic theory as the sample size gets large. In what follows, we shall define various concepts of convergence.

Definition 2.1. A sequence of random variables X_n converges weakly or converges in distribution to X, denoted $X_n \xrightarrow{\mathcal{L}} X$ if as $n \to \infty$

$$P(X_n \leq x) \to P(X \leq x)$$

for all points of continuity x of the cumulative distribution function of X.

Knowing the asymptotic distribution of a test statistic is needed whenever one is interested in hypotheses testing in order to determine approximate critical values.

Definition 2.2. A sequence of random variables X_n converges in probability to X, denoted $X_n \xrightarrow{P} X$ if, as $n \to \infty$

$$P(|X_n - X| > \varepsilon) \to 0$$

for $\varepsilon > 0$.

Convergence in probability is often used to show that an estimator is consistent.

Definition 2.3. A sequence of random variables X_n converges almost surely (a.s.) to X,, denoted $X_n \xrightarrow{a.s.} X$ if

$$P\left(\lim_{n\to\infty} |X_n - X| > \varepsilon\right) = 0$$

for $\varepsilon > 0$.

Convergence almost surely is needed in order to bolster our confidence that an estimator for a parameter will be close to that parameter as the sample size gets large. Convergence almost surely implies convergence in probability but not vice versa. On the other hand, if a sequence of random variables converges in probability, $X_n \xrightarrow{P} X$, then there exists a subsequence of X_n which converges almost surely to X. The following inequality plays a useful role in probability and statistics.

Lemma 2.2. *Markov inequality. Let X be a positive random variable having finite variance. Then for $\varepsilon > 0$.*

$$P(X > \varepsilon) \leq \frac{E[X]}{\varepsilon}$$

Chebyshev's inequality follows from Markov's inequality by considering the variable $|X - \mu|^2$

$$P(|X - \mu| > \varepsilon) \leq \frac{E\left[|X - \mu|^2\right]}{\varepsilon^2}$$

As an application, suppose that $X_1, \ldots, X_n$ is a sequence of independent identically distributed random variables having mean μ and finite variance σ^2. Define $\bar{X}_n = \frac{\sum_{i=1}^n X_i}{n}$.

Then, for $\varepsilon > 0$

$$P\left(\left|\bar{X}_n - \mu\right| > \varepsilon\right) \leq \frac{E\left[\left|\bar{X}_n - \mu\right|^2\right]}{\varepsilon^2} = \frac{\sigma^2}{n\varepsilon^2}$$

from which we conclude that $\bar{X}_n \xrightarrow{P} \mu$ as $n \to \infty$. This is known as the weak law of large numbers.

Definition 2.4. We may define the empirical distribution function:

$$F_n(x) = \frac{1}{n}\sum_{i=1}^{n} I\left[X_i \leq x\right]$$

where $I[A]$ is the indicator function. The empirical distribution function provides an estimator of the distribution function $F(x)$.

The empirical distribution function is an example of a random probability measure because the values it takes depend on the observed random variables drawn from F. We can show that

$$E\left[F_n(x)\right] = F(x), Var\left[F_n(x)\right] = \frac{1}{n}F(x)\left(1 - F(x)\right)$$

Hence, from Chebyshev's inequality, we have $F_n(x) \xrightarrow{P} F(x)$ for every fixed x.

An important result in connection with convergence in distribution is Scheffe's theorem which implies that under certain conditions the point-wise convergence of densities implies the convergence in distribution. This result is used in Bayesian nonparametric statistics.

Theorem 2.2. *(Scheffe) Let μ be a sigma finite measure on a probability space. Let $f, f_1, f_2, \ldots$ be measurable functions on a measure space such that as $n \to \infty$ (i)*

$$f_n(\omega) \to f(\omega)$$

almost everywhere and (ii)

$$\limsup_{n\to\infty} \int |f_n|\, d\mu \leq \int |f|\, d\mu < \infty$$

Then for all measurable sets A

$$\lim_{n\to\infty} \int_A f_n d\mu = \int_A f d\mu$$

Specializing to the case of densities, since $\int f_n d\mu = \int f d\mu = 1$, (ii) is satisfied.

There is as well the strong law of large numbers [Casella and Berger, 2002] which states that if $X_1, \ldots, X_n$ is a sequence of independent identically distributed random variables for which $E|X| < \infty$, then for $\varepsilon > 0$:

$$P\left(\lim_{n\to\infty}\left|\bar{X}_n - \mu\right| > \varepsilon\right) = 1.$$

The single most important result in both probability and statistics is the central limit theorem (C.L.T). In its simplest version, it states that if $\{X_i\}$ is a set of independent identically distributed (i.i.d.) random variables with mean μ and finite variance σ^2, then for large enough n

$$\frac{\sqrt{n}\left(\bar{X}_n - \mu\right)}{\sigma} \xrightarrow{\mathcal{L}} \mathbb{N}(0,1)$$

where $\mathbb{N}(0,1)$ denotes the standard normal distribution with mean 0 and variance 1. The assumptions underlying the Central Limit Theorem are relatively weak. There have been countless applications as in approximations to various events involving the sample mean. As well, it has been used to approximate the probabilities of events involving various discrete distributions such as the binomial, Poisson, and negative binomial. An important companion result is due to Slutsky [Casella and Berger, 2002, p.239].

Theorem 2.3. *(Slutsky's Theorem) Suppose that* $X_n \xrightarrow{\mathcal{L}} X$ *and* $Y_n \xrightarrow{P} c$ *for a constant* $c > 0$. *Then*

$$X_n Y_n \xrightarrow{\mathcal{L}} cX$$

For a direct application of Slutsky's theorem, let $\{X_i\}$ be a sequence of i.i.d. random variables having finite variance σ^2 . Let

$$\bar{X}_n = \frac{\sum_{i=1}^n X_i}{n}, S_n^2 = \frac{\sum\left(X_i - \bar{X}_n\right)^2}{n-1}, n > 1$$

be the sample mean and sample variance, respectively. Then, it can be shown that $S_n^2 \xrightarrow{P} \sigma^2$ and hence

$$\frac{\sqrt{n}\left(\bar{X}_n - \mu\right)}{S_n} = \frac{\sqrt{n}\left(\bar{X}_n - \mu\right)}{\sigma}\frac{\sigma}{S_n} \xrightarrow{\mathcal{L}} \mathbb{N}(0,1)$$

An additional result known as the Delta method enables us to extend the central limit theorem to functions of variables which themselves are asymptotically normally distributed [Casella and Berger, 2002, p.243].

Theorem 2.4. *(Delta Method) Let* $\{X_n\}$ *be a sequence of random variables having mean* θ *and finite variance* σ^2 *for which*

$$\sqrt{n}\left(X_n - \theta\right) \xrightarrow{\mathcal{L}} \mathbb{N}\left(0,\sigma^2\right)$$

Let g *be a differentiable real valued function such that* $g'(\theta) \neq 0$. *Then*

$$\sqrt{n}\left(g\left(X_n\right) - g(\theta)\right) \xrightarrow{\mathcal{L}} \mathbb{N}\left(0, \sigma^2\left[g'(\theta)\right]^2\right)$$

A simple application of the Delta method to the case of Bernoulli variables is given in the next example.

Example 2.5. Let $\{X_n\}$ be a random sample from a discrete distribution for which $X = 1$ with probability θ and $X = 0$ with probability $(1-\theta)$, $0 < \theta < 1$. Let

$$g(\theta) = \theta(1-\theta)$$

The central limit theorem asserts that $\frac{\sqrt{n}(\bar{X}_n - \theta)}{\sqrt{\theta(1-\theta)}} \xrightarrow{\mathcal{L}} \mathbb{N}(0,1)$. Then using the Delta method, the asymptotic distribution of $g\left(\bar{X}_n\right)$ is

$$\sqrt{n}\left(g\left(\bar{X}_n\right) - g(\theta)\right) \xrightarrow{\mathcal{L}} \mathbb{N}\left(0, \theta(1-\theta)\left[g'(\theta)\right]^2\right)$$

provided $g'(\theta) \neq 0$, that is, whenever $\theta \neq \frac{1}{2}$. At $\theta = \frac{1}{2}$, $g'\left(\frac{1}{2}\right) = 0$, and hence the Delta method does not apply directly.

In order to determine the asymptotic distribution of $g\left(\bar{X}_n\right)$ at $\theta = \frac{1}{2}$, we may proceed by using a Taylor expansion about $\frac{1}{2}$

$$\begin{aligned} g\left(\bar{X}_n\right) &= g\left(\frac{1}{2}\right) + g'\left(\frac{1}{2}\right)\left(\bar{X}_n - \frac{1}{2}\right) + \frac{1}{2}g''\left(\frac{1}{2}\right)\left(\bar{X}_n - \frac{1}{2}\right)^2 \\ &= \frac{1}{4} + 0 - \left(\bar{X}_n - \frac{1}{2}\right)^2 \end{aligned}$$

This implies that

$$4n\left(\frac{1}{4} - g\left(\bar{X}_n\right)\right) \xrightarrow{\mathcal{L}} \chi_1^2$$

A generalization of the simple central limit theorem to the case of independent but not necessarily identically distributed random variables is given by the Lindeberg-Feller theorem. This result is particularly useful when dealing with the subject of linear rank statistics.

Theorem 2.5. *(Lindeberg-Feller) Suppose that $\{X_i\}$ are independent random variables with means $\{\mu_i\}$ finite variances $\{\sigma_i^2\}$ and distribution functions $\{F_i\}$. Let $S_n = \sum_{i=1}^n (X_i - \mu_i)$, $B_n^2 = \sum_{i=1}^n \sigma_i^2$ and suppose that as $n \to \infty$*

$$\frac{\sigma_n^2}{B_n^2} \to 0, B_n \to \infty.$$

Then for large enough n

$$\frac{S_n}{B_n} \xrightarrow{\mathcal{L}} \mathbb{N}(0,1)$$

if and only if for every $\epsilon > 0$

$$\frac{\sum_{i=1}^n E\left((X_i - \mu_i)^2 I\left(|X_i - \mu_i| > \epsilon B_n\right)\right)}{B_n^2} \to 0, n \to \infty. \tag{2.6}$$

The Lindeberg condition (2.6) is implied by the Lyapunov condition which states that there exists a $\delta > 0$ *such that*

$$\frac{\sum_{i=1}^{n} E\left|X_i - \mu_i\right|^{2+\delta}}{B_n^{2+\delta}} \to 0, n \to \infty.$$

An important companion result to the central limit theorem is in the form of second order asymptotics due to Berry-Essen. Other results deal with Edgeworth expansions which require finite fourth moment [Casella and Berger, 2002].

Theorem 2.6. *(Berry-Essen) Let* $\{X_i\}$ *be independent identically distributed random variables such that* $E[X_i] = 0, E[X_i^2] = 1$ *and* $E[X_i^3] = \rho < \infty$. *Let* G_n *be the distribution function of* $\frac{1}{n}\sum_{i=1}^{n} X_i$. *Then*

$$\sup_{x \in \mathcal{R}} \left|G_n(x) - \Phi(x)\right| \leq \frac{A\rho}{\sqrt{n}}$$

where A *is a universal constant and* Φ *denotes the standard normal distribution.*

Example 2.6. (Lehmann [1975], p.351) Let $\{Y_n\}$ be a random sample from a discrete distribution for which $Y = 1$ with probability θ and $Y = 0$ with probability $(1 - \theta)$. Set $X_i = iY_i$. We would like to determine the asymptotic distribution of

$$S_n = \sum_{i=1}^{n} X_i$$

We note that the mean and variance of X_i are given respectively by

$$\mu_i = i\theta, \sigma_i^2 = i^2\theta(1-\theta).$$

Consequently,

$$B_n^2 = \sum_{i=1}^{n} \sigma_i^2 = \frac{n(n+1)(2n+1)}{6}\theta(1-\theta)$$

On the other hand, since

$$(X_i - \mu_i)^2 = \begin{cases} i^2(1-\theta)^2 & \text{with probability } \theta \\ i^2\theta^2 & \text{with probability } 1-\theta \end{cases}$$

it follows that for sufficiently large n

$$(X_i - \mu_i)^2 < n^2.$$

and consequently

$$\sum_{i=1}^{n} E\left((X_i - \mu_i)^2 I(|X_i - \mu_i| > \epsilon B_n)\right) = 0.$$

In this example, the Lyapunov condition would not be satisfied.

Finally, we note that demonstrating a test statistic has an asymptotic normal distribution can sometimes be more conveniently done in terms of cumulants by showing $\kappa_{2j+1} = 0, j \geq 1$ and $\lim_{n\to\infty} \frac{\kappa_{2j}}{\kappa_2^j} \to 0, j > 1$ (see, e.g., Kendall and Stuart 1979, p.507). We end this section by pointing that there have been many references in the literature to proving central limit theorems via Stein's lemma. For a new approach, see Ross [2011].

2.3. Multivariate Central Limit Theorem

Definition 2.5. A random m-dimensional vector $\boldsymbol{Y}$ is said to have a multivariate normal distribution with mean vector $\boldsymbol{\mu}$ and variance-covariance matrix Σ, written $\boldsymbol{Y} \sim \mathcal{N}_m(\boldsymbol{\mu}, \Sigma)$, if its density is given by

$$f(\boldsymbol{y}) = (2\pi)^{-m/2} |\Sigma|^{-1/2} \exp -\frac{1}{2} (\boldsymbol{y} - \boldsymbol{\mu})^T \Sigma^{-1} (\boldsymbol{y} - \boldsymbol{\mu})$$

where Σ is a positive definite matrix and $|\Sigma|$ denotes its determinant.

The usual estimators of $(\boldsymbol{\mu}, \Sigma)$ based on a random sample of size n are given respectively by

$$\begin{aligned} \bar{\boldsymbol{Y}}_n &= \frac{1}{n} \sum_{i=1}^n \boldsymbol{Y}_i \\ \hat{\Sigma} &= \frac{1}{n-1} \sum_{i=1}^n \left(\boldsymbol{Y}_i - \bar{\boldsymbol{Y}}_n\right)\left(\boldsymbol{Y}_i - \bar{\boldsymbol{Y}}_n\right)^T \end{aligned}$$

If $\boldsymbol{Y} \sim \mathcal{N}_m(\boldsymbol{\mu}, \Sigma)$ and $\boldsymbol{Z} = A\boldsymbol{Y} + \boldsymbol{b}$ for some $q \times m$ matrix of constants A of rank $q \leq m$, and $\boldsymbol{b}$ is a constant q-vector, then

$$\boldsymbol{Z} \sim \mathcal{N}_q\left(A\boldsymbol{\mu} + \boldsymbol{b}, A\Sigma A^T\right)$$

A companion result is the Cràmer-Wold device which states that an m-dimensional random vector $\mathbf{Y}$ has a multivariate normal distribution if and only if every linear combination $\boldsymbol{c}^T\boldsymbol{Y}$ for every $\boldsymbol{c} \varepsilon R^m$ has a univariate normal distribution. The multivariate generalization of the simple univariate central limit theorem follows.

Theorem 2.7. *Multivariate central limit theorem Let $\boldsymbol{Y}_1, \ldots, \boldsymbol{Y}_n$ be a random sample from some m-variate distribution with mean and variance-covariance matrix $\boldsymbol{\mu}, \Sigma$, respectively. Then, as $n \to \infty$, the asymptotic distribution of $\sqrt{n}\left(\bar{\boldsymbol{Y}}_n - \boldsymbol{\mu}\right)$ is multivariate normal with mean $\boldsymbol{0}$ and variance-covariance matrix Σ.*

Corollary 2.1. *Let A be an $q \times m$ matrix, of rank q. Then $\sqrt{n}\left(A\bar{\boldsymbol{Y}}_n - A\boldsymbol{\mu}\right)$ is asymptotically multivariate normal with mean $\mathbf{0}$ and variance-covariance matrix $A\Sigma A^T$.*

Example 2.7. (Multinomial) Consider a generalization of the binomial distribution which allows for k possible categories. Suppose that at the ith trial we observe the jth category. Let $\boldsymbol{X}_i = (0, \ldots, 0, 1, 0, \ldots, 0)$ be the $k-$*dimensional* vector which records a value equal to 1 at the jth position at trial outcome i. Let the frequency vector $\boldsymbol{N} = (N_1, \ldots, N_k)$ denote the number of occurrences of the categories in n repetitions of the same basic experiment conducted under independent and identical conditions. Let the probability vector $\boldsymbol{\pi} = (\pi_1, \ldots, \pi_k)$ denote the probabilities of observing each of the k categories. It follows that

$$\boldsymbol{N} = \sum_{i=1}^{n} \boldsymbol{X}_i$$

The probability distribution associated with $\boldsymbol{N}$ is given by

$$P(N_i = n_i, i = 1, \ldots, k) = \frac{n!}{n_1 \ldots, n_k} \pi_1^{n_1} \ldots \pi_k^{n_k}, \sum \pi_i = 1, \sum n_i = n$$

and it is called the multinomial distribution. The $\{\boldsymbol{X}_i\}$ are i.i.d. with covariance matrix having (ij) component σ_{ij} given by

$$\sigma_{ij} = \begin{cases} \pi_i (1 - \pi_i) & i = j \\ -\pi_i \pi_j & i \neq j \end{cases}$$

Also the covariance matrix of $\boldsymbol{N}$ is not of full rank and is given in matrix notation by

$$\Sigma = Cov(\boldsymbol{N}) = n \left[(diag(\boldsymbol{\pi})) - \boldsymbol{\pi}\boldsymbol{\pi}^T \right]$$

The multivariate central limit theorem then shows that for $n \to \infty$

$$\frac{1}{\sqrt{n}} (\boldsymbol{N} - n\boldsymbol{\pi}) \xrightarrow{\mathcal{L}} \mathcal{N}(0, \Sigma)$$

The multivariate central limit theorem will be used in Chap. 5 to demonstrate the asymptotic normality of test statistics arising from block designs.

2.4. Markov Chains

In this section, we briefly describe discrete Markov Chains and present some of their more important properties. To motivate the discussion, suppose that we have a collection of discrete valued random variables $X_n, n \geq 0$ taking values in $\mathcal{X} = \{0, 1, \ldots\}$ such that they exhibit the Markov Property, namely,

$$P(X_{n+1} = x_{n+1} | X_i = x_i, i = 0, \ldots, n) = P(X_{n+1} = x_{n+1} | X_n = x_n) \tag{2.7}$$

for all $n \geq 0$. Here, X_n may be thought of as describing the state of a system at time n with (2.7) providing the transition probability of the chain going from x_n at time n to x_{n+1} at time $n+1$.

As an example, suppose that a bird transitions from one branch labeled "0" to another branch labeled "1" with the following transition probabilities

$$
\begin{aligned}
P\left(X_{n+1}=1 \mid X_n=0\right) &= p \\
P\left(X_{n+1}=0 \mid X_n=1\right) &= q
\end{aligned}
$$

Let

$$
\left(P\left(X_n=0\right), P\left(X_n=1\right)\right)^T=\pi_n
$$

and

$$
P=\begin{pmatrix} 1-p & p \\ q & 1-q \end{pmatrix}
$$

Then at time n,

$$
\begin{aligned}
\pi_n &= \pi_{n-1} P \\
&= \pi_{n-2} P^2 \\
\ldots & \\
&= \pi_0 P^n
\end{aligned}
$$

It can be shown that [Hoel et al., 1986, p.3]

$$
\begin{aligned}
P\left(X_n=0\right) &= \frac{q}{p+q}+(1-p-q)^n\left(P\left(X_0=0\right)-\frac{q}{p+q}\right) \\
&\rightarrow \frac{q}{p+q}
\end{aligned}
$$

provided $|1-p-q|<1$. If there exists a distribution π is such that

$$
\pi=\pi P
$$

it is said to be the stationary distribution. In our example, the stationary distribution is given by

$$
\pi=\frac{q}{p+q}
$$

A Markov Chain is said to be irreducible if it is possible to go from one state to every other state in a finite number of steps. It is said to be periodic if the chain can only return to its present state at multiples of some integers greater than 1. A sufficient condition for the chain to be aperiodic is if the return to a state x occurs at irregular times with positive probability. The idea is that an aperiodic chain does not get trapped in a cycle.

To guarantee the existence of a unique stationary distribution, a Markov Chain should be irreducible and aperiodic. In that case, it can be shown that the limiting distribution of the Markov Chain as $n \to \infty$ will be the stationary distribution. We will make use of Markov Chains in connection with Bayesian computation of posterior distributions in Part IV.

3. Basic Concepts in Statistics

In this chapter we review some terminology and basic concepts in classical statistical inference. Statistics is concerned in part with the subjects of estimation and hypothesis testing. The starting point for inference is the collection of data often in the form of a random sample from an underlying distribution. In estimation, one is interested in the data-based point-wise estimation of the parameters of a distribution function as well as in the construction of confidence intervals for them. In testing, one aims to confirm on the basis of the evidence collected, a null hypothesis about a parameter as opposed to some alternative hypothesis. Probability enables us to quantify the level of confidence that an interval constructed on the basis of the random sample will actually contain the true value of a parameter. In Sect. 3.1, we review some important properties of estimators along with methods for obtaining estimators, whereas in Sect. 3.2, we consider hypothesis testing. We conclude with a description of classical Bayesian statistics.

3.1. Parametric Estimation

An important part of statistics is the subject of estimation. Often, one is interested in estimating one or more parameters of a distribution function which characterize a certain physical phenomenon. For example, one may be interested in estimating the average heights of individuals in a population or the number of automobile accidents occurring in a certain period of time. It is natural to collect a random sample from the distribution and to calculate an estimate. Which point estimate should one use and what are its properties? We begin with the definition of the likelihood function which forms the basis of parametric estimation.

Suppose that $X_1, \ldots, X_n$ are random variables having joint density $f(x_1, \ldots, x_n; \theta)$, $\theta \in \mathbb{R}^p$. The likelihood viewed as a function of the parameter θ is defined to be the joint density:

$$L(\theta) = f(x_1, \ldots, x_n; \theta). \tag{3.1}$$

Most statistical inference is concerned with obtaining knowledge about the parameter θ on the basis of the sample collected. It sometimes happens that a statistic provides a summary of the data that is most relevant for drawing inference about θ. The notion of sufficiency formalizes this idea.

Definition 3.1. A statistic $A(X_1, \ldots, X_n)$ is said to be sufficient for θ if the conditional density of $X_1, \ldots, X_n$ given $A(X_1, \ldots, X_n)$ is independent of θ.

The factorization theorem which characterizes this concept states that $A = A(X_1, \ldots, X_n)$ is sufficient if and only if there exists a function h of $A = a$ and θ only and

M. Alvo, *Statistical Inference and Machine Learning for Big Data*, Springer Series in the Data Sciences, https://doi.org/10.1007/978-3-031-06784-6_3

another function g independent of θ such that the joint density factors into the product

$$f(x_1,\ldots,x_n;\theta) = h(a,\theta)\, g(x_1,\ldots,x_n).$$

The concept of sufficiency allows us to focus attention on that function of the data which contains all the pertinent information required for inference on θ.

There are some desirable properties which provide guidance on how to choose among estimators. An estimator $A(X_1,\ldots,X_n)$ is said to be unbiased for the estimation of $g(\theta)$ if for all θ

$$E_\theta [A(X_1,\ldots,X_n)] = g(\theta).$$

Example 3.1. Let $X_1,\ldots,X_n$ be a random sample drawn from a distribution with population mean μ and variance σ^2. It is easy to see from properties of the expectation operator that the sample mean $\bar{X}_n = \frac{1}{n}\sum_{i=1}^n X_i$ is unbiased for μ

$$E\left[\bar{X}_n\right] = \frac{1}{n}\sum_{i=1}^n E[X_i] = \frac{1}{n}\sum_{i=1}^n \mu = \mu$$

whereas the sample variance $S^2 = \frac{1}{n-1}\sum_{i=1}^n (X_i - \bar{X}_n)^2$ is unbiased for σ^2. To see the latter, note that S^2 can be re-expressed as

$$S^2 = \frac{1}{n-1}\left\{\sum_{i=1}^n (X_i-\mu)^2 - n(\bar{X}_n-\mu)^2\right\},$$

Since $E(X_i-\mu)^2 = Var(X_i) = \sigma^2$ and $E(\bar{X}_n-\mu)^2 = Var(\bar{X}_n) = \frac{\sigma^2}{n}$, the result follows.

A desirable property for an estimator for $g(\theta)$ is that it should minimize the mean square error:

$$MSE \equiv E_\theta (A(X_1,\ldots,X_n) - g(\theta))^2. \tag{3.2}$$

The mean square error can be expressed as the sum of the variance and the square of the bias

$$MSE = Var_\theta [A(X_1,\ldots,X_n)] + bias^2$$

where the bias is defined to be

$$bias = \{E_\theta [A(X_1,\ldots,X_n)] - g(\theta)\}$$

In minimizing the MSE, there may at times be a trade-off when the bias is not zero. One may be willing to tolerate a slight bias if it leads to an overall smaller MSE.

A further desirable property of estimators is that of consistency which is a minimal property required of any estimator.

Definition 3.2. An estimator $A(X_1,\ldots,X_n)$ of a parameter $g(\theta)$ is said to be consistent if

$$A(X_1,\ldots,X_n) \xrightarrow{P} g(\theta), as\ n \to \infty.$$

Before considering some examples, we present the method of maximum likelihood developed by R.A. Fisher for finding estimators.

Definition 3.3. An estimator of a parameter θ, denoted $\hat{\theta}_n$, is a maximum likelihood estimator if it maximizes the likelihood function $L(\theta)$ in (3.1).

Provided the derivatives exist, the maximization of the likelihood may sometimes be done by maximizing instead the logarithm of the likelihood since

$$\frac{\partial L(\theta)}{\partial \theta} = \frac{1}{L(\theta)} \frac{\partial L(\theta)}{\partial \theta} = \frac{\partial \log L(\theta)}{\partial \theta} = 0$$

Example 3.2. Let $X_1, \ldots, X_n$ be a random sample from a normal distribution with mean μ and variance σ^2. The log likelihood function is given by

$$\log L\left(\mu, \sigma^2\right) = -\frac{n}{2} \log(2\pi) - \frac{n}{2} \log\left(\sigma^2\right) - \frac{\sum (x_i - \mu)^2}{2\sigma^2}$$

and the maximum likelihood equations are then given by

$$\begin{aligned} \frac{\partial \log L(\mu, \sigma^2)}{\partial \mu} &= \frac{\sum (x_i - \mu)}{\sigma^2} = 0 \\ \frac{\partial \log L(\mu, \sigma^2)}{\partial \sigma^2} &= -\frac{n}{2\sigma^2} + \frac{\sum (x_i - \mu)^2}{2\sigma^{-4}} = 0 \end{aligned}$$

These lead to the estimators $\bar{X}_n$ and $\frac{(n-1)S_n^2}{n}$ for the mean and variance, respectively. The next example demonstrates that it is not always feasible to obtain estimators by differentiating the likelihood function.

Suppose we have a random sample from a uniform distribution on the interval $(0, \theta)$. The likelihood function is given by

$$L(\theta) = \begin{cases} \frac{1}{\theta^n} & 0 < \max_{1 \le i \le n} x_i < \theta \\ & \\ 0 & \textit{elsewhere} \end{cases}$$

Since the range of the density depends on θ, we cannot take the derivative of the likelihood function. Instead we note that the likelihood function decreases with increasing values of θ, and, consequently, the maximum likelihood estimator of θ is $\max_{1 \le i \le n} X_i$.

We now illustrate the properties of estimators in specific distributions.

Example 3.3. Let $X_1, \ldots, X_n$ a random sample from a normal distribution $\mathcal{N}(\mu, \sigma^2)$. Then $\bar{X}_n$ is both unbiased and consistent for μ

$$\begin{aligned} E\left[\bar{X}_n\right] &= \mu \\ \bar{X}_n &\xrightarrow{P} \mu \end{aligned}$$

The consistency follows from the Chebyshev inequality:

$$P\left(\left|\bar{X}_n - \mu\right| > \varepsilon\right) \leq \frac{\sigma^2}{n} \to 0.$$

On the other hand, the maximum likelihood estimator for σ^2 can be shown to be

$$\frac{\sum\left(X_i - \bar{X}_n\right)^2}{n} = \frac{n-1}{n} S_n^2$$

and hence is not unbiased for σ^2. However, from Chebyshev's inequality once again and from the fact that

$$\begin{aligned} Var\left(S_n^2\right) &= \sigma^4\left(\frac{2}{n-1} + \left(\frac{\kappa_4}{\kappa_2^2}\right)\frac{1}{n}\right) \\ &= \frac{2\sigma^4}{n-1} \to 0 \end{aligned}$$

we see that $\frac{\sum\left(X_i - \bar{X}_n\right)^2}{n}$ is consistent.

Example 3.4. Suppose we have a random sample from a uniform distribution on the interval $(0, \theta)$. Then the distribution of $M_n = \max_{1 \leq i \leq n} X_i$ is given by

$$F(t) = \left(\frac{t}{\theta}\right)^n, 0 < t < \theta.$$

The mean and variance are given respectively by $\frac{n}{n+1}\theta$ and $\frac{n}{(n+1)^2(n+2)}\theta^2$. It follows that the estimator $\frac{n+1}{n} M_n$ is unbiased for θ . Its variance is given by $\frac{1}{n(n+2)}\theta^2$, and once again by Chebyshev's inequality the estimator is consistent.

Next, we consider the Rao-Blackwell theorem which allows one to improve on an estimator if given a sufficient statistic. This is an important result, and it represents our first exposure to the concept of projection, a tool that will be used in nonparametric statistics.

Theorem 3.1. *Let $\boldsymbol{X} = (X_1, \ldots, X_n)$. Suppose that $A(\boldsymbol{X})$ is a sufficient statistic for θ and that $h(\boldsymbol{X})$is an unbiased estimator of $g(\theta)$. Then, $h^*(a) = E\left[h(\boldsymbol{X}) | A(\boldsymbol{X}) = a\right]$ is an unbiased estimator of $g(\theta)$ based on A which has variance no larger than $Var_\theta\left[h(\boldsymbol{X})\right]$ unless $h(\boldsymbol{x}) = h^*(A(\boldsymbol{x}))$ a.e. in which case, the variances are equal.*

We illustrate the use of the Rao-Blackwell theorem in the next two examples.

Example 3.5. Given a random sample $X_1, \ldots, X_n$ from some distribution, we would like to estimate the mean $E[X]$. We may begin by proposing the unbiased statistic $h(X) = X_1$. The order statistics

$$A(\boldsymbol{X}) = \left(X_{(1)}, \ldots, X_{(n)}\right)$$

are sufficient. Then the Rao-Blackwell theorem assures that $h^*(a) = E[h(\boldsymbol{X}) | A(\boldsymbol{X}) = \boldsymbol{a}]$ is unbiased and will have variance no larger than $Var_\theta[X_1]$. Now, setting $\boldsymbol{a} = \left(x_{(1)}, \ldots, x_{(n)}\right)$ we have by definition

$$\begin{aligned} E[h(\boldsymbol{X}) | A(\boldsymbol{X}) = \boldsymbol{a}] &= \sum_{i=1}^{n} x_{(i)} P\left(X_1 = x_{(i)} | A(\boldsymbol{x})\right) \\ &= \sum_{i=1}^{n} x_{(i)} \frac{1}{n} \\ &= \bar{x}_n \end{aligned}$$

We conclude that the estimator $\bar{X}_n$ is unbiased and its variance will not exceed the variance of X_1. In fact, we know

$$Var\left[\bar{X}_n\right] = \frac{Var[X_1]}{n}$$

Example 3.6. Suppose that we have a random sample $X_1, \ldots, X_n$ from the uniform distribution on the interval $(0, \theta)$. Let $h(X) = 2\bar{X}_n$. Then $h(X)$ is an unbiased estimator of θ. On the other hand, we know that $A(X) = \max_{1 \le i \le n} X_i$ is sufficient for θ and that

$$E[X | A = a] = \int_0^a \frac{x}{a} dx = \frac{a}{2}$$

We conclude from the Rao-Blackwell theorem that the estimator $\max_{1 \le i \le n} X_i$ will have smaller variance than $2\bar{X}_n$.

The concept of completeness is often used to demonstrate the uniqueness of an estimator which together with the conditions of the important Lehmann-Scheffe theorem leads to unique minimum variance unbiased estimators.

Definition 3.4. A family of probability measures $\{F(x; \theta)\}$ is said to be complete if for any statistic $h(x)$ the equation

$$E_\theta[h(X)] = \int h(x)\, dF(x; \theta) = 0$$

for all θ implies $h(x) = 0$ a.e.

We illustrate this concept in the case of a normal distribution.

Example 3.7. Suppose that we have a random sample $X_1, \ldots, X_n$ from the normal distribution with mean μ and variance σ^2. To show that the family of normal densities indexed by $\theta = (\mu, \sigma^2)$ is complete, we see that the expectation to verify completeness leads us to consider an integral of the form

$$\int h(x)\exp\left(-\frac{n}{2\sigma^2}(x-\mu)^2\right)dx = 0$$

The latter implies that for all θ

$$\int h(x)\exp\left(-\frac{n}{2\sigma^2}x^2\right)\exp\left(\frac{n\mu x}{\sigma^2}\right)dx = 0$$

The uniqueness of the Laplace transform implies

$$h(x)\exp\left(-\frac{n}{2\sigma^2}x^2\right) = 0$$

and hence $h(x) = 0$ which proves completeness.

The Lehmann-Scheffe theorem describes conditions under which an estimator is "best."

Theorem 3.2. *Let $X_1, \ldots, X_n$ be a random sample from a distribution which has density or probability mass function $f(x;\theta)$. Suppose $A(X_1, \ldots, X_n)$ is a sufficient statistic for θ, and let $\{f(x;\theta), \theta \in \Omega\}$ be a complete family. If for some function φ we have $E[\varphi(A)] = \theta$, then $\varphi(A)$ is the unique minimum variance unbiased estimator of θ.*

The Lehmann-Scheffe theorem implies that the usual sample mean and sample variance estimators for the normal distribution are unique minimum variance unbiased estimators. Similarly, in Example 3.4, $\frac{n+1}{n}M_n$ is the unique minimum variance unbiased estimator of θ.

We define next the one parameter family of exponential distributions which includes the normal, exponential, gamma, binomial, negative binomial, and Poisson (see Table 2.2. It does not include the uniform and beta distributions.

Definition 3.5. A one parameter exponential family has densities given by

$$g_\theta(x) = e^{\theta x - K(\theta)}g_0(x), \theta \in \Theta, x \in \mathcal{X}$$

with respect to some measure $m(dx)$ such as the uniform on the real line or a discrete measure on the integers. The sets Θ *and* $\mathcal{X}$ are both subsets of the real line, θ is the natural or canonical parameter, and $g_0(x)$ labeled the carrying density is the null hypothesis density under which $\theta = 0$. The normalizing function $K(\theta)$ satisfies

$$\int g_\theta(x)\,m(dx) = 1$$

Let

$$\mathcal{A} = \left\{\theta : \int e^{\theta x}g_0(x)\,m(dx) < \infty\right\}$$

Then it follows from Holder's inequality that the set $\mathcal{A}$ is convex and hence is a possibly infinite interval in R.

The function $K(\theta)$ is the cumulant or log of the moment generating function computed under $g_0(x)$

$$K(\theta) = \log E[\exp(\theta X)].$$

It is a strictly convex function such that

$$E_\theta[X] = K'(\theta), Var_\theta X = K''(\theta)$$

and $K'(\theta)$ is an increasing function of θ.

In repeated sampling where $\boldsymbol{x} = (x_1, \ldots, x_n)$ are the observed values from a random sample from the one parameter exponential family, the joint density is given by

$$g_\theta(\boldsymbol{x}) = e^{n(\theta\bar{x} - K(\theta))} g_0(\boldsymbol{x}), \theta \in \Theta, \boldsymbol{x} \in \mathcal{X}$$

where

$$\begin{aligned} \bar{x} &= \sum_{i=1}^{n} x_i/n \\ g_0(\boldsymbol{x}) &= \Pi_{i=1}^{n} g_0(x_i) \end{aligned}$$

and $\mathcal{X}$ is interpreted as a set in n-dimensional space. This is once again a member of the exponential family but with different parameters. We may generalize the definition to multiparameter exponential families.

Definition 3.6. A multiparameter exponential family has density given by

$$g_{\boldsymbol{\theta}}(\boldsymbol{x}) = e^{\boldsymbol{\theta}^T\boldsymbol{x} - K(\boldsymbol{\theta})} g_0(\boldsymbol{x}), \boldsymbol{\theta} \in \boldsymbol{\Theta}, \boldsymbol{x} \in \boldsymbol{\mathcal{X}}$$

with respect to some measure $m(d\boldsymbol{x})$ for vectors $\boldsymbol{\theta}, \boldsymbol{x}$.

Specializing to the discrete case, we suppose that

$$\pi_{\boldsymbol{\theta}}(\boldsymbol{x}_j) = \exp\left(\boldsymbol{\theta}^T\boldsymbol{x}_j - K(\boldsymbol{\theta})\right) p_j, j = 1, \ldots, m \tag{3.3}$$

where $\boldsymbol{x}_j$ is the jth value of a $k-$ dimensional random vector $\boldsymbol{X}$ and $\boldsymbol{p} = (p_j)$ denotes the vector of probabilities under the null distribution when $\boldsymbol{\theta} = \boldsymbol{\theta}_0$ such that

$$p_j = P(\boldsymbol{X} = \boldsymbol{x}_j)$$

Here $K(\theta)$ is a normalizing constant for which

$$\sum_j \pi_{\boldsymbol{\theta}}(x_j) = 1$$

Let $A = [x_i, \ldots, x_m]$ be the $k \times m$ matrix of possible vector values of $\boldsymbol{X}$. Then under the distribution specified by $\boldsymbol{p}$,

$$\begin{aligned} \sum \equiv Cov\left(\boldsymbol{X}\right) &= E_p\left(\boldsymbol{X} - E\boldsymbol{X}\right)\left(\boldsymbol{X} - E\boldsymbol{X}\right)^T & (3.4) \\ &= A\left(diag\left(\boldsymbol{p}\right)\right)A^T - \left(A\boldsymbol{p}\right)\left(A\boldsymbol{p}\right)^T & (3.5) \end{aligned}$$

where the expectation is with respect to the model (3.3).

Theorem 3.3. *Cràmer-Rao inequality.Suppose that $X_1, \ldots, X_n$ is a random sample from a distribution having density $f\left(x;\theta\right)$. Under certain regularity conditions which permit the exchange of the order of differentiation and integration, the variance of any estimator $A\left(X_1, \ldots, X_n\right)$ is bounded below by*

$$Var\left(A\left(X_1, \ldots, X_n\right)\right) \geq \frac{\left[b'\left(\theta\right)\right]^2}{I\left(\theta\right)}, \quad (3.6)$$

where $b\left(\theta\right) = E_\theta\left(A\left(X_1, \ldots, X_n\right)\right)$ and

$$I\left(\theta\right) = E_\theta\left[-\frac{\partial^2 \log f\left(X_1, \ldots, X_n;\theta\right)}{\partial\theta^2}\right]. \quad (3.7)$$

The expression in (3.7) is known as the Fisher information, and it plays a key role in estimation and hypothesis testing. The regularity conditions are satisfied by members of the exponential family. Under those conditions, it can be shown that

$$E_\theta\left[\frac{\partial f\left(X_1, \ldots, X_n;\theta\right)}{\partial\theta}\right] = 0$$

and

$$\begin{aligned} I\left(\theta\right) &= E_\theta\left[\left(\frac{\partial \log f\left(X_1, \ldots, X_n;\theta\right)}{\partial\theta}\right)^2\right] \\ &= E_\theta\left[-\frac{\partial^2 \log f\left(X_1, \ldots, X_n;\theta\right)}{\partial\theta^2}\right] \end{aligned}$$

Example 3.8. Suppose we have a random sample from the normal distribution with mean μ and variance σ^2. Then, $\bar{X}_n$ is a consistent and unbiased estimator of the population mean μ whose variance is σ^2/n. The Fisher information can be calculated to be $n\sigma^{-2}$, and hence the $Cr\grave{a}mer - Rao$ lower bound for the variance of any unbiased estimator is σ^2/n. Consequently, the sample mean is best since it has minimal variance.

A multiparameter version of the $Cr\grave{a}mer - Rao$ inequality also exists. Suppose that $\boldsymbol{\theta}$ is a $k-$ dimensional parameter. The $k \times k$ Fisher information matrix $I\left(\boldsymbol{\theta}\right)$ has ij component

$$I_{ij}(\theta) = E_\theta \left[-\frac{\partial^2 \log f(X_1, \ldots, X_n; \theta)}{\partial \theta_i \partial \theta_j} \right].$$

provided the partial derivatives exist.

Let $A(X_1, \ldots, X_n)$ be a vector valued estimator of $\boldsymbol{\theta}$, and let

$$b(\boldsymbol{\theta}) = E_{\boldsymbol{\theta}}(A(X_1, \ldots, X_n)).$$

Then under certain regularity conditions, the multiparameter Cràmer–*Rao* lower bound states that

$$Cov_\theta[(A(X_1, \ldots, X_n))] \geq \left[\frac{\partial b(\theta)}{\partial \theta}\right][I(\theta)]^{-1}\left[\frac{\partial b(\theta)}{\partial \theta}\right]'. \tag{3.8}$$

The above inequality is interpreted to mean that each component on the left-hand side is greater than or equal to the corresponding component on the right-hand side. In general, the regularity conditions require the existence of the Fisher information matrix and demand that either the density function has bounded support and the bounds do not depend on θ or that the density has infinite support and is continuously differentiable and its support is independent of θ.

Example 3.9. Suppose we have a random sample from the normal distribution with mean μ and variance σ^2, and we would like to estimate both the mean μ and the variance σ^2. Let

$$A(X_1, \ldots, X_n) = \left[\bar{X}_n, S_n^2\right]$$

be the vector of the sample mean and sample variance, respectively. We calculate

$$I^{-1} = \begin{pmatrix} \frac{\sigma^2}{n} & 0 \\ 0 & \frac{2\sigma^4}{n} \end{pmatrix}$$

It can be shown that $\bar{X}_n$ and S_n are uncorrelated and that

$$Var\bar{X}_n = \frac{\sigma^2}{n}, VarS_n = \frac{2\sigma^4}{n-1}.$$

Consequently, we conclude that only $\bar{X}_n$ attains the Cr*à*mer Rao lower bound, but S_n does not.

A more detailed introduction to statistical estimation may be found in Casella and Berger [2002] and Lehmann and Romano [2005]. We end this section with a discussion of quantiles and their relationship to the order statistics, a subject which is of importance in nonparametric statistics.

Definition 3.7. The quantile of a distribution function F is given as the inverse

$$F^{-1}(t) = \inf\{x : F(x) \geq t\}, 0 < t < 1.$$

The quantile function is left continuous. When the distribution function F is unknown, it may be approximated by the empirical distribution function F_n defined as

$$F_n(x) = \frac{1}{n}\sum_{i=1}^{n} I[X_i \le x].$$

where $I[A]$ is the indicator function of the set A. Similarly we may define the empirical quantile function for a random sample of size n through its relationship to the order statistics as

$$F_n^{-1}(t) = X_{(i)}, \frac{i-1}{n} < t \le \frac{i}{n} \tag{3.9}$$

It can be shown that as $n \to \infty$

$$\sqrt{n}\left(F_n^{-1}(t) - F^{-1}(t)\right) \underset{\to}{\mathcal{L}} \mathcal{N}\left(0, \frac{t(1-t)}{f^2\left(F^{-1}(t)\right)}\right)$$

A useful result for quantiles is that if U is a uniformly distributed random variable on $(0,1)$, then $X = F^{-1}(U)$ has cumulative distribution F. Hence, given a random sample of uniform random variables $U_1, \ldots, U_n$, it follows that $F^{-1}(U_1), \ldots, F^{-1}(U_n)$ will be a random sample with cumulative distribution function F. By the same token, $F^{-1}\left(U_{(1)}\right), \ldots, F^{-1}\left(U_{(n)}\right)$ will behave like a sample of order statistics from cumulative distribution function F. For further properties of the quantile function, see Van der Vaart [2000, Chapter 21].

3.2. Hypothesis Testing

Suppose that $X_1, \ldots, X_n$ is a random sample from a distribution whose density is given by $f(x;\theta), \theta \in R^p$. In hypothesis testing, it is common to formulate two hypotheses on the parameter θ: the null hypothesis denoted H_0 represents the status quo of no change and the alternative denoted H_1 represents the hypothesis that we are hoping to accept. To illustrate, suppose that we are interested in assessing whether or not a new drug represents a more effective treatment. The null hypothesis indicates that the new drug is not effective whereas the alternative hypothesis states that it is. Since the assessment is made on the basis of evidence contained in a random sample, there is the possibility of error. A type I error results when one falsely rejects the null hypothesis. A type II error occurs when one falsely accepts the null hypothesis. In our example, a type I error means we would incorrectly adopt an ineffective drug and consequently is the more serious error we wish to control. On the other hand, a type II error means we incorrectly forgo an effective drug. Since both types of error cannot simultaneously be controlled, it is customary to prescribe a bound on the type I error probability and then to search for tests that minimize the probability of a type II error.

Formally, we can express the testing problem as

$$H_0 : \theta \in \Theta_0$$

against the alternative hypothesis

$$H_1 : \theta \in \Theta_1$$

where Θ_0, Θ_1 are disjoint sets. In the situation where both Θ_0, Θ_1 consist of single points θ_0 *and* θ_1, respectively, the Neyman-Pearson lemma provides an optimal solution to the hypothesis testing problem. Set $\boldsymbol{x} = (x_1, \ldots, x_n)$, and let $\phi(\boldsymbol{x})$ be a critical function which represents the probability of rejecting the null hypothesis when $\boldsymbol{x}$ is observed. Also let α denote the probability of rejecting the null hypothesis when it is true (also known as the size of the test).

Lemma 3.1. *(Neyman-Pearson Lemma) Suppose that there exists some dominating measure μ with respect to which we have densities $f_0 = f(x; \theta_0)$ and $f_1 = f(x; \theta_1)$. Then the most powerful test of $H_0 : \theta = \theta_0$ against $H_1 : \theta = \theta_1$ is given by*

$$\phi(\boldsymbol{x}) = \begin{cases} 1 & \frac{\prod_{i=1}^{n} f(x_i;\theta_1)}{\prod_{i=1}^{n} f(x_i;\theta_0)} > k \\ \gamma & \frac{\prod_{i=1}^{n} f(x_i;\theta_1)}{\prod_{i=1}^{n} f(x_i;\theta_0)} = k \\ 0 & \frac{\prod_{i=1}^{n} f(x_i;\theta_1)}{\prod_{i=1}^{n} f(x_i;\theta_0)} < k \end{cases}$$

where k, γ are chosen such that $E_{\theta_0}\phi(\boldsymbol{X}) = \alpha$. The power function of a test ϕ is defined to be

$$E_\theta \phi(\boldsymbol{X}) = \int \phi(\boldsymbol{X}) \prod_{i=1}^{n} f(x_i; \theta)\, dx_1 \ldots dx_n$$

and it represents the probability of rejecting the null hypothesis for a given θ.

Example 3.10. Let $X_1, \ldots, X_n$ be a random sample of size n from a $N(\mu, \sigma^2)$ with known variance. Then the uniformly most powerful test of the hypothesis

$$H_0 : \mu = \mu_0$$

against the alternative hypothesis

$$H_1 : \mu = \mu_1 > \mu_0.$$

is given by

$$\phi(\boldsymbol{x}) = \begin{cases} 1 & \bar{x}_n > k \\ 0 & \bar{x}_n < k \end{cases}$$

where k is chosen such that $E_{\theta_0}\phi(\boldsymbol{X}) = \alpha$.

Example 3.11. In the case of a random sample from an exponential distribution with mean θ describing the lifetimes of light bulbs, the uniformly most powerful test of

$$H_0 : \theta = \theta_0$$

against the alternative hypothesis

$$H_1 : \theta = \theta_1 < \theta_0.$$

is given by

$$\phi(\boldsymbol{x}) = \begin{cases} 1 & \bar{x}_n > k \\ \\ 0 & \bar{x}_n < k \end{cases}$$

where k is chosen such that $E_{\theta_0}\phi(\boldsymbol{X}) = \alpha$.

Uniformly most powerful (UMP) tests rarely exist in practice. An exception occurs when the family of densities possesses a monotone likelihood ratio.

Definition 3.8. A family of densities $\{f(x;\theta), \theta \in I\}$ has monotone likelihood ratio if the ratio $\frac{f(x;\theta_2)}{f(x;\theta_1)}$ is non-decreasing in some function $A(x)$ for all $\theta_1 < \theta_2$ in some interval I.

The exponential family of densities has monotone likelihood ratio since

$$\frac{f(x;\theta_2)}{f(x;\theta_1)} = \exp\{(\theta_2 - \theta_1)\, t(x) - (K(\theta_2) - K(\theta_1))\}$$

which is non-decreasing in x provided $t(x)$ is also non-decreasing in x. In the situation where the family of densities has monotone likelihood ratio, it can be shown that there exists uniformly most powerful tests [Lehmann and Romano, 2005]. Specifically, a test ϕ_0 is uniformly most powerful of size α in a class C if it has size α and if its power function satisfies

$$E_\theta \phi_0(\boldsymbol{X}) \geq E_\theta \phi(\boldsymbol{X}), \theta \in \Theta_1$$

for all tests ϕ in class C.

As an application, we consider a test for the mean of a normal distribution

$$H_0 : \mu = \mu_0$$

against the alternative hypothesis

$$H_1 : \mu > \mu_0.$$

The uniformly most powerful test of H_0 against the alternative hypothesis

$$H_1^* : \mu = \mu_1 > \mu_0$$

is given by the Neyman-Pearson lemma. The family has monotone likelihood ratio in x and hence the test becomes

$$\phi(\boldsymbol{x}) = \begin{cases} 1 & \bar{x}_n > k \\ 0 & \bar{x}_n < k \end{cases}$$

The test statistic has non-decreasing power function:

$$\Phi\left(-z_\alpha + \frac{\sqrt{n}(\mu - \mu_0)}{\sigma}\right) \geq \alpha, \mu > \mu_0$$

where z_α is the upper $100(1-\alpha)\%$ point of a standard normal distribution and consequently it is uniformly most powerful.

A generalization of the Neyman-Pearson lemma when both Θ_0, Θ_1 are composite sets is based on the likelihood ratio

$$\Lambda_n = \frac{\sup_{\theta \epsilon \Theta} \prod_{i=1}^n f(x_i;\theta)}{\sup_{\theta \epsilon \Theta_0} \prod_{i=1}^n f(x_i;\theta)}$$

The likelihood ratio test rejects the null hypothesis whenever Λ_n is large. The factorization theorem shows that the likelihood ratio statistic Λ_n is based on the sufficient statistic, and moreover, it is invariant under transformations of the parameter space that leave the hypothesis and alternative hypotheses invariant. For a random sample of size n from the multiparameter exponential family, we have

$$\log \Lambda_n = n \sup_{\theta \in \Theta} \inf_{\theta_0 \in \Theta_0} \left[(\theta - \theta_0)^T \bar{t}_n - K(\theta) + K(\theta_0)\right]$$

where

$$\bar{t}_n = \sum_{i=1}^n t(x_i)/n$$

In certain situations, as in the case of a Cauchy location family, a uniformly powerful test may not exist. On the other hand, a locally most powerful test which maximizes the power function at θ_0 may exist. Provided we can differentiate under the integral sign the power function with respect to θ, the generalized Neyman-Pearson lemma [Lehmann and Romano, 2005] leads to a test of the form

$$\phi(\boldsymbol{x}) = \begin{cases} 1 & \frac{\partial}{\partial\theta} \prod_{i=1}^n f(x_i;\theta_0) > k \prod_{i=1}^n f(x_i;\theta_0) \\ 0 & \frac{\partial}{\partial\theta} \prod_{i=1}^n f(x_i;\theta_0) < k \prod_{i=1}^n f(x_i;\theta_0) \end{cases}$$

which maximizes the power function at θ_0 among all tests of size α.

Definition 3.9. Let $X_1, \ldots, X_n$ be a random sample from some distribution having density $f(x;\theta), \theta \in R$. Let $L(\theta;\boldsymbol{x})$ be the likelihood function where $\boldsymbol{x}^T = (x_1, \ldots, x_n)$ and set

$$l(\theta;\boldsymbol{x}) = logL(\theta;\boldsymbol{x}).$$

The derivative

$$U(\theta;\boldsymbol{x}) = \frac{\partial l(\theta;\boldsymbol{x})}{\partial\theta}$$

is called the score function. The locally most powerful test can be seen to be equivalently based on the score function since the inequality

$$\frac{\partial}{\partial\theta}\prod_{i=1}^{n} f(x_i;\theta_0) > k\prod_{i=1}^{n} f(x_i;\theta_0)$$

can be re-expressed as

$$U \equiv \frac{\partial}{\partial\theta}\left[\log\prod_{i=1}^{n} f(x_i;\theta_0)\right] > k.$$

Lemma 3.2. *For $\theta\epsilon R^p$ define the vector score function $U(\theta;\boldsymbol{x}) = \left(\frac{\partial l(\theta;\boldsymbol{x})}{\partial\theta_i}\right)$. Provided we can differentiate under the integral signs below, the following properties hold*
(i) $E[U(\theta;X)] = 0$ and (ii) $Cov[U(\theta;X)] = I(\theta)$

Proof.

(i)

$$\begin{aligned} E[U(\theta;X)] &= \int \frac{\partial l(\theta;x)}{\partial\theta} f(x;\theta)\,dx \\ &= \int \frac{\partial f(x;\theta)}{\partial\theta}dx \\ &= \frac{\partial}{\partial\theta}\int f(x;\theta)\,dx = 0 \end{aligned}$$

(ii) The (ij) term of the matrix $Cov[U(\theta;X)]$ is given by $E\left[\frac{\partial l(\theta;X)}{\partial\theta_i}\frac{\partial l(\theta;X)}{\partial\theta_j}\right]$

Since

$$\begin{aligned} 0 &= \frac{\partial}{\partial\theta_i}\int \frac{\partial l(\theta;x)}{\partial\theta_j} f(x;\theta)\,dx \\ &= \int \frac{\partial l^2(\theta;x)}{\partial\theta_i\partial\theta_j} f(x;\theta)\,dx + \int \frac{\partial l(\theta;x)}{\partial\theta_j}\frac{\partial}{\partial\theta_i} f(x;\theta)\,dx \\ &= \int \frac{\partial l^2(\theta;x)}{\partial\theta_i\partial\theta_j} f(x;\theta)\,dx + \int \frac{\partial l(\theta;x)}{\partial\theta_j}\frac{\partial l(\theta;x)}{\partial\theta_j}dx \end{aligned}$$

the result follows.

□

There are three distinct possible test statistics for testing a general hypothesis all of which are asymptotically equivalent as the sample size gets large. These are the likelihood ratio test, the Wald test, and the Rao score test. In the lemma below, we outline the proof for the asymptotic distribution of the Rao score test. Let $\hat{\theta}_0$ denote the maximum likelihood estimate of θ under the null hypothesis.

Lemma 3.3. *Rao Score test. Let $X = (X_1, \ldots, X_n)$ be a random sample from some continuous distribution having density $f(x; \theta), \theta \in R^p$, and we wish to test*

$$H_0 : \theta \in \Theta_0$$

against the alternative

$$H_1 : \theta \in \Theta_1.$$

Let $\hat{\theta}_0$ be the maximum likelihood estimate of θ under H_0. Under the null hypothesis, the test statistic

$$U^T\left(\hat{\theta}_0; X\right) I^{-1}\left(\hat{\theta}_0\right) U\left(\hat{\theta}_0; X\right)$$

has asymptotically a χ^2_k as the sample size gets large where k *is the number of constraints imposed by the null hypothesis.*

Proof. The result follows from the multivariate central limit theorem since the score function is a sum of independent identically distributed random vectors

$$\begin{aligned} U(\theta; X) &= \frac{\partial}{\partial \theta}\left[\log \prod_{i=1}^{n} f(X_i; \theta)\right] \\ &= \sum_{i=1}^{n}\left[\frac{\partial}{\partial \theta} \log f(X_i; \theta)\right] \end{aligned}$$

with mean 0 and covariance matrix $I(\theta)$. □

We label the next theorem as the "Three Amigos" because the three test statistics are intimately related.

Theorem 3.4. *Let $X_1, \ldots, X_n$ be a random sample from a continuous distribution having density $f(x; \theta), \theta \in R^p$. Suppose we wish to test the null hypothesis $H_0 : \theta \in \Theta_0$ against the alternative $H_1 : \theta \in \Theta_1$. Then, the likelihood test, the Wald test, and the Rao score test all reject the null hypothesis for large values and are asymptotically distributed as central chi square distributions with k degrees of freedom as $n \to \infty$ where k is the number of constraints imposed by the null hypothesis. Specifically,*

the likelihood ratio test rejects whenever

$$-2 \log\left[\frac{L(\theta_0; X)}{L\left(\hat{\theta}; X\right)}\right] > \chi^2_k(\alpha)$$

The Wald test rejects whenever

$$W = \left(\hat{\theta} - \theta_0\right)^T I\left(\hat{\theta}\right)\left(\hat{\theta} - \theta_0\right) > \chi^2_k(\alpha)$$

The Rao score test rejects whenever

$$U^T\left(\hat{\theta}_0; X\right) I^{-1}\left(\hat{\theta}_0\right) U\left(\hat{\theta}_0; X\right) > \chi_k^2(\alpha)$$

where $\hat{\theta}_0$ *is the maximum likelihood estimate of* θ *under* H_0.

Proof. By expanding the likelihood $L\left(\hat{\theta}; X\right)$ around $\theta \in \Theta_0$ in a second-order Taylor series, it can be shown that these three tests are equivalent [Van der Vaart, 2000, p.231] or [Cox and Hinkley, 1974, p.315 and p.324]. □

We may illustrate the theorem in the following examples.

Example 3.12. Let $X_1, \ldots, X_n$ be a random sample from the Poisson distribution with mean $\theta > 0$. Suppose we wish to test

$$H_0 : \theta = \theta_0$$

against the alternative

$$H_1 : \theta \neq \theta_0.$$

Then the score function evaluated at θ_0 is

$$U\left(\theta_0; \bar{X}\right) = \frac{n}{\theta}\left(\bar{X} - \theta_0\right)$$

whereas the Fisher information is given by $I(\theta) = \frac{n}{\theta}$. Hence the Rao score test rejects whenever

$$\frac{\theta}{n}\left[\frac{n}{\theta}\left(\bar{X} - \theta_0\right)\right]^2 > \chi_1^2(\alpha)$$

The likelihood ratio test and the Wald test have identical forms in this case.

Example 3.13. Let $X_1, \ldots, X_n$ be a random sample from the normal distribution with mean μ and variance σ^2. Suppose we wish to test

$$H_0 : \mu = \mu_0$$

against the alternative

$$H_1 : \mu \neq \mu_0.$$

Then the score function evaluated at μ_0 is

$$U\left(\mu_0; \bar{X}\right) = \frac{n}{\sigma^2}\left(\bar{X} - \mu_0\right)$$

and the Fisher information is given by $I(\theta) = \frac{n}{\sigma^2}$. Hence the Rao score test rejects whenever

$$\frac{\sigma^2}{n}\left[\frac{n}{\sigma^2}\left(\bar{X} - \mu_0\right)\right]^2 > \chi_1^2(\alpha)$$

Since σ^2 is unknown, we may replace it by its maximum likelihood estimator $\frac{n-1}{n}S_n^2$, and hence the score test rejects whenever

$$\left[\frac{n}{n-1}\left(\sqrt{n}\left(\frac{\bar{X}-\mu_0}{S_n}\right)\right)\right]^2 > \chi_1^2(\alpha)$$

Example 3.14. Recall from (3.3) that in the discrete case

$$\pi_{\boldsymbol{\theta}}(\boldsymbol{x}_j) = \exp\left(\boldsymbol{\theta}^T\boldsymbol{x}_j - K(\boldsymbol{\theta})\right)p_j, j = 1,\ldots,m$$

where $\boldsymbol{x}_j$ is the jth value of a k – dimensional random vector $\boldsymbol{X}$ and $\boldsymbol{p} = (p_j)$ denotes the vector of probabilities under the null distribution when $\boldsymbol{\theta} = \boldsymbol{\theta}_0$ such that

$$p_j = P(\boldsymbol{X} = \boldsymbol{x}_j)$$

Here $K(\boldsymbol{\theta})$ is a normalizing constant for which

$$\sum_j \pi_{\boldsymbol{\theta}}(\boldsymbol{x}_j) = 1$$

Suppose we wish to test

$$H_0 : \boldsymbol{\theta} = \boldsymbol{\theta_0}$$

against the alternative

$$H_1 : \boldsymbol{\theta} \neq \boldsymbol{\theta}_0.$$

Letting $\boldsymbol{N}$ denote a multinomial random variable with parameters $(n, \boldsymbol{p})$, we see that the log likelihood as a function of $\boldsymbol{\theta}$ is, apart from a constant, proportional to

$$\begin{aligned}\sum n_j \log(\pi_{\boldsymbol{\theta}}(\boldsymbol{x}_j)) &= \sum n_j\left(\boldsymbol{\theta}^T\boldsymbol{x}_j - K(\boldsymbol{\theta})\right)\\ &= \boldsymbol{\theta}^T\left(\sum n_j\boldsymbol{x}_j\right) - nK(\boldsymbol{\theta})\end{aligned}$$

The score vector under the null hypothesis is then given by

$$\begin{aligned}U(\boldsymbol{\theta}) &= \sum N_j\left(\frac{1}{\pi_{\boldsymbol{\theta}}(\boldsymbol{x}_j)}\frac{\partial\pi_{\boldsymbol{\theta}}(\boldsymbol{x}_j)}{\partial\boldsymbol{\theta}}\right)\\ &= A(\boldsymbol{N} - n\boldsymbol{p})\end{aligned}$$

where $A = [x_1, \ldots, x_m]$

Under the null hypothesis,

$$E[U(\boldsymbol{\theta})] = 0$$

and the score statistic is given by

$$\begin{aligned}\frac{1}{n}\left[A(\boldsymbol{N}-n\boldsymbol{p})\right]^{T}\Sigma^{-1}\left[A(\boldsymbol{N}-n\boldsymbol{p})\right] &= \frac{1}{n}\left[(\boldsymbol{N}-n\boldsymbol{p})\right]^{T}\left(A^{T}\Sigma^{-1}A\right)\left[(\boldsymbol{N}-n\boldsymbol{p})\right] \\ &\xrightarrow{\mathcal{L}} \chi_r^2\end{aligned}$$

where Σ^{-1} is the inverse matrix of Σ and $r = rank\left(A^T\Sigma^{-1}A\right)$.

Here

$$\sum \equiv Cov\left(\boldsymbol{X}\right) = A\left(diag\left(\boldsymbol{p}\right)\right)A^T - \left(A\boldsymbol{p}\right)\left(A\boldsymbol{p}\right)^T$$

We will have occasion to consider this result in Chap. 5 on nonparametric statistics. We conclude this section by a brief discussion of the concept of contiguity which plays an important in connection with the efficiency of tests.

Thus far, we were concerned with determining the asymptotic distribution of the test statistics under the null hypothesis. It is of interest for situations related to the efficiency of tests, to determine the asymptotic distribution of the test statistic under the alternative hypothesis as well. Since most of the tests considered in practice are consistent, they tend to perform well for alternatives "far away" from the null hypothesis. Consequently, the interest tends to focus on alternatives which are "close" to the null hypothesis in a manner to be made more precisely here. The concept of contiguity was introduced by Le Cam in connection with the study of local asymptotic normality. Specifically, the concept enables one to obtain the limiting distribution of a sequence of statistics under the alternative distribution from knowledge of the limiting distribution under the null hypothesis distribution. We follow closely the development and notation in Hajek and Sidak [1967]; Van der Vaart [2000], and we begin with some definitions.

Definition 3.10. Suppose that P and Q are two probability measures defined on the same measurable space $(\Omega, \mathcal{A})$. We shall say that Q is absolutely continuous with respect to P, denoted $Q \ll P$ if $Q(A) = 0$ whenever $P(A) = 0$ for all measurable sets $A\epsilon\mathcal{A}$.

If $Q \ll P$, then we may compute the expectation of a function $f(\boldsymbol{X})$ of a random vector $\boldsymbol{X}$ under Q from knowledge of the expectation under P of the product $\left[f(\boldsymbol{X})\frac{dQ}{dP}\right]$ through the change of measure formula:

$$E_Q\left[f(\boldsymbol{X})\right] = E_P\left[f(\boldsymbol{X})\frac{dQ}{dP}\right]$$

The expression $\left[\frac{dQ}{dP}\right]$ is known as the Radon-Nikodym derivative. The asymptotic version of absolute continuity is the concept of contiguity.

Definition 3.11. Suppose that P_n and Q_n are two sequences of probability measures defined the same measurable space $(\Omega_n, \mathcal{A}_n)$. We shall say that Q_n is contiguous with respect to P_n, denoted $Q_n \lhd P_n$ if $Q_n(A_n) \to 0$ whenever $P_n(A_n) \to 0$ for all measurable sets $A_n \in \mathcal{A}_n$. The sequences $P_n and Q_n$ are mutually contiguous if both $Q_n \lhd P_n$ and $P_n \lhd Q_n$.

Example 3.15. Suppose that under P_n we have a standard normal distribution, whereas under Q_n we have a normal distribution with mean $\mu_n \to \mu$ and variance $\sigma^2 > 0$. Then it follows that P_n and Q_n are mutually contiguous. Note however that if $\mu_n \to \infty$ and then for $A_n = [\mu_n, \mu_n + 1]$, we have that $P_n(A_n) \to 0$, yet $Q_n(A_n) \to$constant.

Le Cam proposed three lemmas which enable us to verify contiguity and to obtain the limiting distribution under one measure given the limiting distribution under another. Suppose that P_n and Q_n admit densities p_n and q_n, respectively, and define the likelihood ratio L_n for typical points $\boldsymbol{x}$ in the sample space as

$$L_n(\boldsymbol{x}) = \begin{cases} \frac{q_n(\boldsymbol{x})}{p_n(\boldsymbol{x})} & p_n(\boldsymbol{x}) > 0 \\ 1 & p_n(\boldsymbol{x}) = q_n(\boldsymbol{x}) = 0 \\ 0 & p_n(\boldsymbol{x}) = 0 < q_n(\boldsymbol{x}) \end{cases}$$

Lemma 3.4. *(Le Cam's First Lemma). Let F_n be the cdf of L_n which converges weakly under P_n at continuity points to a distribution F for which*

$$\int_0^\infty x dF(x) = 1.$$

Then the densities q_n are contiguous to the densities $p_n, n \geq 1$.

As a corollary, if log L_n is under P_n asymptotically normal with mean $-\frac{\sigma^2}{2}$ and variance σ^2, then the densities q_n are mutually contiguous to the densities p_n. This follows from the fact that the moment generating function of a normal (μ, σ^2) evaluated at $t = 1$ must equal 1, that is, for large n,

$$\begin{aligned} E\left[e^{\log L_n}\right] &\to e^{\mu + \frac{\sigma^2}{2}} \\ &= e^{-\frac{\sigma^2}{2} + \frac{\sigma^2}{2}} = 1. \end{aligned}$$

Le Cam's second lemma provides conditions under which a log likelihood ratio is asymptotically normal under the hypothesis probability measure.

Let $\boldsymbol{x} = (x_1, ..., x_n)^T$ and set

$$p_n(\boldsymbol{x}) = \prod_{i=1}^{n} f_{ni}(x_i)$$

$$q_n(\boldsymbol{x}) = \prod_{i=1}^{n} g_{ni}(x_i)$$

Then

$$\log L_n(\boldsymbol{X}) = \sum_{i=1}^{n} \log\left[g_{ni}(X_i) / f_{ni}(X_i)\right]$$

Let

$$W_n = 2\sum_{i=1}^{n} \left\{ \left[g_{ni}\left(X_i\right) / f_{ni}\left(X_i\right) \right]^{\frac{1}{2}} - 1 \right\}$$

Lemma 3.5. *(Le Cam's Second Lemma). Suppose that the following uniform integrability condition holds*

$$\lim_{n\to\infty} \max_{1\le i\le n} P_n\left(\left| g_{ni}\left(X_i\right) / f_{ni}\left(X_i\right) - 1 \right| > \varepsilon \right) = 0, \text{ for all } \varepsilon > 0$$

and W_n is asymptotically $\mathcal{N}\left(-\frac{\sigma^2}{4}, \sigma^2\right)$ under P_n. Then

$$\lim_{n\to\infty} P_n\left(\left| \log\, L_n\left(\boldsymbol{X}\right) - W_n + \frac{\sigma^2}{4} \right| > \varepsilon \right) = 0, \varepsilon > 0$$

and under P_n, $\log\, L_n\left(\boldsymbol{X}\right)$ is asymptotically $\mathcal{N}\left(-\frac{\sigma^2}{2}, \sigma^2\right)$.

The third lemma which is most often utilized establishes the asymptotic distribution under the alternative hypothesis provided the measures are contiguous.

Lemma 3.6. *(Le Cam's Third Lemma). Let P_n, Q_n be sequences of probability measures on measurable spaces $(\Omega_n, \mathcal{A}_n)$ such that $Q_n \lhd P_n$. Let $\boldsymbol{Y}_n$ be a sequence of k – dimensional random vectors such that under P_n*

$$\left(\boldsymbol{Y}_n, \log\, L_n\left(\boldsymbol{X}\right)\right) \xrightarrow{\mathcal{L}} \mathcal{N}_{k+1}\left(\left(\begin{array}{c} \mu \\ -\frac{\sigma^2}{2} \end{array} \right), \left(\begin{array}{cc} \Sigma & \tau \\ \tau' & \sigma^2 \end{array} \right) \right)$$

Then, under Q_n

$$\boldsymbol{Y}_n \xrightarrow{\mathcal{L}} \mathcal{N}_k\left(\mu + \tau, \Sigma\right)$$

We may now consider an important application of Le Cam's lemmas. Suppose that we have two sequences of probability measures defined by densities $p_\theta, p_{\theta+h/\sqrt{n}}$.

Let $\ell\left(\theta; x\right) = \log\, p_\theta\left(x\right)$. Then under suitable regularity conditions, a second-order Taylor series expansion yields under p_θ

$$\log\left[\prod_i \left(\frac{p_{\theta+h/\sqrt{n}}\left(X_i\right)}{p_\theta\left(X_i\right)} \right) \right] = \frac{1}{\sqrt{n}} \sum_{i=1}^{n} h'\dot{\ell}_\theta\left(X_i\right) - \frac{h' I_\theta h}{2} + o\left(1\right) \tag{3.10}$$

where $\dot{\ell}_\theta\left(X_i\right)$ is the k-dimensional vector of partial derivatives. The expansion in (3.10) is known as the local asymptotic normality property (LAN). It follows from the Central Limit Theorem that the right-hand side in (3.10) is asymptotically normal $\mathcal{N}\left(-\frac{h' I_\theta h}{2}, h' I_\theta h\right)$ and consequently, by Le Cam's first lemma, the measures $p_\theta, p_{\theta+h/\sqrt{n}}$ on the left-hand side are contiguous. Next, suppose that we have a sequence of test statistics T_n which are to first order, sums of i.i.d. random variables:

$$\sqrt{n}\left(T_n - \mu\right) = \frac{1}{\sqrt{n}} \sum_{i=1}^{n} \psi\left(X_i\right) + o_{p_\theta}\left(1\right)$$

for some function ψ. Consequently, the joint distribution of $\frac{1}{\sqrt{n}}\sum_i \left(\psi(X_i), \dot{\ell}_\theta(X_i)\right)$ is under p_θ asymptotically multivariate normal

$$\left(\sqrt{n}\,(T_n-\mu), \log\left[\prod_i \left(\frac{p_{\theta+h/\sqrt{n}}(X_i)}{p_\theta(X_i)}\right)\right]\right) \xrightarrow{\mathcal{L}} \mathcal{N}_{k+1}\left(\begin{pmatrix} 0 \\ -\frac{h' I_\theta h}{2} \end{pmatrix}, \begin{pmatrix} \Sigma & \tau \\ \tau' & h' I_\theta h \end{pmatrix}\right)$$

It then follows from Le Cam's third lemma that $\sqrt{n}\,(T_n-\mu)$ is asymptotically normal under $p_{\theta+h/\sqrt{n}}$.

We may demonstrate contiguity in the case of an exponential family of distributions.

Example 3.16. Let $\boldsymbol{X}=(X_1,\dots,X_n)^T$ be a random sample from the exponential family with a real valued parameter θ

$$f(x;\theta)=\exp\left[\theta t(x)-K(\theta)\right]g_0(x)$$

The log likelihood and its derivatives are respectively

$$\begin{aligned}
\ell(\theta;\boldsymbol{X}) &= \left[\theta\sum_{i=1}^n t(X_i)-nK(\theta)\right]\\
\ell'(\theta;\boldsymbol{X}) &= \left[\sum_{i=1}^n t(X_i)-nK'(\theta)\right]\\
\ell''(\theta;\boldsymbol{X}) &= -nK''(\theta)
\end{aligned}$$

It follows that

$$\begin{aligned}
\log\left[\prod_{i=1}^n\left(\frac{p_{\theta+h/\sqrt{n}}(X_i)}{p_\theta(X_i)}\right)\right] &= \frac{h}{\sqrt{n}}\sum_{i=1}^n t(X_i)-n\left[K\left(\theta+\frac{h}{\sqrt{n}}\right)-K(\theta)\right]\\
&= \frac{h}{\sqrt{n}}\sum_{i=1}^n\left[t(X_i)-K'(\theta)\right]-\frac{h^2}{2}K''(\theta)+o(1)
\end{aligned}$$

and, consequently, the measures $p_\theta=f(\boldsymbol{x};\theta)$, $p_{\theta+h/\sqrt{n}}=f\left(\boldsymbol{x};\theta+\frac{h}{\sqrt{n}}\right)$ are contiguous.

The concept of contiguity enables us to obtain the non-null asymptotic distribution of various linear rank statistics to be defined in Chap. 5 on nonparametric statistics.

3.3. Classical Bayesian Statistics

Bayesian methods have been successfully used to incorporate prior information alongside data into the model. This is an important aspect especially in situations where data is scarce. Bayesian methods are straightforward to use and can deal with complex hierarchical models that involve many parameters. In what follows, we shall present classical Bayesian methods and postpone to later chapters the relatively more recent Markov Chain Monte Carlo (MCMC) methods and Bayesian nonparametric paradigms.

The probability distribution function of a random variable X described by a density $f(x;\theta)$ may be a function of a parameter θ often assumed to be an unknown but fixed constant. In the classical Bayesian paradigm, θ is treated as a random variable itself with its own prior distribution function. In that case, $f(x;\theta)$ is considered to be the conditional density of X given θ. If the prior density of θ is given by $\pi(\theta;\alpha)$ where α is a parameter, then the conditional density of θ given $X = x$, labeled the posterior density of θ, is given by Bayes' theorems :

$$\pi(\theta;x,\alpha) = \frac{f(x;\theta)\,\pi(\theta;\alpha)}{\int f(x;\theta)\,\pi(\theta;\alpha)\,d\theta} \tag{3.11}$$

where the marginal density of X is calculated to be

$$\int f(x;\theta)\,\pi(\theta;\alpha)\,d\theta.$$

The use of Bayesian methods enables us to update information about the prior by making use of data embodied in X. The posterior distribution then contains all the information needed to make probability statements about θ. There have been countless applications of classical Bayesian inference in practice, and we refer the reader to Box and Tiao [1973] for further details. We illustrate the computation of the posterior density in a simple example.

Example 3.17. Let $X_1, \ldots, X_n$ be a random sample from a normal distribution with mean θ and known variance σ^2. Assume that θ is itself a random variable having a normal distribution with mean μ and variance τ^2. Then the posterior distribution of θ given the sample data is proportional to

$$\begin{aligned}
\pi(\theta;x,\alpha) &\sim \prod f(x_i;\theta)\,\pi\left(\theta;\mu,\tau^2\right) \\
&\sim \exp -\frac{n\sigma^{-2}}{2}\left(\bar{x}_n-\theta\right)^2 \exp -\frac{\tau^{-2}}{2}\left(\theta-\mu\right)^2 \\
&\sim \exp -\frac{\tau_n^{-2}}{2}\left(\theta-\mu_n\right)^2
\end{aligned}$$

where

$$\begin{aligned}
\mu_n &= \frac{n\sigma^{-2}\bar{x}_n + \tau^{-2}\mu}{n\sigma^{-2}+\tau^{-2}} \\
\tau_n^{-2} &= n\sigma^{-2} + \tau^{-2}
\end{aligned}$$

We recognize this is to be a normal density with mean μ_n and variance τ_n^2. When the sample size is large, the posterior mean μ_n is close to $\bar{x}_n$. Setting

$$s_n^2 = \frac{\sum\left(x_i-\bar{x}_n\right)^2}{n-1}, n>1$$

we can show that the marginal density is proportional to the product

$$\exp -\frac{(n-1)\, s_n^2}{2\sigma^2} \exp -\frac{n\,(\bar{x}_n-\mu)^2}{2\,(n\tau^2+\sigma^2)}$$

from which we deduce that S_n^2 and $\bar{X}_n$ are statistically independent with distributions respectively given by

$$\frac{(n-1)\, s_n^2}{\sigma^2} \sim \chi^2_{n-1}, \bar{X}_n \sim N\left(\mu, \tau^2+\frac{\sigma^2}{n}\right)$$

In any given problem involving Bayesian methods, the specification of the prior and the consequent computation of the posterior may pose some difficulties. In certain problems, this difficulty is overcome if the prior and the posterior come from the same family of distributions in which case the prior is said to be a conjugate prior. In the previous example, both the prior and the posterior distributions were normal distributions but with different parameters. We provide another example below of a conjugate prior.

Example 3.18. Let $X_1, \ldots, X_n$ be a random sample from a Bernoulli distribution with parameter $0 < \theta < 1$:

$$f(x;\theta) = \theta^x (1-\theta)^{1-x}, x = 0, 1$$

If the prior for θ is the Beta distribution with parameters $\alpha > 0, \beta > 0$, then

$$\pi(\theta;\alpha,\beta) = \frac{\Gamma(\alpha+\beta)}{\Gamma(\alpha)\Gamma(\beta)}\theta^{\alpha-1}(1-\theta)^{\beta-1}, 0 < \theta < 1$$

The posterior distribution when $X_i = x_i$ for $i = 1, \ldots, n$ is again a Beta distribution but with parameters

$$\alpha_n = \alpha + \sum x_i, \beta_n = \beta + n - \sum x_i$$

A good exposition of conjugate priors for various distributions is given in DeGroot [1970, Chapter 9] along with the analogue of hypothesis testing in Chap. 11. See also Box and Tiao [1973] for a discussion of improper or noninformative priors. We may now define highest posterior densities regions as in Box and Tiao [1973, p.123].

Definition 3.12. Let $\pi_n(\theta;\boldsymbol{x})$ be the posterior density of θ based on a sample of size n. A highest posterior density region S of content $(1-\alpha)$ for a parameter θ is one for which

(i) $P\{\theta \in S\} = 1 - \alpha$ and
(ii) for $\theta_1 \in S$ and $\theta_2 \in S^c$,

$$\pi_n(\theta_1;\boldsymbol{x}) \geq \pi_n(\theta_2;\boldsymbol{x})$$

A high posterior density region should have the smallest volume within the parameter space of θ.

To illustrate, suppose that $\pi_n(\theta; x)$ is the normal density with mean μ_n and variance σ_n^2. If S is a high posterior density region, then the event $\theta \in S$ is equivalent to saying $\pi_n(\theta; x) > c$ for some positive constant c. Since $\pi_n(\theta; x)$ is a monotonic decreasing function of $\frac{(\theta-\mu_n)^2}{\sigma_n^2}$, it follows that the region S is given by

$$\left\{\theta : \frac{(\theta - \mu_n)^2}{\sigma_n^2} < c\right\}$$

and c is chosen to satisfy the probability requirement. The interval obtained is sometimes called a credible interval.

One of the most important results in classical Bayesian statistics is the Bernstein-von Mises theorem which states that under certain conditions, the posterior distribution may be approximated by a normal distribution. This implies that we may use the machinery related to the normal distribution to obtain credible confidence regions. Our exposition of this famous result closely follows [Ghosh and Ramamoorthi, 2003, Section 1.4 p.33]. We specify the following assumptions.

Let $X_1, X_2, \ldots$ be a sequence of independent identically distributed random variables with density $f(x; \theta)$ with respect to some dominating measure μ and $\theta \in \Theta$ is an open subset of $\mathcal{R}$:

(i) $\{x : f(x; \theta) > 0\}$ is the same for all $\theta \in \Theta$

(ii) $L(\theta, x) = \log f(x; \theta)$ is three times differentiable with respect to $\theta \in (\theta_0 - \delta, \theta_0 + \delta)$ with $E_{\theta_0}\left[\frac{d}{d\theta} L(\theta, X)\right]$ and $E_{\theta_0}\left[\frac{d^2}{d\theta^2} L(\theta, X)\right]$ both finite and

$$\sup_{\theta \in (\theta_0 - \delta, \theta_0 + \delta)} \left|\frac{d^3}{d\theta^3} L(\theta, X)\right| < M(x), E_{\theta_0}[M(X)] < \infty.$$

(iii) The interchange of the order of expectation with respect to θ_0 and differentiation at θ_0 are justified so that

$$\begin{aligned} E_{\theta_0}\left[\frac{d}{d\theta} L(\theta_0, X)\right] &= 0 \\ E_{\theta_0}\left[\frac{d^2}{d\theta^2} L(\theta_0, X)\right] &= -E_{\theta_0}\left[\frac{d}{d\theta} L(\theta_0, X)\right]^2. \end{aligned}$$

(iv) $I(\theta_0) \equiv E_{\theta_0}\left[\frac{d}{d\theta} L(\theta_0, X)\right]^2 > 0$.

(v) For any $\delta > 0$, there exists an $\epsilon > 0$ such that

$$P_{\theta_0}\left\{\sup_{|\theta - \theta_0| > \delta} \frac{1}{n}\left(L_n(\theta) - L_n(\theta_0)\right) \leq -\epsilon\right\} \to 1.$$

(vi) The prior has density $\pi(\theta)$ with respect to Lebesgue measure and is continuous and positive at θ_0.

Theorem 3.5. *(a) Suppose that conditions (i)–(iv) above are satisfied and let $\hat{\theta}_n$ be a consistent solution of the likelihood equation. Then*

$$\sqrt{n}\left(\hat{\theta}_n - \theta_0\right) \underset{\rightarrow}{d} \mathcal{N}\left(0, 1/I\left(\theta_0\right)\right)$$

(b) Let $s = \sqrt{n}\left(\theta - \hat{\theta}_n\right)$ and define $\pi_n^*(s)$ to be the posterior density evaluated at s. If in addition (v) and (vi) are satisfied, then as $n \to \infty$,

$$\int \left| \pi_n^*(s) - \frac{\sqrt{I(\theta_0)}}{\sqrt{2\pi}} e^{-\frac{s^2 I(\theta_0)}{2}} \right| ds \xrightarrow{P_{\theta_0}} 0$$

The Bernstein-von Mises theorem states that the variation distance between the posterior and the approximating Gaussian distribution is a random variable which converges to zero in probability under repeated draws from P_{θ_0}. Lehmann and Casella [1998, p.491] show that the assumptions of the theorem are satisfied in exponential families. See also Van der Vaart [2000, Section 10.2 p.140]. There is also a multiparameter version of the Bernstein-von Mises theorem.

The next result involves the mean of the posterior distribution of θ which is the Bayes estimate under squared error loss.

Theorem 3.6. *Suppose that conditions (i)–(vi) hold. Suppose as well*

$$\int |\theta| \, \pi(\theta) \, d\theta < \infty$$

Let θ_n^* be the mean of the posterior distribution of θ. Then

(a) $\sqrt{n}\left(\hat{\theta}_n - \theta_n^*\right) \xrightarrow{P_{\theta_0}} 0$

(b) $\sqrt{n}\left(\theta_n^* - \theta_0\right) \underset{\rightarrow}{d} \mathcal{N}\left(0, 1/I\left(\theta_0\right)\right)$

As a result, the posterior mean of θ can be approximated by the maximum likelihood estimate up to an error of $o_P\left(n^{-1/2}\right)$. The main theorems remain true if we replace the Fisher information $I(\theta_0)$ by $-\frac{d^2}{d\theta^2} L(\theta, X)$ calculated at $\hat{\theta}_n$, the observed Fisher information per unit observation. We conclude by noting that modern Bayesian methods along with applications are treated in Part IV. Of interest we mention [Alvo and Cabilio, 1982] in connection with the estimation of the difference between two proportions.

4. Multivariate Methods

Multivariate methods are often used to analyze data in the form of vectors. For example, in medical studies, data collected on a single patient may include information on a number of diagnostic measures such as systolic and diastolic blood pressures, weight, height, etc. Since these measurements are all on the same individual, there may exist correlations among the variables which must be taken into account in any statistical analysis. In the next sections, we first present subjects which are generalizations of univariate analysis. Topics such as repeated measure designs, principal component analysis, canonical correlation, and multidimensional scaling which have no counterpart in univariate statistics are treated next. We conclude with the subject of copulas in connection with the simulation of multidimensional data from knowledge of the marginal distributions. We begin with a brief introduction to matrix algebra.

4.1. Matrix Algebra

In the sequel, we shall assume that any vector $\mathbf{v} \in \mathbb{R}^d$ is by default a column vector. The notation $\mathbf{v}^\top$ identifies the transposition of a vector $\mathbf{v}$ and transforms it into a row vector. The multiplication of two vectors $\mathbf{u}, \mathbf{v} \in \mathbb{R}^d$ is defined by

$$\mathbf{u}^\top \mathbf{v} = u_1 v_1 + u_2 v_2 + \cdots + u_d v_d,$$

For matrices $A = \begin{bmatrix} a_{11} & \dots & a_{1n} \\ \dots & \ddots & \dots \\ a_{m1} & \dots & a_{mn} \end{bmatrix}$ and $B = \begin{bmatrix} b_{11} & \dots & b_{1k} \\ \dots & \ddots & \dots \\ b_{n1} & \dots & b_{nk} \end{bmatrix}$, their product is

$$AB = \begin{bmatrix} c_{11} & \cdots & c_{1k} \\ \dots & \ddots & \dots \\ c_{m1} & \cdots & c_{mk} \end{bmatrix},$$

where

$$c_{ij} = \sum_{r=1}^{n} a_{ir} b_{rj}, \quad i = 1, \dots, m, \ j = 1, \dots, k.$$

The *inverse matrix* A^{-1} for a *square non-singular matrix* $A \in \mathbb{R}^{d \times d}$ is defined by the condition

$$A\,A^{-1} = A^{-1} A = \boldsymbol{I}_d,$$

where $\boldsymbol{I}_d \in \mathbb{R}^{d \times d}$ is the *identity matrix* of *rank* d.

M. Alvo, *Statistical Inference and Machine Learning for Big Data*, Springer Series in the Data Sciences, https://doi.org/10.1007/978-3-031-06784-6_4

Definition 4.1. Let A be an $m \times m$ matrix and B and $n \times n$ matrix. The Kronecker product of A and B is the $mn \times mn$ matrix

$$A \otimes B = \begin{bmatrix} a_{11}B & a_{12}B & \dots & a_{1m}B \\ \vdots & \vdots & \ddots & \vdots \\ a_{m1}B & a_{m2}B & \dots & a_{mm}B \end{bmatrix}$$

Some properties of the Kronecker product are

(i) $(A \otimes B)^T = \left(A^T \otimes B^T\right)$

(ii) $(A \otimes B)^{-1} = (A^{-1} \otimes B^{-1})$

(iii) $Tr\,(A \otimes B) = (TrA)\,(TrB)$

(iv) $|A \otimes B| = |A|^n\,|B|^m$

The derivative of a function $f(\boldsymbol{x})$ with respect to $\boldsymbol{x}$ is defined to be the column vector of partial derivatives. We make use of the following expressions for the derivatives.

Lemma 4.1. *The following expressions for derivatives hold*

1. If $f(\boldsymbol{x}) = A\boldsymbol{x}$ for a matrix A, then

$$\frac{\partial}{\partial \boldsymbol{x}} f(\boldsymbol{x}) = A^T, \frac{\partial}{\partial \boldsymbol{x}^T} f(\boldsymbol{x}) = A$$

2a. If $f(\boldsymbol{x}) = \boldsymbol{x}^T A \boldsymbol{x}$, then

$$\frac{\partial}{\partial \boldsymbol{x}} f(\boldsymbol{x}) = \left(A + A^T\right)\boldsymbol{x}, \frac{\partial}{\partial \boldsymbol{x}^T} f(\boldsymbol{x}) = \boldsymbol{x}^T\left(A + A^T\right)$$

2b. For a symmetric matrix $A, A = A^T$ and

$$\frac{\partial}{\partial \boldsymbol{x}} f(\boldsymbol{x}) = 2A\boldsymbol{x}, \frac{\partial}{\partial \boldsymbol{x}^T} f(\boldsymbol{x}) = 2\boldsymbol{x}^T A$$

4.2. Multivariate Analysis as a Generalization of Univariate Analysis

In multivariate analysis, data in the form of p measurements on a single subject is described by a random vector $\boldsymbol{X} = (X_1, X_2, \dots, X_p)^T$ with expected value defined as

$$E[\boldsymbol{X}] = (E[X_1], E[X_2], \dots, E[X_p])^T$$

provided the individual component expectations exist. For two random $p-$dimensional vectors $\boldsymbol{X}_1, \boldsymbol{X}_2$, the covariance is defined to be

$$Cov(\boldsymbol{X}_1, \boldsymbol{X}_2) = E\left[(\boldsymbol{X}_1 - E[\boldsymbol{X}_1])(\boldsymbol{X}_2 - E[\boldsymbol{X}_2])^T\right]$$

In a sample of size N, data for the ith subject consisting of p measurements may be written as

$$\boldsymbol{Y}_i^T = [Y_{i1}, Y_{i2}, \ldots, Y_{ip}]$$

whereas for the entire sample, it may be represented by an $N \times p$ matrix

$$\boldsymbol{Y} = \begin{bmatrix} Y_{11} & Y_{12} & \ldots & Y_{1p} \\ Y_{21} & Y_{22} & \ldots & Y_{2p} \\ \vdots & \vdots & \ddots & \vdots \\ Y_{N1} & Y_{N2} & \ldots & Y_{Np} \end{bmatrix} \tag{4.1}$$

The expectation of the matrix Y is defined as the matrix of expectations

$$E[\boldsymbol{Y}] = \begin{bmatrix} E[Y_{11}] & E[Y_{12}] & \ldots & E[Y_{1p}] \\ E[Y_{21}] & E[Y_{22}] & \ldots & E[Y_{2p}] \\ \vdots & \vdots & \ddots & \vdots \\ E[Y_{N1}] & E[Y_{N2}] & \ldots & E[Y_{Np}] \end{bmatrix}$$

The following lemma provides expressions for the mean and variance of quadratic forms involving multivariate normal distributions. Quadratic forms play an important role in problems of regression and analysis of variance.

Lemma 4.2. *Suppose that* $\boldsymbol{X} \sim \mathcal{N}_p(\boldsymbol{\mu}, \Sigma)$. *Then for a symmetric matrix* A

(i) $E\left[\boldsymbol{X}^T A \boldsymbol{X}\right] = Tr[A\Sigma] + \boldsymbol{\mu}^T A \boldsymbol{\mu}$

(ii) $Var\left[\boldsymbol{X}^T A \boldsymbol{X}\right] = 2Tr[A\Sigma A\Sigma] + 4\boldsymbol{\mu}^T A\Sigma A \boldsymbol{\mu}$

Next, we define the Wishart distribution which plays an analogous role in multivariate analysis as the chi square does in univariate statistics.

Definition 4.2. Suppose that $\boldsymbol{Y}_1, \ldots, \boldsymbol{Y}_N$ are a random sample from a multivariate normal $\mathcal{N}_p(0, \Sigma)$. Define

$$W \equiv \boldsymbol{Y}^T \boldsymbol{Y} = \sum_{i=1}^{N} \boldsymbol{Y}_i^T \boldsymbol{Y}_i.$$

Then W is said to have a central Wishart distribution with N degrees of freedom denoted $W \sim \mathcal{W}_p(N, \Sigma)$. If $\boldsymbol{Y}_i \sim \mathcal{N}_p(\mu_i, \Sigma)$,$i = 1, \ldots, N$, then W has a non-central Wishart distribution with N degrees of freedom and non-centrality parameter $\Upsilon = E\left[\boldsymbol{Y}^T\right] E[\boldsymbol{Y}]$ denoted $W \sim \mathcal{W}_p(N, \Sigma, \Upsilon)$.

When $p = 1$, $\Sigma = \sigma^2$, we have $W \sim \sigma^2 \chi^2_N$. In practice, the functional form of the Wishart distribution is seldom utilized. An important result for Wishart distributions is Cochran's theorem which provides the basis for the distribution of quadratic forms in regression and in the analysis of variance.

Theorem 4.1. *Let $\boldsymbol{Y}_1, \ldots, \boldsymbol{Y}_N$ be a random sample from $\mathcal{N}_p(0, \Sigma)$. Let*

$$\mathbf{Y} = \begin{bmatrix} Y_{11} & Y_{12} & \ldots & Y_{1p} \\ Y_{21} & Y_{22} & \ldots & Y_{2p} \\ \vdots & \vdots & \ddots & \vdots \\ Y_{N1} & Y_{N2} & \ldots & Y_{Np} \end{bmatrix}$$

and suppose we may write

$$\mathbf{Y}^T\mathbf{Y} = \sum_{i=1}^{k} \mathbf{Y}^T A_i \mathbf{Y}$$

where rank $A_i = r_i$ and $I_{N\times 1} = \sum_{i=1}^{k} A_i$. Then $\mathbf{Y}^T A_i \mathbf{Y} \sim \mathcal{W}_p(r_i, \Sigma)$ and statistically independent for all i if and only if $\sum_{i=1}^{k} r_i = N$.

For an application, suppose we have a random sample $\boldsymbol{Y}_i \sim N_p(\mu, \Sigma), i = 1, \ldots, N$. Define the sample variance S

$$(N-1)\, S = \sum_{i=1}^{N} \left(\mathbf{Y}_i - \bar{\mathbf{Y}}\right)\left(\mathbf{Y}_i - \bar{\mathbf{Y}}\right)^T$$

As in the univariate case, Cochran's theorem implies that $\bar{\mathbf{Y}}$ and S_N are statistically independent, and moreover

$$(N-1)\, S_N \sim \mathcal{W}_p(N-1, \Sigma)$$

The density of S_N is given by

$$f(S) \sim |S|^{\frac{N-2p-1}{2}} |\Sigma|^{-\frac{N}{2}} \exp -\left(Ntr\Sigma^{-1}S\right)$$

The multivariate extension of the Student t distribution is Hotelling's T^2 statistic. Set

$$T^2 = N\left(\bar{\mathbf{Y}} - \boldsymbol{\mu}\right)^T S^{-1}\left(\bar{\mathbf{Y}} - \boldsymbol{\mu}\right)$$

Then it can be shown that

$$\frac{N-p}{p}\frac{T^2}{N-1} \sim F(p, N-p) \tag{4.2}$$

Some further distribution results useful for drawing inference in multivariate settings are:

Lemma 4.3. *Suppose $\mathbf{Y} \sim N_p(\mu, \sigma^2 I)$ and A is a symmetric matrix of rank r. Then*

$$\frac{\mathbf{Y}^T A \mathbf{Y}}{\sigma^2} \sim \chi_r^2(\delta)$$

a non-central chi-square distribution with r degrees of freedom if and only if the matrix

A is idempotent, that is, $A = A^2$. *The non-centrality parameter is*

$$\delta = \frac{\boldsymbol{\mu}^T A \boldsymbol{\mu}}{\sigma^2}$$

An extension of the previous lemma is the following lemma.

Lemma 4.4. *Suppose* $\boldsymbol{Y} \sim N_p(\boldsymbol{\mu}, \Sigma)$ *and* A *is a symmetric matrix of rank* r. *Then*

$$\boldsymbol{Y}^T A \boldsymbol{Y} \sim \chi_r^2(\delta)$$

a non-central chi-square distribution with r *degrees of freedom if and only if the matrix* $A\Sigma$ *is idempotent, that is,* $A\Sigma A = A$. *The non-centrality parameter is*

$$\delta = \boldsymbol{\mu}^T A \boldsymbol{\mu}$$

The next result is useful in regression analysis. Suppose that the vector U can be partitioned into two random vectors $\boldsymbol{X}_1, \boldsymbol{X}_2$ such that

$$U = \begin{bmatrix} \boldsymbol{X}_1 \\ \boldsymbol{X}_2 \end{bmatrix} \sim N_{p+q}\left(\begin{bmatrix} \boldsymbol{\mu}_1 \\ \boldsymbol{\mu}_2 \end{bmatrix}, \Sigma = \begin{bmatrix} \Sigma_{11} & \Sigma_{12} \\ \Sigma_{21} & \Sigma_{22} \end{bmatrix} \right)$$

Then,

$$\boldsymbol{X}_2 \sim N_q(\boldsymbol{\mu}_2, \Sigma_{22})$$

whereas the conditional density of $\boldsymbol{X}_1$ given $\boldsymbol{X}_2 = \boldsymbol{x}_2$ is

$$N_p(\boldsymbol{\mu}_{1.2}, \Sigma_{1.2})$$

with

$$\begin{aligned} \boldsymbol{\mu}_{1.2} &= \boldsymbol{\mu}_1 + \Sigma_{12}\Sigma_{22}^{-1}(\boldsymbol{x}_2 - \boldsymbol{\mu}_2) \\ \Sigma_{1.2} &= \Sigma_{11} - \Sigma_{12}\Sigma_{22}^{-1}\Sigma_{21} \end{aligned}$$

4.2.1. The General Linear Model

The multivariate analogue of the general linear model is

$$Y = XB + \varepsilon$$

where Y is an $N \times p$ matrix given by (4.1), $X = 1_{N\times q}$, $B_{q\times p}$ is a matrix of unknown parameters, and ε is an $N \times p$ matrix of random error terms.

We assume the error terms in ε have mean equal to 0 so that $E[Y] = XB$ and

$$\begin{aligned} Var[Y] &= I_N \otimes \Sigma \\ &\equiv \begin{pmatrix} \Sigma & 0 & \ldots & 0 \\ 0 & \Sigma & \ldots & 0 \\ \vdots & \vdots & \ddots & \vdots \\ 0 & 0 & \ldots & \Sigma \end{pmatrix} \end{aligned}$$

The method of least squares to estimate the parameters in the matrix B consists of minimizing

$$Tr\left[(Y - XB)^T (Y - XB)\right]$$

and this leads to the normal equations

$$\left(X^T X\right) B = X^T Y$$

4.2.2. One Sample Problem

We may illustrate the general linear model in the one sample problem formulated as

$$\boldsymbol{Y} = X\boldsymbol{B} + \boldsymbol{\varepsilon}$$

where $\boldsymbol{Y}$ is given by (4.1), $X = 1_{N\times 1}$ and

$$\boldsymbol{B} = [\mu_1, \mu_2, \ldots, \mu_p]$$

We assume the column error vector $\boldsymbol{\varepsilon}$ has mean 0 and

$$\begin{aligned} Var[\boldsymbol{Y}] &= I_N \otimes \Sigma \\ &\equiv \begin{pmatrix} \Sigma & 0 & \ldots & 0 \\ 0 & \Sigma & \ldots & 0 \\ \vdots & \vdots & \ddots & \vdots \\ 0 & 0 & \ldots & \Sigma \end{pmatrix} \end{aligned}$$

In general, one is interested in testing a hypothesis on the mean vector

$$\begin{aligned} H_0 &: \boldsymbol{\mu} = \boldsymbol{\mu}_0 \\ vs & \\ H_1 &: \boldsymbol{\mu} \neq \boldsymbol{\mu}_0 \end{aligned}$$

It can be shown that the likelihood ratio test leads to Hotelling's test statistic

$$T_0^2 = N\left(\bar{Y} - \boldsymbol{\mu}_0\right)^T S^{-1}\left(\bar{Y} - \boldsymbol{\mu}_0\right) \tag{4.3}$$

which rejects H_0 whenever

$$\frac{N-p}{p}\frac{T_0^2}{N-1} \geq F(\alpha;p,N-p)$$

Here $F(\alpha;p,N-p)$ is the upper α percentage point of the F distribution with degrees of freedom $p, N-p$. In fact, we may calculate the likelihood ratio test whereby we reject H_0 whenever

$$\Lambda \equiv \frac{\max\limits_{H_0} L(\mu,\Sigma)}{\max L(\mu,\Sigma)} < c$$

for some constant c to be determined to guarantee a type I error. Some algebra shows that

$$\Lambda^{-(2/N)} = 1 + \frac{T_o^2}{N-1}$$

Hotelling's statistic is a generalization of the univariate student t statistic. The development here may be used to derive simultaneous confidence intervals for components of μ. Specifically, a $100(1-\alpha)\%$ confidence region for the mean is given by the set of all $\boldsymbol{\mu}$ such that

$$N\left(\bar{\boldsymbol{Y}}-\boldsymbol{\mu}\right)^T S^{-1}\left(\bar{\boldsymbol{Y}}-\boldsymbol{\mu}\right) \leq \frac{p(N-1)}{(N-p)}F(\alpha;p,N-p)$$

Other test statistics due to Wilks, Roy, and Lawley-Hotelling are described in Timm [1975].

4.2.3. Two-Sample Problem

In the two-sample problem, we suppose that we have two independent sets of random samples

$$\boldsymbol{Y}_i \sim \begin{cases} \mathcal{N}_p(\boldsymbol{\mu}_1,\Sigma_1) & i=1,\ldots,N_1 \\ \\ \mathcal{N}_p(\boldsymbol{\mu}_2,\Sigma_2) & i=N_1+1,\ldots,N \end{cases}$$

and we wish to test

$$\begin{aligned} H_0 &: \boldsymbol{\mu}_1 = \boldsymbol{\mu}_2 \\ vs& \\ H_1 &: \boldsymbol{\mu}_1 \neq \boldsymbol{\mu}_2 \end{aligned}$$

The two-sample problem may be formulated as a special case of the general linear model

$$\boldsymbol{Y} = X\boldsymbol{B} + \boldsymbol{\varepsilon}$$

where $\boldsymbol{Y}$ is given by (4.1),

$$X = \begin{pmatrix} 1 & 0 \\ \vdots & \vdots \\ 1 & 0 \\ 0 & 1 \\ \vdots & \vdots \\ 0 & 1 \end{pmatrix}_{(N_1+N_2)\times 2}$$

and

$$B = \begin{pmatrix} \mu_{11} & \mu_{12} & \cdots & \mu_{1p} \\ \mu_{21} & \mu_{22} & \cdots & \mu_{2p} \end{pmatrix}$$

We shall make the additional assumption that the covariance matrices are equal but unknown so that

$$Var\,[\mathbf{Y}] = I_{(N_1+N_2)} \otimes \Sigma.$$

Set

$$\begin{aligned} \bar{Y}_{i.} &= \frac{\sum_{j=1}^{N_i} Y_{ij}}{N_i} \\ S_i &= \frac{\sum_{j=1}^{N_i} \left(Y_{ij} - \bar{Y}_{i.}\right)\left(Y_{ij} - \bar{Y}_{i.}\right)^T}{N_i - 1} \\ S_{pooled} &= \frac{(N_1 - 1)\, S_1 + (N_2 - 1)\, S_2}{N_1 + N_2 - 2}. \end{aligned}$$

The two-sample Hotelling test statistic

$$T_0^2 = \frac{N_1 N_2}{N_1 + N_2}\left(\bar{Y}_{1.} - \bar{Y}_{2.}\right)^T S_{pooled}^{-1}\left(\bar{Y}_{1.} - \bar{Y}_{2.}\right) \tag{4.4}$$

rejects H_0 whenever

$$\frac{N_1 + N_2 - p - 1}{p\,(N_1 + N_2 - 2)} \frac{T_0^2}{N - 1} \geq F\,(\alpha; p, N_1 + N_2 - p - 1)$$

where $F\,(\alpha; p, N_1 + N_2 - p - 1)$ is the upper α percentage point of the F distribution with degrees of freedom $p, N_1 + N_2 - p - 1$.

This assumption that the populations have a common covariance matrix is rather strong. In the case where the covariance matrices are not known to be equal but the sample sizes are large, we may replace the estimate $\left(\frac{N_1+N_2}{N_1 N_2}\right) S_{pooled}$ in (4.4) by the separate estimate

$$S_{separate} = \frac{S_1}{N_1} + \frac{S_2}{N_2}.$$

The multivariate analysis of variance then proceeds in a straightforward fashion. We refer the reader to Johnson and Wichern [2008, ch.6] in general and for more extensive coverage to Timm [1975, ch.5] for details on problems of profile analysis and growth curve analysis. Problems in connection with the nested, Latin square and split plot designs are also discussed. We postpone a discussion of regression analysis to Part III of this book where machine learning is discussed.

4.3. Structure in Multivariate Data Analysis

In Sect. 4.2 we briefly described the analogues of one- and two-sample problems in the multivariate case. In the next section, we consider issues related to the structure of multivariate data for which there is no counterpart in univariate statistics. These include the topics of principal components analysis, factor analysis, canonical correlation, linear discriminant analysis, and multidimensional scaling.

4.3.1. Principal Component Analysis

How can we interpret a high-dimensional multivariate data set on p variables? Conceptually, a plot of multivariate data appears as a cloud of points in an axis system defined by usually correlated variables. In that context, it is difficult to interpret the overall meaning of the variables. Principal component analysis is a multivariate technique that transforms the cloud into a new axis system defined by uncorrelated (orthogonal) variables.

Let $\boldsymbol{Y}$ be random p-dimensional vector having mean $\boldsymbol{\mu}$ and variance-covariance matrix Σ. The simplest transformation is the one consisting of linear combinations

$$X_i = \boldsymbol{w}_i^T \boldsymbol{Y}, i = 1, \ldots, p$$

where $\boldsymbol{w}_i$ is a p-dimensional scalar vector. We know that

$$\begin{aligned} E\left[X_i\right] &= \boldsymbol{w}_i^T \boldsymbol{\mu} \\ Var\left[X_i\right] &= \boldsymbol{w}_i^T \Sigma \boldsymbol{w}_i \end{aligned}$$

We would first like to determine the vector w_1 so as to maximize $Var\left[X_1\right]$ while restricting w_1 to have length equal to 1. Using the method of Lagrange multipliers, define

$$G \equiv V\left[X_1\right] - \lambda_1 \left(w_1^T w_1 - 1\right) = w_1^T \Sigma w_1 - \lambda_1 \left(w_1^T w_1 - 1\right)$$

Differentiating with respect to w_1

$$\frac{\partial G}{\partial w_1} = 2\Sigma w_1 - 2\lambda_1 w_1 = 0.$$

A nontrivial solution exists if and only if

$$|\Sigma - \lambda_1 I| = 0$$

Hence, w_1, λ_1 are the first eigenvector and eigenvalue, respectively, of the matrix Σ, and X_1 is called the first principal component. The second principal component is the solution to

$$(\Sigma - \lambda_2 I)\, w_2 = 0$$

subject to $w_2^T w_2 = 1$ and to the orthogonality condition

$$w_1^T w_2 = 0.$$

We know that the eigenvectors of the symmetric matrix Σ are orthogonal and hence w_2, λ_2 correspond to the second eigenvector and eigenvalue, respectively, of the matrix Σ. Proceeding in this way, we may determine all the p principal components. Since,

$$Var\,[X_i] = w_i^T \Sigma w_i = \lambda_i$$

the ratios

$$\frac{\lambda_i}{\sum_{i=1}^{p} \lambda_i}, i = 1, \ldots, p$$

measure the contribution of each component relative to the sum of the variances. In practice, the analysis is usually performed on the sample estimate of the matrix Σ.

It is common practice to interpret the data set in terms of a few components which account for the "lion's" share of the total variance. Principal component analysis has been used as a dimension reduction technique as well as in problems of clustering. No distributional assumptions are made. The main difficulty in using principal components lies in the interpretation of the components. We illustrate this technique on Fisher's Iris data.

Example 4.1. (Fisher's Iris Data) The Fisher Iris data available in R has been analyzed extensively in the literature [Fisher, 1936]. There are 150 objects consisting of 50 *Iris setosa*, 50 Iris versicolor, and 50 Iris virginica. Four variables are measured on each object: sepal length (SL), sepal width (SW), petal length (PL), and petal width (PW). Initially, it was thought that Iris versicolor was a hybrid of *Iris setosa* and Iris virginica. A correlation plot is shown in Fig. 4.1 from which it can be seen that sepal length and sepal width are strongly correlated.

Analyzing *Iris setosa* separately, we have the following results for the principal components.

It is seen that the first two components explain 76.5% and 11.95%, respectively, of the total variation. From the eigenvector, the first component deals with the sepal length and width. The second component is more difficult to interpret. It appears to contrast the combined sepal and petal lengths with sepal width. Using the first two principal components, we see from Fig. 4.2 that there is a clear separation between the three varieties. A principal component analysis appears as an exercise in Morrison [1990, p.349].

In a second example, we consider body measurements on individuals as a help in designing clothing.

Figure 4.1.: Correlation plots for Iris data

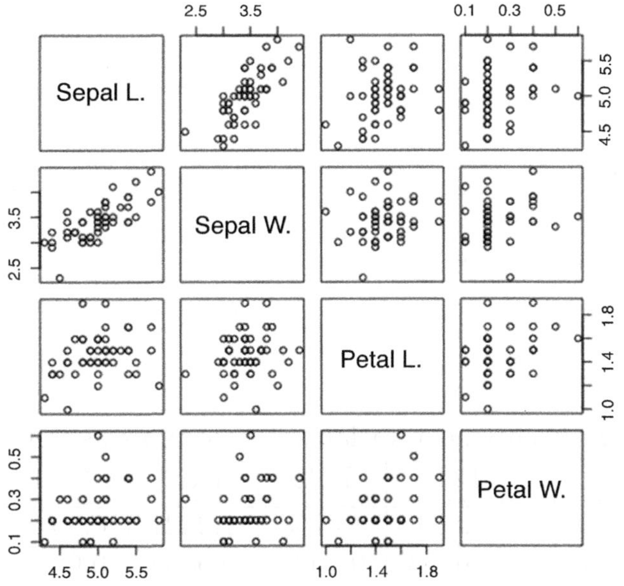

Table 4.1.: Principal component analysis of Iris data

	PC1	PC2	PC3	PC4
Sepal length	0.6691	0.5979	−0.4400	0.0361
Sepal width	0.7341	−0.6207	0.2746	0.0196
Petal length	0.0965	0.4901	0.8324	0.2399
Petal width	0.0636	0.1309	0.1951	−0.9699
Eigenvalues	11.59	1.81	1.31	0.44
Standard deviation	0.4863	0.1921	0.1637	0.950
Proportion of total variance	76.5	11.95	8.64	2.90
Cumulative proportion	76.5	0.884	0.971	1.000

Example 4.2. Consider a data set contains 23 body measurements on 507 physically active individuals, along with their gender and age. A full description of each measurement can be found by importing the openintro package and using the help operator on the bdims data set. The data set is available in the "openintro" package. The authors cite that "knowing the distributions of various body dimensions is helpful in designing clothing (commercial and military uniforms, for example), and furniture in the home and the work place."

Figure 4.2.: Plot of the first two principal components for Iris data

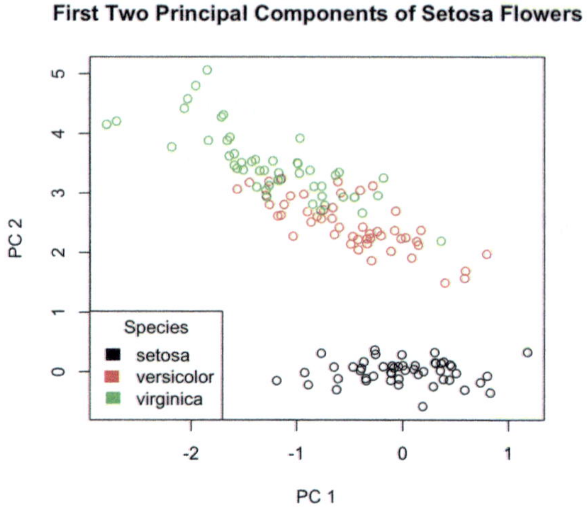

Table 4.2.: Component variance for body measurements

	PC1	PC2	PC3	PC4	PC5	PC6
Standard deviation	3.864	1.5758	1.0321	0.9548	0.6934	0.6585
Proportion of total variance	0.6493	0.1080	0.0463	0.0396	0.0209	0.0189
Cumulative proportion	0.6493	0.7572	0.8036	0.8432	0.8641	0.8830

The first two principal components combined maintain approximately 76% of the original variation in the data set. These components are weighted averages of the body measurements. Retaining only the first two components, we may plot the data by gender of the individual as shown in color coded in Fig. 4.3. There is a clear separation on the basis of gender. Components of the variance are shown in Table 4.2.

4.3.2. Factor Analysis

In principal component analysis, it is frequently difficult to interpret the meaning of the components. In various studies in the social sciences, it is common to make use of the principal components as a first step in a factor analysis. Factor analysis has been used in the literature in order to group together into "factors" variables that are highly correlated among themselves, the goal being to better explain the underlying data. It should be emphasized that the factors into which the variables are grouped are unobserved or latent factors. As an example, in the Wechsler adult intelligence study, scores were assigned to a variety of variables including comprehension, arithmetic skills, vocabulary extent, picture completion and picture arrangement, etc. A factor analysis might group together in one factor, variables related to performance while placing variables related to cognitive ability into another factor.

Figure 4.3.: Principal components for body measurements

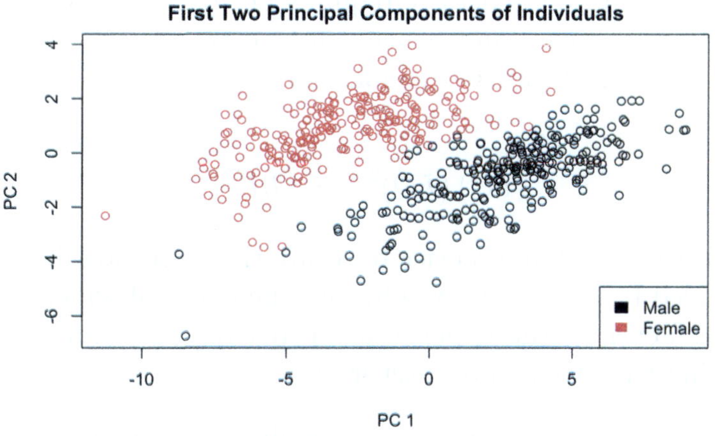

The factor model may be represented in matrix notation as

$$\boldsymbol{Y} - \boldsymbol{\mu} = LF + \boldsymbol{\varepsilon}$$

where $\boldsymbol{Y}$ is the observed p dimensional random vector having mean $\boldsymbol{\mu}$ and covariance Σ, L is the $p \times m$ matrix of factor loadings, F denotes the m dimensional vector of latent factors, and $\boldsymbol{\varepsilon}$ is a p dimensional unobserved error term. For this model, it is assumed that

- F and $\boldsymbol{\varepsilon}$ are independent, both having mean 0
- $\text{Cov}(F) = I$, $\text{Cov}(\boldsymbol{\varepsilon}) = \Psi$, a diagonal matrix

Under these assumptions it can be shown that we may express the covariance matrix Σ as the sum

$$\Sigma = LL^T + \Psi$$

with

$$Cov\left(X, F^T\right) = L$$

The goal in factor analysis is to determine estimates for the factors F and their loadings L. In practice, a principal component analysis is the starting point. These components now serve as the factors. The estimation process as described in Morrison [1990, p.358] begins with the likelihood function for the sample variance S given previously by a Wishart density:

$$f\left(S\right) \sim |S|^{\frac{N-2p-1}{2}} |\Sigma|^{-\frac{N}{2}} \exp - \left(Ntr\Sigma^{-1}S\right)$$

Taking the logarithm of $f\left(S\right)$ and substituting for Σ, we have

$$\log f\left(S\right) \sim \frac{N-2p-1}{2} \log|S| - \frac{N}{2} \log\left|LL^T + \Psi\right| - Ntr\left(LL^T + \Psi\right)^{-1} S$$

The development proceeds to then find the maximum likelihood estimates of L and Ψ by differentiation. The solution generally requires an iterative process. It is important to note that the factor loadings (and hence the factors) are unique up to an orthogonal transformation. This fact can be demonstrated by noting that for an orthogonal matrix, Q

$$(LQ)\,(LQ)^T \;=\; LL^T$$

As a consequence, it is common in practice to rotate the factor loadings by an orthogonal matrix. This is referred to as a varimax solution, the effect of which is to increase some of the loadings in a factor at the expense of decreasing the smaller loadings in other factors, thereby easing the problem of interpretation.

Example 4.3. The Big5 data set contains personality traits most recognized within the academic psychology community. These are extroversion, neuroticism, agreeableness, conscientiousness, and openness to experience. The data contains the responses from a 50 question test. Each factor in the "Big5" data set has ten questions. The data set contains the responses of over a million people, and it is available from the Open-Source Psychometrics Project at https://openpsychometrics.org/_rawdata/. The full list of questions can be found on the openpsychometrics website.

The correlations between questions aimed at measuring the same personality trait are moderate to high for the most part, while most of the correlations between questions measuring different traits are low.

It can be seen from Fig. 4.4 that the five factors correspond fairly well to the five traits as one would expect. This analysis can be used to determine which questions isolate the personality trait well. The AGR10 question is "I make people feel at ease". This question was meant to measure agreeableness but also had almost as much influence on the extroversion factor. The CSN4 question is "I make a mess of things". This question was meant to measure conscientiousness but had a noticeable influence on the neuroticism factor.

4.3.3. Canonical Correlation

Suppose we have two sets of variables: one which measures verbal ability and the other which measures nonverbal ability. In canonical correlation, the goal consists of finding which linear combinations of variables from the first set best correlate with linear combinations from the second set in such a way that previous linear combinations are uncorrelated. Another example may involve a set of ability variables and a set of personality variables. A canonical correlation study would consist of finding which linear combination of ability variables best correlates with personality variables.

Figure 4.4.: Correlation plot for Big5 data

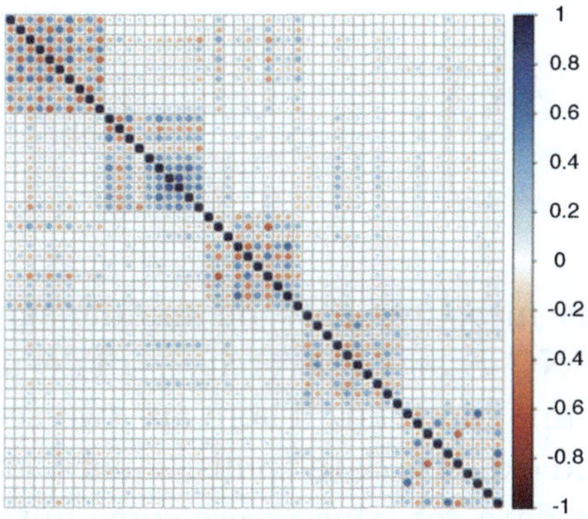

```
##
## Loadings:
##     Factor1 Factor2 Factor3 Factor4 Factor5
## E1    0.669
## E2   -0.677          -0.118
## E3    0.655  -0.251   0.257   0.134
## E4   -0.697   0.156
## E5    0.733           0.212   0.100
## E6   -0.567          -0.149          -0.220
## E7    0.739  -0.102   0.158
## E8   -0.552
## E9    0.606                           0.119
## E10  -0.650   0.188
## N1   -0.110   0.670
## N2    0.114  -0.489
## N3   -0.138   0.597   0.161
## N4    0.139  -0.316           0.100
## N5            0.539          -0.112  -0.122
## N6            0.741
## N7            0.723          -0.154
## N8            0.751          -0.160
## N9            0.708  -0.171
## N10  -0.244   0.610          -0.168
## A1                   -0.425
## A2    0.349           0.529
## A3            0.244  -0.401  -0.195
## A4                    0.790
## A5   -0.143          -0.651
## A6            0.156   0.595
## A7   -0.318   0.106  -0.610
## A8    0.131           0.578
## A9    0.112   0.114   0.708
## A10   0.341  -0.119   0.384   0.153   0.100
## C1           -0.103           0.599   0.119
## C2            0.102          -0.528   0.126
## C3                            0.401   0.260
## C4            0.359          -0.556
## C5                            0.624
## C6            0.168          -0.582
## C7                            0.535
## C8            0.223  -0.150  -0.476
## C9                            0.624
## C10                           0.469   0.231
## O1                                    0.576
## O2            0.209                  -0.537
## O3            0.105                   0.533
## O4            0.118  -0.108          -0.464
## O5    0.203                   0.167   0.607
## O6                                   -0.504
## O7           -0.127           0.194   0.490
## O8                   -0.110           0.539
## O9   -0.134   0.164   0.175           0.336
## O10   0.187                           0.679
##
```

Let $\boldsymbol{X} = (X_1, \ldots, X_q)^T$ and $\boldsymbol{Y} = (Y_1, \ldots, Y_p)^T$ denote the two sets of variables, $p \leq q$. We may partition the variance-covariance matrix of the $(p+q)$ variates $\left[\boldsymbol{Y}^T | \boldsymbol{X}^T\right]$ as

$$\Sigma = \begin{pmatrix} \Sigma_{11} & \Sigma_{12} \\ \Sigma_{21} & \Sigma_{22} \end{pmatrix}$$

We assume Σ is of full rank and Σ_{11}, Σ_{22} are both non-singular. To determine the first most highly correlated linear combinations, let $U = \boldsymbol{a}^T \boldsymbol{Y}$, $V = \boldsymbol{b}^T \boldsymbol{X}$, and maximize

$$Cov\,(U, V) = \boldsymbol{a}^T \Sigma_{12} \boldsymbol{b}$$

subject to $\boldsymbol{a}^T \Sigma_{11} \boldsymbol{a} = \boldsymbol{b}^T \Sigma_{22} \boldsymbol{b} = 1$. It can be shown Timm [1975] or Johnson and Wichern [2008] for $\boldsymbol{a}$ we need to solve the eigenvalue problem

$$\left| \Sigma_{12} \Sigma_{22}^{-1} \Sigma_{21} - \rho^2 \Sigma_{11} \right| = 0$$

whereas for $\boldsymbol{b}$ we need to solve the eigenvalue problem

$$\left| \Sigma_{21} \Sigma_{11}^{-1} \Sigma_{12} - \rho^2 \Sigma_{22} \right| = 0$$

Let $\rho_1^2 \geq \rho_2^2 \geq \cdots \geq \rho_p^2$ be the eigenvalues and $\boldsymbol{a}_1, \boldsymbol{a}_2, \ldots, \boldsymbol{a}_p$ be the corresponding eigenvectors. It can be shown in general that the nonzero positive roots $|\rho_i|$ are the canonical correlations between $U_i = \boldsymbol{a}_i^T \boldsymbol{Y}$ and $V_i = \boldsymbol{b}_i^T \boldsymbol{X}$, $i = 1, \ldots, p$ and

$$\begin{aligned} Cov\left(U_i, U_j\right) &= Cov\left(V_i, V_j\right) = \delta_{ij} \\ Corr\left(U_i, V_j\right) &= \left|\rho_i\right| \delta_{ij} \end{aligned}$$

where $\delta_{ij} = \begin{cases} 1 & i = j \\ 0 & i \neq j \end{cases}$.

In practice, a canonical correlation analysis is performed on the sample variance-covariance matrix S which is partitioned as

$$S = \begin{pmatrix} S_{11} & S_{12} \\ S_{21} & S_{22} \end{pmatrix}$$

Example 4.4. The data considered contains measurements on middle-aged men in a health fitness club (Dr. A. C. Linnerud, NC State University). The data set is available in the plsdepot package under the name linnerud. There are two sets of data: physiological data (weight, waist, and pulse) and exercise data (pulls, squats, and jumps). The correlations are shown in Fig. 4.5. In that data set, there were some outliers which were removed as they had a large impact on the correlations. The resulting correlations for the entire data set are shown in Fig. 4.5. A canonical correlation was performed resulting in factor loadings for the two sets as follows:

Physiological data

Weight (−0.029371464), waist (0.653772378), pulse (0.003056312)

exercise data

Pulls (−0.03611262), squats (−0.02200047) , jumps (0.02883732)

There is an overall canonical correlation of approximately 0.862 between the two sets showing that physiological and exercise data are highly correlated.

Figure 4.5.: Correlations for fitness data

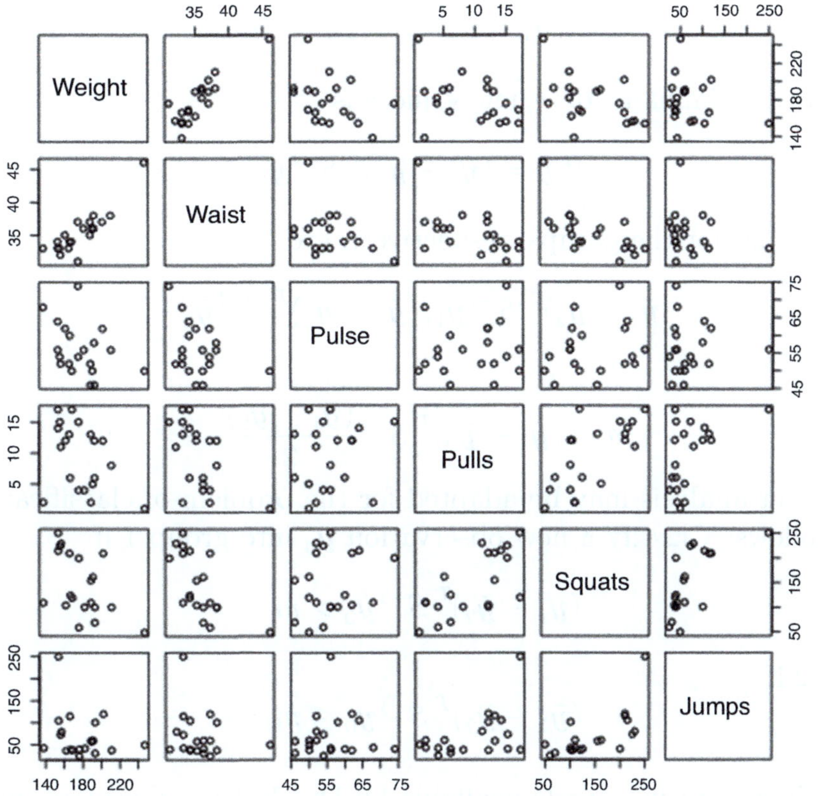

4.3.4. Linear Discriminant Analysis

The subject of linear discriminant analysis may be presented in different ways [Johnson and Wichern, 2008]. We will use Fisher's original approach in this section. Suppose that we wish to place a subject or object into one of two groups. The decision of which group the subject is to be placed will be made on the basis of a vector of measurements $\boldsymbol{y}$ on each subject. We assume that we have a priori random samples from each group for which we have calculated the sample means and sample variances $\bar{\boldsymbol{y}}_i, S_i^2$ for $i = 1, 2$. Fisher's original idea was to linearize the problem and to look for that vector l which maximizes the ratio

$$\frac{\left(\boldsymbol{l}^T\bar{\boldsymbol{y}}_1 - \boldsymbol{l}^T\bar{\boldsymbol{y}}_2\right)^2}{\boldsymbol{l}^T\Sigma_1\boldsymbol{l} + \boldsymbol{l}^T\Sigma_2\boldsymbol{l}}$$

Assume a common covariance $\Sigma_1 = \Sigma_2 = \Sigma$ for the two groups, and let S be the pooled estimate of the common covariance. The above ratio may be viewed as the square of the distance between groups relative to the variability within the two groups. This is an eigenvalue problem. It can be shown that $\boldsymbol{l}$ satisfies

$$\boldsymbol{l} = S^{-1}\left(\bar{\boldsymbol{y}}_1 - \bar{\boldsymbol{y}}_2\right)$$

and the maximum of that ratio is given by

$$\lambda = (\bar{\boldsymbol{y}}_1 - \bar{\boldsymbol{y}}_2)^T S^{-1} (\bar{\boldsymbol{y}}_1 - \bar{\boldsymbol{y}}_2)$$

Hence, the linear discriminant for a new value y is

$$\boldsymbol{l}^T \boldsymbol{y} = (\bar{\boldsymbol{y}}_1 - \bar{\boldsymbol{y}}_2)^T S^{-1} \boldsymbol{y}$$

The mean values for the groups, respectively, are

$$(\bar{\boldsymbol{y}}_1 - \bar{\boldsymbol{y}}_2)^T S^{-1} \bar{\boldsymbol{y}}_1, (\bar{\boldsymbol{y}}_1 - \bar{\boldsymbol{y}}_2)^T S^{-1} \bar{\boldsymbol{y}}_2$$

with midpoint

$$m = (\bar{\boldsymbol{y}}_1 - \bar{\boldsymbol{y}}_2)^T S^{-1} \frac{(\bar{\boldsymbol{y}}_1 + \bar{\boldsymbol{y}}_2)}{2}$$

Linear discriminant analysis may be adapted for the problem of classification involving two groups as follows: Classify a new observation $\boldsymbol{y}_0$ into group 1 if

$$(\bar{\boldsymbol{y}}_1 - \bar{\boldsymbol{y}}_2)^T S^{-1} \boldsymbol{y}_0 > m$$

and into group 2 if

$$(\bar{\boldsymbol{y}}_1 - \bar{\boldsymbol{y}}_2)^T S^{-1} \boldsymbol{y}_0 \leq m$$

Example 4.5. The data was made available by Dr. Kristen Gorman and the Palmer Station Long-Term Ecological Research Program. It contains data on Adélie, Chinstrap, and Gentoo penguins observed on islands near Palmer Station, Antarctica, from 2007 to 2009 which can be found in the "palmerpenguins" package. The correlations for the penguin data are shown in Fig. 4.6.

As seen in the head function, the data has missing values which were removed before proceeding. A random seed is set, and the data was split 4:1 into training and test sets. The model is then used to predict the species of the penguins in the test set. A confusion matrix can be used to view the results. Only one penguin in the test set was incorrectly classified as shown in the table below.

		Classified	
Originally	Adelie	Chinstrap	Gentoo
Adelie	33	1	0
Chinstrap	0	12	0
Gentoo	0	0	23

4.3.5. Multidimensional Scaling

Inhaber and Alvo [1978] viewed world science as an input-output system. In the example to be discussed, an input-output model of the citation patterns of the 18 leading

Figure 4.6.: Correlations for penguin data

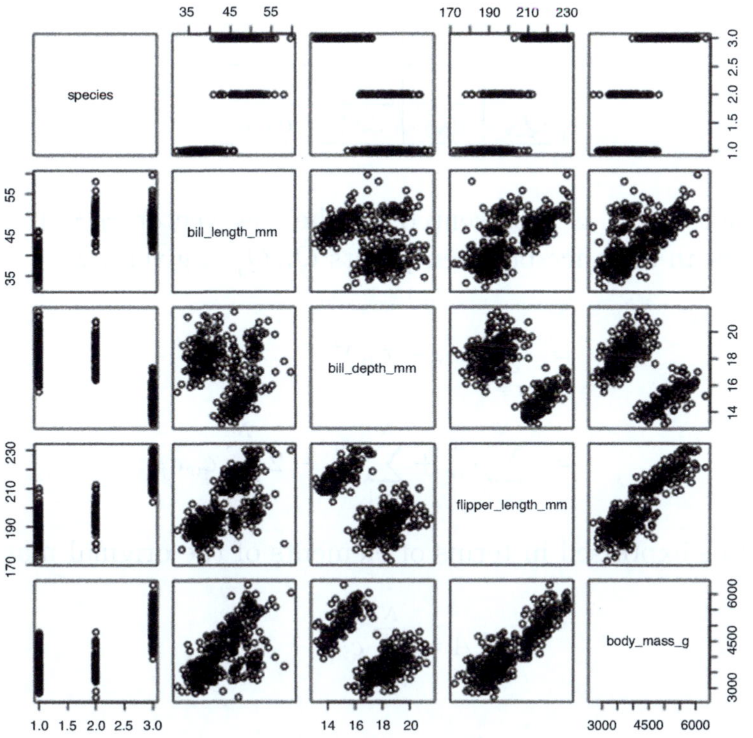

countries in international science was constructed. These countries produce most of the world's science. Multidimensional scaling is a multivariate technique which can be used to graphically display the position of countries with respect to each other in science citations. Such information is useful in discussing transnational interactions in science.

Perhaps the simplest method of determining how closely the vectors, rows, or columns of a matrix are related to each other is the multidimensional scaling system described by Gower [1966, 1967]; Kruskal [1964a,b]. The scaling calculates a "stress", which is a measure of the mathematical distance between vectors. Subject to certain constraints, the stress is minimized by an iteration process until it reaches its lowest possible value. The advantage of this method in terms of understanding is that the stresses can be considered as being reduced to two dimensions and so can be plotted. In our example, the points representing each nation indicate the relative position it occupies with respect to an arbitrary origin. If, for example, two nations are close in this plot, then their citation patterns should be similar.

To explain the theory underlying multidimensional scaling, suppose that we have a symmetric matrix $A = (a_{ij})$ of order N whose eigenvalues $\{\lambda_i\}$ and corresponding eigenvectors are displayed below

	λ_1	...		λ_N
Q_1	c_{11}	...		c_{1N}
...				...
...				...
Q_N	c_{N1}	...		c_{NN}

The point denoted, Q_i, is represented by the row vector $c_i = (c_{i1}, \ldots, c_{iN})$. The square of the Euclidean distance between points Q_i, Q_j is given by

$$\begin{aligned} d_{ij}^2 &= \sum_{r=1}^{N} (c_{ir} - c_{jr})^2 \\ &= \sum_{r=1}^{N} c_{ir}^2 + \sum_{r=1}^{N} c_{jr}^2 - 2\sum_{r=1}^{N} c_{ir}c_{jr} \end{aligned}$$

This distance may be expressed in terms of elements of the original matrix A by noting that

$$A = \sum_{i=1}^{N} c_i c_i^T$$

and hence

$$a_{ii} = \sum_{r=1}^{N} c_{ir}^2, \, a_{ij} = \sum_{r=1}^{N} c_{ir}c_{jr}$$

and

$$d_{ij}^2 = a_{ii} + a_{jj} - 2a_{ij}.$$

If the original matrix A is either a similarity matrix or a product moment correlation matrix, $a_{ii} = 1$ so that

$$d_{ij}^2 = 2\,(1 - a_{ij})$$

One way to facilitate the calculation of the coordinates of the points Q_i is to force the matrix A to have diagonal elements equal to 0. This can be done by a process of double centering of rows and columns. Centering the a_{ij} about their rows and columns, we define

$$b_{ij} = a_{ij} - \left(\bar{a}_{i.} - \bar{\bar{a}}_{..}\right) - \left(\bar{a}_{.j} - \bar{\bar{a}}_{..}\right) + \bar{\bar{a}}_{..}$$

where

$$\begin{aligned} \bar{a}_{i.} &= \frac{\sum_{j=1}^{N} a_{ij}}{N}, \, \bar{a}_{.j} = \frac{\sum_{i=1}^{N} a_{ij}}{N} \\ \bar{\bar{a}}_{..} &= \frac{\sum_{i=1}^{N}\sum_{j=1}^{N} a_{ij}}{N^2} \end{aligned}$$

This may be represented in matrix notation

$$B = -\frac{1}{2} CDC$$

where $D = \left(d_{ij}^2\right)$,

$$C = I - \frac{J_N}{N}$$

I is the $N \times N$ identity matrix and J_N is the $N \times N$ matrix of ones. The centering matrix is symmetric and idempotent. (A similar centering, as we shall see in Chap. 5, is done for Hoeffding's combinatorial central limit theorem). For Euclidean distance, the matrix B can be shown to be positive semi definite.

Hence, we have for $i \neq j$

$$a_{ij} = -\frac{d_{ij}^2}{2}$$

We may proceed to find the coordinates of the Q_i after the double centering.

Next, we note that it is not always possible to define a notion of distance between two objects. An example described in Young and Hamer [1987] involves a study in developmental psychology to investigate the nature of a child's conception of the body, and how this conception evolves in time. In multidimensional scaling, we begin with a matrix A along with a similarity measure δ_{ij} to compare all possible pairs M among N objects. These may be arranged in increasing order of similarity as

$$\delta_{i_1 j_1} < \cdots < \delta_{i_M j_M} \tag{4.5}$$

Let $v_j, j = 1, \ldots, k$ be the resulting normalized eigenvectors of the matrix A, each of dimension N and set $w_i^T = (c_{i1}, \ldots, c_{ik}), i = 1, \ldots, N$. The Euclidean distances in this space are now defined in terms of distances between the $k \times 1$ vectors $\{w_i\}$

$$d_{ij}^{(k)} = \|w_i - w_j\|$$

The ordered pairs $\left(d_{ij}^{(k)}, \delta_{ij}\right)$ form a cloud of points in $k-$dimensional space. In order to ensure that the distances $d_{ij}^{(k)}$ match the relationship defined by (4.5), Kruskal [1964b] defined a measure of "stress" to assess the extent to which the configuration falls short of the perfect match:

$$Stress\,(k) = \left\{ \frac{\sum_{i<l} \left(d_{ij}^{(k)} - \hat{d_{ij}}^{(k)} \right)^2}{\sum_{i<l} \left(d_{ij}^{(k)} \right)^2} \right\}^{1/2} \tag{4.6}$$

where, the $\hat{d_{ij}}^{(k)}$ are determined so as to reduce (4.6) subject to preserving the order prescribed in (4.5). Various software packages including R may be used to perform the calculations and to plot the resulting configuration.

In another application, [Inhaber and Alvo, 1978] applied multidimensional scaling in a study of scientific citations using data in the 1973 edition of the Science Citation Index. Suppose that there are N countries. Let p_{ij} be the proportion of scientific citations made by scientists in country i to scientific journals published in country j. Then,

$$\sum_{j=1}^{N} p_{ij} = 1.$$

Let x_{ij} be the observed number of scientific citations made by country i to country j in a sample of n_i citations. Then,

$$(x_{i1}, \ldots, x_{iC}) \sim \mathit{Multinomial}\,(n_i; p_{ij}, j = 1, \ldots, C)$$

Set

$$\hat{p}_{ij} = \frac{x_{ij}}{n_i}.$$

A measure of similarity between two countries i, j may be the test statistic we would use to test the null hypothesis that the proportions $p_{il} = p_{jl}$ for all $l = 1, \ldots, N$

$$\delta_{ij} = n_i \sum_{l=1}^{N} \frac{(\hat{p}_{il} - q_{il})^2}{q_{il}} + n_j \sum_{l=1}^{N} \frac{(\hat{p}_{jl} - q_{jl})^2}{q_{jl}}$$

where

$$\begin{aligned} q_{il} &= q_{jl} = \frac{(x_{il} + x_{jl})}{n_i + n_j}, l \neq i, l \neq j \\ q_{ij} &= q_{ji} = \frac{(x_{ij} + x_{ji})}{n_i + n_j} \\ q_{ii} &= q_{jj} = \frac{(x_{ii} + x_{jj})}{n_i + n_j} \end{aligned}$$

A measure of distance between two countries i, j may be

$$d_{ij} = \left\{ (\hat{p}_{ij} - \hat{p}_{ji})^2 + (\hat{p}_{ii} - \hat{p}_{jij})^2 + \sum_{l \neq i, l \neq j} (\hat{p}_{il} - \hat{p}_{jl})^2 \right\} \tag{4.7}$$

In Fig. 4.7 we show the relationship of nations in terms of their publishing patterns. Those nations whose publishing effort is confined mostly to their own nationals are predominantly in the upper right-hand corner. For example, the USSR has 98.7% of its output from residents of that country. Corresponding values for India are 95.0%; for Poland, 92.3%. Nations like the USA and the UK occupy an intermediate position. Denmark, Switzerland, and the Netherlands are clearly at the opposite end of the scale from countries which publish almost exclusively the work of their own nationals.

In Fig. 4.8, the data was modified by noting that papers published in a given country

Figure 4.7.: Relationship of nations by publishing patterns

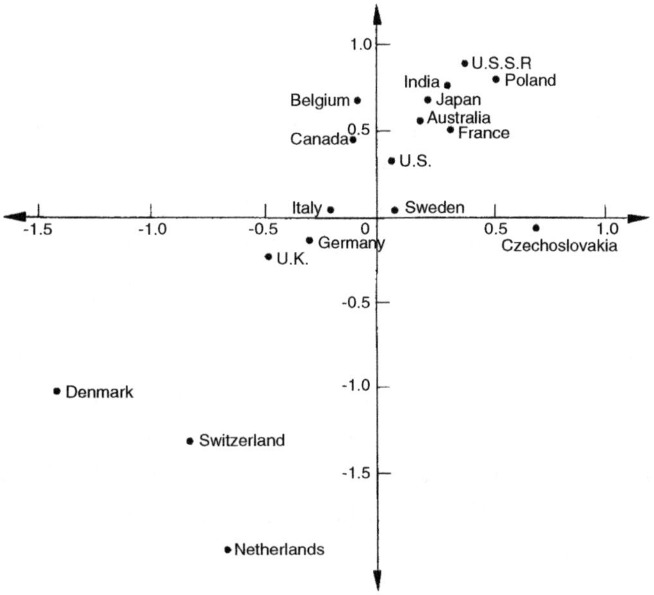

Figure 4.8.: International citation patterns, corrected for publishing patterns

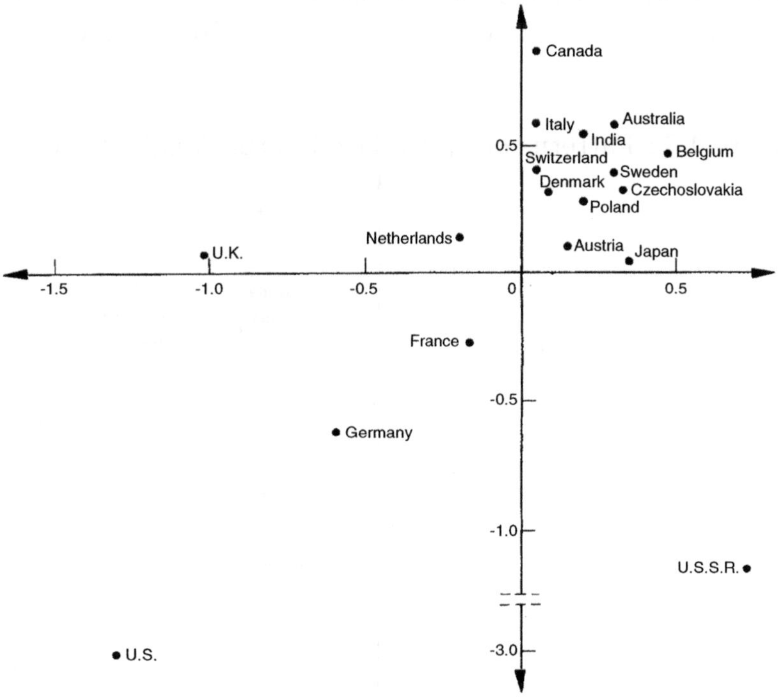

do not always necessarily originate from the nationals of that country. The location of the USA is far from other countries. However, now we can see more of a pattern emerging among the other nations. Germany and the UK, with their relatively strong dependence on U.S. science, are the countries closest to it. The original data, not displayed here, showed that the UK had more citations to US science than it did to British papers. The proportion of German citations to U.S. science to that of German papers is also about one. As well, US science cites papers from the UK and Germany fairly strongly.

Canada appears far removed from the USA in Fig. 4.8. This may be due to the fact that Canada refers much more to US publications than the other way around so that the distribution of citations to and references from is highly asymmetrical between the two countries. This produces a larger distance between them in Fig. 4.8. The same reasoning can be applied to other pairs of countries. The relationship of the USSR to the USA in Fig. 4.8 deserves comment. Their relative closeness arises from the fact that papers in the USSR cite US papers frequently. The ratio of Soviet citations to US papers to that of their own papers is about one-third. Almost all papers published in the USSR are by Soviet nationals. The ratio of Soviet to the USA/the USA to Soviet citations is about 5 in. The corresponding ratio for Canada and the USA is about 10, implying less similarity of citation patterns.

Finally, we can make a graphical allowance for the fact that some countries have many more journals and thus more citations to these journals than other countries. In Fig. 4.9 we compiled data to reflect this. One of the major differences between this distribution and the previous ones is that the USA is now much closer to other countries. In effect, the consequences of the large number of citations it receives have been removed. The USA cited about 3400 journals in the network and received references from about 3800. A second consequence is a clustering around the USA of three countries: the UK,

Figure 4.9.: Patterns of up-to-date international citations

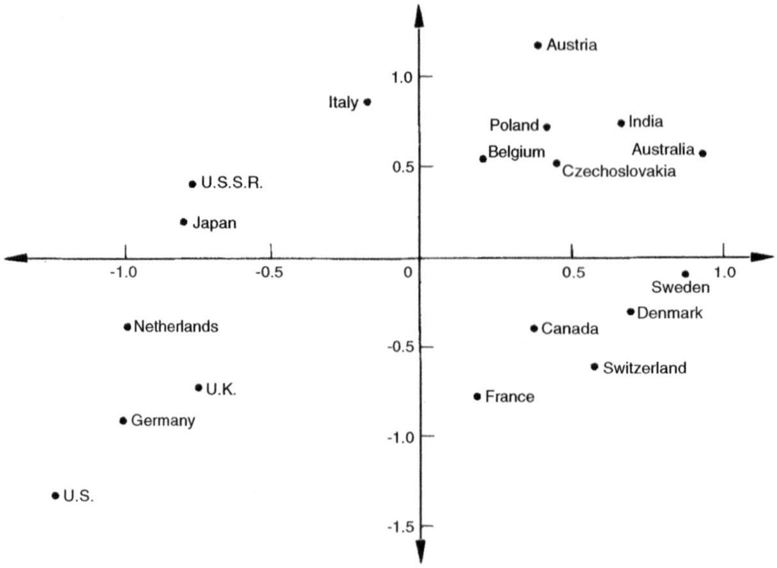

Germany, and the Netherlands. The first two are well-known associates of the USA in scientific enterprises. The location of the Netherlands is due to a large proportion of its publications (about 35%) originating from US authors. Many of these US authors obviously cite Dutch publications and vice versa.

In other works, [Morrison, 1990, p.391] uses multidimensional scaling to compare Latin manuscripts whereas [Alvo and Yu, 2014, p.11] apply these methods to movie rating data.

4.3.6. Copula Methods

Copula methods play an important role in statistics because they provide convenient ways to simulate multivariate data from knowledge of the marginal distributions and to compute expectations through Monte Carlo integration. There are applications in various fields of science, engineering, medicine, and quantitative finance. In the following presentation, we shall restrict attention to the bivariate case.

Definition 4.3. A two-dimensional copula $C(u,v)$ is a joint distribution of two uniformly distributed random variables $U = F(X).V = G(Y)$ where $F\ and\ G$ are the marginal cumulative distributions of $X\ and\ Y$, respectively. Setting $I = [0,1]$, a copula therefore satisfies the usual properties:

1. $C(u,0) = 0 = C(0,v), (u,v) \in (I \times I)$
2. For $u_1 \leq u_2, v_1 \leq v_2, in\ I$

$$C(u_2,v_2) - C(u_2,v_1) - C(u_1,v_2) + C(u_1,v_1) \geq 0$$

The principal result in the study of copulas is Sklar's (1959) theorem [Nelsen, 2006, p.18] which we quote:

Theorem 4.2. *Let X and Y be random variables with distribution functions F and G, respectively, and joint distribution function H. Then there exists a copula C such that for all x,y in the plane*

$$H(x,y) = C(F(x), G(y)).$$

If F and G are continuous, then C is unique; otherwise, C is uniquely determined on the space Range F × Range G.

Conversely, if C is a copula and F and G are distribution functions, then the function H is a joint distribution function with marginal distributions F and G for X and Y, respectively.

This theorem is remarkable in that it characterizes the joint distribution of two variables as a function of the marginals. Consequently, knowing the copula function along with the marginal distributions enables one to generate the joint distribution of the variables. It can be shown that the copula function is invariant under strictly increasing transformations of the variables $X\ and\ Y$. The copula function is given by

$$C(u,v) = H\left(F^{-1}(u), G^{-1}(v)\right)$$

Example 4.6. The bivariate normal density $f(x, y) =$

$$\frac{1}{2\pi\sigma_1\sigma_2\sqrt{1-\rho^2}} \exp \frac{-1}{2(1-\rho^2)} \left[\left(\frac{x-\mu_1}{\sigma_1}\right)^2 + \left(\frac{y-\mu_2}{\sigma_2}\right)^2 - 2\rho \left(\frac{x-\mu_1}{\sigma_1}\right)\left(\frac{y-\mu_2}{\sigma_2}\right) \right]$$

may be expressed through a change of variables, as

$$f(u, v) = \frac{1}{\sqrt{|R|}} \exp \frac{-1}{2} \left(\left[\Phi^{-1}(u), \Phi^{-1}(v)\right] \left(R^{-1} - I_2\right) \left[\Phi^{-1}(u), \Phi^{-1}(v)\right]^T \right)$$

where $||$ denotes determinant, I_2 is the 2×2 unit matrix, and

$$\begin{aligned} \Phi^{-1}(u) &= \frac{x-\mu_1}{\sigma_1}, \Phi^{-1}(v) = \frac{y-\mu_2}{\sigma_2} \\ R &= \begin{pmatrix} 1 & \rho \\ \rho & 1 \end{pmatrix} \end{aligned}$$

Here, $\Phi()$ is the cumulative distribution function of a standard normal. The Gaussian copula is then seen to be

$$\begin{aligned} \int_{-\infty}^{x} \int_{-\infty}^{y} \frac{1}{2\pi\sqrt{1-\rho^2}} \exp \frac{-1}{2(1-\rho^2)} \left(s^2 - 2\rho st + t^2\right) dsdt &= \\ &= \int_0^u \int_0^v f(s, t)\, dsdt \\ &= C(u, v). \end{aligned}$$

We encounter the Gaussian copula in the context of Bayesian analysis. Specifically, suppose that

$$f(x|\theta) \sim N\left(\theta, \sigma^2\right), \pi(\theta) \sim N\left(\mu, \tau^2\right).$$

Then the marginal density of X is $N(\mu, \sigma^2 + \tau^2)$. The joint density of x and θ is the bivariate normal in the integrand above with correlation

$$\rho = \sqrt{\frac{\tau^2}{\sigma^2 + \tau^2}}$$

and

$$\begin{aligned} \mu_1 &= \mu_2 = \mu \\ \sigma_1^2 &= \sigma^2 + \tau^2 \\ \sigma_2^2 &= \tau^2. \end{aligned}$$

Calculating , $F^{-1}(u) = -\ln(1-u), G(y) = -\ln(1-v)$, we have

$$C_\theta(u,v) = u + v - 1 + (1-u)(1-v)\, e^{-\theta \ln(1-u)\ln(1-v)}$$

When $\theta = 0$, we have independence of the variables.

Example 4.9. (Bivariate Extreme Value Logistic or Gumbel-Hougaard) Suppose the joint distribution function is

$$H_\theta(x,y) = \exp\left[e^{-\theta x} + e^{-\theta y}\right]^{-1/\theta}, \theta \geq 1$$

Then we can show the copula is given by

$$C_\theta(u,v) = \exp\left\{-\left[(-\ln u)^\theta + (-\ln v)^\theta\right]^{1/\theta}\right\}$$

with generator

$$\varphi_\theta(t) = (-\ln t)^\theta$$

When $\theta \to 1$, we have independence; when $\theta \to \infty$, we have perfect dependence $C_\theta(u,v) = \min(u,v)$.

Example 4.10. (Clayton Copula) The Clayton copulas are defined by

$$C_\theta(u,v) = \left(u^{-\theta} + v^{-\theta} - 1\right)^{-1/\theta}, \theta > -1$$

The generator is given by

$$\varphi_\theta(t) = \frac{t^{-\theta} - 1}{\theta}$$

When $\theta \to 0$, we have independence; when $\theta \to \infty$, we have perfect dependence $C_\theta(u,v) = \min(u,v)$.

Example 4.11. (Frank Copula) The Frank copulas are defined by

$$C_\theta(u,v) = \frac{-1}{\theta}\ln\left(1 + \frac{\left(e^{-\theta u} - 1\right)\left(e^{-\theta v} - 1\right)}{e^{-\theta} - 1}\right)^{-1/\theta}, \theta \neq 0$$

The generator is given by

$$\varphi_\theta(t) = -\ln\frac{e^{-\theta t} - 1}{e^{-\theta} - 1}$$

Example 4.12. (Gumbel-Barnett) These are the survival copulas associated with Gumbel's bivariate exponential distribution:

$$C_\theta(u,v) = uve^{(-\theta(\ln u)(\ln v))}, 0 < \theta \leq 1$$

The generator is given by

$$\varphi_\theta(t) = \ln(1 - \theta \ln t)$$

Copulas may be estimated from a random sample of n pairs (x_i, y_i) as

$$\hat{C}_n(u,v) = \frac{\sum I_{[F_n(X_i) \le u, G_n(Y_i) \le v]}}{n} \tag{4.8}$$

$$= \frac{\sum_{i=1}^{n} I_{\left[\frac{R_i}{n} \le u, \frac{S_i}{n} \le v\right]}}{n} \tag{4.9}$$

where R_i is the rank of X_i among the $X's$ and S_i is the rank of Y_i among the $Y's$.

In the study of nonparametric statistics, we shall consider the relationship between the important Spearman and Kendall measures of correlation. For now, we define the population concept of concordance.

Definition 4.4. Let (X_1, Y_1) and (X_2, Y_2) be a pair of independent identically distributed random vectors. The vectors are said to be concordant whenever $X_1 < X_2$ we have $Y_1 < Y_{2.} or$ $X_1 > X_2$ we have $Y_1 > Y_{2.}$. Otherwise, we say the pairs are discordant.

Kendall's tau as a measure of concordance, is defined to be

$$\tau = P[(X_1 < X_2)(Y_1 < Y_{2.}) > 0] - P[(X_1 < X_2)(Y_1 < Y_{2.}) < 0]$$

Let $C(u,v)$ be the copula of (X, Y). It can be shown that [Nelsen, 2006, Theorem 5.1.3, p.161]

$$\tau = 4 \int_0^1 \int_0^1 C(u,v)\, dC(u,v) - 1$$

Moreover, if $C(u,v)$ is Archimedean with generator $\varphi(t)$, then [Nelsen, 2006, Corollary 5.1.4, p.163]

$$\tau = 1 + 4 \int_0^1 \frac{\varphi(t)}{\varphi'(t)} dt$$

The importance of this result is that for Archimedean copulas, the Kendall coefficient can be computed from a single integral involving the generator.

Another measure of concordance is due to Spearman. Given three independent identically distributed random vectors (X_1, Y_1), (X_2, Y_2), and (X_3, Y_3), the Spearman coefficient is defined to be

$$\rho_{XY} = 3P[(X_1 < X_2)(Y_1 < Y_{3.}) > 0] - P[(X_1 < X_2)(Y_1 < Y_{3.}) < 0]$$

It can be shown that [Nelsen, 2006, Theorem 5.1.6, p.167]

$$\rho_{XY} = 12 \int_0^1 \int_0^1 C(u,v)\, dudv - 3$$

Notes. 1. Genest and Rivest [1993] examined the problem of selecting an Archimedean copula to provide a suitable representation of the dependence structure between two variates X and Y. Genest and Neslehova [2007] presented an introduction to copulas for count data. For other works on copulas, see Bentoumi et al. [2019]; Joe [2001].

2. For additional theoretical details on multivariate methods and their applications, we refer the reader to Timm [1975] and to Johnson and Wichern [2008]. The R package https://cran.r-project.org/web/views/Multivariate.html can be used to facilitate the implementation of the various procedures.

3. Applications of multivariate analysis to *big data* abound. As an example we cite [Camacho et al., 2018] whereby multivariate methods are utilized for detecting anomalies in industrial monitoring and networking in cyber security. The paper also discusses various software used.

5. Nonparametric Statistics

We begin with a quote W.J. Conover's description of the difference between nonparametric and parametric statistics. "Nonparametric methods use approximate solutions to exact problems, while parametric methods use exact solutions to approximate problems."

Nonparametric statistics is concerned with the treatment of statistical problems when the underlying distribution of the data is not explicitly known. The traditional problems of estimation and hypothesis testing in parametric statistics are usually formulated in terms of parameters of distributions. In testing a hypothesis on the mean lifetime of microchips when the lifetime is described by an exponential distribution, we can make use of the likelihood ratio test. Nonparametric statistics differs markedly from parametric statistics in that fewer such assumptions are made and it is not possible to write down a likelihood function. Problems in parametric statistics often have their counterpart in nonparametric statistics. In Table 5.1, we list various parametric problems and their counterparts in nonparametric statistics. Hence, for example, a test on the mean of a normal distribution may be reformulated as a test on the median of an otherwise unspecified distribution. The various Kolmogorov-Smirnov tests have no parametric counterpart as they look for any differences between distribution functions.

In testing problems, a violation of the assumptions results in a "lack of power." leading one to more likely commit a type II error and failing to detect a difference when one truly exists. Assuming a normal distribution for skewed data may result in improper confidence intervals. Often times, a small sample size limits our ability to ascertain whether or not the data are normally distributed prohibits an appeal to asymptotic methods.

The type of data analyzed in nonparametric statistics tests includes ordinal as well as ranked data which are less sensitive to the influence of outliers. Clustering and classification problems involve the use of algorithms such as K-means, nearest neighbor, and support vector machine methods. Nonparametric statistics has a wider range of applicability and is considered more robust than parametric statistics.

We begin with the study of goodness-of-fit tests whereby we wish to ascertain whether or not the data follow a specified distribution. We then proceed with an introduction to the basic tools in nonparametric statistics. These include the use of ranks and U-statistics. We apply these in the study of tests of location and scale. In Sect. 5.9 we outline a unified theory of hypothesis testing and apply it in problems of testing for trend in proportions and in tests for umbrella alternatives. Density estimation is briefly discussed. We conclude the chapter with the important subject of spatial data analysis and its application to COVID-19 data.

M. Alvo, *Statistical Inference and Machine Learning for Big Data*, Springer Series in the Data Sciences, https://doi.org/10.1007/978-3-031-06784-6_5

Table 5.1.: Parametric and nonparametric problems

Parametric problems	Nonparametric problems
One-sample T-test	Median test
Two-sample T-test	Two-sample Wilcoxon or Mann-Whitney test
One way analysis of variance	Kruskal-Wallis; Mood's median test
Factorial design with a factor and a blocking variable	Friedman randomized block design
Two-way design for dichotomous data	Cockran's Q test
Matched pair T-test	Sign or Wilcoxon signed rank tests
Two-sample scale	Mood's test; Ansari-Bradley; Siegel Tukey test
Levene or Brown-Forsythe tests for equality of variances	Conover squared ranks test
Parametric regression	Nonparametric regression methods
Test of trend or independence	Spearman or, Kendall correlation tests; runs test
Classification: LDA, logistic regression, neural networks	Nearest neighbor, support vector machines, CART

5.1. Goodness-of-Fit Tests

The main question to ask is when should one resort to using nonparametric methods. Karl Pearson addressed this question in (1900) when he formulated the chi-squared goodness-of-fit test. Suppose that we observe data $x_1, \ldots, x_n$ and we would like to test if these come from a specific distribution $F(x)$. The test proceeds by selecting k disjoint intervals $\{I_i\}$ spanning the real line and counting the number of $x's$ say $\{o_i\}$ occurring in each of the k intervals. Then, using the specified distribution F, we calculate the probabilities $\{p_i\}$ of observing a value in each interval. The expected number of $x's$ is given by $e_i \equiv np_i$. The test statistic

$$\chi^2 = \sum_{i=1}^{k} \frac{(o_i - e_i)^2}{e_i}$$

rejects the hypothesis that the observed data come from F whenever $\chi^2 > \chi^2_{k-s}(\alpha)$ where $\chi^2_{k-s}(\alpha)$ is the upper $100(1-\alpha)\%$ point of a chi-squared distribution with $(k-s)$ degrees of freedom. Here s is the number of parameters which are estimated from F in order to compute the p_i. The number of intervals k is left unspecified but should be reasonably large, say 10, and the e_i should be at least 5 for each interval. The test is simple to implement and works fairly well when the sample size n is large.

In cases when the sample size is small, we may use the Kolmogorov-Smirnov test. We first define the empirical distribution function which serves to estimate the theoretical distribution itself.

Definition 5.1. The empirical distribution function is defined as

$$F_n(x) = \frac{1}{n}\sum_{i=1}^{n} I(X_i \leq x), \infty < x < \infty \tag{5.1}$$

where $I(A)$ is the indicator function of the set A.

The importance of the empirical distribution function of X is underlined by the Glivenko-Cantelli theorem [Van der Vaart, 2000, p.266] which asserts that for a random sample of size n,

$$D_n = \sup_x |F_n(x) - F(x)| \to 0, a.s.$$

The Kolmogorov–Smirnov test statistic D_n is an omnibus test which is sensitive simultaneously to changes in both location and spread. Specifically, we may test the null hypothesis that the underlying cumulative distribution of X is given by $F(t)$. The test is conducted by calculating $\sqrt{n}D_n$ and rejecting at level α the null hypothesis that the data come from $F(t)$ whenever

$$\sqrt{n}D_n > d_\alpha$$

where d_α is the upper $100(1-\alpha)\%$ point of the variable D which has cumulative distribution function:

$$P(D \leq d) = 1 - 2\sum_{k=1}^{\infty} (-1)^{k-1} e^{-2k^2 d^2}$$

The test can also be adapted to test for the equality of two distributions. In the two-sample case, we may use the statistic

$$D_{n,m} = \sup_x |F_n(x) - G_m(x)|$$

where F_n, G_m are the two empirical distributions based on sample sizes n, m respectively. Writing

$$H_0 : F(x) = G(x), \textit{for all } x$$

vs

$$H_1 : F(x) \neq G(x), \textit{for some } x$$

the two-sample test rejects the null hypothesis H_0 at level α whenever

$$D_{n,m} > \sqrt{-\left(\ln\frac{\alpha}{2}\right)\left(\frac{n+m}{2nm}\right)}.$$

Arguably both the one-sample and two-sample tests lack power since they are designed to guard against all departures from the null hypotheses. There has been a large body of work on various aspects of these statistics [Van der Vaart, 2000]. A different class of tests is based on the statistic

$$A_n = n\int_{-\infty}^{\infty} (F_n(x) - F(x))^2 w(x)\, dF(x)$$

Here A_n may be viewed as a measure of separation between F_n and F, but it is not a metric since it is not symmetric. For a discussion of metrics (distances) on probability measures, see Dudley (2002, pages 393–398) and Gibbs and Su (2002). If $w(x) = 1$ for all x, we have the Cràmer–von Mises statistic, whereas if

$$w(x) = \frac{1}{F(x)(1 - F(x))}$$

we have the Anderson–Darling statistic which places more weight on the tails of the distribution and thus is more sensitive to extreme values. The statistic A_n allows more flexibility in testing as it can assign different weights to $F(x)$. The actual calculation of the test statistic can be simplified by using the probability integral transformation whereby $Y = F(X)$ has a uniform distribution. We note that it is frequently the case that parameters of the distribution $F(x)$ will need to be estimated on the basis of the sample. For further references to empirical distribution based tests, see D'Agostino and Stephens [1986]; Stephens [1974].

Various studies have found that, even in this corrected form, the test based on A_n is less powerful for testing normality than the Shapiro-Wilk test [Shapiro and Wilk, 1965]. For instance, the test based on A_n is known not to work well in samples with many identical values. Other tests have their own disadvantages. In parametric statistics, when the sample size is large, the use of the normal distribution for tests on the mean is quite common in light of the central limit theorem. However, tests on the variance need to be treated differently in non-normal situations given the dependence of the sample variance on the kurtosis Table 2.1 (see Kendall and Stuart 1979, p.494). In those cases, nonparametric methods are to be preferred. In the next section, we introduce the important subject of linear rank statistics.

5.2. Linear Rank Statistics

There are a number of important tools in nonparametric statistics, two of which are linear rank statistics and U statistics. In our development, test statistics will often appear as either sums of i.i.d. random variables, U statistics, or linear rank statistics. U statistics were introduced by Hoeffding [1948], whereas the subject of linear rank statistics was developed by Hajek and Sidak [1967]. A study of statistical methods based on ranks is given by Alvo and Yu [2014], whereas a novel parametric approach to nonparametric statistics is described in Alvo and Yu [2018].

In this section, we mention briefly some of the important properties of linear rank statistics which arise in various contexts in nonparametric hypothesis testing. Let $X_1, \ldots, X_N$ constitute a random sample from some continuous distribution $F(x)$. The order statistics are defined as

$$X_{(1)} \leq X_{(2)} \leq \cdots \leq X_{(N)}$$

and indicate which are the smallest, second smallest, etc. Hence,

$$X_{(1)} = \min\{X_1, \ldots, X_N\}, X_{(N)} = \max\{X_1, \ldots, X_N\}$$

An important measure of location for a distribution is the median defined as that value m for which

$$F(m) \geq 0.5 \text{ and } 1 - F(m) \geq 0.5.$$

The sample median serves to estimate the population median, and it is defined to be the middle value if the sample size n is an odd integer and the average of the two middle values if n is even. The median can be used to test for symmetry in a distribution.

Definition 5.2. The rank of X_i denoted R_i is the relative position among the $\{X_1, \ldots, X_N\}$ when ordered from smallest to largest.

For example, among the values 1, 5, 7, and 10, the value 7 has rank 3. We may view the rank of X_i in terms of the empirical distribution function and write

$$F_N(X_i) = \frac{R_i}{N}.$$

A working definition for continuous data can be given as

$$R_i = \sum_{j=1}^{N} sgn[X_i - X_j]$$

where

$$sgn[u] = \begin{cases} 1 & u > 0 \\ 0 & u < 0 \end{cases}$$

The ranks in a random sample are invariant under any monotone transformation of the data. They, along with the order statistics constitute the basic tools in nonparametric statistics. We begin with the definition of a linear rank statistic.

Definition 5.3. Let $X_1, \ldots, X_N$ constitute a random sample from some continuous distribution and let $R_1, \ldots, R_N$ represent the corresponding ranks. A statistic of the form

$$S_N = \sum_{i=1}^{N} a(i, R_i)$$

where the $\{a(i, j)\}$ represents a matrix of values is called a linear rank statistic. The statistic is called a simple linear rank statistic if

$$a(i, R_i) = c_i a(R_i)$$

where $\{c_1, \ldots, c_N\}$ and $\{a(1), \ldots, a(N)\}$ are given vectors of real numbers. The $\{c_i\}$ are called regression coefficients, whereas the $\{a_i\}$ are called scores.

To illustrate why the $\{c_i\}$ are called regression coefficients, suppose that $X_1, \ldots, X_N$ are a set of independent random variables in the simple linear regression model

$$X_i = \mu + \beta c_i + \varepsilon_i, i = 1, \ldots, N$$

where $\{\varepsilon_i\}$ are i.i.d. continuous random variables with median 0 and cumulative distribution function F. To test hypotheses about the slope β, consider the ranks of the differences $X_i - \beta c_i$ denoted by $R_1, \ldots, R_N$. Assuming that $c_1 \leq c_2 \leq \cdots \leq c_N$, then the usual sample correlation coefficient for the pairs $(c_i, a(R_i))$is

$$\frac{\sum (c_i - \bar{c})(a(R_i) - \bar{a})}{\sqrt{\left[\sum (c_i - \bar{c})^2 \sum (a(R_i) - \bar{a})^2\right]}}$$

where $a(.)$ is an increasing function and $\bar{a}, \bar{c}$ are averages, respectively. Specializing $c_i = i, a(R_i) = R_i$, we obtain the Spearman rank correlation statistic

$$\rho = \frac{\sum \left(i - \frac{N+1}{2}\right)\left(R_i - \frac{N+1}{2}\right)}{\left(\frac{N(N^2-1)}{12}\right)} \tag{5.2}$$

This is as an example of a linear rank statistic which can be used to test for a linear trend. It correlates the ordered time scale with the corresponding observed ranks of the data.

Simple linear rank statistics arise as well when we have a random sample of size m from a population and a random sample of size n from another. We are interested in testing the null hypothesis that the two populations are the same against the alternative that the first population is shifted to the right of the second. We may rank all the observations together and retain only the ranks of one of the populations by defining the regression coefficients

$$c_i = \begin{cases} 1 & i = 1, \ldots, m \\ \\ 0 & i = m+1, \ldots, m+n. \end{cases}$$

The resulting linear rank statistic due to Wilcoxon consists of the sum of the ranks from the first population. Next, we consider next some useful properties of linear rank statistics.

Lemma 5.1. *Suppose that the ranks $R_1, \ldots, R_N$ are uniformly distributed over the set of $N!$ permutations of the integers $1, 2, \ldots, N$. Let $S_N = \sum_{i=1}^N c_i a(R_i)$. Then,*

(i) for $i = 1, \ldots, N$

$$E(R_i) = \frac{N+1}{2}, Var(R_i) = \frac{(N^2-1)}{12}$$

and for $i \neq j$

$$Cov\left(R_i, R_j\right) = -\frac{N+1}{12}$$

(ii) $ES_N = N\bar{c}\bar{a}$ *where* $\bar{c} = \sum_{i=1}^{N} c_i/N$ *and* $\bar{a} = \sum_{i=1}^{N} a_i/N$
(iii) $Var\ S_N = \frac{1}{N-1}\sum_{i=1}^{N}\left(c_i - \bar{c}\right)^2 \sum\left(a_i - \bar{a}\right)^2$
(iv) Let $Q_N = \sum_{i=1}^{N} d_i b\left(R_i\right)$ *be another linear rank statistic. Then*

$$cov\left(S_N, Q_N\right) = \sigma_{ab}\sum_{i=1}^{N}\left(c_i - \bar{c}\right)\left(d_i - \bar{d}\right)$$

where

$$\sigma_{ab} = \frac{1}{N-1}\sum_{i=1}^{N}\left(a_i - \bar{a}\right)\left(b_i - \bar{b}\right)$$

and

$$\bar{b} = \sum_{i=1}^{N} b_i/N, \bar{d} = \sum_{i=1}^{N} d_i/N.$$

Proof. The proof of this lemma is straightforward and may be found in Hajek and Sidak [1967, p.57]. □

The lemma implies that the correlation between two rankings is

$$\frac{Cov\left(R_i, R_j\right)}{Var\left(R_i\right)} = \frac{-\frac{(N+1)}{12}}{\frac{(N^2-1)}{12}} = -\frac{1}{N-1} \downarrow 0, as\ N \to \infty.$$

It can be shown that for the coefficient in (5.2), the first four moments are given by

$$\begin{aligned} E\rho &= 0, Var\rho = \frac{1}{N-1} \\ E\rho^3 &= 0 \\ E\rho^4 &= \frac{3}{\left(N^2-1\right)}\left\{1 + \frac{12\left(N-2\right)\left(N-3\right)}{25N\left(N-1\right)^2}\right\} \end{aligned}$$

Knowledge of the first four moments enables us to approximate the distribution of ρ using the Pearson system of distributions. See Kendall and Stuart [1979, Section 3.1 p.504] for reference to tables of the exact distribution of ρ.

Under certain conditions, linear rank statistics are asymptotically normally distributed. We shall consider square integrable functions ϕ defined on the interval $(0, 1)$ which have the property that they can be written as the difference between two non-decreasing functions and which satisfies

$$0 < \int_0^1 \left[\phi\left(u\right) - \bar{\phi}\right]^2 du < \infty \tag{5.3}$$

where $\bar{\phi} = \int_0^1 \phi(u)\, du$. These functions will be used to define the score functions. The next theorem demonstrates a linear algebraic result that will prove useful in demonstrating the equivalence between the Spearman and Kendall tau statistics.

Theorem 5.1. *(Projection Theorem) Let $T = T(R_1, \ldots, R_N)$ be an arbitrary rank statistic and set*

$$\hat{a}(i,j) = E[T|R_i = j]$$

Then the projection of T into the family of linear rank statistics is given by

$$\hat{T} = \frac{N-1}{N} \sum_{i=1}^{N} \hat{a}(i, R_i) - (N-2)\, ET.$$

Proof. The proof of this theorem is given in Hajek and Sidak [1967, p.59]. □

The projection theorem reflects the geometric flavor of the subject of linear rank statistics.

Theorem 5.2. *Suppose that $(R_1, \ldots, R_N)$ are uniformly distributed over the set of $N!$ permutations of the integers $1, 2, \ldots, N$. Let the score function be given by*

$$a(i) = \phi\left(\frac{i}{N+1}\right)$$

where ϕ is a square integrable function on $(0,1)$. Let $S_N = \sum_{i=1}^{N} c_i a(R_i)$. Then, provided the Noether condition holds as $N \longrightarrow \infty$

$$\frac{\sum_{i=1}^{N} (c_i - \bar{c})^2}{\max_{i \leq N} x_i (c_i - \bar{c})^2} \longrightarrow \infty \tag{5.4}$$

it follows that

$$\frac{S_N - N\bar{c}\bar{a}}{\sigma_N} \xrightarrow{\mathcal{L}} N(0,1)$$

where

$$\sigma_N^2 = \frac{1}{N-1} \sum_{i=1}^{N} (c_i - \bar{c})^2 \sum (a_i - \bar{a})^2$$

Proof. The details of the proof of this theorem, given in Hajek and Sidak [1967, p.160], consist in showing that the $\left\{\frac{R_i}{N+1}\right\}$ behave asymptotically like a random sample of uniformly distributed random variables. Hence, S_N can be shown to be equivalent to a linear combination of independent random variables to which we can apply the Lindeberg-Feller central limit theorem. □

Does [1982] obtained higher-order asymptotic results which provide rates of convergence along with Berry-Essen theorems under the null hypothesis. He also provided third order approximations for the distribution functions of simple linear rank statistics in the form of Edgeworth expansions. We do not discuss these topics further in this book.

We sometimes encounter rank statistics that are not linear rank statistics, and we wish nonetheless to demonstrate that they are asymptotically normally distributed. One approach is to "project" the rank statistic into the family of linear rank statistics for which normality can be demonstrated. Then, proceed by showing that the "difference" between the two statistics is asymptotically negligible. An example to which this approach is applied is the Kendall rank correlation:

$$\tau = \frac{\sum_{i \neq j} sgn\,(i-j)\, sgn\,(R_i - R_j)}{N\,(N-1)} \tag{5.5}$$

whose projection into the family of linear rank statistics is

$$\begin{aligned} \hat{\tau} &= \frac{8\sum\left(i-\frac{N+1}{2}\right)\left(R_i-\frac{N+1}{2}\right)}{N^2\,(N-1)} \\ &= \frac{2}{3}\left(\frac{N+1}{N}\right)\rho \end{aligned} \tag{5.6}$$

It can be shown that

$$Var\,(\tau) = \frac{2\,(2N+5)}{9N\,(N-1)},\, Var\,(\hat{\tau}) = \frac{4}{9\,(N-1)}\left(\frac{N+1}{N}\right)^2$$

and hence in view of the projection theorem, the Kendall and Spearman statistics are asymptotically equivalent. Kendall and Stuart [1979, p.512] have shown that both Spearman's and Kendall's correlation coefficients have an efficiency of $\left(\frac{3}{\pi}\right)^{1/3} = 0.98$ as tests of randomness against normal alternatives and an efficiency of $\left(\frac{3}{\pi}\right)^2 = 0.91$ as tests of independence against normal alternatives. See also Van der Vaart [2000, Chapters 8,14].

The usual Pearson correlation coefficient between two random variables X, Y is defined as

$$Corr\,(X,Y) = \frac{Cov\,(X,Y)}{\sqrt{Var\,(X)\,Var\,(Y)}}.$$

This coefficient is not invariant under monotone increasing transformations of the variables. Moreover, even if the variables are perfectly correlated, its value may not be equal to 1. A simple example is to the case where $Y = X^3$ where X a standard normal variable.

The previous results dealt with the asymptotic distribution of linear rank statistics under the null hypothesis. For comparisons of power, it is also important to determine the asymptotic distribution under alternatives to the null hypothesis. To this end, we consider the two-sample problem whereby $X_1, \ldots, X_m \sim f\,(x-\Delta)$ and $Y_1, \ldots, Y_n \sim f\,(x)$, with f absolutely continuous so that if $\Delta > 0$, it implies the $X's$ are shifted to the right of the $Y's$. In that case, one would expect the ranks of the $X's$ to be larger.

We adopt the notation in Hajek and Sidak [1967] for ease of reference. Suppose that we are interested in testing

$$H_0 \quad : \quad \Delta = 0$$

$$H_1 \quad : \quad \Delta > 0$$

Let $U_N^{(1)} < \cdots < U_N^{(N)}$ be the order statistics for a random sample of size N from a uniform distribution on $(0,1)$.

Definition 5.4. The function

$$\varphi(u, f) = -\frac{f'\left(F^{-1}(u)\right)}{f\left(F^{-1}(u)\right)}, 0 < u < 1$$

where f is the density function corresponding to the cumulative distribution function F is called the optimal score function.

Given that $F^{-1}\left(U_N^{(1)}\right), \ldots, F^{-1}\left(U_N^{(N)}\right)$ behave like a sample of order statistics from a cumulative distribution function F, we have the following equivalence in distribution:

$$X_{(k)} \stackrel{d}{=} F^{-1}\left(U_N^{(k)}\right).$$

Hence, for the optimal expected value scores function for location problems, we have the following equivalences:

$$\begin{aligned} a_N(i, f) &\equiv E\left[\varphi\left(U_N^{(i)}, f\right)\right] & (5.7) \\ &= E\left[-\frac{f'\left(F^{-1}\left(U_N^{(i)}\right)\right)}{f\left(F^{-1}\left(U_N^{(i)}\right)\right)}\right], 1 \le i \le N & (5.8) \\ &= E\left[-\frac{f'\left(X_{(i)}\right)}{f\left(X_{(i)}\right)}\right] \\ &= N\tbinom{N-1}{i-1} \int_0^1 \varphi(u, f)\, u^i (1-u)^{N-i}\, du. \end{aligned}$$

Alternatively, for N sufficiently large and φ smooth in a neighborhood of $\frac{i}{N+1}$, we may define the approximations

$$\begin{aligned} a_N(i, f) &\cong \varphi\left(EU_N^{(i)}, f\right) & (5.9) \\ &= -\frac{f'\left(F^{-1}\left(EU_N^{(i)}\right)\right)}{f\left(F^{-1}\left(EU_N^{(i)}\right)\right)} & (5.10) \\ &= -\frac{f'\left(F^{-1}\left(\frac{i}{N+1}\right)\right)}{f\left(F^{-1}\left(\frac{i}{N+1}\right)\right)} & (5.11) \end{aligned}$$

The optimality of rank tests has been discussed by Hoeffding. Specifically, Let $Z_1, \ldots, Z_N$ be a sequence of independent random variables with densities $f_1, \ldots, f_N$ respectively. Suppose that the null hypothesis postulates that the variables are identically distributed. Let $\boldsymbol{R} = (R_1, \ldots, R_N)$ denote the vector of ranks where R_i is the rank of Z_i among $Z_1, \ldots, Z_N$. The following lemma is due to Hoeffding.

Lemma 5.2. *(Hoeffding's Formula) Assume that $V_1 < \cdots < V_N$ are the order statistics of a sample of size N from a density h. Then*

$$P(\boldsymbol{R} = \boldsymbol{r}) = \int \cdots \int_{(v_1 < \cdots < v_N)} f_1(v_1) \cdots f_N(v_N)\, dv_1 \ldots dv_N \tag{5.12}$$

$$= \int \cdots \int_{(v_1 < \cdots < v_N)} \frac{f_1(v_1) \ldots f_N(v_N)}{N! h(v_1) \cdots h(v_N)} N! \left[\prod_{i=1}^{N} h(v_i)\right] dv_1 \ldots dv_N \tag{5.13}$$

$$= \frac{1}{N!} E_h \left[\frac{f_1\left(V_{(r_1)}\right) \cdots f_N\left(V_{(r_N)}\right)}{h\left(V_{(r_1)}\right) \cdots h\left(V_{(r_N)}\right)}\right] \tag{5.14}$$

To illustrate the use of Lemma 5.2, consider the two-sample problem where $X_1, \ldots, X_m$ constitute a random sample from a cdf F having density f and $Y_1, \ldots, Y_n$ are a random sample from a cdf G having density g. As well, assume that $f = 0$ implies $g = 0$. Substitute in (5.14)

$$f_i = \begin{cases} f & i = 1, \ldots, m \\ g & i = m+1, \ldots, m+n \end{cases}$$

and set $h = g$

$$P(R = r) = \frac{1}{N!} E_g \left[\frac{f\left(V_{(r_1)}\right) \ldots f\left(V_{(r_m)}\right)}{g\left(V_{(r_1)}\right) \ldots g\left(V_{(r_m)}\right)}\right]$$

It follows that if $R_1, \ldots, R_m$ denote the ranks of variables in the first population:

$$\begin{aligned} P\{R_1 = r_1, \ldots, R_m = r_m\} &= \sum P(R_1 = r_1, \ldots, R_m = r_m, R_{m+1}, \ldots, R_{m+n}) \\ &= \frac{1}{N!} E_g \left[\frac{f\left(V_{(r_1)}\right) \ldots f\left(V_{(r_m)}\right)}{g\left(V_{(r_1)}\right) \ldots g\left(V_{(r_m)}\right)}\right] \left[\sum 1\right] \\ &= \frac{m!n!}{(m+n)!} E_g \left[\frac{f\left(V_{(r_1)}\right) \ldots f\left(V_{(r_m)}\right)}{g\left(V_{(r_1)}\right) \ldots g\left(V_{(r_m)}\right)}\right] \\ &= E_g \left[\frac{f\left(V_{(r_1)}\right) \ldots f\left(V_{(r_m)}\right)}{g\left(V_{(r_1)}\right) \ldots g\left(V_{(r_m)}\right)}\right] / \binom{m+n}{m} \qquad (5.15) \end{aligned}$$

where $V_{(1)} < \cdots < V_{(N)}$ are the order statistics from a random sample of N uniform variables on $(0, 1)$ and the second summation $[\sum 1]$ is over all possible permutations of the first and second samples among themselves. The expectation E_g is with respect to the probability measure under which the n observations are i.i.d. with common density function g, assuming that g is positive whenever f is. In the next two examples, we obtain locally most powerful tests for location and for scale in the two-sample problem.

Theorem 5.3. *Assume $\int |f'(x|)\, dx < \infty$. The test with critical region*

$$\sum_{i=1}^{N} c_i \left[-\frac{f'\left(X_{(i)}\right)}{f\left(X_{(i)}\right)} \right] \geq k$$

is the locally most powerful test for testing

$$H_0 : f(x_1, \ldots, x_N) = \prod_{i=1}^{N} f(x_i)$$

against

$$H_1 : f(x_1, \ldots, x_N) = \prod_{i=1}^{N} f(x_i - \Delta c_i), \Delta > 0$$

where the $\{c_i\}$ are regression constants indicating group membership and are given by

$$c_i = \begin{cases} 1 & i = 1, \ldots, m \\ 0 & i = m+1, \ldots, m+n \end{cases}$$

Proof. See Hajek and Sidak [1967, p.69]. □

Similar results are also valid for tests against scale alternatives. Let

$$\varphi_1(u, f) = -1 - F^{-1}(u) \frac{f'\left(F^{-1}(u)\right)}{f\left(F^{-1}(u)\right)}, 0 < u < 1$$

and define the scores

$$a_{1N}(i, f) \cong \varphi_1\left(EU_N^{(i)}, f\right) = \varphi_1\left(\frac{i}{N+1}, f\right)$$

Theorem 5.4. *Assume $\int_{-\infty}^{\infty} |xf'(x)\, dx| < \infty$. The test with critical region*

$$\sum_{i=1}^{N} c_i \left[X_{(i)} \frac{f'\left(X_{(i)}\right)}{f\left(X_{(i)}\right)} \right] \geq k$$

is the locally most powerful test for testing

$$H_0 : f(x_1, \ldots, x_N) = \prod_{i=1}^{N} f_X(x_i)$$

against

$$H_1 : f(x_1, \ldots, x_N) = \exp\left(-\Delta \sum_{i=1}^{N} c_i\right) \prod_{i=1}^{N} f_X(x_i - \mu) e^{-\Delta c_i}, \Delta > 0$$

where the $\{c_i\}$ *are known regression constants and* μ *is an arbitrary real number.*

Proof. See Hajek and Sidak [1967, p.69]. □

To illustrate these results, we consider some examples.

Example 5.1. Let $f(x)$ is the density of a standard normal random variable. Then

$$\begin{aligned}\varphi(u, f) &= -\frac{f'(F^{-1}(u))}{f(F^{-1}(u))} \\ &= \Phi^{-1}(u)\end{aligned}$$

and

$$\varphi_1(u, f) = \left[\Phi^{-1}(u)\right]^2 - 1$$

Consequently for the location problem, the van der Waarden test statistic uses the approximate score function (5.9) and is given by

$$\sum_{i=1}^{m} \Phi^{-1}\left(\frac{R_i}{N+1}\right)$$

Example 5.2. If $f(x)$ is the density of a logistic random variable, then

$$f(x) = F(x)(1 - F(x))$$

and hence

$$\begin{aligned}\varphi(u, f) &= -\frac{f'(F^{-1}(u))}{f(F^{-1}(u))} \\ &= 2u - 1\end{aligned}$$

Therefore, the score function is

$$\begin{aligned}a_N(i, f) &\cong \varphi\left(EU_N^{(i)}, f\right) \\ &= \frac{i}{N+1}\end{aligned}$$

The Mann-Whitney-Wilcoxon statistic is then

$$\sum_{i=1}^{m} \varphi\left(\frac{R_i}{N+1}, f\right) \sim \sum_{i=1}^{m}\left(\frac{R_i}{N+1}\right)$$

For the scale alternative, we have

$$\varphi_1 (u, f) = (2u - 1) \log \left[\frac{u}{1-u} \right] - 1$$

Example 5.3. In the case of the double exponential,

$$f (x) = \frac{1}{2} \exp - |x| ,$$

$$-\frac{f' (x)}{f (x)} = sgn (x)$$

and the score functions are

$$\begin{aligned} \varphi (u, f) &= sgn (2u - 1) \\ \varphi_1 (u, f) &= - \log (1 - |2u - 1|) \end{aligned}$$

Example 5.4. As a further example for the scale alternatives, we cite the Ansari-Bradley test which assigns small scores to the extreme ranks as follows:

$$a_{1N} (i, f) = \frac{N+1}{2} - \left| i - \frac{N+1}{2} \right|$$

Hence, populations with a larger dispersion would tend to have smaller ranks.

The asymptotic distribution of various linear rank statistics under the alternative was obtained using the concept of contiguity [Hajek and Sidak, 1967, Section 6.12]. See also Hajek [1968] for a more general result. In this section, we have cited the optimal scores for both the location and scale two-sample problems. Unfortunately, these results rely on knowing the underlying distributions, and as a result we find ourselves in a dilemma. To partially circumvent this difficulty, [Alvo, 2021] proposed instead the use of the Pearson system of distributions defined in (2.2) whereby for members of this family of distributions, we have the ratio

$$\frac{f' (x)}{f (x)} = \frac{x - a}{b_0 + b_1 x + b_2 x^2}$$

The constants are functions of the first four central moments which exist for most distributions, the Cauchy distribution being an important exception. Now set

$$h (x) = -\frac{f' (x)}{f (x)}$$

and define a new score function

$$a(i) = h\left(\frac{i}{N+1}\right)$$

We may appeal to Theorem 5.2 for the asymptotic normality of the linear rank statistic provided $h(u)$ is square integrable. A simulation study was conducted to assess the power of the following linear rank statistic:

$$S = \sum_{i=1}^{N} c_i h\left(\frac{R_i}{N+1}\right)$$

We note that this new statistic takes into account both the moments and the rankings of the actual data. For a simple linear rank statistic $T = \sum_{i=1}^{N} c_i a(R_i)$, the mean and variance are given respectively by

$$\mu_T = N\bar{a}\bar{c} \tag{5.16}$$

$$\sigma_T^2 = \frac{1}{N-1}\sum_{i=1}^{N}(c_i - \bar{c})^2 \sum_{i=1}^{N}(a_i - \bar{a})^2 \tag{5.17}$$

where the $\{c_i\}$ are regression constants indicating group membership given by

$$c_i = \begin{cases} 1 & i = 1, \ldots, m \\ 0 & i = m+1, \ldots, N \end{cases}$$

Note

$$\bar{c} = \frac{m}{m+n}$$

$$\sum_{i=1}^{m+n}(c_i - \bar{c})^2 = \frac{mn}{m+n} = n\bar{c}$$

Below, we report on a limited simulation study involving the following three linear rank statistics. Here, T_N and T_L are the locally most powerful tests when the underlying distributions are normal and logistic, respectively.

1. $T_N = \sum c_i \Phi^{-1}\left(\frac{R_i}{N+1}\right)$
2. $T_L = \sum c_i \left(\frac{R_i}{N+1}\right)$
3. $T_{new} = \sum c_i h\left(\frac{R_i}{N+1}\right)$

Table 5.2.: Power of tests for location alternative

						Δ					
Distribution	Statistic	0	0.1	0.2	0.4	0.6	0.8	1	1.25	1.5	1.75	2
Normal	T_N	0.0263	0.054	0.117	0.328	0.6156	0.854	0.963	0.996	1	1	1
	T_L	0.027	0.054	0.1152	0.319	0.602	0.844	0.958	0.996	1	1	1
	T_{new}	0.0273	0.054	0.115	0.319	0.601	0.843	0.958	0.995	0.998	1	1
Logistic	T_N	0.0254	0.0371	0.063	0.1348	0.247	0.405	0.574	0.759	0.899	0.959	0.989
	T_L	0.024	0.038	0.063	0.141	0.252	0.415	0.588	0.773	0.898	0.965	0.991
	T_{new}	0.0254	0.038	0.063	0.140	0.2527	0.414	0.588	0.771	0.897	0.964	0.989
Laplace	T_N	0.0258	0.049	0.091	0.220	0.439	0.6643	0.8423	0.947	0.988	0.997	1
	T_L	0.028	0.053	0.095	0.247	0.484	0.7116	0.879	0.967	0.994	0.999	1
	T_{new}	0.0286	0.052	0.094	0.2449	0.478	0.707	0.874	0.963	0.992	0.997	1

The means and variances were numerically calculated in accordance with (5.16) and (5.17). For T_{new}, the moments of the data were first obtained. In general, the linear rank statistics considered when standardized have an asymptotic standard normal distribution as $\min(m, n) \to \infty$, that is,

$$S = \frac{T - \mu_T}{\sigma_T} \xrightarrow{d} N(0, 1)$$

The power results reported in Table 5.2 were obtained in a simulation study with $m = n = 30$ and for a nominal level for rejecting the null hypothesis under H_0 equal to 0.025. All the test statistics satisfy the nominal level requirement. It is seen that the new test statistic performs well irrespective of the underlying distribution.

In general the test statistics which have an asymptotic standard normal distribution as $\min(m, n) \to \infty$ take the standardized form

$$S = \frac{T - \mu(T)}{\sigma_T}$$

The following results were obtained in a simulation study with $m = n = 30$.

Similar results are also valid for tests against scale alternatives. Define the scores

$$a_{1N}(i, f) \cong \varphi_1\left(EV_N^{(i)}, f\right) = \varphi_1\left(\frac{i}{N+1}, f\right)$$

where

$$\varphi_1(v, f) = -1 - F^{-1}(v)\frac{f'\left(F^{-1}(v)\right)}{f\left(F^{-1}(v)\right)}, 0 < v < 1.$$

Assume $f(x)$ is absolutely continuous with $\int_{-\infty}^{\infty} |xf'(x)|\,dx < \infty$. Hajek and Sidak [1967, p.94] showed that the test statistic with critical region

$$\sum_{i=1}^{N} c_i a_{1N}(R_i f) \geq k$$

is the locally most powerful test for testing

$$H_0 : f(x_1, \ldots, x_{m+n}) = \prod_{i=1}^{m+n} f_X(x_i)$$

against

$$H_1 : f(x_1, \ldots, x_{m+n}) = \exp(-m\Delta) \prod_{i=1}^{m} f_X\left[(x_i - \mu)e^{-\Delta}\right] \prod_{i=m+1}^{m+n} f_X(x_i - \mu), \Delta > 0.$$

Under the alternative, both densities have the same location parameter μ, considered a nuisance parameter.

In the special case where F is a normal distribution with mean 0 and variance σ^2, then the locally most powerful rank test is based on the sum

$$T_{CK} = \sum_{i=1}^{N} c_i \left[\Phi^{-1}\left(\frac{R_i}{N+1}\right)\right]^2$$

It is known as the Capon-Klotz test. On the other hand, if $f(x) = e^{-x}, x > 0$, then the locally most powerful rank test is based on the sum of the approximate Savage scores given by

$$T_S = -\sum_{i=1}^{N} c_i \left[\log\left(1 - \frac{R_i}{N+1}\right)\right].$$

A new statistic is proposed as follows. Rank the combined data by their absolute values and denote these by $\{R_i^*\}$. Then compute

$$T_{new}^* = \sum_{i=1}^{N} c_i \left(\frac{R_i^*}{N+1}\right) h\left(\frac{R_i^*}{N+1}\right)$$

A simulation study was conducted, and the results are reported in Table 5.3 with $m = n = 30$ and for a nominal level for rejecting the null hypothesis under H_0 equal to 0.025. As expected, the Capon-Klotz statistic performs well when the underlying distribution is normal but not otherwise, while the Savage test performs well when the underlying distribution but not otherwise. In all cases the new statistic performs well.

Table 5.3.: Power of tests for scale alternative

						Δ				
Distribution	Statistic	0	0.25	0.5	0.75	1	1.25	1.5	1.75	2
Normal	T_{CK}	0.022	0.224	0.666	0.934	0.994	1	1	1	1
	T_S	0.022	0.058	0.119	0.182	0.256	0.320	0.369	0.406	0.434
	T^*_{new}	0.024	0.201	0.617	0.902	0.989	1	1	1	1
Exponential	T_{CK}	0.025	0.047	0.074	0.096	0.097	0.087	0.062	0.041	0.027
	T_S	0.026	0.143	0.449	0.780	0.951	0.994	1	1	1
	T^*_{new}	0.025	0.125	0.360	0.660	0.872	0.967	0.994	1	1

5.3. U Statistics

The subject of U statistics (the letter U is for unbiased) was introduced by Hoeffding [1948]. Let $X_1, \ldots, X_n$ be a random sample from some distribution F, and let $h(x_1, \ldots, x_m), m \leq n$ be a real valued measurable function such that

$$E(h(X_1, \ldots, X_m)) = \theta_F$$

The smallest integer m with this property is called the degree of θ_F. A U statistic is defined to be the sum

$$U_n = \binom{n}{m}^{-1} \sum h(X_{i_1}, \ldots, X_{i_m})$$

where the summation is taken over all combinations of m integers $(i_1 < i_2 < \cdots < i_m)$ chosen from $(1, 2, \ldots, n)$.

A U statistic is a minimum variance unbiased estimator of θ_F based on the random sample of size n. This follows from the Lehmann-Scheffe theorem since the statistic is an unbiased estimator and depends on the data only through the order statistics which are complete and sufficient statistics. Examples of U statistics are the moments of a distribution, the variance, and the serial correlation coefficient [Hajek and Sidak, 1967].

Example 5.5. The moments of a univariate distribution are given by the choice $h(x) = x^r$. The sample variance is obtained from the choice of $h(x_i, x_j) = \frac{(x_i - x_j)^2}{2}$. The Gini mean difference statistic popular in economics is obtained from $h(x_i, x_j) = |x_i - x_j|$. The unscaled statistic

$$\frac{n \sum_i \left(X_i - \bar{X}\right)^3}{(n-1)(n-2)}, n \geq 3$$

appearing in the estimator of skewness is an example of a U statistics [Maesono, 2004].

U statistics are symmetric, and in this connection we cite the article by Lai and Wang [1993] on Edgeworth expansions for symmetric statistics. In the next theorem, we indicate a general expression for the variance of a U statistic. Define for $c = 0, 1, \ldots, m$

$$h_c(x_1, \ldots, x_c) = h(x_1, \ldots, x_c, X_{c+1}, \ldots, X_m)$$

Theorem 5.5. *$Var(U_n) = \binom{n}{m}^{-1} \sum_{c=1}^{m} \binom{m}{c} \binom{n-m}{m-c} \sigma_c^2$ where*

$$\begin{aligned} \sigma_c^2 &= Cov\left(h\left(X_{i_1}, \ldots, X_{i_m}\right), h\left(X_{j_1}, \ldots, X_{j_m}\right)\right) \\ &= E\left[\left(h_c\left(X_1, \ldots, X_c\right) - \theta_F\right)\left(h_c\left(X_1, \ldots, X_c\right) - \theta_F\right)\right] \end{aligned}$$

The variances are non-decreasing

$$\sigma_1^2 \leq \sigma_2^2 \leq \cdots \leq \sigma_m^2$$

and moreover, for large n, if $\sigma_m^2 < \infty$,

$$Var\left(U_n\right) \sim m^2 \sigma_1^2 / n$$

Proof. See Ferguson [1996]. □

Definition 5.5. A U statistic has a degeneracy of order k if $\sigma_i^2 = 0$ for $i \leq k$ and $\sigma_{k+1}^2 > 0$.

The asymptotic distribution of a U statistic depends on whether or not there is a degeneracy as we see in the next theorem.

Theorem 5.6. *Let $\sigma_m^2 < \infty$.*

(i) Suppose that $\sigma_1^2 > 0$. Then for large n

$$\sqrt{n}\left(U_n - \theta_F\right) \xrightarrow{\mathcal{L}} \mathcal{N}\left(0, m^2 \sigma_1^2\right)$$

(ii) If $\sigma_1^2 = 0$ and hence the $U-$statistic has a degeneracy of order 1, then

$$n\left(U_n - \theta_F\right) \xrightarrow{\mathcal{L}} \sum_{j=1}^{\infty} \lambda_j \left(Z_j^2 - 1\right)$$

where the Z_js are i.i.d. $N(0,1)$ and the $\{\lambda_j\}$ are the eigenvalues satisfying

$$h\left(x_1 x_2\right) - \theta_F = \sum \lambda_k \varphi_k\left(x_1\right) \varphi_k\left(x_2\right)$$

for orthonormal functions $\{\varphi_k(x)\}$ for which

$$E\left(\varphi_k(X) \varphi_j(X)\right) = \delta_{kj}$$

Proof. See Ferguson [1996]. □

In the study of block designs, the usual Friedman statistic is a U statistic which under the null hypothesis of uniformly distributed rankings has a degeneracy of order 1. In the section on block designs, we shall study other examples whose behavior under the null distribution is similar. See Van der Vaart [2000, p.167–171] for additional examples of degenerate U statistics.

Next we define two-sample U statistics and state a result on their variance.

Definition 5.6. Consider two independent random samples $X_1, \ldots, X_{n_1}$ from F and $Y_1, \ldots, Y_{n_2}$ from G. Let $h(x_1, \ldots, x_{m_1}, y_1, \ldots, y_{m_2})$ be a kernel such that

$$E\left[h\left(X_1, \ldots, X_{m_1}, Y_1, \ldots, Y_{m_2}\right)\right] = \theta_{F,G}$$

A two-sample U statistic is defined to be

$$U_{n_1,n_2} = \binom{n_1}{m_1}^{-1}\binom{n_2}{m_2}^{-1}\sum h\left(X_{i_1}, \ldots, X_{i_{m_1}}, Y_{j_1}, \ldots, Y_{j_{m_2}}\right)$$

where the sum is taken over all subscripts $1 \leq i_1 < \cdots < i_{m_1} \leq n_1$ chosen from $1, 2, \ldots, n_1$ and subscripts $1 \leq j_1 < \cdots < j_{m_2} \leq n_2$ chosen from $1, 2, \ldots, n_2$, respectively.

Theorem 5.7. *Var* $\left[U_{n_1,n_2}\right] = \binom{n_1}{m_1}^{-1}\binom{n_2}{m_2}^{-1}\sum_{i=1}^{m_1}\sum_{j=1}^{m_2}\binom{m_i}{i}\binom{n_1-m_1}{m_1-i}\binom{m_2}{j}\binom{n_2-m_2}{m_2-j}\sigma_{ij}^2$
where $\sigma_{ij}^2 =$

$$Cov\left(h\left(X_1, .., X_i, X_{i+1}, \ldots, X_{m_1}, Y_1, \ldots, Y_j Y_{j+1,}, \ldots, Y_{m_2}\right), h\left(X_1, \ldots, X_i, X'_{i+1}, \ldots, X'_{m_1}, Y_1, \ldots, Y_j Y'_{j+1,}, \ldots, Y'_{m_2}\right)\right)$$

If $\sigma_{m_1 m_2}^2 < \infty$ *and* $\frac{n_1}{n_1+n_2} \to p$, $0 < p < 1$ *as* $n_1 + n_2 \to \infty$, *then*

$$\sqrt{n_1+n_2}\left(U_{n_1,n_2} - \theta_{F,G}\right) \xrightarrow{\mathcal{L}} N\left(0, \sigma^2\right)$$

where

$$\sigma^2 = \frac{m_1^2}{p}\sigma_{10}^2 + \frac{m_2^2}{1-p}\sigma_{01}^2$$

Proof. See Ferguson [1996] □

Modern detailed accounts on U-statistics and further results may be found in Ferguson [1996]; Lee [1990].

5.4. Hoeffding's Combinatorial Central Limit Theorem

Let $(Y_{n1}, \ldots, Y_{nn})$ be a random vector which takes each of the $n!$ permutations of $(1, \ldots, n)$ with equal probability $\frac{1}{n!}$. Set

$$S_n = \sum_{i=1}^{n} a_n(i)\, b_n(Y_{ni})$$

and for $c_n(i,j) = a_n(i)\, b_n(j)$

$$d_n(i,j) = c_n(i,j) - \frac{1}{n}\sum_{g=1}^{n} c_n(g,j) - \frac{1}{n}\sum_{h=1}^{n} c_n(i,h) + \frac{1}{n^2}\sum_{g=1}^{n}\sum_{h=1}^{n} c_n(g,h)$$

Recall that a similar type of transformation was used in multidimensional scaling (see Sect. 4.3.5) to doubly center a matrix. Hoeffding [1951] showed that under some conditions on $\{a_n(i)\}$ and $\{b_n(i)\}$, S_n has an asymptotic normal distribution.

Theorem 5.8. *Hoeffding [1951] The distribution of S_n is asymptotically normal with mean*

$$ES_n = \frac{1}{n}\sum_i\sum_j c_n(i,j)$$

and variance

$$VarS_n = \frac{1}{n-1}\sum_i\sum_j d_n^2(i,j)$$

provided

$$\lim_{n\to\infty} n\frac{\max(a_n(i)-\bar{a}_n)^2}{\sum(a_n(i)-\bar{a}_n)^2}\frac{\max\left(b_n(i)-\bar{b}_n\right)^2}{\sum\left(b_n(i)-\bar{b}_n\right)^2} = 0$$

An immediate application of Hoeffding's theorem is to the Spearman footrule

$$S_n = \sum_{i=1}^{n}|R_i - i| \tag{5.18}$$

It can be shown that

$$\begin{aligned} E(S_n) &= \frac{n^2}{3} + O(n) \\ Var(S_n) &= \frac{2n^3}{45} + O\left(n^2\right) \end{aligned}$$

and hence a standardized S_n is asymptotically normal [Diaconis and Graham, 1977].

In related works, [Alvo and Charbonneau, 1997] consider the use of Spearman's footrule in tests of trend when the data are incomplete. Genest et al. [2010] provides a review along with some related results on copulas. A further reference which provides a connection of the Spearman footrule to a linear combination of a Markov Chain is due to Sen et al. [2011]. A surprising result stated in the theorem below relates the empirical copula function (Chap. 4) to the Spearman correlation [Nelsen, 2006, Theorem 5.6.2, p.220].

Theorem 5.9. *Let $\hat{C}_n$ denote the empirical copula for the sample (x_i, y_i). Then Spearman's rho statistic can be expressed as*

$$\sum_{i=1}^{n}\sum_{j=1}^{n}\hat{C}_n\left(\frac{i}{n},\frac{j}{n}\right) = \frac{1}{n}\sum_{i=1}^{n} iR_i$$

where $\{R_i\}$ denote the ranks of the y_i when the ranks of the $x_i^{\prime}s$ are in natural order. Similar results hold for Kendall's tau in terms of the empirical copula frequency function.

Goodness-of-fit tests were useful in order to determine if nonparametric tests should be used in further analysis. Having decided to proceed with nonparametric tests, we may now consider the study of one sample, paired samples, and two-sample problems. In the next several subsections, we briefly describe some of the main problems in nonparametric statistics. The one and two way design layouts follow.

5.5. Nonparametric Tests

We shall consider nonparametric tests in relation to the types of sampling data available. Table 5.4 displays the types of data along with the corresponding tests of location and scale.

5.5.1. One-Sample Tests of Location

In the simplest one-sample problem, we are interested in testing a hypothesis about the median of a distribution.

Suppose that we have a random sample $X_1, \ldots, X_n$ from a population having a (not necessarily continuous) distribution $F_X(x)$ with a unique median M (i.e., $F_X(M) = 0.5$). We would like to test the hypotheses

$$H_0 : M = M_0 \text{ versus } H_1 : M > M_0. \tag{5.19}$$

Alvo and Yu [2018] developed a parametric paradigm, in which they first define a score function, sensitive to changes in the median. Let $Y = \text{sgn}\,(X - M_0)$ and define the change of measure by

$$\pi_\theta\,(y) = \exp\left(\theta y - K(\theta)\right) g(y), \quad y = -1, 0, 1, \tag{5.20}$$

where $g(y)$ represents the null probability distribution of Y. Let $g(1) = p_+, g(0) = p_0$ and $g(-1) = p_-$ such that $p_+ + p_0 + p_- = 1$. It is natural to assume that $p_+ = p_-$. Here $K\,(\theta)$ satisfies

$$p_+ e^\theta + p_- e^{-\theta} + p_0 = e^{K(\theta)}.$$

Since $p_+ + p_- + p_0 = 1$ and $p_+ = p_-$, it follows that $K\,(0) = K'\,(0) = 0$. It is easy to see that testing the hypotheses in (5.19) is the same as testing

$$H_0 : \pi\,(1;\theta) = \pi\,(-1;\theta)$$

versus

$$H_1 : \pi\,(1;\theta) > \pi\,(-1;\theta)$$

or equivalently

$$H_0 : \theta = 0$$

versus

$$H_1 : \theta > 0$$

Table 5.4.: Nonparametric problems

Types of samples	Test for location	Tests for dispersion
One sample	Sign test	
	Wilcoxon signed rank test	
	Tests for trend [Alvo and Berthelot, 2012; Alvo and Cabilio, 1994]	
Multi samples	Median tests	Moses' test
	Mann-Whitney, Wilcoxon tests	Ansari-Bradley; Siegel Tukey
	Kruskal-Wallis	Alvo-Pan tests [Alvo and Pan, 1997]
	Jonckheere-Terpestra tests for ordered alternatives [Alvo and Cabilio, 1995b]	Conover squared ranks test
	Test for ordered alternatives [Alvo and Pan, 1997]	
	Test for umbrella alternatives [Alvo, 2008]	
Matched pairs	Sign test	
	Wilcoxon matched pairs signed test	
Block designs	Two-way randomized block (Friedman, Durbin, [Alvo et al., 1982])	
	Incomplete block designs [Alvo and Cabilio, 1995b; Alvo and Yu, 2014]	
	Page, Joncheere tests [Alvo and Cabilio, 1995b]	
	Test for interaction	

For a sample of size n, let n_+, n_-, n_0 denote the observed number of cases where $y = 1, -1, 0$ respectively. Note that under H_0, the score function is given by $U = (n_+ - n_-)$ and the Fisher information is

$$I = nK''(0) = n(1 - \hat{p}_0) = n - n_0$$

The score test statistic for testing

$$H_0 : \theta = 0 \text{ versus } H_1 : \theta \neq 0$$

is

$$S_{sign} = \frac{U^2}{I} = \frac{(n_+ - n_-)^2}{n - n_0} = \frac{4\left(n_+ + \frac{n_0}{2} - \frac{n}{2}\right)^2}{n - n_0} = \frac{4\left(n_+ - \frac{n - n_0}{2}\right)^2}{n - n_0}. \quad (5.21)$$

Consequently large values of S_{sign} reject the null hypothesis H_0. For a large sample size n, the score test statistic S_{sign} has a χ_1^2 distribution under H_0.

When no ties are present (i.e., $n_0 = p_0 = 0$), n_+ has a binomial distribution with probability p_+. Under H_0, we have $p_+ = 0.5$. When $M > M_0$, $p_+ > 0.5$, and hence large values of n_+ lead to rejecting H_0. The p-value of the sign test is then $\Pr(B \geq n_+^*)$ where $B \sim Binomial(n, 0.5)$ and n_+^* is the observed value n_+.

Similarly, when $M < M_0$, the p-value is $\Pr(B \leq n_+^*)$ whereas when $M \neq M_0$, the p-value is $2 \times \min\{\Pr(B \geq n_+^*), \Pr(B \leq n_+^*)\}$. We illustrate these computations in the following example.

Example 5.6. The weekly sales of new mobile phones at a mobile phone shop in a mall are collected over the past 12 weeks. The number of phones sold is recorded for each of 12 weeks and is given below:

45 32 39 29 64 55 38 212 187 124 320 188

Last year, the *median* weekly sales was 50 units. Is there sufficient evidence to conclude that median sales this year are higher than last year? Test at $\alpha = 0.05$.
Solution. Let M be true median weekly sales this year and formulate the hypotheses

$$H_0 : M = 50 \text{ versus } H_1 : M > 50,$$

We calculate $n = 12$ and $n_+ = 7$. The p-value for the test is given by

$$\text{pvalue} = \Pr(B \geq 7 | p = 0.5)) = \sum_{i=7}^{12} \binom{12}{i} 0.5^i (1 - 0.5)^{12-i} = \sum_{i=7}^{12} \binom{12}{i} 0.5^{12} = 0.3872.$$

Since the p-value > 0.05, we do not have enough evidence to reject H_0 at the 5% level of significance. Thus, there are no grounds upon which to conclude that median sales are now higher than 50 units per week.

Example 5.7. A food product is advertised to contain 75 mg of sodium per serving, while preliminary studies indicate that servings may contain more than that amount.

We can formulate this problem as a test for the median M of the amount of sodium preserving:

$$H_0 : M = 75 \text{ versus } H_1 : M > 75$$

We are given that 40 packages of a company's food product are examined and there are 26 packages whose amount of sodium per serving are greater than 75. Making use of the normal approximation, the p-value of sign test is

$$\begin{aligned}
\Pr(B \geq 26) &= \Pr(B > 25.5) \quad \text{(continuity correction)} \\
&= \Pr(Z > \frac{25.5 - 20}{\sqrt{0.25 \times 40}}) \quad (Z \sim N(0,1) \text{ under } H_0) \\
&= \Pr(Z > 1.7393) = 0.041
\end{aligned}$$

Since the p-value < 0.05, we conclude that the median amount of sodium per serving is greater than 75 mg at the 5% level of significance.

The sign test gets its name from the fact that the test depends only on the sign of the differences $X_i - M_0$ but not their magnitude. This kind of test has the advantage that it can still be performed for data with only signs available. It also protects well against outliers, as only the sign of the outlier is of interest no matter how far it is away from M_0. When either $p_0 = 0$ or $n_0 = 0$, we obtain the usual sign test [Hajek and Sidak, 1967, p.110].

5.5.2. Confidence Interval for the Median

Definition 5.7. A confidence interval is called *distribution-free* if

$$\Pr(L < \phi < U) = 1 - \alpha$$

independent of the distribution F where $L = f_1(X_1, \ldots, X_n)$ and $U = f_2(X_1, \ldots, X_n)$ are statistics and ϕ is a parameter to be estimated.

Suppose the distribution of the data $F(x)$ is continuous. The procedure to construct a symmetric two-sided distribution-free confidence interval for the median M with level of confidence $(1 - \alpha)$ is given below:

(a) Find c_α such that $\Pr(c_\alpha \leq B^* \leq n - c_\alpha) = 1 - \alpha$, with $B^* \sim \mathit{Binomial}(n, 0.5)$.

(b) Let $X_{(1)} < X_{(2)} < \cdots < X_{(n)}$ be the order statistics.

(c) Then $(X_{(c_\alpha)}, X_{(n+1-c_\alpha)})$ is a $100(1-\alpha)\%$ confidence interval for M since it satisfies $\Pr(X_{(c_\alpha)} < m < X_{(n+1-c_\alpha)}) = 1 - \alpha$.

To show $\Pr(X_{(c_\alpha)} < m < X_{(n+1-c_\alpha)}) = 1 - \alpha$, note that

$$\begin{aligned} B^* \geq c_\alpha &\Leftrightarrow \text{ number of } (X_i > M) \geq c_\alpha \\ &\Leftrightarrow M < \text{ at least } c_\alpha \text{ of } X_i \\ &\Leftrightarrow M < X_{(n+1-c_\alpha)} < \cdots < X_{(n)} \\ &\Leftrightarrow M < X_{(n+1-c_\alpha)}. \end{aligned}$$

Similarly,

$$\begin{aligned} B^* \leq n - c_\alpha &\Leftrightarrow \text{ number of } (X_i > M) \leq n - c_\alpha \\ &\Leftrightarrow M < \text{ at most } n - c_\alpha \text{ of } X_i \\ &\Leftrightarrow M > \text{ at least } c_\alpha \text{ of } X_i \\ &\Leftrightarrow X_{(1)} < \cdots < X_{(c_\alpha)} < M \\ &\Leftrightarrow M > X_{(c_\alpha)}. \end{aligned}$$

Therefore, the confidence interval is

$$X_{(c_\alpha)} < M < X_{(n+1-c_\alpha)}$$

5.5.3. Wilcoxon Signed Rank Test

The sign test can be derived from the theory of U-statistics by defining the kernel function $h(Z_i)$ on the sign of the difference $Z_i = X_i - M$:

$$h(Z_i) = \text{sgn}(Z_i), \quad i = 1, 2, \ldots, n.$$

The Wilcoxon signed ranked test, used in matched pairs situations, takes into account both the rank and the sign. It is more powerful than the sign test, and it is used to test the hypothesis that the distribution of X is symmetric about M.

Consider the two-variable kernel:

$$Y_{ij} = h(Z_i, Z_j) = I(Z_i + Z_j > 0),\ i \leq j,\ i, j = 1, 2, \ldots, n$$

where $(Z_i + Z_j)$ are the Walsh sums.

As we saw, the sign test can be derived from the theory of U-statistics by defining the kernel function on the differences $z = x - m$:

$$h(z) = I(z > 0).$$

The Wilcoxon signed ranked test which takes into account both the rank and the sign is more powerful. It is used to test the hypothesis that the distribution of X is symmetric about m. Consider the two-variable kernel

$$h(z_i, z_j) = I(z_i + z_j > 0)$$

We may construct U statistics U_{1n} and U_{2n} for each of these kernels, respectively, along with

$$\begin{aligned} W_n^+ &= \sum_{i \le j} I(Z_i + Z_j > 0) \\ &= \sum_{i=1}^{n} I(Z_i > 0) + \sum_{i<j} I(Z_i + Z_j > 0) \end{aligned}$$

Though itself not a U-statistic, W_n^+ is, in fact, the sum of two U-statistics:

$$\begin{aligned} W_n^+ &= \sum_{i \le j} I(Z_i + Z_j > 0) \\ &= \sum_{i} I(Z_i > 0) + \sum_{i<j} I(Z_i + Z_j > 0) \\ &= nU_{1n} + \binom{n}{2} U_{2n} \end{aligned} \tag{5.22}$$

with respective kernels $I(Z_i > 0)$ and $I(Z_i + Z_j > 0)$. Under the null hypothesis, we have that

$$E\left(\sum_i I(Z_i > 0)\right) = \frac{n}{2}, \quad E\left(\sum_{i<j} I(Z_i + Z_j > 0)\right) = \frac{n(n-1)}{4}$$

and hence

$$E\left(W_n^+\right) = \frac{n(n+1)}{4}.$$

As well,

$$\begin{aligned} Var(W_n^+) &= Var(nU_{1n}) + Var\left(\binom{n}{2} U_{2n}\right) + 2Cov\left(nU_{1n}, \binom{n}{2} U_{2n}\right) \\ &= \frac{n}{4} + \frac{n(n^2-1)}{12} + \frac{n(n-1)}{8} \\ &= \frac{n(n+1)(2n+1)}{24} \end{aligned}$$

It can be shown that for large n, the asymptotic distribution of W_n^+ is determined by the distribution of U_{2n} the second term (5.22) [Lehmann, 1975, p.124 and p.337]. The test rejects the null hypothesis for large values of W_n^+ whose distribution is as $n \to \infty$:

$$\frac{W_n^+ - n(n+1)/4}{\sqrt{n(n+1)(2n+1)/24}} \Rightarrow_{\mathcal{L}} N(0,1).$$

Alternatively, we may derive a test statistic from a parametric point of view. Define the two smooth models:

$$\begin{aligned}\pi_1(z_i;\theta_1) &= \exp\left[\theta_1 I(z_i>0) - K_1(\theta_1)\right] g_Z(z_i),\\ \pi_2(z_i,z_j;\theta_2) &= \exp\left[\theta_2 I(z_i+z_j>0) - K_2(\theta_2)\right] g_Z(z_i) g_Z(z_j)\end{aligned}$$

where g_Z is the density of Z assumed to be symmetric around 0 under the null hypothesis:

$$H_0 : \theta_1 = \theta_2 = 0.$$

Although the subject of composite likelihood is deferred to Chap. 9, it is instructive in the present context to derive a score test statistic from the composite likelihood function for models $\pi_1 and \pi_2$. The logarithm of the composite likelihood is proportional to

$$\ell(\theta_1,\theta_2) \sim \left[\theta_1 \sum_{i=1}^{n} I(z_i>0) - nK_1(\theta_1)\right] + \left[\theta_2 \sum_{i<j} I(z_i+z_j>0) - \binom{n}{2} K_2(\theta_2)\right].$$

The score vector is

$$U = \left(\sum_{i=1}^{n} I\left(Z_i>0\right), \sum_{i<j} I\left(Z_i+Z_j>0\right)\right)^T$$

with the variance-covariance of U is given by the matrix

$$\Sigma = \begin{pmatrix} \frac{n}{4} & \frac{n(n-1)}{16} \\ \frac{n(n-1)}{16} & \frac{n\left(n^2-1\right)}{12} \end{pmatrix}$$

The inverse is thus

$$\Sigma^{-1} = \frac{1}{\left(\frac{n^2(n^2-1)}{48} - \left(\frac{n(n-1)}{16}\right)^2\right)} \begin{pmatrix} \frac{n\left(n^2-1\right)}{12} & -\frac{n(n-1)}{16} \\ -\frac{n(n-1)}{16} & \frac{n}{4} \end{pmatrix}.$$

It follows from the theory of multivariate U-statistics that as $n \to \infty$,

$$\left(U - E(U)\right)^T \Sigma^{-1} \left(U - E(U)\right) \xrightarrow{\mathcal{L}} \chi^2_2$$

Note that the usual Wilcoxon signed rank statistic, SR_+, is defined as

$$SR_+ = \sum_{i=1}^{n} R_i^+ I(Z_i>0),$$

where R_i^+ be the rank of $|Z_i|$ among $\{|Z_j|, j=1,\ldots,n\}$. The following lemma shows that W_n^+ and SR_+ are equal.

Lemma 5.3. $W_n^+ = SR_+$.

Proof. Suppose without loss in generality that the $Z_i^{\prime}s$ are ordered in absolute value:

$$|Z_1| < \cdots < |Z_n|.$$

Fix j and consider all the pairs $(Z_i, Z_j), i \leq j$. The sum $Z_i + Z_j > 0$ if and only if $Z_j > 0$. The number of indices for which this is true is equal to j, the rank of $|Z_j|$. □

An adjustment for the variance can be made in the case of ties which are defined as follows.

Definition 5.8. A tied ordering of t objects is a partition into e sets, $1 \leq e \leq t$, each containing g_i objects, $g_1+g_2+\cdots+g_e = t$, so that the g_i objects in each set share the rank $\sum_{j=1}^{i-1} g_j + g_i(g_i+1)/2, 1 \leq i \leq e$. Such a tie pattern is denoted by $\delta = (g_1, g_2, \ldots, g_e)$.

It can be shown that in the case of ties, the variance of W_n^+ above may be replaced by

$$n(n+1)(2n+1)/24 - \frac{\sum g_l^3 - \sum g_l}{48}$$

5.6. Multi-Sample Tests

There are several applications of the sign test for paired comparisons whereby the null hypothesis is that the distribution of the difference $Z = Y - X$ is symmetric about 0. As a special case, we can consider the shift model where $Z = (Y - \Delta) - X$ so that the treatment adds a shift of size Δ to the control. The asymptotic properties of the sign test are discussed in several textbooks. See, for example, [Lehmann, 1975, Chapter 4]. There are occasions when bivariate data is collected. For example, we may be interested in testing the effectiveness of a diet. Individuals are weighed before and after. This is an example of a matched pair design. Other examples are concerned with the shrinkage of material after washing and drying or the computer lab usage in 2 separate months under identical circumstances. Both the sign and Wilcoxon signed rank tests can be used on the paired differences.

In the more general two-sample problem, we are presented with two independent random samples, $X_1, \ldots, X_n$ and $Y_1, \ldots, Y_m$ with continuous cumulative distributions F and G, respectively. Such would be the case if one is interested in assessing the effect of a treatment on a group of patients. One sample would serve as the control group and the other as the treatment group. In general, we would like to test the null hypothesis

$$H_0 : G = F$$

against the alternative

$$H_0 : G \neq F$$

The alternative may be that the treatment effect is additive, in which case the alternative becomes

$$H_1 : G(x) = F(x - \Delta), \varDelta > 0.$$

If the treatment effect depends on x, then a more general alternative might be

$$H_1 : G(x) \leq F(x) \forall x$$

which indicates that small values are less likely under G than under F. In that case, we say that Y is stochastically larger than X. We consider in the next section a test for this general alternative.

5.6.1. Two-Sample Tests for Location

Let X and Y be two independent random variables with cumulative continuous distributions F and G, respectively. Suppose that we are interested in testing the hypotheses:

$$H_0 : G(x) = F(x) \text{ for all } x \text{ vs } H_1 : G(x) \leq F(x) \tag{5.23}$$

for all x and strict inequality for some x.

Example 5.8. If $X \sim \mathcal{N}(\mu_X, \sigma^2)$ and $Y \sim \mathcal{N}(\mu_Y, \sigma^2)$, then the hypotheses become

$$H_0 : \mu_Y = \mu_X \text{ vs } H_1 : \mu_Y \geq \mu_X$$

so that the alternative points to a shift in the mean of the distribution of Y.

Alvo and Yu [2018] showed that a *parametric* approach to this problem leads to the U-statistic:

$$\begin{aligned} U &= \sum_{j=1}^{m} \sum_{i=1}^{n} I(X_i < Y_j) \\ &= \# \{(X_i, Y_j) : X_i < Y_j\} \end{aligned}$$

which rejects for large values. This is the two-sided Mann Whitney statistic. It can be shown

$$U = \sum_{j=1}^{m} R(Y_j) - \frac{n(n+1)}{2}.$$

The sum

$$\sum_{j=1}^{m} R(Y_j) \tag{5.24}$$

is known as the Wilcoxon rank sum statistic and it is equivalent to the Mann Whitney statistic. Both tests reject the null hypothesis for large values of the sum of the ranks from the Y population. Properties of the Wilcoxon test under the alternative can be derived from the properties of U-statistics. The following theorem provides the theoretical basis for the test statistic.

Theorem 5.10. *Let X, X', Y, Y' be independent random variables with X, X' having distribution F and Y, Y' having distribution G. Then*

(a) $E(U) = mnq_1$, $Var(U) = mnq_1(1-q_1)+mn\,(n-1)\,(q_2-q_1^2)+mn\,(m-1)\,(q_3-q_1^2)$ *where*

$$\begin{aligned} q_1 &= P\,(X<Y)\,, \\ q_2 &= P\,(X<Y, X<Y')\,, \\ q_3 &= P\,(X<Y, X'<Y)\,. \end{aligned}$$

(b) *if* $\min(n,m) \to \infty$, *with* $m/n \to \lambda > 0$, *then*

$$\frac{U - mnq_1}{\sqrt{VarU}} \xrightarrow{\mathcal{L}} \mathcal{N}\,(0,1)\,.$$

(c) *Under the null hypothesis whereby* $G\,(x) = F\,(x)$ *for all* x,

$$\begin{aligned} q_1 &= \frac{1}{2} \\ q_2 = q_3 &= \frac{1}{3} \end{aligned}$$

and

$$\frac{U - \frac{mn}{2}}{\sqrt{mn\left(\frac{n+m+1}{12}\right)}} \xrightarrow{\mathcal{L}} \mathcal{N}\,(0,1)\,.$$

The case of ties can be dealt with by assigning the mid rank to tied rankings. In that case, the variance under the null hypothesis becomes

$$Var(U) = mn\left(\frac{n+m+1}{12}\right) - \frac{mn\left(\sum_{i=1}^{e} g_i^3 - g_i\right)}{12\,(m+n)\,(m+n-1)}$$

where e represents the number of distinct values and g_i represents the number of items tied at value i [Lehmann, 1975, p.355].

5.6.2. Multi-Sample Test for Location

The nonparametric analogue of the one-way analysis of variance is given by the Kruskal-Wallis test. A random sample $X_{ij}, j = 1, \ldots, n_i$ is observed for each population $i, i = 1, \ldots, k$. It is desired to test the null hypothesis that the medians of the populations are equal against the alternative that at least two are different. Setting $n = \sum_{i=1}^{k} n_i$, the test is conducted by first ranking all the observations together assuming no ties and then computing the statistic

$$KW = \frac{12}{n\,(n+1)} \sum_{i=1}^{k} n_i \left(\bar{R}_i - \frac{n+1}{2}\right)^2 \tag{5.25}$$

where $\bar{R}_i$ is the average of the ranks for the ith population. The statistic KW weights the squared deviations of the individual population average ranks from the overall average rank $\frac{n+1}{2}$. It can be shown that

$$KW = \frac{12}{n(n+1)} \sum_{i=1}^{k} n_i \bar{R}_i^2 - 3(n+1)$$

Under the null hypothesis, the Kruskal-Wallis statistic KW has an asymptotic chi-square distribution with $(k-1)$ degrees of freedom.

$$KW \xrightarrow{\mathcal{L}} \chi^2_{k-1}.$$

The null hypothesis is rejected for large values of KW.

5.6.3. Tests for Dispersion

Suppose that we have two independent random samples $X_1, \ldots, X_{n_1}$ and $Y_1, \ldots, Y_{n_2}$ from normal distributions $\mathcal{N}(\mu_1, \sigma_1^2)$ and $\mathcal{N}(\mu_2, \sigma_2^2)$, respectively. We would like to test the null hypothesis

$$H_0 : \sigma_1 = \sigma_2$$

against the alternative

$$H_1 : \sigma_1 \neq \sigma_2$$

The test for the equality of the two variances makes use of the Fisher distribution statistic

$$F = \frac{S_1^2}{S_2^2}$$

where for $n_1 > 1, n_2 > 1$,

$$S_1^2 = \frac{\sum_{i=1}^{n_1} \left(X_i - \bar{X}_{n_1}\right)^2}{n_1 - 1}, S_2^2 = \frac{\sum_{j=1}^{n_2} \left(Y_j - \bar{Y}_{n_2}\right)^2}{n_2 - 1}$$

The statistic F which, under the null distribution has a Fisher distribution with $(n_1 - 1, n_2 - 1)$ degrees of freedom, rejects H_0 when its values are either too large or too small. The distribution of the statistic is particularly sensitive to non normality of the underlying distributions [Miller, 1968].

In the nonparametric setting, the Ansari-Bradley test [Ansari and Bradley, 1960] assumes the two populations have equal medians and any differences between them will occur in their dispersion. The combined data from the two samples is first ordered from smallest to largest. Ranks are then assigned from both ends starting with unity and working toward the center. We then have the set of ranks:

$$\begin{cases} 1, 2, \ldots, \frac{n_1+n_2}{2}, \frac{n_1+n_2}{2}, \ldots, 2, 1 & n_1 + n_2 \textit{ even} \\ \\ 1, 2, \ldots, \frac{n_1+n_2+1}{2}, \ldots, 2, 1 & n_1 + n_2 \textit{ odd} \end{cases}$$

The test statistic denoted AB consists of the sum of the ranks of the first sample. Intuitively, the sample with the greater dispersion will have observations on both ends and hence will receive the smaller ranks compared to the population with the smaller dispersion. It can be shown that for large sample sizes n_1, n_2

$$AB \sim \mathcal{N}\left(\mu, \sigma^2\right)$$

where

$$\mu = \begin{cases} \frac{n_1(n+2)}{4} & n \textit{ even} \\ \\ \frac{n_1(n+1)^2}{4n} & n \textit{ odd} \end{cases}$$

and

$$\sigma^2 = \begin{cases} \frac{n_1 n_2 (n+2)(n-2)}{48(n-1)} & n \textit{ even} \\ \\ \frac{n_1 n_2 (n+1)\left(3+n^2\right)}{48n^2} & n \textit{ odd} \end{cases}$$

For a two-sided alternative, the test rejects whenever the values of AB are either too small or too large. Tables for small sample sizes are found in Daniel [1990] along with a description of another test due to Moses's test. A different test called the squared ranks test which mimics the F test is due to Conover [1999].

5.7. Compatibility

Alvo and Cabilio [1993] introduced the notion of compatibility defined below to deal with the situation when not all items in an experiment are ranked. We begin with some notation and then proceed with some applications.

Notation Incomplete ranks will be denoted by "-," and corresponding incomplete rankings will be written with an upper script "*."

For example, the ranking $\nu^* = (2, -, 3, 4, 1)^T$ indicates that object 2 is not ranked among the 5 objects presented. Object 1 is given rank 2, object 3 is assigned rank 3, and so on. We define the following definition of compatibility for rankings.

Definition *The complete ranking ν of t objects is said to be compatible with an incomplete ranking ν^* of a subset of k of these objects, $2 \leq k \leq t$, if the relative ranking of every pair of objects ranked in ν^* coincides with their relative ranking in ν.*

An incomplete ranking gives rise to a class of order preserving complete rankings. As an example, denoting by $C(\nu^*)$ the set of complete rankings compatible with $\nu^* = (2, -, 3, 4, 1)^T$, we have that

$$C(\nu^*) = \left\{ (2,5,3,4,1)^T, (2,4,3,5,1)^T, (2,3,4,5,1)^T, (3,2,4,5,1)^T, (3,1,4,5,2)^T \right\}$$

The total number of complete rankings of t objects compatible with an incomplete ranking of a subset of k objects is given by $t!/k!$. This follows from the fact that there are $\binom{t}{k}$ ways of choosing k integers for the ranked objects, one way in placing them to preserve the order and then $(t-k)!$ ways of rearranging the remaining integers. The product is thus

$$a = \binom{t}{k}(t-k)! = t!/k!$$

The notion of compatibility establishes a connection between an incomplete ranking and the class of complete rankings from which the incomplete ranking could have arisen.

Example 5.9. Suppose that $t = 3, k = 2$. In that case, the possible incomplete rankings are denoted by

$$\begin{aligned} \nu_{11}^* &= (1,2,-)^T, \nu_{12}^* = (2,1,-)^T, \nu_{21}^* = (1,-,2)^T, \\ & \nu_{22}^* = (2,-,1)^T, \nu_{31}^* = (-,1,2)^T, \nu_{32}^* = (-,2,1)^T \end{aligned}$$

We may associate with every incomplete ranking a ($t!$ x 1) compatibility vector, also denoted by $C(\nu^*)$, whose ith component is 1 or 0 according to whether ν_i is compatible with ν^*. A summary can be provided by a compatibility matrix as follows:

$$C = \begin{array}{c|cccccc} & \nu_{11}^* & \nu_{12}^* & \nu_{21}^* & \nu_{22}^* & \nu_{31}^* & \nu_{32}^* \\ \nu_1 & 1 & 0 & 1 & 0 & 1 & 0 \\ \nu_2 & 1 & 0 & 1 & 0 & 0 & 1 \\ \nu_3 & 0 & 1 & 1 & 0 & 1 & 0 \\ \nu_4 & 1 & 0 & 0 & 1 & 0 & 1 \\ \nu_5 & 0 & 1 & 0 & 1 & 1 & 0 \\ \nu_6 & 0 & 1 & 0 & 1 & 0 & 1 \end{array}$$

5.8. Tests for Ordered Alternatives

Consider a randomized block experiment given by the model

$$X_{ij} = b_i + \tau_j + e_{ij},\ i = 1, \ldots, n, j = 1, \ldots, t.$$

where b_i is the ith block effect, τ_j is the jth treatment effect, and the e_{ij} are independent identically distributed error terms having a continuous distribution. We wish to test the hypothesis of no treatment effect

$$H_0 : \tau_1 = \tau_2 = \cdots = \tau_t$$

against the ordered alternative

$$H_1 : \tau_1 \leq \tau_2 \leq \cdots \leq \tau_t$$

with at least one inequality strict. This problem has been considered in the literature by Jonckheere [1954] and Page [1963] where the data in each block is complete. It often happens in experiments, however, that one or more observations are missing from one or more blocks. The example below describes such a situation.

Example 5.10. As an example, we consider the lymph heart pressure data in mm Hg taken over a 24-hour period at 6-hour intervals on 8 toads during an induced dehydration period in Table 5.5. The data shows that some cells are empty. If we identify the 6 hour intervals as treatments and the toads with the blocks, then we have $n = 8, t = 4, k_1 = k_2 = k_6 = k_7 = k_8 = 4, k_3 = k_5 = 3, k_4 = 2$. We wish to test against the alternative $H_1 : \tau_1 \geq \tau_2 \geq \tau_3 \geq \tau_4$ with at least one inequality strict. The data shows that some cells are empty.

Table 5.5.: Lymph heart pressure in mm Hg

		Time			
Toad ID	**Block**	6 **hr**	12 **hr**	18 **hr**	24 **hr**
21	1	11.865	9.832	7.567	10.168
22	2	5.601	4.892	4.032	3.126
23	3		14.415	14.185	7.800
24	4	13.267			9.953
25	5	8.006	7.793		7.582
27	6	17.692	16.644	15.327	11.573
28	7	9.027	7.973	11.855	6.820
29	8	9.789	7.967	7.758	7.849

Alvo and Cabilio [1995b] extended the Jonchheere and Page test statistics to the incomplete ranking case by making use of the concept of compatibility. To this end, define o_{il} to be the label of the lth object ranked in the ith block and set

$$A_S^*(i) = \frac{(t+1)}{(k_i+1)} \sum_{l=1}^{k_i} \left(o_{il} - \frac{t+1}{2}\right) \left(\mu_i^*(o_{il}) - \frac{k_i+1}{2}\right)$$

where k_i is the number of objects ranked in block i. This statistic represents a measure of the association between the observed ranking μ_i^* for the treatments labeled $\{o_{il}\}$ and the complete criterion ranking specified by the alternative chosen to be $\{1, 2, \ldots, k\}$. It can be interpreted as the sum of a naive statistic which ignores gaps and a correction term:

$$\begin{aligned} A_S^*(i) &= \frac{(t+1)}{(k_i+1)} \sum_{l=1}^{k_i} \left(l - \frac{t+1}{2} \right) \left(\mu_i^*(o_{il}) - \frac{k_i+1}{2} \right) + \\ & \frac{4}{(k_i+1)} \sum_{l=1}^{k_i} (o_{il} - l) \left(\mu_i^*(o_{il}) - \frac{k_i+1}{2} \right) \end{aligned}$$

The term $(o_{il} - l)$ represents the number of missing observations to the left of o_{il} and is a weighting for the observed rank. The proposed test which generalizes the Page statistic is the sum

$$L_S^* = \sum_{i=1}^{n} A_S^*(i)$$

A further informative expression may be had by defining

$$\begin{aligned} u_{ij} &= \mu_i^*(j)\,\delta_{ij} + \frac{k_i+1}{2}(1-\delta_{ij}) \\ V_i &= \sum_i j u_{ij} \end{aligned}$$

where $\delta_{ij} = 1$ if treatment j is ranked in block i and $= 0$ otherwise. Note that for each block i,

$$\sum_{j=1}^{t} \frac{(t+1)}{(k_i+1)} u_{ij} = \frac{t(t+1)}{2}$$

Then,

$$A_S^*(i) = \frac{t+1}{k_i+1} V_i - \frac{t(t+1)^2}{4}$$

and

$$L_S^* = \sum_{i=1}^{n} \frac{t+1}{k_i+1} V_i - \frac{nt(t+1)^2}{4}.$$

In this form, we see that instead of taking the sum of the ranks assigned to treatment j as we would in the complete case, we calculate a weighted sum of the scores assigned to that treatment, which is the sum of the ranks in a block in the complete case. Large values of L_S^* are consistent with the alternative.

Turning attention to the Kendall statistic, let

$$a_i(l,m) = \begin{cases} sgn(\mu_i^*(o_{im}) - \mu_i^*(o_{il})) & \text{if } \delta_{il} = \delta_{im} = 1 \\ 1 - \frac{2\mu_i^*(o_{il})}{(k_i+1)} & \text{if } \delta_{il} = 1, \delta_{im} = 0 \\ \frac{2\mu_i^*(o_{im})}{(k_i+1)} - 1 & \text{if } \delta_{il} = 0, \delta_{im} = 1 \\ 0 & \text{otherwise} \end{cases}$$

where $\delta_{il} = 1$ or 0 according to whether or not the lth treatment response is observed in the ith block. We may now write the Kendall statistic:

$$\begin{aligned} A_K^*(i) &= \sum_{l<m}^{k_i} a_i(l,m) \\ &= \sum_{l<m}^{k_i} sgn\left(\mu_i^*(o_{im}) - \mu_i^*(o_{il})\right) + \\ &\quad \frac{4}{(k_i+1)} \sum_{l=1}^{k_i} (o_{il} - l)\left(\mu_i^*(o_{il}) - \frac{k_i+1}{2}\right) \end{aligned}$$

The extended Jonckheere statistic can then be defined to be

$$L_K^* = \sum_{i=1}^{n} A_K^*(i)$$

Alternatively, setting $U_i^*(l,m) = \frac{1}{2}(1 + a_i(l,m))$ we obtain the extended Jonckheere statistic:

$$L_K{}^* = 2\sum_{i=1}^{n} \sum_{l<m} U_i^*(l,m) - \frac{nt(t-1)}{2}$$

The statistics L_S^*, L_K^* are respectively generalizations of the Page and Jonckheere statistics for ordered alternatives. Under H_0 and conditional on the pattern of missing observations, the following theorem was shown to be true.

Theorem 5.11. *(a)* $\frac{L_S^*}{\sigma_S} \xrightarrow{\mathcal{L}} N(0,1)$ *as* $n \to \infty$, $k_i \geq 2$, *where* $\sigma_S^2 = \sum_{i=1}^{n} \sigma_S^2(i)$,

$$\sigma_S^2(i) = \frac{k_i(t+1)^2}{12(k_i+1)} \sum_{l=1}^{k_i} (o_{il} - \bar{o}_l)^2,$$

and $\bar{o}_l = \sum_{l=1}^{k_i} o_{il}/k_i$

(b) $\frac{L_K^*}{\sigma_K} \xrightarrow{\mathcal{L}} N(0,1)$ *as* $n \to \infty$ $k_i \geq 2$, *where* $\sigma_K^2 = \sum_{i=1}^{n} \sigma_K^2(i)$ *and*

$$\sigma_K^2(i) = \frac{16}{(t+1)^2}\sigma_S^2(i) + \frac{5k_i(k_i-1)}{18} + \frac{8}{3(k_i+1)} \sum_{l=1}^{k_i} (o_{il} - l)\left(l - \frac{k_i+1}{2}\right)$$

Proof. The proofs consist of demonstrating the Lyapunov condition holds and then applying the Lindeberg-Feller central limit theorem. See [Alvo and Cabilio, 1995b] for the details. □

These methods were applied to the data in 5.10. Ranking from largest to smallest, we calculate $L_S^* = 26.75$. It was seen in Alvo and Cabilio [1995b] that the p-value of the test is $P(L_S^* \geq 26.75) = 0.00006$.

5.9. A Unified Theory of Hypothesis Testing

In this section, we present a general approach based on the ranks of the data for testing various hypotheses. The approach consists of first defining two sets of rankings, one set most in agreement with the data and the other set most in agreement with the alternative. Next, measure of the separation between the two sets by the average "distance" over all pairs of rankings, one from each set. Small values of this statistic are consistent with the alternative hypothesis and consequently lead to rejection of the null hypothesis. This general methodology can be applied various testing problems. Our notation in this chapter differs somewhat from the notation of previous chapters. It will be convenient to denote observed ranks by μ since we will be constructing a set of permutations generated by it.

The following unified approach was described in Alvo and Pan [1997]. Let H_0, H_1 be the null and alternative hypotheses, respectively, in a typical testing situation. Let $\mathcal{P} = \{\eta\colon [\eta(1), \ldots, \eta(\mathrm{n})]\}$ be the set of all permutations of the integers $1, \ldots, n$, and let $d(\mu, \nu)$ be a measure of distance between permutations μ, ν.

- Step 1: Let $X_1, \ldots, X_n$ be a collection of random variables from some continuous distribution, and let $\mu(i)$ be the rank of X_i among the $X's$. The continuity assumption ensures that with probability one, ties occur with probability 0.
- Step 2: Define $\{\mu\}$ to be the set of all rankings which are "equivalent" to the observed ranking μ in the sense that ranks occupied by identically distributed random variables are exchangeable.
- Step 3: Define E to be the class of extremal rankings which are "most in agreement" with the alternative. The set E is not data dependent but rather depends on the alternative hypothesis.
- Step 4: Define the distance between the set of rankings $\{\mu\}$ and the set E

  $$d(\{\mu\}, E) = \sum_{\mu \in \{\mu\}} \sum_{\nu \in E} d(\mu, \nu).$$

  Small values of $d(\{\mu\}, E)$ are consistent with the alternative and lead to the rejection of the null hypothesis.

We shall illustrate the use of the unified theory in two specific instances: tests for umbrella alternatives and tests of trend for proportions.

5.9.1. Umbrella Alternatives

Let $X_{i(1)}, \ldots, X_{i(m_i)}, i = 1, \ldots, k$ be k independent random samples with $X_{i(l)}, l = 1, \ldots, m_i$ having an absolutely continuous distribution function $F_i(x)$. In the parametric case, we may have $F_i(x) \equiv F(x - \theta_i)$, where F has median zero. We shall be concerned with testing the hypothesis of no treatment effect against the alternative that there is a monotone treatment effect subject to a change in direction. Alternatives of this type arise in situations where the treatment effect changes in direction after reaching a peak. For example, the effectiveness of a drug may change with time, the effectiveness in learning as a function of age may peak at a certain stage, the reaction to increasing levels of a drug dosage may peak at a certain point and decrease thereafter, etc. Formally, letting F_p be the distribution where the turning point occurs, the hypothesis and the alternative are respectively:

$$H_0 \ : \ F_1(x) = \cdots = F_k(x) \tag{5.26}$$

$$H_1 \ : \ F_1(x) \geq \cdots \geq F_{p-1}(x) \geq F_p(x) \leq F_{p+1}(x) \leq \cdots \leq F_k(x), \tag{5.27}$$

with at least one strict inequality for some x. Equivalently, in the parametric case, the hypotheses become

$$H_0 \ : \ \theta_1 = \cdots = \theta_k \tag{5.28}$$

$$H_1 \ : \ \theta_1 \leq \cdots \leq \theta_{p-1} \leq \theta_p \geq \theta_{p+1} \geq \cdots \geq \theta_k. \tag{5.29}$$

with at least one strict inequality in H_1. We note that an umbrella alternative contains as special cases, the ordered alternatives corresponding to $p = k$ or $p = 1$.

In the following section, we construct test statistics based on the ranks of the observations for the situation when the location of the peak is known. Test statistics using both the Spearman and Kendall distances are determined. We obtain the asymptotic null distributions for these test statistics under the assumption that the minimum of the sample sizes gets large. We propose an algorithm to estimate the location of peak when it is unknown. A limited simulation study shows that this algorithm performs well.

In arriving at test statistics, we will assume that there is a single peak and that its location is known. Let $n_i = \sum_{h=1}^{i} m_h, i = 1, \ldots, k,\ n = n_k,\ \hat{n} = n - m_p, n_0 = 0$. Let $\mathcal{P} = \{\mu : [\mu(1), \ldots, \mu(n)]\}$ be the set of all permutations of the integers $1, 2, \ldots, n$, and let $d(\mu, \nu)$ be a distance function between permutations μ and ν. In view of the unified theory, we propose the following steps.

Step 1: Rank all the observations together so that the smallest gets rank 1, the next smallest rank 2, etc. Let the $n-$dimensional vector

$$\pi = (\pi(1), \ldots, \pi(m_1) \,|\, \pi(m_1 + 1), \ldots, \pi(m_1 + m_2) \,|\, \ldots, \pi(n)) \tag{5.30}$$

represent the ranks of the $\left\{X_{i(l)}\right\}, i = 1, \ldots, k;\ l = 1, \ldots, m_i$ and grouped by populations. Since distributions are assumed continuous, ties among the observations occur with probability zero.

Step 2: Define $\{\pi\}$ to be the subclass of permutations π in the sense that ranks occupied by identically distributed random variables are exchangeable. This subclass consists of all permutations π where the rankings within each population are permuted among themselves only. The cardinality of $\{\pi\}$ is given by the product $(\Pi m_h!)$.

Step 3: Define E to be an extremal subclass of $\mathcal{P}$ with H_1. The extremal set E does not necessarily correspond to the entire critical region but rather consists of those permutations which provide the strongest evidence in favor of the alternative. In the present context, permutations in E are such that ranks occupied by observations from F_i are always less than those from $F_{i'}$ if $i < i' \leq p$, whereas the reverse is true if $p \leq i < i'$. Moreover, ranks attributed to a distribution consist of consecutive integers. Hence the cardinality of E is equal to $c\,(\Pi m_i!)$ where $c = \binom{k-1}{p-1}$.

The enumeration of the extremal set E is a two-stage procedure. First, choose the relative order of the $(p-1)$ populations $F_1, \ldots, F_{p-1}$ among $F_1, \ldots F_{p-1}, F_{p+1}, \ldots, F_k$. This can be done in $c = \binom{k-1}{p-1}$ ways. Then partition the integers $1, \ldots, n$ in accordance with the prescribed ordering of the populations while taking into account corresponding sample sizes. The extremal set E is finally obtained by permuting the integers within each population. Population F_p is always allocated the last m_p integers, namely, $\hat{n} + 1, \ldots, n$. The cardinality of E is therefore equal to $c\,(\Pi m_i!)$.

Step 4: Let $d\,(\mu, \nu)$ be a distance function between two permutations $\mu and \nu$, and define the distance between the two sets $\{\pi\}$ and an extremal set E:

$$d\left(\{\pi\}, E\right) = \sum_{\mu \in \{\pi\}} \sum_{\nu \in E} d\,(\mu, \nu) \tag{5.31}$$

Small values of $d\left(\{\pi\}, E\right)$ are inconsistent with the null hypothesis and consequently lead to rejection of H_0.

Consider the distances between permutations defined by Spearman and Kendall, respectively, as

$$\begin{aligned} d_S\,(\mu, \nu) &= \frac{1}{2} \sum_{i=1}^{k} \sum_{l=1}^{m_i} \left[\mu\,(i\,(l)) - \nu\,(i\,(l))\right]^2 \\ &= \frac{n\,(n^2-1)}{12} - \sum_{i=1}^{k} \sum_{l=1}^{m_i} \left[\nu\,(i\,(l))\right] \left[\mu\,(i\,(l)) - \frac{n+1}{2}\right] \end{aligned}$$

and

$$\begin{aligned} d_K\,(\mu, \nu) &= \sum_{i_1(l) < i_2(l')} \left\{1 - sgn\left[\mu\,(i_1\,(l)) - \mu\,(i_2\,(l'))\right] sgn\left[\nu\,(i_1\,(l)) - \nu\,(i_2\,(l'))\right]\right\} \\ &= \frac{n\,(n-1)}{2} \\ &\quad - \sum_{i_1(l) < i_2(l')} \left\{sgn\left[\mu\,(i_1\,(l)) - \mu\,(i_2\,(l'))\right] sgn\left[\nu\,(i_1\,(l)) - \nu\,(i_2\,(l'))\right]\right\} \end{aligned}$$

The Spearman and Kendall distances are both right invariant in the sense that they are invariant with respect to a change in the relabeling of the objects ranked. It was shown in Alvo [2008] that the test statistic corresponding to Spearman distance when the location of the peak is known is given by

$$d_S\left(\{\pi\}, E\right) = \frac{n\left(n^2-1\right)}{12} c\left(\Pi m_i!\right)^2 - c\left(\Pi m_i!\right)^2 S \tag{5.32}$$

where

$$v_i = \begin{cases} c^{-1} \sum_{j=i}^{k+i-p} \left(a_{ij} + \frac{1+m_i}{2}\right) \binom{j-1}{i-1} \binom{k-1-j}{p-1-i} & i < p \\ \left(\hat{n} + \frac{1+m_p}{2}\right) & i = p \\ c^{-1} \sum_{j=k-i+1}^{k-i+p} \left(b_{ij} + \frac{1+m_i}{2}\right) \binom{j-1}{k-i} \binom{k-1-j}{i-1-p} & i > p \end{cases} \tag{5.33}$$

and

$$S = \sum_{i=1}^{k} m_i v_i \left[\bar{\pi}_i - \frac{n+1}{2}\right],$$

where $\bar{\pi}_i$ represents the average of the ranks for the ith population.

As well

$$\begin{aligned} a_{ij} &= \left(\sum_{h=1}^{i-1} m_h\right) + \left(\sum_{h=k+i-j+1}^{k} m_h\right) \\ &= n_{i-1} + n - n_{k+i-j} \end{aligned}$$

and

$$\begin{aligned} b_{ij} &= \left(\sum_{h=i+1}^{k} m_h\right) + \left(\sum_{h=1}^{j+i-k-1} m_h\right) \\ &= n - n_i + n_{j+i-k-1} \end{aligned}$$

In the special case $m_i = m$ where an equal number of observations is taken from each population, $a_{ij} = b_{ij} = (j-1)\, m$ and

$$v_i = \begin{cases} c^{-1} \sum_{j=i}^{k+i-p} \left(jm + \frac{1-m}{2}\right) \binom{j-1}{i-1} \binom{k-1-j}{p-1-i} & i < p \\ km + \left(\frac{1-m}{2}\right) & i = p \\ c^{-1} \sum_{j=k-i+1}^{k-i+p} \left(jm + \frac{1-m}{2}\right) \binom{j-1}{k-i} \binom{k-1-j}{i-p-1} & i > p \end{cases} \tag{5.34}$$

and hence

Table 5.6.: Wechsler Adult intelligence scale scores

		Age		
16 − 19	20 − 34	35 − 54	55 − 69	> 70
8.62	9.85	9.98	9.12	4.80
9.94	10.43	10.69	9.89	9.18
10.06	11.31	11.40	10.57	9.27

$$v_i = \begin{cases} n\frac{i}{p} + \left(\frac{1-m}{2}\right) & i \leq p \\ n\frac{(k+1-i)}{k+1-p} + \left(\frac{1-m}{2}\right) & i > p \end{cases} \tag{5.35}$$

It follows that the test statistic becomes

$$S = mn\left\{\sum_{i\leq p}\frac{i}{p}\left[\bar{\pi}_i - \frac{n+1}{2}\right] + \sum_{i>p}\frac{(k+1-i)}{(k+1-p)}\left[\bar{\pi}_i - \frac{n+1}{2}\right]\right\} \tag{5.36}$$

Assume that $\min(m_i) \to \infty$ in such a way as

$$\frac{m_i}{n} \to \lambda_i > 0. \tag{5.37}$$

and define

$$\sigma_p^2 = (n+1)^2 \frac{\sum_{i=1}^k m_i (v_i - \bar{v})^2}{12} \tag{5.38}$$

Then the test statistic $S = \sum_{i=1}^{k} m_i v_i \left[\bar{\pi}_i - \frac{n+1}{2}\right]$ corresponding to the Spearman distance is asymptotically normal with mean 0 and variance equal to σ_p^2. The proof consists of representing S as a linear rank statistic. The case when the peak is unknown is discussed in Alvo [2008]. As well, it is shown that the Kendall and Spearman test statistics are asymptotically equivalent under (5.37).

In order to illustrate the Spearman test, we consider the data in Table 5.6 on adult intelligence scale scores on males by age groups where we assume that the peak is located in the $35 - 54$ age group. We calculate $S = 155.79, \sigma_p^2 = 4486.464$ and a standardized test statistic $\frac{S}{\sigma_p} = 2.326$. The resulting p-value is 0.01.

5.9.2. Tests for Trend in Proportions

There are several instances in practice when one is interested in testing for a trend in proportions. For instance, one may be interested in the trend in birth rates, in mortality rates, or in the incidence of a certain disease. As an example, we will consider the mortality statistics in South Africa during the period 2000–2008. One may ask if there is an increasing trend in the mortality rates (Table 5.7).

Table 5.7.: Mortality statistics for South Africa 2000–2008

Year	Number of deaths	Population size
2000	416,155	43,789,115
2001	454,882	43,997,828
2002	502,050	44,187,637
2003	556,779	44,344,136
2004	576,709	42,718,530
2005	598,131	42,768,678
2006	612,778	43,647,658
2007	603,094	43,586,097
2008	592,073	43,421,021

In such problems, one usually observes at time values $t_1 < t_2 < \cdots < t_k$ a random sample of n_i binary variables $\{y_{ij}\}$ with y_{ij} taking values 1 or 0 with unknown probabilities θ_i and $1 - \theta_i$ respectively. The observed data may be viewed as:

Time	t_1	t_2	$\ldots$	t_k	Total
	y_1	y_2	$\cdots$	y_k	y
	$n_1 - y_1$	$n_2 - y_2$	$\cdots$	$n_k - y_k$	$n - y$
Total	n_1	n_2	$\cdots$	n_k	n

where $y_i = \sum_j y_{ij}$, $y = \sum_i y_i$. Let $\bar{\theta}_i = y_i/n_i$ and $\bar{\theta} = y/n$. Without loss in generality, we shall be interested in detecting a monotone increasing trend in θ_i. Let $\{x_i\}$ represent arbitrary preselected scores, $x_1 < x_2 < \cdots < x_k$ which mimic a monotone increasing time trend. The linear model regression of y_{ij} expressed by

$$\theta_i = \alpha + \beta\left(x_i - \bar{x}_k\right) \tag{5.39}$$

and subject to the constraint that $\sum_i n_i\left(x_i - \bar{x}_k\right) = 0$ where $\bar{x}_k = \sum n_i x_i/n$ yields estimates

$$\hat{\alpha} = \frac{\sum_i n_i \bar{\theta}_i}{\sum_i n_i} = \bar{\theta}, \quad \hat{\beta} = \frac{\sum_i n_i\left(x_i - \bar{x}_k\right)\left(\bar{\theta}_i - \bar{\theta}\right)}{\sum_i n_i\left(x_i - \bar{x}_k\right)^2}.$$

The hypothesis of homogeneity is

$$H_0 : \theta_i = \theta, i = 1, ..k$$

or equivalently $\beta = 0$. Possible alternatives are

$$\begin{aligned} H_1 &: \theta_1 \leq \theta_2 \leq \cdots \leq \theta_k.\text{with at least one strict inequality} \\ H_2 &: \theta_i \neq \theta_j, \text{ for at least one pair } (i, j) \end{aligned}$$

Hypothesis H_2 can also be expressed as

$$\sum_{i<j} (\theta_i - \theta_j) = \sum_i (k + 1 - 2i)\,\theta_i \neq 0 \tag{5.40}$$

Under the null hypothesis of homogeneity, the estimate of variance of $\hat{\beta}$ is given by

$$V\left(\hat{\beta}\right) = \frac{\bar{\theta}\left(1 - \bar{\theta}\right)}{\sum_i n_i\left(x_i - \bar{x}_k\right)^2}$$

and consequently we reject H_0 in favor of H_1 (or equivalently $\beta > 0$) for large values of the statistic

$$\hat{\beta}/\sqrt{V\left(\hat{\beta}\right)} = \frac{\sum_i n_i\left(x_i - \bar{x}_k\right)\left(\bar{\theta}_i - \bar{\theta}\right)}{\sqrt{\bar{\theta}\left(1 - \bar{\theta}\right)}\sqrt{\sum_i n_i\left(x_i - \bar{x}_k\right)^2}} \tag{5.41}$$

which for large samples has a standard normal. If we suppose further that $\bar{\theta}$ is small (of the order of $1\% - 2\%$) so that $\bar{\theta}^2$ is negligible, then the statistic becomes

$$\frac{\sum_i n_i\left(x_i - \bar{x}_k\right)\left(\bar{\theta}_i - \bar{\theta}\right)}{\sqrt{\sum_i n_i\bar{\theta}\left(x_i - \bar{x}_k\right)^2}}.$$

The differences between observed and expected frequencies $n_i\left(\bar{\theta}_i - \bar{\theta}\right)$ are multiplied by the score effect $\left(x_i - \bar{x}_k\right)$ in the numerator, whereas the square of the score effects are weighted by the expected frequencies in the denominator. The statistic provides a test of trend in frequencies as opposed to a test of trend on the proportions. Since the sample correlation between $\left\{\bar{\theta}_i - \bar{\theta}\right\}$ and $\left\{x_i - \bar{x}_k\right\}$ is given by

$$\begin{aligned} &\frac{\sum_i n_i\left(x_i - \bar{x}_k\right)\left(\bar{\theta}_i - \bar{\theta}\right)}{\sqrt{\sum_i n_i\left(\bar{\theta}_i - \bar{\theta}\right)^2\sum_i n_i\left(x_i - \bar{x}_k\right)^2}} \\ &= \hat{\beta}\sqrt{\frac{\sum_i n_i\left(x_i - \bar{x}_k\right)^2}{\sum_i n_i\left(\bar{\theta}_i - \bar{\theta}\right)^2}} \end{aligned}$$

we may view the test of monotonicity as equivalent to a test that the correlation is 0.

Alternatively, the test of homogeneity may be conducted by treating the data as coming from a $2 \times k$ contingency table. That test rejects H_0 in favor of H_2 for large values of the statistic

$$\sum_i \frac{\left(n_i\bar{\theta}_i - n_i\bar{\theta}\right)^2}{n_i\bar{\theta}} + \sum_i \frac{\left(n_i\left(1-\bar{\theta}_i\right) - n_i\left(1-\bar{\theta}\right)\right)^2}{n_i\left(1-\bar{\theta}\right)} \tag{5.42}$$

$$= \frac{\sum_i n_i\left(\bar{\theta}_i - \bar{\theta}\right)^2}{\bar{\theta}\left(1-\bar{\theta}\right)} \tag{5.43}$$

which for large samples $n_i \to \infty$ has a chi-square distribution with $(k-1)$ degrees of freedom. In that case there is no need to define scores. We note that even though H_1 is included in H_2, one-sided tests which focus strictly on H_1 will in general be more powerful. In the next section, we make use of the general theory of hypothesis testing to construct a nonparametric test statistic based on the ranks of the observations to test H_0 against H_2.

We now proceed with the construction of the test statistics. Recall the following measures of similarities due to Spearman, Kendall, and Hamming, respectively, for rankings $\mu = (\mu(1), \ldots, \mu(n))', \nu = (\nu(1), \ldots, \nu(n))'$:

$$\begin{aligned}
\mathcal{A}_S(\mu,\nu) &= \sum_{i=1}^{n}\left(\mu(i) - \frac{n+1}{2}\right)\left(\nu(i) - \frac{n+1}{2}\right) \\
\mathcal{A}_K(\mu,\nu) &= \sum_{i<j} sgn\left(\mu(j) - \mu(i)\right) sgn\left(\nu(j) - \nu(i)\right) \\
\mathcal{A}_H(\mu,\nu) &= \sum_{i=1}^{n}\sum_{j=1}^{n}\left(I\left[\mu(i) = j\right] - \frac{1}{n}\right)\left(I\left[\nu(i) = j\right] - \frac{1}{n}\right)
\end{aligned}$$

where $sgn(x)$ is either 1 or -1 depending on whether $x > 0$ or $x < 0$ and where $I[\cdot]$ is the indicator function taking values 1 or 0 depending on whether the statement in brackets holds or not. In the model considered, the ranking which describes the time points may be viewed as a tied ranking with tie pattern:

$$\delta_1 = (n_1, n_2, \ldots, n_k) \tag{5.44}$$

and with ordering

$$\left\langle (1, \ldots, n_1)\,(n_1+1, \ldots, n_1+n_2) \ldots \left(\sum_{i=1}^{k-1} n_i + 1, \ldots, \sum_{i=1}^{k} n_i\right)\right\rangle \tag{5.45}$$

On the other hand, the binary variables y_{ij} have with $e = 2$ the simple tie pattern:

$$\delta_2 = (y, n-y) \tag{5.46}$$

To compute the test statistic corresponding to Spearman similarity, note that

$$\begin{aligned}&\mathcal{A}_S\left(\mu_{\delta_1},\nu_{\delta_2}\right)\\&=\sum_{i=1}^{n}E\left[\left(\mu(i)-\frac{n+1}{2}\right)|C\left(\mu\right)\right]E\left[\left(\nu(i)-\frac{n+1}{2}\right)|C\left(\nu\right)\right]\end{aligned}$$

The average of the compatible ranks at time t_1 is

$$g_1=\left(\frac{n_1+1}{2}\right)$$

at time t_2 it is

$$g_2=n_1+\left(\frac{n_2+1}{2}\right)$$

and so on. In general, at time t_l, the average rank is

$$g_l=\sum_{i=1}^{l-1}n_i+\left(\frac{n_l+1}{2}\right),l=1,..,k$$

Clearly,

$$\sum_{i=1}^{k}n_ig_i=\frac{n\left(n+1\right)}{2}$$

Hence, the conditional expectation

$$E\left[\nu(l)|C\left(\nu\right)\right]=g_l,l=1,\ldots,n_i$$

Turning attention now to the binary observations, the average rank for the $n-y$ observations which take value 0 is $l_1=\frac{n-y+1}{2}$, whereas the average rank for the y observations which take value 1 is

$$l_2=n-y+\left(\frac{y+1}{2}\right)=n-\frac{y}{2}+\frac{1}{2}$$

It follows that at time t_i

$$E\left[\mu(l)|C\left(\mu\right)\right]=\begin{cases}l_1 & l=1,\ldots,n_i-y_i\\ l_2 & l=n_i-y_i+1,\ldots,n_i\end{cases},$$

Since $l_2-l_1=\frac{n}{2}$ and $\sum_{i=1}^{k}n_ig_i=\frac{n(n+1)}{2}$, we have that

$$\begin{aligned}\mathcal{A}_S\left(\mu_{\delta_1},\nu_{\delta_2}\right)&=\sum_{i=1}^{k}\left[g_i-\frac{n+1}{2}\right]\left[l_2y_i+\left(n_i-y_i\right)l_1\right]\\&=\frac{n}{2}\sum_{i=1}^{k}\left[g_i-\frac{n+1}{2}\right]y_i.\end{aligned}$$

Hence, we may define the Spearman statistic to be

$$\begin{aligned} S &= \sum_{i=1}^{k}\left[g_i - \frac{n+1}{2}\right] y_i \\ &= \sum_{i=1}^{k} n_i c_i \left(\overline{\theta_i} - \overline{\theta}\right) \end{aligned}$$

where $c_i = g_i - \frac{n+1}{2}$ and $\sum_{i=1}^{k} n_i c_i = 0$.

We now consider the test statistic corresponding to Kendall's similarity measure. Note that within a given time period, the difference between ties is zero, and hence there is no contribution to the distance. Between different time periods, we have

$$\begin{aligned} &\sum_{i<j}^{n} E\left[sgn\left(\mu(j) - \mu(i)\right)|C\left(\mu\right)\right] E\left[sgn\left(\nu(j) - \nu(i)\right)|C\left(\nu\right)\right] \\ &= \sum_{i<j}^{k}\left[\left(n_i - y_i\right) y_j - \left(n_j - y_j\right) y_i\right] \\ &= \sum_{i<j}^{k}\left[n_i y_j - n_j y_i\right] \end{aligned}$$

Now

$$\begin{aligned} \sum_{j=i+1}^{k} n_j &= \left[n - n_i - \sum_{j=1}^{i-1} n_j\right] \\ &= \left[n - g_i - \frac{n_i - 1}{2}\right] \end{aligned}$$

Hence,

$$\begin{aligned} \mathcal{A}_K\left(\mu_{\delta_1}, \nu_{\delta_2}\right) &= \sum_{i=1}^{k} y_i\left[g_i - \frac{n_i+1}{2}\right] - \sum_{i=1}^{k} y_i\left[n - g_i - \frac{n_i - 1}{2}\right] \\ &= 2\sum_{i=1}^{k}\left[g_i - \frac{n+1}{2}\right] y_i \end{aligned}$$

and it is seen that the Kendall and Spearman statistics are equivalent.

The asymptotic distribution of the Spearman test statistic under the null hypothesis may be determined under two different assumptions. Let $\{y_i\}$ be k independent binomials (n_i, θ_i), and suppose we would like to test H_0 vs H_1. In most applications the asymptotic situation of interest occurs when

$$n_i \to \infty, \text{ with } \frac{n_i}{n} \to \lambda_i > 0,\ i = 1, \ldots, k. \tag{5.47}$$

We show next that the Spearman statistic has an asymptotic normal distribution under either (5.47) or whenever the $\{n_i\}$ are bounded while $k \to \infty$. The following result was shown in Alvo and Berthelot [2012].

Theorem 5.12. *Suppose that both $y \to \infty$ and $n - y \to \infty$ as $n \to \infty$. Under H_o the Spearman test statistic has asymptotically a standard normal distribution, i.e.,*

$$\frac{\sum_{i=1}^{k} n_i c_i \left(\bar{\theta}_i - \bar{\theta}\right)}{\sqrt{\sum_{i=1}^{k} c_i^2 n_i \theta_i \left(1 - \theta_i\right)}} \xrightarrow{\mathcal{L}} N(0, 1)$$

under either (i) (5.47) or (ii) the $\{n_i\}$ are bounded and $k \to \infty$.

We may estimate θ_i either by $\bar{\theta}_i$ or by $\bar{\theta}$. In the first case, the test rejects whenever

$$\frac{\sum_{i=1}^{k} c_i y_i}{\sqrt{\sum_{i=1}^{k} c_i^2 n_i \bar{\theta}_i \left(1 - \bar{\theta}_i\right)}} \geq z_\alpha$$

where z_α is the upper $100(1-\alpha)\,\%$ percentage point from a standard normal distribution. The expression for the estimate of the asymptotic power becomes

$$1 - \Phi\left(z_\alpha - \frac{\sum_{i=1}^{k} n_i c_i \theta_i}{\sqrt{\sum c_i^2 n_i \theta_i \left(1 - \theta_i\right)}}\right) = \Phi\left(\frac{\sum_{i=1}^{k} n_i c_i \theta_i}{\sqrt{\sum_{i=1}^{k} c_i^2 n_i \theta_i \left(1 - \theta_i\right)}} - z_\alpha\right)$$

It is seen that the power converges to 1 with increasing n.

Alternatively, we may use the statistic

$$\frac{\sum_{i=1}^{k} n_i c_i \left(\bar{\theta}_i - \bar{\theta}\right)}{\sqrt{\bar{\theta}\left(1 - \bar{\theta}\right)}\sqrt{\sum_{i=1}^{k} n_i c_i^2}} \tag{5.48}$$

which in the simulation studies reported appears to more closely attain the prescribed significance level. In the case of equal sample sizes, $n_i = n_0$ say, the test statistic (5.48) takes the simpler form:

$$\frac{\sum_{i=1}^{k} \left(i - \frac{k+1}{2}\right)\left(\bar{\theta}_i - \bar{\theta}\right)}{\sqrt{n_0}\sqrt{\bar{\theta}\left(1 - \bar{\theta}\right)}\sqrt{\frac{k(k^2-1)}{12}}}$$

Example 5.11. Returning to the data on mortality rates in South Africa, we calculated values of 260.1 for the Spearman which yielded a p-value $< 10^{-4}$.

5.10. Randomized Block Designs

The subject of experimental designs has been addressed in several textbooks, and we shall not attempt to provide an in-depth review in this book. Some general references are

John and Williams [1995]; Kendall et al. [1983]. In this section, we consider applications to ranking data.

The randomized block design is a two-way layout where the experimental units consist of "blocks." Within each block, the treatments are randomly allocated to the units inside each block. When all treatments appear at least once in each block, we have a completely randomized block design. Otherwise, we have an incomplete randomized block design. The use of blocks serves to eliminate the effect of nuisance variables. Randomization within the blocks allows one to test for treatment effects.

As an example, animal litters may be considered as blocks which account for genetic differences. Farmlands may be divided into parcels which are homogeneous within. Here, a block consists of contiguous plots of land that share the same characteristics (moisture, fertility, acidity, etc.). If, for example, we want to test the difference between different fertilizers on crop yield, we can apply a different fertilizer (the treatment) at random to different plots in the block (therefore controlling for the nuisance factors). In genetic studies, twins may be considered as constituting a block. In an experiment on people or animals, age, weight, and state of health are continuous variables which are often suitable for determining blocks. To be in the same block, individuals do not have to have exactly the same weight. Instead weight ranges can be chosen so that blocks have a suitable size. Similarly, severity of disease can be used to block patients in a clinical trial.

The basic model involving a single nuisance variable is given by

$$X_{ij} = \mu + \alpha_i + \beta_j + \varepsilon_{ij}, i = 1, \ldots, t; j = 1, \ldots, n$$

where X_{ij} is the response arising from block j treatment i, μ is the general location parameter, α_i is the ith treatment effect, β_j is the jth effect from block j, and ε_{ij} is a random error term. This is equivalent to a two-factor analysis of variance model without replication. In nonparametric statistics, the responses within each are replaced by their ranks within that block. The Friedman test statistic is then given by

$$G = \frac{12n}{t\,(t+1)} \sum_i \left(\bar{R}_i - \frac{t+1}{2} \right)^2$$

and the test rejects the null hypothesis of no treatment effect whenever $G \geq \chi^2_{t-1}(\alpha)$. Here, $\bar{R}_i$ is the average of the ranks assigned to the ith treatment. A result in Alvo and Yu [2014] relates G to the average of all pairwise correlations between blocks

$$G = (t-1)\left\{(n-1)\,\bar{\rho}_n + 1\right\}$$

In some situations, it may not always be possible to apply a treatment for every unit in a block. This leads naturally to the study of incomplete block designs of which there are several. As an example, the popular balanced incomplete block design imposes the following conditions:

$$bk = rt, \lambda\,(t-1) = r\,(k-1)$$

where b is the number of blocks, k is the number of treatments compared in each block, r is the number of blocks for each treatment, t is the number of treatments, and λ is the number of pairs of treatments being compared in each block. Then, Durbin Durbin [1951] has shown that

$$G = \frac{12}{n\lambda t(k+1)} \sum_{i=1}^{t} \left(R_i - \frac{nr(k+1)}{2} \right)^2 \xrightarrow{\mathcal{L}} \chi^2_{t-1}$$

Here, R_i is the sum of the ranks assigned to the ith treatment. Alvo and Cabilio [1991] presented a distance-based approach which yields a variety of test statistics for different block designs which include both the Friedman and Durbin statistics. A parametric approach to the problem due to Alvo [2015] leads to an alternative derivation of the Friedman statistic.

An interesting application was to The Sound of Music and Word of Mouth: hearing music and hearing about music [Geelhoed, 2001]. Here, subjects were asked to rank order a list of 13 items, which related to how they became aware of the music that they acquired. The goal of the study was to cluster the subjects if possible in some manner. The study involved 70 volunteers, 35 women (mean age 29.97 years), and 35 men (mean age 31.11 years). Using a computerized version of the Youden squares incomplete block design, the 13 items were presented in 13 blocks of four items to rank at a time. After each session, individuals were given feedback on how they rank ordered the list of 13 items, and without exception subjects were in agreement with that particular rank order. Use of the Durbin statistic indicated a high level of agreement among the subjects. At the top of the list was radio (mean rank 3.4), as the medium via which people become aware of new music. In second place was "Someone playing music to you," becoming aware in a social setting (mean rank 4.3). Film and video (mean rank 5.1) are also important carriers of new music, closely followed by recommendation (mean rank 5.3) and in the fifth place, TV (mean rank 5.5).

5.11. Density Estimation

In most classification problems, the underlying population densities are generally not known. However, an estimate of the density function can be constructed from the observed data. Such an estimate may then be used in the Bayes classification rule. The estimate will be more reliable with increasing numbers of observations. The optimality of the classifier will not only depend on the number of observations available but equally upon the method utilized to obtain the estimate of the density.

There are many techniques available to estimate a density function. The kernel density method of estimation is the one most commonly applied. It has been widely researched and has been shown to perform well in many classification problems. Silverman [1988] provides a thorough study of this subject in both univariate and multivariate cases. In what follows, we present a brief description of kernel density estimation and then report on a simulation study based on the work in Leger [2009].

5.11.1. Univariate Kernel Density Estimation

Let $X_1, \ldots, X_n$ be a random sample from a given univariate population whose true density is $f(x)$. The kernel density estimate is given by

$$\hat{f}(x) = \frac{1}{nh} \sum_{i=1}^{n} K\left(\frac{x - X_i}{h}\right) \tag{5.49}$$

where $K(x)$ is a kernel function to be specified and h is the bandwidth or smoothing parameter. A necessary condition for $\hat{f}$ to be a proper density function is

$$\int K(x)\, dx = 1$$

Likewise, if $K(x) \geqslant 0$, $\forall x$ and $K(x)$ has derivatives of all order, then $\hat{f}(x)$ will possess the same characteristics. Furthermore, Eq. (5.49) indicates that the density estimate will be influenced by the smoothing parameter. Hence, the quality and accuracy of the estimator $\hat{f}$ will depend on the selection of the kernel as well as on the bandwidth.

In order to select the best kernel and smoothing parameter, a criterion must be used to evaluate the estimator $\hat{f}$ in comparison to the true density, f. The most commonly used method of measuring the overall accuracy of $\hat{f}$ is to use its mean integrated square error, (MISE) defined as

$$\begin{aligned} \text{MISE} &= E\left[\int \left\{\hat{f}(x) - f(x)\right\}^2 dx\right] \\ &= \int \operatorname{Var}(\hat{f}(x))\, dx + \int \left[\operatorname{bias}(\hat{f}(x))\right]^2 dx. \end{aligned} \tag{5.50}$$

We seek the kernel function and bandwidth which minimize (5.50). Note that $\sum_{i=1}^{n} K\left(\frac{x-X_i}{h}\right)$ is a sum of independent, identically distributed variables. Let Y represent a variable with the same distribution as $X_1, \ldots, X_n$. Then the variance and expected value of $\hat{f}(x)$ are given by

$$\begin{aligned} E\left[\hat{f}(x))\right] &= E\left[\frac{1}{nh} \sum_{i=1}^{n} K\left(\frac{x - X_i}{h}\right)\right] \\ &= \int \frac{1}{h} K\left(\frac{x - y}{h}\right) f(y)\, dy \end{aligned} \tag{5.51}$$

and

$$
\begin{aligned}
\operatorname{Var}\left[\hat{f}(x)\right] &= \operatorname{Var}\left[\frac{1}{nh}\sum_{i=1}^{n} K\left(\frac{x-X_i}{h}\right)\right] \\
&= \frac{1}{n}\int \frac{1}{h^2}\left[K\left(\frac{x-y}{h}\right)\right]^2 f(y)\,dy - \\
&\quad \frac{1}{n}\left[\int \frac{1}{h}K\left(\frac{x-y}{h}\right) f(y)\,dy\right]^2 \qquad (5.52)
\end{aligned}
$$

Substituting the expressions found in (5.51) and (5.52) into Eq. (5.50), we obtain an exact formula for the mean integrated square error of $\hat{f}(x)$. However, in many cases, minimizing the exact MISE in order to find the best kernel function and smoothing parameter can be difficult, computationally intensive, and impractical. To rectify this, [Silverman, 1988] suggests using approximate values for the variance and bias. This leads to a simpler, approximate formula for the mean integrated square error. He further indicates that this approximation is quite satisfactory since minimizing it will give results that coincide closely to the results that would have been obtained if the exact expression of the MISE had been used instead, even for small sample sizes.

Theorem 5.13. *Suppose the kernel $K(t)$ is a symmetric function that satisfies*

$$\int K(t)\,dt = 1, \int tK(t)\,dt = 0$$

and

$$\int t^2 K(t)\,dt = b \neq 0$$

Then, the approximate mean integrated square error is given by

$$MISE \approx \frac{1}{4}h^4 b^2 \int \left(f''(x)\right)^2\,dx + \frac{1}{nh}\int (K(t))^2\,dt$$

Using this simpler expression for the mean integrated square error, we can proceed to find the kernel function and bandwidth that will minimize it. We begin with the smoothing parameter, h. It can be shown that the optimal bandwidth, h_{opt}, which minimizes the approximate mean integrated square error is given by

$$h_{opt} = b^{-2/5}\left\{\int (K(t))^2\,dt\right\}^{1/5}\left\{\int (f''(x))^2\,dx\right\}^{-1/5} n^{-1/5}$$

which depends on the kernel, $K(t)$, and the true density $f(x)$. This can be problematic since in practice no information is known about f. Nevertheless, [Silverman, 1988] proposes a solution to this problem as we will see below.

To find the optimal kernel function, $K(t)$, substitute h_{opt} in the expression for the approximate MISE:

$$\begin{aligned}
\text{MISE} &\approx \frac{1}{4}h_{opt}^4 b^2 \int \left(f''(x)\right)^2 dx + \frac{1}{nh_{opt}} \int (K(t))^2 \, dt \\
&= \frac{1}{4}b^{-8/5}\left\{\int (K(t))^2\, dt\right\}^{4/5}\left\{\int (f''(x))^2\, dx\right\}^{-4/5} b^2 n^{-4/5}\left\{\int (f''(x))^2\, dx\right\} + \\
&\quad n^{-1}b^{2/5}\left\{\int (K(t))^2\, dt\right\}^{-1/5}\left\{\int (f''(x))^2\, dx\right\}^{1/5} n^{1/5}\left\{\int (K(t))^2\, dt\right\} \\
&= \frac{1}{4}b^{2/5}\left\{\int (K(t))^2\, dt\right\}^{4/5}\left\{\int (f''(x))^2\, dx\right\}^{1/5} n^{-4/5} + \\
&\quad b^{2/5}\left\{\int (K(t))^2\, dt\right\}^{4/5}\left\{\int (f''(x))^2\, dx\right\}^{1/5} n^{-4/5} \\
&= \frac{5}{4}b^{2/5}\left\{\int (K(t))^2\, dt\right\}^{4/5}\left\{\int (f''(x))^2\, dx\right\}^{1/5} n^{-4/5}
\end{aligned}$$

Since $\frac{5}{4}\left\{\int (f''(x))^2\, dx\right\}^{1/5} n^{-4/5}$ will have the same value for all kernels, then the optimal kernel will be the one which minimizes $b^{2/5}\left\{\int (K(t))^2\, dt\right\}^{4/5}$ where $b = \int t^2 K(t)\, dt$. It can be shown [Hallin and Tribel, 2000] that the optimal kernel is the Epanechnikov kernel defined as

$$K_e(t) = \begin{cases} \frac{3}{4\sqrt{5}}\left(1 - \frac{1}{5}t^2\right) & \text{if } -\sqrt{5} \leqslant t \leqslant \sqrt{5} \\ 0 & \text{otherwise.} \end{cases}$$

It is important to note that there exist many other choices for the kernel function. Table 5.8 lists four other possible choices and further indicates the efficiency of each kernel in comparison to the Epanechnikov kernel. We notice from this table that the efficiencies are quite high and so we conclude that the kernels do not differ greatly in terms of the mean integrated square error. Hence, as [Silverman, 1988] states, "it is perfectly legitimate, and indeed desirable, to base the choice of the kernel on other considerations, for example the degree of differentiability required or the computational effort involved." In fact, it is precisely for these reasons that we will choose to use a Gaussian kernel in the Monte Carlo simulation study.

The problem of choosing the smoothing parameter remains. Recall that

$$h_{opt} = b^{-2/5}\left\{\int (K(t))^2\, dt\right\}^{1/5}\left\{\int (f''(x))^2\, dx\right\}^{-1/5} n^{-1/5}.$$

Table 5.8.: Some kernels and their efficiencies

Kernel	$K(t)$	Efficiency
Epanechnikov	$\begin{cases} \frac{3}{4\sqrt{5}}\left(1-\frac{1}{5}t^2\right) & \text{if } -\sqrt{5}\leqslant t\leqslant\sqrt{5} \\ 0 & \text{otherwise} \end{cases}$	1
Biweight	$\begin{cases} \frac{15}{16}(1-t^2)^2 & \text{if } \lvert t\rvert<1 \\ 0 & \text{otherwise} \end{cases}$	$\left(\frac{3087}{3125}\right)^{1/2} \approx 0.9939$
Triangular	$\begin{cases} 1-\lvert t\rvert & \text{if } \lvert t\rvert<1 \\ 0 & \text{otherwise} \end{cases}$	$\left(\frac{243}{250}\right)^{1/2} \approx 0.9859$
Gaussian	$\frac{1}{\sqrt{2\pi}}\,\mathrm{e}^{-(1/2)t^2}, \quad -\infty<t<\infty$	$\left(\frac{36\pi}{125}\right)^{1/2} \approx 0.9512$
Rectangular	$\begin{cases} \frac{1}{2} & \text{if } \lvert t\rvert<1 \\ 0 & \text{otherwise} \end{cases}$	$\left(\frac{108}{125}\right)^{1/2} \approx 0.9295$

Once the kernel has been selected, b and $\int (K(t))^2\,dt$ can easily be computed. Silverman [1988] suggests using "a standard family of distributions to assign a value to the term $\int (f''(x))^2\,dx$" such as a normal distribution with variance σ^2. In this case, if also a Gaussian kernel is utilized, then h_{opt} is found to be

$$h_{opt} = 1.06\sigma n^{-1/5}. \tag{5.53}$$

The population variance, σ^2, is usually unknown and so σ can be replaced in (5.53) by the estimated standard deviation of the observations $X_1, \ldots, X_n$. For a more robust estimate of the spread, it has been suggested to use

$$h^*_{opt} = 1.06\mathrm{A}n^{-1/5}$$

where A =min (standard deviation, interquartile range/1.34). This choice of smoothing parameter works very well for unimodal densities $f(x)$. It is therefore this smoothing parameter that will be used when conducting the simulations. In this Monte Carlo simulation study, all simulated data were obtained from unimodal densities, and so this choice of bandwidth seems reasonable and appropriate.

In cases where the population does not appear to have a unimodal density and no additional information is available about the true density $f(x)$, then h^*_{opt} is no longer the best choice of bandwidth. Instead, there exist other methods, such as least squares cross-validation or likelihood cross-validation, which lead to a better selection of the smoothing parameter. Additionally, it is possible to select the bandwidth by minimizing a criterion other than the mean integrated square error.

5.11.2. The Rank Transform

Another problem with Fisher's classifier as described in Sect. 5.3 is that the rule depends on the means. These values can be easily influenced by extreme observations. To minimize the effect of such outliers, Conover and Iman [1981] propose applying Fisher's discriminant function to the ranks of the observations instead. Hence, we transform the data into a new set of observations which are the ranks of the original data. We then apply Fisher's classification rule using these transformed observations. Therefore, $\bar{x}$ will now be replaced by the average rank of the observations of the first population. Similarly, $\bar{y}$ will be replaced by the average rank of the observations of the second population, and S will now be a pooled estimate of the variance of the ranks of the observations.

Not only is this rank transform robust to outliers, but such a transformation may also help when the population densities are not normal or when the populations have unequal covariance matrices. This stems from the fact that we are no longer dealing with the population densities but rather with the distribution of the ranked data. Finally, it is important to note that Akritas [1990] and Thompson [1991] both express that the rank transform is most effective when populations differ in location. When only a change in scale is present, the rank transform is less useful. Thankfully, in most classification problems, the populations differ in location, and so the rank transform is often applicable.

We report on a limited simulation study in Table 5.9. The results for the Bayes and kernel methods are comparable. The cases where the rank transform method performs poorly are indicated by *. The following abbreviations are utilized in the table for the density functions:

$N(\mu, \sigma^2)$ is a normal distribution with mean μ and variance σ^2.

$C(\mu, \sigma)$ is a Cauchy distribution with location μ and scale σ.

$Log(\mu, \sigma)$ is a logistic distribution with location μ and scale σ.

$Exp(\mu, \sigma)$ is an exponential distribution with location μ and scale σ.

5.11.3. Multivariate Kernel Density Estimation

The results for univariate density estimation can be extended to multivariate populations. Let $X_1, \ldots, X_n$ be a random sample from a $p-$dimensional population with density function $f(x)$. The multivariate kernel density estimate of the density function, $f(x)$, constructed from the observations is given by

$$\hat{f}(x) = \frac{1}{nh^p} \sum_{i=1}^{n} K\left(\frac{x - X_i}{h}\right)$$

where $K(x)$ is the kernel function and h is the smoothing parameter. As in the univariate case, we can find an expression for the approximate mean integrated square error of $\hat{f}(x)$ and minimize it in order to find the optimal bandwidth and kernel function.

Table 5.9.: Cross-validation error rate (%)

Population 1: $n_1 = 25$	Population 2: $n_2 = 50$	Bayes	Kernel	Rank transform
$N(0,1)$	$N(2,1)$	15.07	15.51	18.55
$N(0,1)$	$N(2,2)$	29.73	28.4	25.27
$C(0,1)$	$C(2,1)$	25.05	26.37	26.51
$C(0,1)$	$C(2,2)$	29.55	34.77	31.4
$Exp(0,1)$	$Exp(2,1)$	5.63	6.24	5.91
$Exp(0,1)$	$Exp(2,2)$	5.52	6.03	16.19 *
$Log(0,1)$	$Log(2,1)$	25.89	28.6	27.64
$Log(0,1)$	$Log(2,2)$	32.6	33.89	33.92
$C(0,1)$	$Log(2,1)$	25.17	26.03	27.01
$C(0,1)$	$Log(3,2)$	27.15	29.01	27.55
$Log(0,1)$	$N(2,1)$	18.52	18.29	24.23 *
$Log(0,3)$	$N(3,1)$	12.48	12.24	32.05 *
$C(0,1)$	$N(1,1)$	26.93	27.11	35.48 *
$C(0,1)$	$N(3,2)$	23.16	20.17	23.43

Let $K(t)$ be a radially symmetric probability density function, and let

$$\alpha = \int t_1^2 K(t)\, dt$$

and

$$\beta = \int (K(t))^2\, dt$$

Then combining multivariate Taylor series expansions, we have the following approximations:

$$\text{bias}(\hat{f}(x)) \approx \frac{1}{2} h^2 \alpha \nabla^2 f(x) \tag{5.54}$$

and

$$\text{Var}(\hat{f}(x)) \approx n^{-1} h^{-p} \beta f(x). \tag{5.55}$$

Substituting (5.54) and (5.55) into (5.50) gives the following approximate value of the mean integrated square error:

$$\text{MISE} \approx \frac{1}{4} h^4 \alpha^2 \int \left\{ \nabla^2 f(x) \right\}^2 dx + n^{-1} h^{-p} \beta. \tag{5.56}$$

Hence the optimal bandwidth is

$$h_{opt}^{p+4} = p\beta\alpha^{-2}\left\{\int \left(\nabla^2 f(x)\right)^2 dx\right\}^{-1} n^{-1}. \tag{5.57}$$

Moreover, it can be shown that the kernel function which minimizes the approximate MISE is a multivariate version of the Epanechnikov kernel and is defined as

$$K_e(x) = \begin{cases} \frac{1}{2}v_p^{-1}(p+2)(1-x'x)^2 & \text{if } x'x < 1 \\ 0 & \text{otherwise} \end{cases}$$

where v_p is the volume of the unit $p-$dimensional sphere.

As in the univariate case, many other kernel functions are available. Three other possible multivariate kernels that can be used when estimating a density function are listed in Table 5.10. The kernel $K_g(x)$ is the multivariate extension of the Gaussian kernel. The selection of the kernel is often based on criteria such as the differentiability requirements and computational considerations.

Table 5.10.: Three multivariate kernels

Kernel	$K(x)$
Multivariate Gaussian kernel	$K_g(x) = (2\pi)^{-p/2}\exp\left(-\frac{1}{2}x'x\right)$
Multivariate kernel 2	$K_2(x) = \begin{cases} 3\pi^{-1}(1-x'x)^2 & \text{if } x'x < 1 \\ 0 & \text{otherwise} \end{cases}$
Multivariate kernel 3	$K_3(x) = \begin{cases} 4\pi^{-1}(1-x'x)^3 & \text{if } x'x < 1 \\ 0 & \text{otherwise} \end{cases}$

Once the kernel has been selected, values for α and β can readily be calculated and substituted in (5.57) to find h_{opt}. However, an exact value for this smoothing parameter still cannot be obtained since h_{opt} depends on the true population density, $f(x)$. To solve this problem, we proceed as in the univariate case and select a family of distributions in order to obtain a value for the term $\int \{\nabla^2 f(x)\}^2 dx$. A natural choice is to take $f(x)$ to be the standard $p-$dimensional normal density. In this case,

$$\int \left\{\nabla^2 f(x)\right\}^2 dx = (2\sqrt{\pi})^{-p}\left(\frac{2p+p^2}{4}\right).$$

Then, if a multivariate Gaussian kernel is also used, we have

$$h_{opt} = \{4/(2p+1)\}^{1/(p+4)}\, n^{-1/(p+4)}.$$

Silverman [1988] adds that if instead the true population density, $f(x)$, is multivariate normal with covariance matrix Σ, then h_{opt} can be modified to take this fact into account. Let S be a, possibly robust, estimate of the true covariance matrix Σ, and let s_{ii} be the ith diagonal element of S. Then a more appropriate smoothing parameter might be σh_{opt} where

$$\sigma^2 = \frac{1}{p}\sum_{i=1}^{p} s_{ii}.$$

This is the smoothing parameter which will be used for the multivariate simulations. This bandwidth was selected since all multivariate data sets utilized in the Monte Carlo simulation study were generated from various multivariate normal distributions.

It is important to note that as in the univariate situation, the smoothing parameter, $\sigma h_{opt} = \sigma \{4/(2p+1)\}^{1/(p+4)} n^{-1/(p+4)}$, should be used with caution. Although this approach provides a quick and easy way of obtaining an initial value of the bandwidth, it is not the optimal value for all situations. Other univariate methods, such as least squares cross-validation and likelihood cross-validation, can be extended to the multivariate case and may yield a more ideal value for the smoothing parameter. Also, in some cases, a vector of bandwidths may even be more appropriate than a single value.

We report in Table 5.11 on a limited simulation study for which we assumed bivariate normal densities for the populations. The means utilized in the multivariate simulations are

$$\mu_1 = \begin{bmatrix} 0 \\ 0 \end{bmatrix}, \mu_2 = \begin{bmatrix} 2 \\ 2 \end{bmatrix}, \mu_3 = \begin{bmatrix} 1 \\ -3 \end{bmatrix}, \text{ and } \mu_4 = \begin{bmatrix} 5 \\ -5 \end{bmatrix}$$

The results for the Bayes and kernel methods are again comparable. The cases where the rank transform method performs poorly are indicated by *.

Finally, an additional concern presents itself when doing multivariate kernel density estimation. We must now consider the sample size required in order to obtain an accurate estimate of the true population density, $f(x)$. Evidently, as the number of dimensions (or variables) increases, the amount of data needed to produce an adequate estimate will also augment. The problem is that the number of observations required can increase alarmingly fast. As an example, [Silverman, 1988] studies the following problem. "Suppose that the true density f is unit multivariate normal, and that the kernel is normal. Also, suppose that it is of interest to estimate f at the point 0, and that the bandwidth h has been chosen to minimize the mean square error at this point. A not unreasonable aim would be to ensure that the relative mean square error $E\{\hat{f}(0) - f(0)\}^2/f(0)^2$ is fairly small, say less than 0.1."

Table 5.12 gives the approximate sample sizes, accurate to about three figures, required to meet the criteria of this problem for various dimensions.

It should be noted that for classification problems, we are more interested in determining if the ratio $\hat{f}_1(z)/\hat{f}_2(z)$ is greater than p_2/p_1 rather than correctly estimating the value of $f_1(z)$ or $f_2(z)$. In this case, an incorrect value of $f_1(z)$ or $f_2(z)$ may still lead to a correct classification. Hence, the sample sizes needed to correctly classify a new

Table 5.11.: Cross-Validation Error Rate (%) for Multivariate simulations

Population 1: $n_1 = 25$	Population 2 $n_2 = 50$	Bayes	Kernel	Rank transform
$\mu_1, \Sigma = \begin{bmatrix} 1 & 0 \\ 0 & 1 \end{bmatrix}$	$\mu_2, \Sigma = \begin{bmatrix} 1 & 0 \\ 0 & 1 \end{bmatrix}$	7.88	8.61	9.65
$\mu_1, \Sigma = \begin{bmatrix} 1 & 0 \\ 0 & 1 \end{bmatrix}$	$\mu_3, \Sigma = \begin{bmatrix} 1 & 0 \\ 0 & 1 \end{bmatrix}$	5.16	6.61	8.85
$\mu_1, \Sigma = \begin{bmatrix} 1 & 0 \\ 0 & 1 \end{bmatrix}$	$\mu_4, \Sigma = \begin{bmatrix} 1 & 0 \\ 0 & 1 \end{bmatrix}$	0.04	0.03	2.49 *
$\mu_1, \Sigma = \begin{bmatrix} 1 & -0.5 \\ -0.5 & 1 \end{bmatrix}$	$\mu_2, \Sigma = \begin{bmatrix} 1 & -0.5 \\ -0.5 & 1 \end{bmatrix}$	2.11	2.8	3.36
$\mu_1, \Sigma = \begin{bmatrix} 1 & -0.5 \\ -0.5 & 1 \end{bmatrix}$	$\mu_3, \Sigma = \begin{bmatrix} 1 & -0.5 \\ -0.5 & 1 \end{bmatrix}$	6.55	7.45	10.39 *
$\mu_1, \Sigma = \begin{bmatrix} 1 & -0.5 \\ -0.5 & 1 \end{bmatrix}$	$\mu_4, \Sigma = \begin{bmatrix} 1 & -0.5 \\ -0.5 & 1 \end{bmatrix}$	0.25	0.49	4.93 *
$\mu_1, \Sigma = \begin{bmatrix} 4 & 3 \\ 3 & 4 \end{bmatrix}$	$\mu_2, \Sigma = \begin{bmatrix} 4 & 3 \\ 3 & 4 \end{bmatrix}$	30.04	33.17	31.09
$\mu_1, \Sigma = \begin{bmatrix} 4 & 3 \\ 3 & 4 \end{bmatrix}$	$\mu_3, \Sigma = \begin{bmatrix} 4 & 3 \\ 3 & 4 \end{bmatrix}$	7.55	10.63	9.45
$\mu_1, \Sigma = \begin{bmatrix} 4 & 3 \\ 3 & 4 \end{bmatrix}$	$\mu_4, \Sigma = \begin{bmatrix} 4 & 3 \\ 3 & 4 \end{bmatrix}$	0.03	0.05	0.15
$\mu_1, \Sigma = \begin{bmatrix} 1 & 0.3 \\ 0.3 & 2 \end{bmatrix}$	$\mu_3, \Sigma = \begin{bmatrix} 1 & 0.3 \\ 0.3 & 2 \end{bmatrix}$	10.47	11.65	11.16
$\mu_1, \Sigma = \begin{bmatrix} 1 & 0.3 \\ 0.3 & 2 \end{bmatrix}$	$\mu_4, \Sigma = \begin{bmatrix} 1 & 0.3 \\ 0.3 & 2 \end{bmatrix}$	0.03	0.05	1.59
$\mu_1, \Sigma = \begin{bmatrix} 1 & 0 \\ 0 & 1 \end{bmatrix}$	$\mu_2, \Sigma = \begin{bmatrix} 2 & 0 \\ 0 & 2 \end{bmatrix}$	11.12	14.23	13.93
$\mu_1, \Sigma = \begin{bmatrix} 1 & 0 \\ 0 & 1 \end{bmatrix}$	$\mu_3, \Sigma = \begin{bmatrix} 2 & 0 \\ 0 & 2 \end{bmatrix}$	9.53	10.91	11.77
$\mu_1, \Sigma = \begin{bmatrix} 1 & 0 \\ 0 & 1 \end{bmatrix}$	$\mu_4, \Sigma = \begin{bmatrix} 2 & 0 \\ 0 & 2 \end{bmatrix}$	0.23	0.32	2.12

observation may indeed be smaller than those given in Table 5.12. However, although the sample sizes may be smaller, they are nevertheless still large and continue to increase with the dimensions.

Gingras et al. [1995] considered density estimation to analyze flood data via regression. For further references see Van der Vaart [2000, chapter 24].

Table 5.12.: Sample sizes required for increasing dimension

Dimension	Required sample size
1	4
2	19
3	67
4	223
5	768
6	2790
7	10700
8	43700
9	187000
10	842000

5.12. Spatial Data Analysis

Environmental data is of major concern today as new policies are enacted to tackle the adverse effects of climate change. There are examples of spatial data in epidemiology, geology, forestry, atmospheric science, and any area of science where measurements are made in geographical locations. Clark et al. [1988] reported on the evaluation of regional long-term sulfur deposition models used to calculate mean air concentrations of SO_2 and amounts of S wet and dry deposition for monthly, seasonal, and annual periods across North America. Recently, there has been an interest in developing new spatial interpolation methods to model disease such as the COVID-19 pandemic [Goovaerts, 2005, 2017; Schwab, 2015]. In what follows, we shall describe the mathematics behind spatial analysis and then apply the theory to real data. We refer the reader to Cressie [2015] for several references as well as for a more extensive development of the subject. We have utilized the same notation for ease of reference.

Suppose that $s \in R^d$ is a point in $d-$ dimensional space and $Z(s)$ represents data collected at s. As s varies in a domain D, we can generate a random field.

Definition 5.9. A random process $\{Z(s), s \in D\}$ is intrinsically stationary if

$$\begin{aligned} E[Z(s+h) - Z(s)] &= 0 \\ Var[Z(s_1) - Z(s_2)] &= 2\gamma(s_1 - s_2) \end{aligned} \tag{5.58}$$

The quantity $\gamma(h)$ is known as the semi-variogram. Under intrinsic stationarity, the semi-variogram is a function only of the separation $s_1 - s_2$ between realizations at two sites. If moreover,

$$\gamma(s_1 - s_2) = \gamma^*(||s_1 - s_2||)$$

then the semi-variogram is said to be isotropic. The semi-variogram is anisotropic if its behavior is a function of the direction in space.

Suppose observations are made at sites $\{s_1, \ldots, s_n\}$. Then, the method moments estimator of the semi-variogram is

$$\hat{\gamma}(h) = \frac{1}{2|N(h)|} \sum [Z(s_i) - Z(s_j)]^2$$

where the summation is over pairs of points s_i, s_j a separation h apart

$$N(h) = \{(i,j); s_i - s_j = h\}$$

and $|N(h)|$ is the number of distinct elements in $N(h)$. Other estimators are possible and are described in Cressie [2015, p.74]. Normally, the semi-variogram is 0 whenever $h = 0$. If $\gamma(h) \to \gamma_0 > 0$, then we say that there is a nugget effect. In theory, a nugget effect is not possible for processes for which $E[Z(s+h) - Z(s)]^2 \to 0, =$ as $\|h\| \to 0$. It represents a small scale variability or measurement error. For purposes of drawing inference, we shall also define the concept of second-order stationarity.

Definition 5.10. A random process is said to be second-order stationary if

$$\begin{aligned} E[Z(s)] &= \mu, \text{for all } s \in D. \\ Covar[Z(s_1) - Z(s_2)] &= C(s_1 - s_2), \text{for all } s_1 s_2 \in D. \end{aligned}$$

One of the characteristic properties of a second-order stationary process is that the covariogram C is positive semi-definite:

$$\sum_{i=1}^{n} \sum_{j=1}^{n} a_i a_j C(s_i - s_j) \geq 0$$

for sites $\{s_1, \ldots, s_n\}$ and real numbers $\{a_i\}$ for which $\sum a_i = 0$. Such processes are also intrinsically stationary and

$$\gamma(h) = C(0) - C(h)$$

The above condition may be expressed in terms of the semi-variogram:

$$\sum_{i=1}^{n} \sum_{j=1}^{n} a_i a_j \gamma(s_i - s_j) \leq 0$$

In practice, the isotropic semi-variogram needs to be modeled and calibrated on the basis of the observations at the sites. Two popular models are

Linear in R^2

$$\gamma(h; c_0, b) = \begin{cases} 0 & h = 0 \\ c_0 + b_1 h & h \neq 0 \end{cases}$$

Exponential in R^2

$$\gamma(h; c_0, c_e, a) = \begin{cases} 0 & h = 0 \\ c_0 + c_e(1 - \exp(-h/a)) & h \neq 0 \end{cases}$$

Generally, though not always, the semi-variogram increases with $\|h\|$ and flattens out after some point $\gamma(\infty)$ called a sill which is simply the variance

$$\gamma(\infty) = Var[Z(x)] = C(0)$$

5.12.1. Spatial Prediction

The principal objective in spatial analysis is to predict the value $Z(s_0)$ at site s_0 from data $\{Z(s_i), i = 1, \ldots, n\}$ at sites $\{s_1, \ldots, s_n\}$. Historically, this problem originated in the context of geostatistics, and the methodology used was given the name kriging after the South African mining engineer D.G. Krige who initiated an empirical development of the subject; see Matheron [1971]. The idea in ordinary kriging is to estimate $Z(s_0)$ by a weighted average $\sum_{i=1}^{n} \lambda_i Z(s_i)$ of the observed data in such a way as to minimize the mean square error. The result is stated in the following theorem.

Theorem 5.14. *Suppose that we observe data $\{Z(s_i), i = 1, \ldots, n\}$ at sites $\{s_1, \ldots, s_n\}$ for which $E[Z(x)] = \mu$. Let $\{\lambda_i, i = 1, \ldots, n\}$ be a set of weights for which $\sum_{i=1}^{n} \lambda_i = 1$. Let s_0 be a given site. Then the optimal choice of weights for which*

$$E\left[Z(s_0) - \sum_{i=1}^{n} \lambda_i Z(s_i)\right]^2$$

is minimized is given by the solution to

$$-\sum_{j=1}^{n} \lambda_j \gamma(s_i - s_j) + \gamma(s_0 - s_j) - L = 0, i = 1, \ldots, n$$

subject to $\sum_{i=1}^{n} \lambda_i = 1$. The mean squared prediction error is

$$2\sum_{j=1}^{n} \lambda_j \gamma(s_0 - s_j) - \sum_{i=1}^{n}\sum_{j=1}^{n} \lambda_i \lambda_j \gamma(s_i - s_j)$$

Proof. The proof proceeds by noting that

$$\begin{aligned}\left[Z(s_0) - \sum_{i=1}^{n} \lambda_i Z(s_i)\right]^2 &= -\frac{1}{2}\sum_{i=1}^{n}\sum_{j=1}^{n} \lambda_i \lambda_j (Z(s_i) - Z(s_j))^2 \\ &+ \sum_{j=1}^{n} \lambda_i (Z(s_0) - Z(s_j))^2\end{aligned}$$

See Cressie [2015, p.121]. □

Universal kriging is a generalization of ordinary kriging whereby the mean μ in ordinary kriging is now modeled by

$$\mu(s) = \sum_{j=1}^{p+1} \beta_{j-1} f_{j-1}(s)$$

Here $\{\beta_j, j = 0, \ldots, p\}$ are parameters and $\{f_j(s), j = 0, \ldots, p\}$ are given auxiliary variables which are functions of the sites. The optimal linear predictor for $Z(s_0)$ is obtained in a similar manner as for ordinary kriging [Cressie, 2015, p.153]. There are different variants of kriging such as indicator function kriging, co-kriging and block kriging. We do not pursue these subjects in this book. We conclude by noting that the calculations in practice can be performed by a variety of software including R.

Spatial analyses on environmental data were conducted in a series of articles [Alvo and Voldner, 1999; Clark et al., 1988; Voldner and Alvo, 1989, 1993]. A Bayesian kriging approach was described in Alvo and Voldner [1999]. We conclude this section with a current application of spatial analysis on COVID-19 data (Mu [2021]).

Example 5.12. (COVID-19) The pandemic caused by the coronavirus disease 2019 (COVID-19) has led to an unprecedented number of disruptions in society and in the world at large. The use of geostatistical methods allows us to monitor the spread of the virus and to assess the effectiveness of various interventions and policy changes. In her M.Sc. thesis, [Mu, 2021] made use of two Bayesian spatial-temporal models labeled, ATA Poisson kriging, and ATP Poisson kriging to track how the virus spreads over the province of Ontario in Canada. Her data consisted of age adjusted infection rates provided by Statistics Canada. In her analysis, she included policy factors to take into account control measures as well as auxiliary variables such as average household size, proportion of self-employed workers in all class workers, and proportion of self-employed workers in all class workers. The general spatial-temporal model was formulated as follows:

$$\begin{aligned} y_{it} &\sim Poisson(\mu_{it}) \\ \log \mu_{it} &= \log e_i + \theta_{it} \\ \theta_{it} &= \alpha + \beta X + u_i + v_i + \gamma_t + \phi_t + \delta_{it} \end{aligned}$$

where y_i is the age-adjusted number of infections in area i, e_i and θ_i are respectively the corresponding population sizes and the log of the infected risk. The auxiliary variables are represented by X, u_i is a spatially structured Markov random effect, and v_i is a spatially unstructured random error term expressed as

$$
\begin{aligned}
u_i|u_k, k &\neq i \sim N\left(\frac{\sum u_{k\neq i} u_k \omega_{ki}}{\sum_{k\neq i}\omega_{ki}}, \frac{\tau_u^2}{\sum_{k\neq i}\omega_{ki}}\right) \\
v_i &\sim N\left(0, \tau_v^2\right) \\
\gamma_t|\gamma_{t-1}, \gamma_{t-2} &\sim N\left(2\gamma_{t-1} + \gamma_{t-2}, \sigma_\gamma^2\right) \\
\phi_t &\sim N\left(0, \sigma_\phi^2\right)
\end{aligned}
$$

The indicator variables ω_{ki} are defined as

$$
\omega_{ki} = \begin{cases} 1 & k, i \text{ adjacent} \\ 0 & otherwise \end{cases}
$$

The term δ takes into account the space-time interaction. A Laplace approximation was used to fit this model. Figure 5.1 exhibits the evolution of the infected risks at the public health unit level. In addition, she defined a Poisson kriging model to take into account the different shapes and sizes of the geographical. For a given number N of geographical units v_α (e.g., counties, public health units) denote the number of age adjusted recorded infection cases by $d(v_\alpha)$. Let $\mathbf{u}_\alpha = (\mathbf{x}_\alpha, \mathbf{y}_\alpha)$ be the vector of spatial coordinates for the centroid of the unit v_α. The observed infection rate is then denoted as $z(\mathbf{u}_\alpha) = d(\mathbf{u}_\alpha)/n(\mathbf{u}_\alpha)$, where $n(\mathbf{u}_\alpha)$ is the population size in unit v_α. At each location $\mathbf{u}_\alpha$, the disease counts $d(\mathbf{u}_\alpha)$ can be interpreted as realizations of random variables $D(\mathbf{u}_\alpha)$ that follow a Poisson distribution with parameter (expected number of counts) that is the product of the population size $n(\mathbf{u}_\alpha)$ by the local risk $R(\mathbf{u}_\alpha)$:

$$
D(\mathbf{u}_\alpha)|R(\mathbf{u}_\alpha) = \text{Poisson}(n(\mathbf{u}_\alpha)R(\mathbf{u}_\alpha)) \quad \alpha = 1, \ldots, \text{N}
$$

Given the risk $R(\mathbf{u}_\alpha)$, the infection count variables $D(\mathbf{u}_\alpha)$ are assumed to be conditionally independent. Hence any spatial correlation among the counts is influenced by spatial trends in either the population sizes or in the local risks. The risk variable $R(\mathbf{u}_\alpha)$ itself can be modeled as a stationary random field with mean m, variance σ_R^2, and variance function $C_R(\mathbf{h})$ [Goovaerts, 2017]. The conditional mean and variance of the rate variable $Z(\mathbf{u}_\alpha)$ are defined, respectively, as

$$
\begin{aligned}
E[Z(\mathbf{u}_\alpha)|R(\mathbf{u}_\alpha)] &= E[\frac{D(\mathbf{u}_\alpha)}{n(\mathbf{u}_\alpha)}|R(\mathbf{u}_\alpha)] \\
&= \frac{1}{n(\mathbf{u}_\alpha)} E[D(\mathbf{u}_\alpha)|R(\mathbf{u}_\alpha)] \\
&= R(\mathbf{u}_\alpha)
\end{aligned}
$$

$$\begin{aligned}\mathrm{Var}[Z(\mathbf{u}_\alpha)|R(\mathbf{u}_\alpha)] &= \mathrm{Var}[\frac{D(\mathbf{u}_\alpha)}{n(\mathbf{u}_\alpha)}|R(\mathbf{u}_\alpha)] \\ &= \frac{1}{n(\mathbf{u}_\alpha)^2}\mathrm{Var}[D(\mathbf{u}_\alpha)|R(\mathbf{u}_\alpha)] \\ &= \frac{R(\mathbf{u}_\alpha)}{n(\mathbf{u}_\alpha)}\end{aligned}$$

For the covariance expression, the conditional independence of observations, infection cases, at different sites leads to

$$\begin{aligned}E[D_iD_j|R] &= \mathrm{Cov}[D_i, D_j|R] + E[D_i|R_i]E[D_j|R_j] \\ &= \delta_{ij}n_iR_i + n_in_jR_iR_j\end{aligned}$$

Figure 5.1.: Estimated cases per 100,000

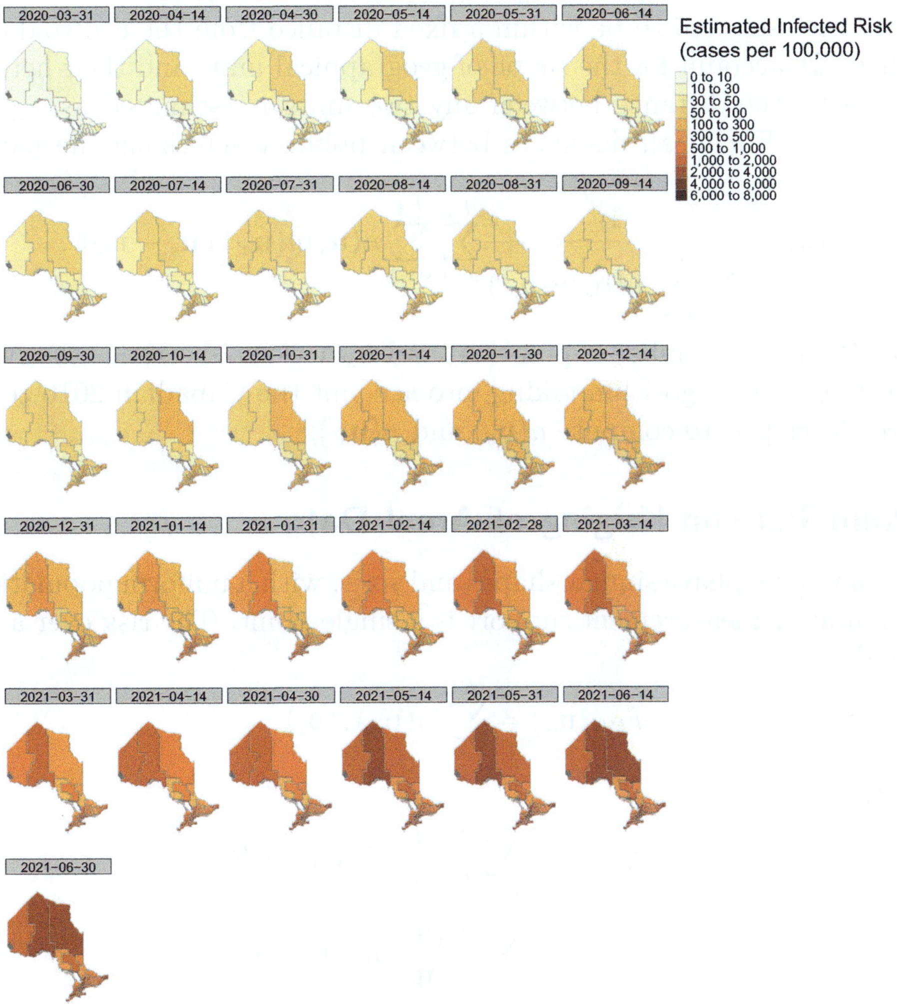

whereas the unconditional mean rate observed in region α, $Z(\mathbf{u}_\alpha)$ is

$$E_R[Z(\mathbf{u}_\alpha)] = E_R E[Z(\mathbf{u}_\alpha)|R(\mathbf{u}_\alpha)] = E_R[R(\mathbf{u}_\alpha)] = m,$$

Similarly, the unconditional variance of $Z(\mathbf{u}_\alpha)$ is given by

$$\begin{aligned}\text{Var}[Z(\mathbf{u}_\alpha)] &= E_R[\text{Var}((Z(\mathbf{u}_\alpha)|R(\mathbf{u}_\alpha))] + \text{Var}_R[E(Z(\mathbf{u}_\alpha)|R(\mathbf{u}_\alpha))] \\ &= E_R[R(\mathbf{u}_\alpha)/n(\mathbf{u}_\alpha)] + \text{Var}_R[R(\mathbf{u}_\alpha)] \\ &= m/n(\mathbf{u}_\alpha) + \sigma_R^2\end{aligned}$$

The estimator can be formulated as a linear combination of K neighboring rates or functions of those rates:

$$\hat{r}_{PK}(\mathbf{u}_\alpha) = \sum_{i=1}^{K} \lambda_i(\mathbf{u}_\alpha) z(\mathbf{u}_i),$$

where $\lambda_i(\mathbf{u}_\alpha)$ is the weight assigned to the rate $z(\mathbf{u_i})$ when estimating the risk at $\mathbf{u}_\alpha$, the geographical centroid of the unit $\mathbf{v}_\alpha$.

Neighbors were selected to be within a fixed distance from the centroid of the area to be smoothed. To account for the shape of geographical units and their heterogeneous population density, the distance between any two units is estimated as a population-weighted average of Euclidean distances between points discretizing the pair of units:

$$\text{Dist}(v_\alpha, v_\beta) = \frac{1}{\sum\limits_{s=1}^{P_\alpha}\sum\limits_{s'=1}^{P_\beta} n(\mathbf{u}_s)n(\mathbf{u}_{s'})} \sum_{s=1}^{P_\alpha}\sum_{s'=1}^{P_\beta} n(\mathbf{u}_s)n(\mathbf{u}_{s'}) \left\|\mathbf{u}_s - \mathbf{u}_{s'}\right\|, \tag{5.59}$$

where P_α and P_β are the number of points $\mathbf{u}_s$ and $\mathbf{u}_{s'}$ used to discretize the two units v_α and v_β, respectively. An algorithm taking into account the Canadian 2016 census block level data was developed to compute $n(\mathbf{u}_s)$ and $n(\mathbf{u}_{s'})$.

5.12.2. Point Poisson Kriging of Areal Data

Assume that all units $\mathbf{v}_\alpha$ have similar shapes and sizes, with a uniform population density and that each unit or measurement support is a single point. The risk over a given unit v_α is estimated by

$$\hat{r}_{PK}(\mathbf{u}_\alpha) = \sum_{i=1}^{K} \lambda_i(\mathbf{u}_\alpha) z(\mathbf{u}_i)$$

with expectation:

$$\begin{aligned}E[\hat{r}_{PK}(\mathbf{u}_\alpha)|R] &= \sum_{i=1}^{K} \frac{\lambda_i(\mathbf{u}_\alpha)}{n_i(\mathbf{u}_\alpha)} E[D(\mathbf{u}_i)|R(\mathbf{u}_i)] \\ &= \sum_{i=1}^{K} \frac{\lambda_i(\mathbf{u}_\alpha)}{n(\mathbf{u}_i)} n_i(\mathbf{u}_i) R(\mathbf{u}_i) \\ &= \sum_{i=1}^{K} \lambda_i(\mathbf{u}_\alpha) R(\mathbf{u}_i)\end{aligned} \tag{5.60}$$

$$\begin{aligned} E[\hat{r}_{PK}(\mathbf{u}_\alpha)] &= \sum_{i=1}^{K} \lambda_i(\mathbf{u}_\alpha) E[R(\mathbf{u}_i)] \\ &= m \sum_{i=1}^{K} \lambda_i(\mathbf{u}_\alpha) = 1 \end{aligned} \tag{5.61}$$

for unbiasedness. Similarly, the conditional mean square error of prediction is

$$\begin{aligned} E[(\hat{r}_{PK}(\mathbf{u}_\alpha) - R(\mathbf{u}_\alpha))^2|R] &= E[(\sum_{i=1}^{K} \lambda_i(\mathbf{u}_\alpha)\frac{D(\mathbf{u}_i)}{n(\mathbf{u}_i)} - R(\mathbf{u}_\alpha))^2|R] \\ &= \sum_{i=1}^{K}\sum_{j=1}^{K} \frac{\lambda_i(\mathbf{u}_\alpha)}{n(\mathbf{u}_i)}\frac{\lambda_j(\mathbf{u}_\alpha)}{n(\mathbf{u}_j)} E[D(\mathbf{u}_i)D(\mathbf{u}_j)|R] + R^2(\mathbf{u}_\alpha) \\ &\quad - 2R(\mathbf{u}_\alpha)\sum_{i=1}^{K} \frac{\lambda_i(\mathbf{u}_\alpha)}{n(\mathbf{u}_i)} E[D(\mathbf{u}_i)|R] \\ &= \sum_{i=1}^{K}\sum_{j=1}^{K} \lambda_i(\mathbf{u}_\alpha)\lambda_j(\mathbf{u}_\alpha)R(\mathbf{u}_i)R(\mathbf{u}_j) + \sum_{i=1}^{K} \frac{\lambda_i^2(\mathbf{u}_\alpha)}{n(\mathbf{u}_i)} R(\mathbf{u}_\alpha) + R^2(\mathbf{u}_\alpha) \\ &\quad - 2R(\mathbf{u}_\alpha)\sum_{i=1}^{K} \lambda_i(\mathbf{u}_\alpha)R(\mathbf{u}_i) \end{aligned} \tag{5.62}$$

Then, unconditionally it can be shown to be

$$E[(\hat{r}_{PK}(\mathbf{u}_\alpha) - R(\mathbf{u}_\alpha))^2] = E_R E[(\hat{r}_{PK}(\mathbf{u}_\alpha) - R(\mathbf{u}_\alpha))^2|R]$$

$$\begin{aligned} &= \sigma_R^2 + \sum_{i=1}^{K} \frac{\lambda_i^2(\mathbf{u}_\alpha)}{n(\mathbf{u}_i)} m + \sum_{i=1}^{K}\sum_{j=1}^{K} \lambda_i(\mathbf{u}_\alpha)\lambda_j(\mathbf{u}_\alpha) C_R(\mathbf{u}_i, \mathbf{u}_j) \qquad (5.63) \\ &- 2\sum_{i=1}^{K} \lambda_i(\mathbf{u}_\alpha) C_R(\mathbf{u}_i, \mathbf{u}_\alpha) \end{aligned}$$

The problem thus reduces to the ordinary kriging optimization process which requires modeling the estimate of the semi-variogram:

$$\hat{\gamma}_R(\mathbf{h}) = \frac{1}{2\sum_{\alpha=1}^{N(\mathbf{h})} \frac{n(\mathbf{u}_\alpha n(\mathbf{u}_{\alpha+h})}{n(\mathbf{u}_\alpha)+n(\mathbf{u}_{\alpha+h})}} \sum_{\alpha=1}^{N(\mathbf{h})} \{\frac{n(\mathbf{u}_\alpha)n(\mathbf{u}_{\alpha+h})}{n(\mathbf{u}_\alpha)+n(\mathbf{u}_{\alpha+h})}[z(\mathbf{u}_\alpha) - z(\mathbf{u}_{\alpha+h})]^2 - m*\} \tag{5.64}$$

For the Area-to-Point Poisson kriging system, the covariances $\bar{C}_R(v_i, v_\alpha)$ are replaced by Area-to-Point covariances $\bar{C}_R(v_i, \mathbf{u}_s)$ approximated as follows:

$$\bar{C}_R(v_i, \mathbf{u}_s) = \frac{1}{\sum_{s'=1}^{P_i} w_{s's}} \sum_{s'=1}^{P_i} w_{s's} C(\mathbf{u}_{s'}, \mathbf{u}_s)$$

where P_i is the number of points used to discretize the area v_i and the weights $w_{s's}$ are computed as

$$w_{s's} = n(\mathbf{u}_{s'}) \times n(\mathbf{u}_s)$$

The Area-to-Point kriging estimator variance is given as

$$\sigma^2_{PK}(\mathbf{u}_s) = C_R(0) - \sum_{i=1}^{K} \lambda_i(\mathbf{u}_s)\bar{C}_R(v_i, \mathbf{u}_s) - \mu(\mathbf{u}_s)$$

See Mu [2021] for further details.

5.13. Efficiency

We conclude this chapter with a discussion of the efficiency of tests. Knowledge of the distribution of the test statistics under both the null and the alternative hypotheses permits us to compare tests. In the usual cases such as for the sign and the Wilcoxon tests, the calculations are straightforward. However, in the multi-sample situations, the calculations are more complicated, and for these, we appeal to Le Cam's lemmas. This is illustrated in the case of test statistics involving both the Spearman and Hamming distances. The smooth embedding approach is useful in that for a given problem, it leads to test statistics whose power function can be determined. The latter is then assessed against the power function of the optimal test statistic derived from Hoeffding's formula for any given underlying distribution of the data. We begin with the definition of Pitman efficiency.

5.13.1. Pitman Efficiency

Suppose that one is interested in testing a null hypothesis H_0 against an alternative H_1. Consider two competing tests $T_{n,1}, T_{n,2}$ both of which are at level α and are consistent, that is, the power function converges to 1 as the sample size gets large. We may choose between them only if we can define a suitable criterion that takes into account the size as well as the power function of the test. The criteria proposed by Pitman is defined for local alternatives close to the null hypothesis and provides a simple overall measure of the comparison between two tests. However, the power function provides a measure of performance for a full range of alternatives and in that sense is more informative. The presentation below follows Lehmann [1975].

Let $X_1, \ldots, X_n$ be a random sample from some cumulative distribution function F having density $f(x;\theta)$ for some parameter θ. Let Θ_0, Θ_1 denote the spaces where θ takes its values under the null and alternative hypotheses, respectively. The power function of a test statistic T_n is defined to be the probability of rejecting the null hypothesis and is given by

$$\beta_n(T_n;\theta) = P_\theta(T_n \text{ rejects } H_0)$$

The size of the test is defined to be

$$\alpha_n = sup_{\theta\epsilon\Theta_0}\beta_n\left(T_n;\theta\right)$$

In the specific situation where

$$H_0 : \theta = 0$$

against

$$H_1 : \theta > 0$$

we shall be concerned with smooth local alternatives of the form $\theta_n = \frac{h}{\sqrt{n}}$. These alternatives make it more difficult for a test to choose between the hypotheses as n gets large. We shall suppose that for both tests the following three conditions are satisfied:

(i) the test statistic T_n is asymptotically normal under P_{θ_n}

$$P_{\theta_n}\left(\frac{\sqrt{n}\left(T_n - \mu\left(\theta_n\right)\right)}{\sigma\left(\theta_n\right)} \leq x\right) \longrightarrow \Phi\left(x\right)$$

where it is convenient to let $\mu\left(\theta_n\right)$ denote the mean of T_n and $\frac{\sigma^2(\theta_n)}{n}$ denote the variance of T_n.

(ii) $\mu\left(\theta_n\right)$ is differentiable at 0. $\sigma\left(\theta_n\right) \longrightarrow \sigma\left(0\right)$ as $\theta_n \rightarrow 0$ and

(iii) the test rejects for large values of the test statistic, that is,

$$\frac{\sqrt{n}\left(T_n - \mu\left(0\right)\right)}{\sigma\left(0\right)} > z_\alpha$$

where z_α is the upper $100\left(1-\alpha\right)\%$ point of a standard normal distribution.

Under these conditions, the corresponding power function converges to

$$\begin{aligned}
\beta_n\left(\theta_n\right) &= P_{\theta_n}\left(\frac{\sqrt{n}\left(T_n - \mu\left(\theta_n\right)\right)}{\sigma\left(\theta_n\right)} \geq z_\alpha\frac{\sigma\left(0\right)}{\sigma\left(\theta_n\right)} - \frac{\sqrt{n}\left(\mu\left(\theta_n\right) - \mu\left(0\right)\right)}{\sigma\left(\theta_n\right)}\right) \\
&= P_{\theta_n}\left(\frac{\sqrt{n}\left(T_n - \mu\left(\theta_n\right)\right)}{\sigma\left(\theta_n\right)} \geq z_\alpha\frac{\sigma\left(0\right)}{\sigma\left(\theta_n\right)} - \frac{h}{\sigma\left(\theta_n\right)}\frac{\left(\mu\left(\theta_n\right) - \mu\left(0\right)\right)}{\theta_n}\right) \\
&\longrightarrow \Phi\left(hc - z_\alpha\right)
\end{aligned}$$

where

$$\begin{aligned}
c &= lim_{n\rightarrow\infty}\left(\frac{1}{\sigma\left(\theta_n\right)}\frac{\left(\mu\left(\theta_n\right) - \mu\left(0\right)\right)}{\theta_n}\right) \\
&= \left(\frac{\mu'\left(0\right)}{\sigma\left(0\right)}\right)
\end{aligned}$$

The quantity c is called the slope of the test. The larger the slope, the more powerful the test. A comparison of two tests $T_{n,1}$ *and* $T_{n,2}$ may then be made in terms of the ratio of the slopes denoted by c_1 *and* c_2, respectively:

$$\frac{c_1}{c_2} = \left[\frac{\mu_1'(0)}{\sigma_1(0)}\right] / \left[\frac{\mu_2'(0)}{\sigma_2(0)}\right] \tag{5.65}$$

Hence, $T_{n,1}$ is preferred to $T_{n,2}$ if $\frac{c_1}{c_2} > 1$.

Example 5.13. One-sample tests comparisons for the sign test, the T-test, and the Wilcoxon signed rank test

To calculate the slope of the sign test, we have a random sample of size n from some distribution having absolutely continuous cumulative distribution function $F(x-\theta)$ which is symmetric around its median θ. We would like to test the null hypothesis $H_0 : \theta = 0$ and the one-sided alternative $H_1 : \theta > 0$.

Let $T_{n,1}$ be the proportion of $X_i > 0$. The sign test rejects the null hypothesis for large values of $T_{n,1}$. Set

$$p = P(X > 0) = 1 - F(-\theta)$$

Then, $T_{n,1}$ has a binomial distribution with parameters n and probability of success p. Hence,

$$\begin{aligned} \mu(\theta) = E_\theta[T_{n,1}] &= p, \\ Var_\theta[T_{n,1}] &= \frac{p(1-p)}{n} \end{aligned}$$

It follows that $\sigma^2(\theta) = p(1-p)$ and the slope is

$$\frac{\mu'(0)}{\sigma(0)} = \frac{2f(0)}{\sigma} \tag{5.66}$$

where $f(x) = F'(x)$.

Next, consider the usual T-test for H_0 vs H_1 which rejects the null hypothesis for large values of the sample mean $T_{n,2} = \bar{X}_n$. Hence,

$$\begin{aligned} \mu(\theta) &= E[T_{n,2}] \\ &= \int x dF(x-\theta) \\ &= \theta \end{aligned}$$

Since the test statistic is given by

$$\frac{\sqrt{n}\bar{X}_n}{S_n}$$

where S_n^2 represents the sample variance, then as $n \to \infty$

$$S_n^2 \to \sigma^2 = \int x^2 dF(x)$$

It follows that

$$\frac{\mu'(0)}{\sigma(0)} = \frac{1}{\sigma} \tag{5.67}$$

Hence, when $f(x)$ is the standard normal density, the asymptotic relative efficiency of the sign test to the T-test is

$$(2f(0))^2 = \left(\frac{2}{\sqrt{2\pi}}\right)^2 = 0.637$$

The Wilcoxon signed rank statistic rejects for large values of

$$T_{n,3} = \sum_{i=1}^{n} R_i^+ I(X_i > 0)$$

where as in Sect. 5.2, R_i^+ is the rank of $|X_i|$ among the $\{|X_i|, i = 1, \ldots, n\}$. It follows that under the null hypothesis

$$\begin{aligned} E_0[T_{n,3}] &= \frac{n(n+1)}{4} \\ Var_0[T_{n,3}] &= \frac{n(n+1)(2n+1)}{24} \end{aligned}$$

The calculations of the mean and variance under the alternative for the Wilcoxon signed rank statistic are more involved and are given by the following theorem:

Theorem 5.15. *Let*

$$\begin{aligned} p_1 &= P(X > 0) \\ p_2 &= P(X_1 + X_2 > 0) \\ p_3 &= P(X_1 + X_2 > 0, X_1 > 0) = \frac{(p_2 + p_1^2)}{2} \\ p_4 &= P(X_1 + X_2 > 0, X_1 + X_3 > 0) \end{aligned}$$

Then,

$$E[T_{n,3}] = np_1 + \frac{n(n-1)}{2} p_2$$

and

$$\begin{aligned} Var[T_{n,3}] = {} & np_1(1-p_1) + \frac{n(n-1)}{2} p_2(1-p_2) \\ & + 2n(n-1)(p_3 - p_1 p_2) + n(n-1)(n-2)\left(p_4 - p_2^2\right) \end{aligned}$$

Proof. See Lehmann [1975, p.47]. □

The probabilities in the theorem above can be approximated for values of θ close to 0 so that

$$\begin{aligned} p_1 &= 1 - F(-\theta) \\ &\approx 1 - [F(0) - \theta f(0)] \\ &= \frac{1}{2} - \theta f(0) \end{aligned}$$

Also,

$$\begin{aligned} p_4 &= P(X_1 + X_2 > 0, X_1 + X_3 > 0) \\ &= \left(\int 1 - F(-x) \right)^2 f(x)\, dx \\ &= \frac{1}{3} \end{aligned}$$

and

$$\begin{aligned} p_2 &= P(X_1 + X_2 > 0) \\ &\approx 1 - F*(-2\theta) \\ &= \frac{1}{2} + 2\theta f*(0) \end{aligned}$$

where $F*$ is the convolution distribution and $f*$ is its density. Using the symmetry of f whereby $f(x) = f(-x)$, we have that

$$f*(0) = \int_{-\infty}^{\infty} f^2(x)\, dx$$

It follows that for small values of θ

$$Var_0[T_{n,3}] \approx \frac{n(n+1)(2n+1)}{24}$$

In order to compute the slope for the Wilcoxon signed rank test, we note that

$$\mu(\theta) = E[T_{n,3}] = np_1 + \frac{n(n-1)}{2} p_2$$

so that

$$\mu'(0) = nf(0) + n(n-1) \int_{-\infty}^{\infty} f^2(x)\, dx$$

and

$$\frac{\mu'(0)}{\sigma(0)} = \sqrt{12} \int_{-\infty}^{\infty} f^2(x)\, dx \tag{5.68}$$

The calculations of the slopes in (5.66), (5.67), and (5.68) enable us to compute the asymptotic relative efficiencies.

Example 5.14. Two-sample tests comparisons: Mann-Whitney-Wilcoxon test and T-test

Suppose that we have two independent random samples of sizes n and m, respectively, from distributions $F(x)$,G$(x-\Delta)$ having densities f, g. Recalling the Mann-Whitney-Wilcoxon test, it was seen that the test statistic

$$T_{m,n} = \frac{1}{mn}\sum_{j=1}^{m}\sum_{i=1}^{n} I(X_i < Y_j)$$

rejects the null hypothesis for large values. Then

$$\begin{aligned}\mu(\Delta) &= E[T_{m,n}] \\ &= P(X<Y) \\ &= \int_{-\infty}^{\infty} F(y) f(y-\Delta)\,dy \\ &= \int_{-\infty}^{\infty} F(y+\Delta)\,dF(y)\end{aligned}$$

and provided we can differentiate under the integral sign

$$\mu'(0) = \int_{-\infty}^{\infty} f^2(y)\,dy.$$

The variance of $T_{m,n}$, given from Theorem 5.15, is asymptotically

$$\begin{aligned}Var[T_{m,n}] &= \frac{1}{(mn)^2}\left[mnq_1(1-q_1) + mn(n-1)(q_2-q_1^2) + mn(m-1)(q_3-q_1^2)\right] \\ &\approx \left[o(1) + m^{-1}(q_2-q_1^2) + n^{-1}(q_3-q_1^2)\right]\end{aligned}$$

so that for $\frac{m}{m+n} \to \lambda > 0$,

$$\begin{aligned}(m+n)\,Var[T_{m,n}] &\approx \lambda^{-1}(q_2-q_1^2) + (1-\lambda)^{-1}(q_3-q_1^2) \\ &= \frac{1}{12\lambda(1-\lambda)}\end{aligned}$$

It follows that the slope is given by

$$\frac{\mu'(0)}{\sigma(0)} = \sqrt{12\lambda(1-\lambda)}\int_{-\infty}^{\infty} f^2(x)\,dx \tag{5.69}$$

By way of comparison, we may also compute the slope for the two-sample T-test using the statistic which rejects the null hypothesis for large values. It is given by the ratio

Table 5.13.: Relative efficiency of the Wilcoxon two-sample test relative to the T-test

Distribution	Relative efficiency of the Wilcoxon test to the t test
Normal	$3/\pi = 0.954$
Logistic	$\pi^2/9 = 1.097$
Uniform	1

$$\left(\frac{\bar{Y}_m - \bar{X}_n}{S_{n,m}\sqrt{m^{-1}+n^{-1}}}\right)$$

where $\bar{X}_n$ *and* $\bar{Y}_m$ represent the corresponding sample means and $S^2_{n,m}$ represents the pooled sample variance estimate of the common variance σ^2. It is straightforward to show that the slope of the two-sample T-test is

$$\frac{\mu'(0)}{\sigma(0)} = \frac{\sqrt{\lambda(1-\lambda)}}{\sigma} \tag{5.70}$$

We record in Table 5.13 the relative efficiency of the Wilcoxon test relative to the T-test for various underlying distributions.

5.13.2. Application of Le Cam's Lemmas

The calculation of the asymptotic distribution under the alternative is not always as straightforward as in the previous section. In what follows we shall consider examples where it is necessary to calculate the asymptotic relative efficiency of tests by making use of Le Cam's contiguity concepts and lemmas described in Sect. 3.2. We consider the simple example due to Van der Vaart [2000, Section 14.11] involving the median test. Other calculations of efficiency involving the general multi-sample problem are described in greater detail in Alvo and Yu [2018].

Example 5.15. (Median Test) Consider the two-sample test with m observations in the first sample and n in the second. Set $N = m + n$. The median test rejects the null hypothesis that the medians in the two samples are the same for large values of

$$T_N = \frac{1}{N}\sum_{i=m+1}^{N} I\left(R_i \leq \frac{n+1}{2}\right)$$

where $R_1, \ldots, R_N$ are the ranks of the complete data and $I(A)$ is the indicator function of the set A.. Under the null hypothesis

$$\sqrt{N}\left(T_N - \frac{n}{2N}\right) = \frac{1}{N\sqrt{N}}\left(-n\sum_{i=1}^{m} I\left(F(X_i) \leq 1/2\right) + m\sum_{j=1}^{n} I\left(F(Y_i) \leq 1/2\right)\right) + o_P(1) \tag{5.71}$$

$$\xrightarrow{L} \mathcal{N}\left(0, \frac{\lambda(1-\lambda)}{4}\right) \tag{5.72}$$

On the other hand, for alternatives $\theta_N = \frac{h}{\sqrt{N}}$, the log-likelihood ratio given by

$$log \frac{\prod f(X_i) \prod g(Y_j - \theta_N)}{\prod f(X_i) \prod g(Y_j)} = -\frac{h\sqrt{1-\lambda}}{\sqrt{n}} \sum_{i=1}^{n} \frac{g'}{g}(Y_i) - \frac{h^2(1-\lambda) I_g}{2} + o_P(1) \tag{5.73}$$

and hence is asymptotically normal. Moreover the joint distribution of the linear parts is multivariate normal. Slutsky's lemma implies a similar result for the right-hand sides. Hence by Le Cam's third lemma, under alternatives $\theta_N = \frac{h}{\sqrt{N}}$,

$$\sqrt{N}\left(T_n - \frac{n}{2N}\right) \underset{\rightarrow}{\mathcal{L}} \mathcal{N}\left(\tau(h), \frac{\lambda(1-\lambda)}{4}\right)$$

where

$$\tau(h) = -h\lambda(1-\lambda) \int_{F(y) \le 1/2} \left(\frac{f'(y)}{f(y)}\right) dF(y)$$

is the asymptotic covariance. The slope of the median test is then given by

$$-2\sqrt{\lambda(1-\lambda)} \int_0^{1/2} \left(\frac{f'(v)}{f(v)}\right) \left(F^{-1}(v)\right) dv \tag{5.74}$$

5.14. Permutation Methods

Permutation or randomization tests as they are sometimes called provide an effective though computationally intensive method of determining the sampling distribution of a test statistic under the null hypothesis of "no association." The basic idea which originated with R.A. Fisher consists of permuting the labels of the observed data and computing the value of the test statistic. This operation is repeated a large number of times thereby creating a histogram of values that approximates the null distribution of the test statistic. A p-value may then be calculated by counting the number of cases which yield values at least as extreme as the one obtained from the observed data. Since the permutation test is conditional on the observed data, it is distribution-free. Permutation tests are particularly useful when the data sets are small and when the underlying distribution of the test statistic is uncertain.

Example 5.16. Let $X_1, \ldots, X_{n_1}$ be a random sample from a continuous cumulative distribution function $F(x)$ and let $Y_1, \ldots, Y_{n_2}$ be an independent random sample from $F(x - \Delta)$, where Δ is finite. Let $n = n_1 + n_2$.

We would like to test the hypotheses

$$H_0 : \Delta = 0$$

against

$$H_1 : \Delta > 0.$$

Suppose that the test statistic to be used is given by the difference in the sample means

$$T = \bar{X} - \bar{Y}$$

Let t_0 be the observed value of T. Under the null hypothesis, the combined sample of n observations is assumed to come from the same cumulative distribution function F. Ignoring the labels which identify the distribution from which the observations came, there would be $\binom{n}{n_1}$ possible samples of n_1 observations assigned to the first population and n_2 observations assigned to the second. This is precisely the number of possible permutations of the labels, and each is equally likely. For each of these permutations, we may compute a value of the statistic T thereby creating a reference distribution (which also includes the observed t_0) and then count how many of these exceed the observed t_0 in order to calculate the p-value. An interesting discussion is given in Kendall and Stuart [1979, p.501].

Notes We close this chapter by noting that there have been a number of important developments in nonparametric statistics as pertaining to *big data*. We mention in particular the work on random forests as developed by Breiman [2001]. See also Genuer et al. [2017] for a review of random forests as applied to *big data*. As well, there is the Apache Hadoop open-source software programmed in Java for performing *big data* calculations.

6. Exponential Tilting and Its Applications

In a landmark paper, Neyman [1937] considered the nonparametric goodness-of-fit problem and introduced the notion of smooth tests of fit by proposing a parametric family of densities to model the alternative to the null hypothesis. The type of embedding proposed by Neyman and further elaborated by Rayner et al. [2009] focused on global properties of the data. We begin with a description of the Neyman approach and then proceed by defining an embedding which focuses on local properties in line with the notion of exponential tilting. Alvo [2015] and Alvo and Yu [2018] also exploited the notion of exponential tilting to develop a parametric approach to nonparametric statistics.

6.1. Neyman Smooth Tests

Let $X_1, \ldots, X_n$ be a random sample from a continuous cumulative distribution function $F(x)$ with continuous density $f(x)$. We would like to test the null hypothesis

$$H_0 : F(x) = F_0(x), \, for \ all \ x$$

for known F_0 against the alternative

$$H_1 : F(x) \neq F_0(x), \, for \ some \ x.$$

Since the random variable $Y = F_0(X)$ has a uniform distribution, we may instead test the null hypothesis that we have a random sample $Y_1, \ldots, Y_n$ from the uniform distribution on the interval $[0, 1]$. If the alternative were specified more precisely, one could use the Neyman-Pearson lemma to arrive at a uniformly most powerful test. More often than not, however, the alternative hypothesis is vague. Various tests such as the Kolmogorov-Smirnov and Anderson-Darling test statistics have been proposed for this problem [Van der Vaart, 2000, p.277]. In most cases, however, such tests have little power. Moreover, if the null hypothesis of uniformity is rejected, it is not clear what the alternative hypothesis should be. Neyman [1937] considered this problem of testing for goodness of fit by embedding the uniform distribution into a larger class of alternatives

M. Alvo, *Statistical Inference and Machine Learning for Big Data*, Springer Series in the Data Sciences, https://doi.org/10.1007/978-3-031-06784-6_6

defined by

$$\pi_\theta(x) = \exp\left(\sum_{j=1}^{d}\theta_j h_j(x) - K(\theta)\right), 0 < x < 1 \tag{6.1}$$
$$= \exp\left(\theta^T h(x)\right) - K(\theta)$$

where

$$h(x) = (h_1(x), \ldots, h_d(x))^T$$

and $K(\theta)$ is the normalizing constant depending on θ chosen to make $\pi(x;\theta)$ a proper density. The functions $\{h_j(x)\}$ consist of orthonormal polynomials that satisfy

$$\int_0^1 h_i(x)\,dx = 0 \tag{6.2}$$

$$\int_0^1 h_i(x)\,h_j(x)\,dx = \delta_{ij} = \begin{cases} 0 & i \neq j \\ 1 & i = j \end{cases} \tag{6.3}$$

The orthonormal polynomials may be selected in order to detect different alternatives, and these lead to different test statistics. Some common choices of orthogonal functions (denoted with an upper asterisk) are listed in Table 6.1.

Under the model in (6.1), the test for uniformity then becomes a test of

$$H_0 : \theta = 0 \text{ vs } H_1 : \theta \neq 0. \tag{6.4}$$

so that under H_0, X has a uniform distribution on the interval $(0, 1)$. For small values of θ_j close to 0, we see that $\pi_\theta(x)$ is close to the uniform density. Hence, the model in (6.1) provides a "smooth" transition from the null hypothesis and as we shall later see, leads to tests with optimal properties. The next theorem specifies the test statistic.

Theorem 6.1. *Let $X_1, \ldots, X_n$ be a random sample from the density in (6.1). The score test rejects the null hypothesis whenever*

$$\sum_{j=1}^{d} U_j^2 > c_\alpha$$

where

$$U_j = \frac{1}{\sqrt{n}}\sum_{i=1}^{n} h_j(X_i)$$

and c_α satisfies $P\left(\chi_d^2 > c_\alpha\right) = \alpha$.

Proof. The log likelihood for the density in (6.1) is

$$l(\theta) = \sum_{j=1}^{d}\theta_j\sum_{i=1}^{n} h_j(x_i) - nK(\theta)$$

Table 6.1.: Orthogonal functions for various densities

Density	Polynomials	Orthogonal function
Normal: $\frac{1}{\sqrt{2\pi}} \exp\left(-\frac{x^2}{2}\right)$	Hermite	$h_j^*(x) = \sum_{t=0}^{[j/2]} \frac{(-1/2)^t j!}{t!(j-2t)!} x^{j-2t}$
Poisson: $\exp^{-\lambda} \lambda^x / x!$	Poisson-Charlier	$h_j^*(x;\lambda) = \sqrt{\lambda^j / j!} \sum_{t=0}^{j} (-1)^{j-t} \binom{x}{t} ! \lambda^{-t} \binom{j}{t}$
Exponential: e^{-x}		$h_j^*(x) = \sum_{t=0}^{j} \binom{j}{t} (-x)^t / t!$
Gamma: $x^{\alpha-1} \frac{\exp(-x/\beta)}{\Gamma(\alpha)\beta^\alpha}$	Laguerre	$h_j^*(x) = \frac{e^x}{j!} \frac{d^j}{dx^j} \left(e^{-x} x^j\right)$
Hypergeometric $\frac{\binom{M}{x}\binom{N-M}{n-x}}{\binom{N}{n}}$	Chebyshev	$h_0(x,n) = 1, h_1(x,n) = 1 - \frac{2x}{n}, x = 0, 1, \ldots, n$ $(i+1)(n-i)h_{i+1}(x,n) = (2i+1)(n-2x)h_i(x,n) - i(n+i+1)h_{i-1}(x,n)$

which yields the score vector

$$U(\theta) = \frac{\partial l(\theta)}{\partial \theta}$$

with jth component

$$\frac{\partial l(\theta)}{\partial \theta_j} = \sqrt{n} u_j - n \frac{\partial K(\theta)}{\partial \theta_j}$$

Differentiating with respect to θ_j and evaluating the derivative at $\theta = 0$

$$\frac{\partial}{\partial \theta_j} \int_0^1 \pi_\theta(x)\, dx = 1$$

leads to

$$\frac{\partial K(\theta)}{\partial \theta_j} = 0$$

and consequently

$$\frac{\partial l(\theta)}{\partial \theta_j} = \sqrt{n} u_j$$

Also, the (i, j) component of the information matrix $I_n(\theta)$ evaluated at $\theta = 0$ is

$$\frac{\partial^2 K(\theta)}{\partial \theta_i \partial \theta_j} \quad = \quad n\delta_{ij}$$

It follows that the score test statistic at $\theta = 0$ is

$$U^T(\theta) I_n^{-1}(\theta) U(\theta) = \sum_{j=1}^{d} U_j^2$$

An alternative direct proof makes use of the fact that the $U_i's$ are sums of i.i.d. random variables. □

The density in (6.1) motivates the notion of exponential tilting to describe many common nonparametric statistics. Suppose that a random variable X has density given by

$$\pi_\theta(x) = \exp(\theta x - K(\theta))\, g_0(x)$$

where $K(\theta)$ is a normalizing function and $g_0(x)$ is the carrying density whenever $\theta = 0$. We record in Table 6.2 various examples of the change of measure for different choices of carrying densities $g_0(x)$. The normalizing constant $K(\theta)$ is the cumulant generating function of $g_0(x)$ computed as the logarithm of its moment generating function:

$$K(\theta) = \log E_0 [\exp(\theta X)] .$$

We note that $K(\theta)$ is a strictly convex function such that

$$E_\theta[X] = K'(\theta), Var_\theta[X] = K''(\theta)$$

Example 6.1. Consider a random sample of size n from a normal distribution with mean μ and variance σ^2. We would like to test a null hypothesis on the mean μ. Under the smooth alternative formulation, the log of the likelihood function as a function of θ is proportional to

$$\left[n\theta\left(\bar{x}_n - \mu\right) - \frac{1}{2}\sigma^2\theta^2\right]$$

where $\bar{x}_n$ is the sample mean. Setting the derivative of the log with respect to θ equal to 0 shows that the maximum likelihood estimate of θ is

$$\hat{\theta} = \frac{n\left(\bar{x}_n - \mu\right)}{\sigma^2}.$$

Consequently, $\hat{\theta}$ represents the shift of the sample mean from the null hypothesis mean μ as specified by the density $g_0(x)$. The score statistic is given by the derivative of the log likelihood evaluated at $\theta = 0$:

$$n\left[\bar{x}_n - \mu\right]$$

Moreover, from the above equation, the information function is given by $K''(0) = n\sigma^2$. Hence, the Rao score statistic becomes

$$W = n^2\left[\bar{X}_n - \mu\right]\left(K''(0)\right)^{-1}\left[\bar{X}_n - \mu\right] \tag{6.5}$$

$$= \frac{n}{\sigma^2}\left[\bar{X}_n - \mu\right]^2 \tag{6.6}$$

The null hypothesis is rejected for large values of W which asymptotically as $n \to \infty$, has a chi-square distribution with one degree of freedom. The advantage of the Rao score test statistic (6.5) is that all the derivatives are computed under $\theta = 0$. Asymptotically, it is equivalent both to the generalized likelihood ratio test and to the Wald test. It is seen then that (6.5) is the usual two-sided test statistic for the mean of a normal. We may generalize the previous sections to the case of composite hypotheses.

6.2. Smooth Models for Discrete Distributions

The smooth alternative model previously described also extends to discrete distributions where we may utilize the orthogonal Chebyshev polynomials in (6.1). We may thus test the null hypothesis of uniformity over the integers $0, 1, \ldots, n$ against an alternative defined in terms of the Chebyshev polynomials. To motivate their use, we indicate below that the Chebyshev polynomials were used to generate the hypergeometric distribution [Alvo and Cabilio, 2000].

The Chebyshev polynomials over the integers $x = 0, 1, \ldots, n$ may be expressed in descending factorial form as

$$h_j(x, n) = \sum_{m=0}^{j}(-1)^m\binom{j}{m}\binom{j+m}{m}\binom{x}{m}\binom{n}{m}^{-1} \tag{6.7}$$

Table 6.2.: Change of measure distributions

Density $g_0(x)$	$\pi(x;\theta)$	$K(\theta)$
$\mathcal{N}(\mu,\sigma^2)$	$\mathcal{N}(\mu+\theta\sigma^2,\sigma^2)$	$\frac{\theta^2\sigma^2}{2}+\theta\mu$
$\beta\exp-\beta x, x>0,\beta>0$	$(\beta-\theta)\exp-(\beta-\theta)x,\theta<\beta$	$-\log\left(1-\frac{\theta}{\beta}\right)$
$\binom{n}{p}p^x(1-p)^{n-x}, x=0,\ldots,n$	$\binom{n}{p}p_\theta^x(1-p_\theta)^{n-x}, x=0,\ldots,n, 0<p_\theta<1, p_\theta=\frac{pe^\theta}{1-p+pe^\theta}$	$-\log(1-p_\theta)$
$\frac{e^{-\mu}\mu^x}{x!},\mu>0,x=0,1,\ldots$	$\frac{e^{-\mu_\theta}\mu_\theta^x}{x!},\mu_\theta=\mu e^\theta,x=0,1,\ldots$	$\mu(e^\theta-1)$
$\frac{\beta^\alpha}{\Gamma(\alpha)}x^{\alpha-1}e^{-\beta x},\alpha,\beta>0$	$\frac{(\beta-\theta)^\alpha}{\Gamma(\alpha)}x^{\alpha-1}e^{-(\beta-\theta)x},\alpha,\beta-\theta>0$	$-\alpha\log\left(1-\frac{\theta}{\beta}\right)$
$\mathcal{N}(\mu,\Sigma)$	$\mathcal{N}(\mu+\Sigma\theta,\Sigma)$	$\frac{\theta^2\Sigma}{2}+\theta\mu$
$\frac{1}{2^{r/2}\Gamma(r/2)}x^{r/2-1}e^{-x/2},x,r>0$	$\frac{(1/2-\theta)^{r/2}}{\Gamma(r/2)}x^{r/2-1}e^{-(1/2-\theta)x},r>0,\theta<1/2$	$(-r/2)\log(1-2\theta),\theta<\frac{1}{2}$
$(1-p)^{x-1}p,x=1,2,\ldots;p>0$	$(1-p_\theta)^{x-1}p_\theta,x=1,2,\ldots;p_\theta=(1-\theta-\log(1-p))>0$	$\log\left[\frac{pe^\theta}{1-(1-p)e\theta}\right]$

where $0 \leq j \leq n$. For these polynomials we have the recursions

$$\begin{aligned} h_0(x,n) &= 1, h_1(x,n) = 1 - \frac{2x}{n}, \\ (j+1)(n-j)h_{j+1}(x,n) &= (2j+1)(n-2x)h_j(x,n) - j(n+j+1)h_{j-1}(x,n). \end{aligned}$$

For every n, the vectors

$$\boldsymbol{\varepsilon}_j = (h_j(0,n), \ldots, h_j(n,n))^T \text{ for } j = 0, 1, \ldots, n$$

form a basis in $(n+1)$-dimensional space. Hence, any function $\{g(x)\}$ defined over the integers $x = 0, \ldots, n$, can be expressed in vector notation:

$$\boldsymbol{g} = \sum_{j=0}^{n} g_j \boldsymbol{\varepsilon}_j$$

where the vector $\boldsymbol{g} = (g(0), g(1), \ldots, g(n))^T$ and the g_j are obtained from the relation

$$g_j = \frac{\boldsymbol{g}^T \boldsymbol{\varepsilon}_j}{||\boldsymbol{\varepsilon}_i||^2}.$$

Equivalently, for each x

$$g(x) = \sum_{j=0}^{n} g_j h_j(x,n) \text{ for } x = 0, 1, \ldots, n.$$

The Chebyshev polynomials have been particularly useful in quickly and easily generating complete tables of values of the hypergeometric distribution given by

$$p(x; M, N) = \frac{\binom{M}{x}\binom{N-M}{n-x}}{\binom{N}{n}} \text{ for } \max(0, n-N+M) \leq x \leq \min(n, M).$$

The connection between the Chebyshev polynomials and the hypergeometric distribution is based on the following theorem proven in Alvo and Cabilio [2000].

Theorem 6.2. *The expected value of a function of a hypergeometric variable with parameters (M, N, n) is equal to a linear combination of the first n Chebyshev polynomials in parameters M, N as*

$$E[g(X)] = \sum_{j=0}^{n} g_j h_j(M,N).$$

Proof. The proof proceeds by showing that

$$\sum_{x=0}^{n} h_j(x,n) p(x; M, N) = h_j(M,N) \text{ for all } j = 0, 1, \ldots, n, \text{and } M = 0, 1, \ldots, N.$$

As a corollary, we have that if $g(x) = I(X = x)$, then[1]

$$p(x; M, N) = \sum_{j=0}^{n} g_j h_j(M, N) \tag{6.8}$$

□

The model in (6.1) can also be adapted to apply to categorical data. Suppose we have m categories and that

$$\pi_\theta(\boldsymbol{x}_j) = \exp\left(\theta^T \boldsymbol{x}_j - K(\theta)\right) p_j, j = 1, \ldots, m \tag{6.9}$$

where $\boldsymbol{x}_j$ is the jth value of a $k-$ dimensional random vector $\boldsymbol{X}$ and $\boldsymbol{p} = (p_j)$ denotes the vector of probabilities under the null distribution when $\theta = \theta_0$. Here $K(\theta)$ is a normalizing constant for which

$$\sum_j \pi_\theta(\boldsymbol{x}_j) = 1$$

Let $A = [\boldsymbol{x}_i, \ldots, \boldsymbol{x}_m]$ be the $k \times m$ matrix of possible vector values of X. Then under the distribution specified by $\boldsymbol{p}$,

$$\begin{aligned} \sum \equiv Cov(\boldsymbol{X}) &= E_p(\boldsymbol{X} - E\boldsymbol{X})(\boldsymbol{X} - E\boldsymbol{X})^T & (6.10) \\ &= A(diag(\boldsymbol{p}))A^T - (A\boldsymbol{p})(A\boldsymbol{p})^T & (6.11) \end{aligned}$$

where the expectation is with respect to the model (6.9)

Define

$$\boldsymbol{\pi}(\theta) = (\pi_\theta(\boldsymbol{x}_1), \ldots, \pi_\theta(\boldsymbol{x}_m))^T$$

and suppose that the probability vector $\boldsymbol{\pi}(\theta)$ is a function of a vector parameter θ as in (6.9). We would like to test

$$H_0 : \theta = 0 \; vs \; H_1 : \theta \neq 0$$

Letting $\boldsymbol{N}$ denote a multinomial random variable with parameters $(n, \boldsymbol{\pi}(\theta))$, we see that the log likelihood as a function of θ is, apart from a constant, proportional to

$$\begin{aligned} \sum n_j \log(\pi_\theta(\boldsymbol{x}_j)) &= \sum n_j \left(\theta^T \boldsymbol{x}_j - K(\theta)\right) \\ &= \theta' \left(\sum n_j \boldsymbol{x}_j\right) - nK(\theta) \end{aligned}$$

The score vector under the null hypothesis is then given by

$$\begin{aligned} U(\theta) &= \sum N_j \left(\frac{1}{\pi_j(\theta)} \frac{\partial \pi_j(\theta)}{\partial \theta}\right) \\ &= T(\boldsymbol{N} - n\boldsymbol{p}) \end{aligned}$$

[1]The R package hypersampleplan with subroutine hypergeotable (N, n) can be used to compute tables of hypergeometric probabilities simultaneously for all values of x=$0, 1, \ldots, n$ and $M = 0, 1, \ldots, N$.

Under the null hypothesis,

$$E\left[U\left(\theta\right)\right]=0$$

and the score statistic is given by

$$\begin{aligned}\frac{1}{n}\left[A(\boldsymbol{N}-n\boldsymbol{p})\right]^{T}\Sigma^{-1}\left[A(\boldsymbol{N}-n\boldsymbol{p})\right] &= \\ \frac{1}{n}\left[(\boldsymbol{N}-n\boldsymbol{p})\right]'\left(A^{T}\Sigma^{-1}A\right)\left[(\boldsymbol{N}-n\boldsymbol{p})\right] &\xrightarrow{\mathcal{L}} \chi_{r}^{2}\end{aligned} \tag{6.12}$$

where Σ^{-1} is the inverse matrix of Σ and $r=rank\left(A^{T}\Sigma^{-1}A\right)$.

Alvo [2015] exploited the model in (6.9) to obtain an alternative derivation of the Friedman statistic where the categories are identified as the $t!$ rankings of the integers $1,\ldots,t$. See Alvo et al. [2018].

Figure 6.1.: Rejection sampling

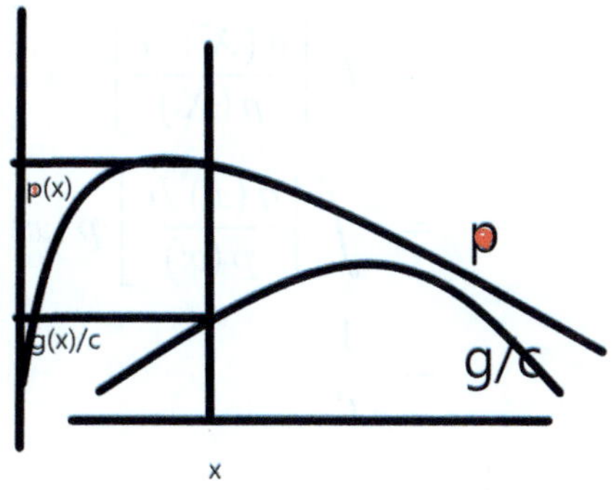

6.3. Rejection Sampling

The acceptance-rejection algorithm attributed to Herman Rubin [Casella et al., 2004] is designed to simulate values from a target posterior density $g\left(x\right)$, whose form is known but otherwise difficult to simulate from. We may instead make use of a proposal density $p\left(x\right)$ from which it is easier to simulate. At each iteration, the acceptance-rejection algorithm begins by randomly selecting a point x from $p\left(x\right)$ and sampling independently a random variable U from a uniform on $(0,1)$. We accept that value x as coming from $g\left(x\right)$ whenever

$$\begin{aligned}U &< \frac{g\left(x\right)/c}{p\left(x\right)} \\ &= \frac{L}{c}\end{aligned}$$

where $L=\frac{g}{p}$ is the likelihood ratio and the constant c is chosen such that $\frac{g(x)/c}{p(x)} < 1$ for all x as shown in Fig. 6.1. We see that points below $g(x)/c$ more likely belong to the target density. In fact, a relative frequency plot of such points will result in the reconstruction of the target density itself.

In the next theorem, we demonstrate why this algorithm works.

Theorem 6.3. $P\left[U < \frac{g(x)/c}{p(x)}\right] = \frac{1}{c}$

Proof. The probability of accepting a value X is

$$\begin{aligned} P\left[U < \frac{g(X)/c}{p(X)}\right] &= E\left\{E\left[I_{\left(U<\frac{g(x)/c}{p(x)}\right)}|X=x\right]\right\} \\ &= EP\left[U < \frac{g(x)/c}{p(x)}|X=x\right] \\ &= E\left[\frac{g(X)/c}{p(X)}\right] \\ &= \int \left[\frac{g(x)/c}{p(x)}\right] p(x)\,dx \\ &= \frac{1}{c} \end{aligned}$$

Consequently, we must choose $c \geq 1$. □

Generally, the proposal density should be chosen to resemble the target density in terms of location and spread but which is easier to sample from. The number of samples required to obtain an accepted value follows a geometric distribution with probability $1/c$ of success, and it follows that the average number of samples required is therefore c. One important question is how should one choose the proposal density. One approach is to embed the target density $g(x)$ into an exponential family of proposal densities $p_\theta(x)$ defined by

$$p_\theta(x) = \exp(\theta x - K(\theta))\, g(x).$$

This is another use of exponential tilting. The likelihood ratio in that case is given by

$$\begin{aligned} L(\theta) &= \frac{g(x)}{p_\theta(x)} \\ &= \exp(-\theta x + K(\theta)) \end{aligned}$$

We interpret $K(\theta)$ to be the cumulant generating function of the target density, that is,

$$\exp[K(\theta)] = \int \exp(\theta x)\, g(x)\, dx$$

In that connection, the moment generating function of the proposal density is given by

$$\begin{aligned}\int \exp(\eta x)\, p_\theta(x)\, dx &= \int \exp(\eta x) \exp(\theta x - K(\theta))\, g(x)\, dx \\ &= \int \exp((\eta+\theta)x - K(\theta))\, g(x)\, dx \\ &= \exp[K(\eta+\theta) - K(\theta)]\end{aligned}$$

From this expression, we may derive important properties of the proposal distribution as we shall see in the examples below. We consider an example whereby we wish to sample from a truncated normal.

Example 6.2. (Wikipedia) Let

$$g\left(x;\mu,\sigma^2\right) = N\left(\mu,\sigma^2\right)$$

and suppose X has a truncated $N(\mu,\sigma^2)$ where $X \geq b > \mu$. Hence, the density of X is given by

$$g\left(x;\mu,\sigma^2\right)\frac{I_{(x\geq b)}}{P(X\geq b)}$$

The moment generating function of the proposal density $p_\theta(x)$ and from which we will sample is

$$\begin{aligned}\psi_\theta(\eta) &= \exp[K(\eta+\theta) - K(\theta)] \\ &= \eta\left(\mu+\theta\sigma^2\right) + \frac{\sigma^2\eta^2}{2}\end{aligned}$$

which is recognized to be $N(\mu+\theta\sigma^2,\sigma^2)$. We may choose the parameter θ in a convenient way so as to center the proposal density at the mean $b = \mu+\theta\sigma^2$. Hence, we choose

$$\theta^* = \frac{b-\mu}{\sigma^2}$$

Note that

$$P(X\geq b) = P\left(Z \geq \frac{b-\mu}{\sigma}\right) \leq \sqrt{2\pi}\frac{e^{-\left(\frac{b-\mu}{\sigma}\right)^2/2}}{\frac{b-\mu}{\sigma}}$$

The likelihood ratio is for all x:

$$\begin{aligned} L(\theta^*) &= \frac{g(x;\mu,\sigma^2)}{g(x;\mu+\theta^*\sigma^2,\sigma^2)} \frac{I_{(x\geq b)}}{P(X\geq b)} \\ &= \frac{\exp(-\theta^* x + K(\theta^*))}{P(X\geq b)} I_{(x\geq b)} \\ &\leq \frac{\exp(-\theta^* b + K(\theta^*))}{P(X\geq b)} \\ &\leq \left(\frac{b-\mu}{\sigma}\right) \equiv c \end{aligned}$$

since $\theta^* > 0$. Hence, the simulation consists of sampling x from $N(b,\sigma^2)$ and independently $U \sim U(0,1)$. Accept x as coming from the truncated normal whenever

$$U \leq \frac{L(\theta^*)}{c}$$

The average number of iterations required under this approach is of order $c = \left(\frac{b-\mu}{\sigma}\right)$, whereas under the traditional approach, the number of iterations is

$$\frac{1}{P(X\geq b)} = ce^{\left(\frac{b-\mu}{\sigma}\right)^2/2}$$

which is much larger. Hence, the change of measure approach reduces the number of iterations considerably. We may use the Learn Bayes package [Albert, 2009] available in R to implement the acceptance-rejection algorithm.

Example 6.3. In this example, we develop the acceptance-rejection algorithm for a truncated gamma distribution. Suppose that we would like to sample from a truncated *Gamma* (α,β), where $X \geq b > \alpha/\beta$ and α and β are the shape and rate parameters, respectively. The usual *Gamma* (α,β) has density

$$g(x;\alpha,\beta) = \frac{\beta^\alpha}{\Gamma(\alpha)} x^{\alpha-1} e^{-\beta x}, x>0, \alpha>0, \beta>0$$

and moment generating function

$$\exp[K(\theta)] = \left(1-\frac{\theta}{\beta}\right)^{-\alpha}, \theta < \beta.$$

The proposal density is given by

$$p_\theta(x) = e^{\theta x - K(\theta)} g(x)$$

which has moment generating function

$$\psi_\theta(\eta) = \left(1-\frac{\theta+\eta}{\beta}\right)^{-\alpha} - \left(1-\frac{\theta}{\beta}\right)^{-\alpha}, \theta > \beta$$

The mean and variance of the proposal density are given respectively by

$$\left(\frac{\alpha_0}{\beta_0}\right), \quad \frac{\alpha_0(\alpha+1)}{\beta_0^2} - \left(\frac{\alpha_0}{\beta_0}\right)^2 \tag{6.13}$$

where $\theta < \beta$ and

$$\alpha_0 = \alpha\left(1-\frac{\theta}{\beta}\right)^{-\alpha}, \beta_0 = \beta\left(1-\frac{\theta}{\beta}\right)$$

The mean and variance in 6.13 suggest using $Gamma(\alpha_0, \beta_0)$ as the proposal density. In that case, choose θ so that

$$\begin{aligned} b &= \alpha_0/\beta_0, \\ &= \frac{\alpha}{\beta\left(1-\frac{\theta}{\beta}\right)^{\alpha+1}} \end{aligned}$$

that is

$$\theta^* = \beta\left(1-\left(\frac{\alpha}{b\beta}\right)^{\frac{1}{\alpha+1}}\right) > 0.$$

Now in order to compute the bound c for all x, we need an expression for $P(X \geq b)$. Since

$$\int_b e^{-x\beta} x^{\alpha-1} dx \sim \beta^{-1} e^{-b\beta} b^{\alpha-1},$$

we have

$$\begin{aligned} \frac{g(x)}{p_\theta(x) P(X \geq b)} I_{(x\geq b)} &\leq \frac{\Gamma(\alpha_0)}{\beta_0^{\alpha_0} e^{-x\beta_0} x^{\alpha_0-1}} \frac{e^{-x\beta} x^{\alpha-1}}{\int_b e^{-x\beta} x^{\alpha-1} dx} \\ &\leq \frac{\Gamma(\alpha_0)}{\beta_0^{\alpha_0}} \frac{e^{-x(\beta-\beta_0)} x^{\alpha-\alpha_0}}{\int_b e^{-x\beta} x^{\alpha-1} dx} \end{aligned}$$

Since $\beta > \beta_0$, $\alpha > \alpha_0$, the right-hand side is a decreasing function of x which takes its maximum at $x = b$ and we have

$$\frac{\Gamma(\alpha_0)}{\beta_0^{\alpha_0}} \frac{e^{-x(\beta-\beta_0)} x^{\alpha-\alpha_0}}{\int_b e^{-x\beta} x^{\alpha-1} dx} \leq \frac{\beta}{g(b; \alpha_0, \beta_0)} \equiv c$$

Hence, the simulation consists of sampling x from $Gamma(\alpha_0, \beta_0)$ and independently

$U \sim U(0,1)$. Accept x as coming from the truncated Gamma whenever

$$U \leq \frac{L(\theta^*)}{c} I_{(x \geq b)}$$

The average number of iterations is of order c, whereas under the naive method, the average number of iterations needed is

$$\frac{1}{P(X \geq b)} = \frac{\beta}{e^{-b\beta} b^{\alpha-1}} c$$

which is much larger. Hence, once again, the change of measure approach reduces the number of iterations considerably.

An interesting exercise would consist of developing an accept-rejection algorithm for sampling for the Laplace density given by

$$g(x) = \frac{\alpha}{2} \exp -\alpha |x|, \alpha > 0, |x| < \infty$$

Johnson et al. [1994, 153].

Rejection sampling does have a serious limitation because finding a reasonable bound c over the entire space may not be possible. Moreover, a very large value of c makes the method inefficient since the acceptance probability would be small.

6.4. Tweedie's Formula: Univariate Case

Empirical Bayes does not require knowledge of the prior or hyperprior distributions and hence may be viewed as an alternative to hierarchical Bayes. In this section, we consider Tweedie's formula which has played an important role in empirical Bayes methodology. It provides an expression for the posterior mean and posterior variance as a function of the marginal distribution.

Consider the special case of exponential families:

$$\pi_\theta(y) = e^{\theta y - K(\theta)} g_0(y), \theta \in \Theta, y \in \mathcal{Y}$$

and suppose that $\pi(\theta)$ is a prior density on θ with respect to Lebesgue measure on Θ. We do not assume knowledge of the prior. The posterior density of θ is given by

$$\begin{aligned} \pi_y(\theta) &= \frac{e^{\theta y - K(\theta)} \pi(\theta) g_0(y)}{g_Y(y)} \\ &= e^{\theta y - K(\theta) - \lambda(y)} \pi(\theta) \end{aligned}$$

where

$$\lambda(y) = \log\left(\frac{g_Y(y)}{g_0(y)}\right)$$

The posterior is seen to be once again a member of the exponential family but with cumulant function $\lambda(y)$. The marginal density is given by

$$g_Y(y) = \int \pi_\theta(y)\, \pi(\theta)\, d\theta.$$

Set

$$l_0(y) \equiv \log g_0(y),\, l(y) \equiv \log g_Y(y).$$

Then, a direct differentiation of the marginal with respect to y yields Tweedie's formula for the posterior mean and posterior variance in terms of derivatives of $l_0(y)$ and $l(y)$:

$$E[\theta|y] = l'(y) - l_0'(y) \tag{6.14}$$

and

$$Var[\theta|y] = l''(y) - l_0''(y) \tag{6.15}$$

Efron [2011] made use of (6.14) and (6.15) in an empirical Bayes application in genetics as described in the next example.

Example 6.4. The following application to micro-arrays appeared in Efron [2011].

Suppose $y \sim N(\mu, \sigma^2), \sigma^2$ known, and $\mu \sim \pi(\mu)$. Noting that

$$\pi_\mu(y) = \exp\left(\frac{\mu y}{\sigma^2} - \frac{\mu^2}{2\sigma^2}\right)\left[\frac{1}{\sqrt{2\pi}\sigma}\exp\left(-\frac{y^2}{2\sigma^2}\right)\right]$$

so that $\theta = \frac{\mu}{\sigma^2}$ and we have

$$g_0(y) = \frac{1}{\sqrt{2\pi}\sigma}\exp\left(-\frac{y^2}{2\sigma^2}\right)$$

Hence,

$$l_o'(y) = -\frac{y}{\sigma^2}$$

and

$$E\left(\frac{\mu}{\sigma^2}|y\right) = l'(y) + \frac{y}{\sigma^2} \tag{6.16}$$

Similarly, the posterior variance of μ is given by

$$Var\left[\frac{\mu}{\sigma^2}|y\right] = l''(y) + \frac{1}{\sigma^2} \tag{6.17}$$

The quantity $\sigma^2 l'(y)$ for the posterior mean is called the Bayes bias correction with respect to the maximum likelihood estimator. Generally, if the prior is not specified, the quantity $l(y)$ which depends on the marginal distributions is unknown. However, in the empirical Bayes context, it can be estimated from the data $y_1, \ldots, y_N$ labeled features, which often requires N to be large.

As an example, in a study of prostate cancer for men, gene expression levels were recorded for each patient for $N = 6033$ genes. There were $n_1 = 50$ men serving as controls and $n_2 = 52$ male patients who had prostate cancer. Let t_i be the two-sample t statistic comparing patients with controls for gene i, $i = 1, \ldots, N$. Under the null hypothesis that there is no difference in gene expression level between the two groups, $t_i \sim T_{100}$ distribution with $n_1 + n_2 - 2 = 100$ degrees of freedom. For ease of computation, we may convert to normal statistics and set

$$z_i = \Phi^{-1}\left(F_{100}\left(t_i\right)\right), i = 1, \ldots, N$$

In the exponential model formulation described above, y corresponds to z. The z's can now be used to produce a histogram from which the marginal density can be estimated using Lindsey's method [Efron, 2011]. In this study, it was reasonably assumed as a consequence that $z \sim N(\mu, 1)$. Under the null hypothesis, the prostrate effect $\delta = 0$.

A more accurate estimation of the derivatives of the logarithms of the marginal densities $l'(y), l''(y)$, can be obtained by recalling the Pearson system of distributions (2.2) whereby a density $f(x)$ may be determined from the first four central moments, namely,

$$\frac{d \log f(x)}{dx} = \frac{1}{f(x)} \frac{df(x)}{dx} = \frac{x - a}{b_0 + b_1 x + b_2 x^2}$$

where the constants are defined

$$\begin{aligned} b_1 &= a = -\frac{\sqrt{\mu_2}\sqrt{\beta_1}\left(\beta_2 + 3\right)}{A} \\ b_0 &= -\frac{\mu_2\left(4\beta_2 - 3\beta_1\right)}{A} \\ b_2 &= -\frac{\left(2\beta_2 - 3\beta_1 - 6\right)}{A} \end{aligned}$$

with

$$A = 10\beta_2 - 12\beta_1 - 18$$

and

$$\beta_1 = \frac{\mu_3^2}{\mu_2^3}$$

$$\beta_2 = \frac{\mu_4}{\mu_2^2}$$

The application of the Pearson system to the estimation of the marginal distribution in Tweedie's formula is then clear: given the first four central moments, we estimate $l'(y)$ by

$$\hat{l}'(y) = -\frac{y-a}{b_0 + b_1 y + b_2 y^2}$$

As well,

$$\hat{l}''(y) = \frac{b_2 y^2 - 2ab_2 y - (ab_1 + b_0)}{(b_0 + b_1 y + b_2 y^2)^2}$$

In this way, there is no need to identify the approximating distribution. The use of the Pearson system has three distinct advantages. First, it eliminates the need to determine the entire marginal distribution. Second, it makes use of the first four moments of a distribution to determine the derivative of the logarithm of the density. The latter is important in situations where the variance of a distribution depends on the kurtosis, and third, the number of features (or genes in our application) need not be large. Alvo [2021] illustrates this application on Efron's prostrate data from which the following values were calculated:

$$\mu_1 = 0.0030, \mu_2 = 1.2885, \beta_1 = 0.0017, \beta_2 = 3.6445$$

Hence,

$$\hat{a} = 0.017116, \hat{b}_0 = 1.019168, \hat{b}_1 = 0.017116, \hat{b}_2 = 0.069679, A = 18.42417$$

Figure 6.2.: Posterior mean and variance for the prostrate data

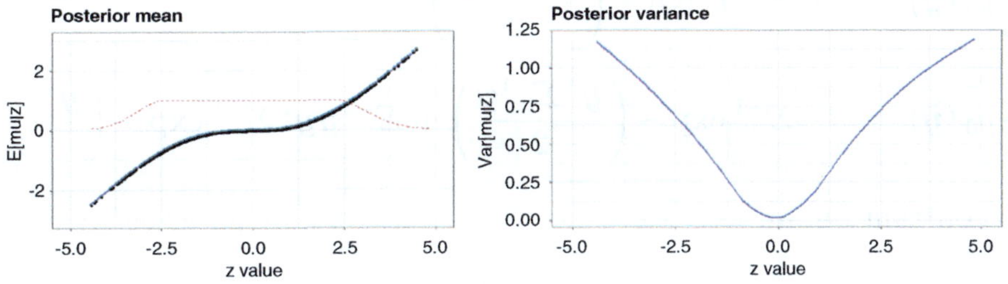

Plots of the posterior mean and variance against z are exhibited in Fig. 6.2. The plot of the posterior mean is flatter compared with Efron's plot and the values are closer to the overall mean 0. There are some absolute values which exceed 2 and those point to genes that require further investigation. The dashed curve is an estimate of the local false discovery rate which is the conditional probability of the null distribution being true given z. This rate decreases from a value of 1 near $z = 0$ to a value of 0 at the extremes. The plot of the posterior variance is smallest at 0 and grows on either end as z gets further away from 0. The variance is a measure of uncertainty/error.

6.5. Tweedie's Formula: Multivariate Case

The multivariate form of Tweedie's formula for the posterior mean is obtained in a straightforward manner. We make use of (4.1) in what follows.

Suppose that $\boldsymbol{Y} \sim N_p(\boldsymbol{\mu}, \Sigma), \Sigma$, known and let $\boldsymbol{\mu}$ have prior $\pi(\boldsymbol{\mu})$. Set

$$\begin{aligned} g_0(\boldsymbol{y}) &= \frac{|\Sigma|^{-1/2}}{(2\pi)^{p/2}} \exp-\left(\frac{\boldsymbol{y}^T\Sigma^{-1}\boldsymbol{y}}{2}\right) \\ g_{\boldsymbol{Y}}(\boldsymbol{y}) &= \frac{|\Sigma|^{-1/2}}{(2\pi)^{p/2}} \int \exp-\frac{1}{2}\left((\boldsymbol{y}-\boldsymbol{\mu})^T\Sigma^{-1}(\boldsymbol{y}-\boldsymbol{\mu})\right)\pi(\boldsymbol{\mu})\,d\boldsymbol{\mu} \\ l(\boldsymbol{y}) &= \log g_{\boldsymbol{Y}}(\boldsymbol{y}) \end{aligned}$$

Note that

$$\begin{aligned} l'(\boldsymbol{y}) &= \frac{g'_{\boldsymbol{Y}}}{g_{\boldsymbol{Y}}} \\ l''(\boldsymbol{y}) &= \frac{g''_{\boldsymbol{Y}}}{g_{\boldsymbol{Y}}} - \left(\frac{g'_{\boldsymbol{Y}}}{g_{\boldsymbol{Y}}}\right)\left(\frac{g'_{\boldsymbol{Y}}}{g_{\boldsymbol{Y}}}\right)^T \\ &= \frac{g''_{\boldsymbol{Y}}}{g_{\boldsymbol{Y}}} - l'(\boldsymbol{y})\,(l'(\boldsymbol{y}))^T \end{aligned}$$

Ignoring the constant when differentiating and noting that Σ is symmetric, we have

$$\begin{aligned} \frac{\partial}{\partial \boldsymbol{y}} g_0(\boldsymbol{y}) &= -\Sigma^{-1}\boldsymbol{y}\exp-\left(\frac{\boldsymbol{y}^T\Sigma^{-1}\boldsymbol{y}}{2}\right) \\ \frac{\partial^2}{\partial \boldsymbol{y}\partial \boldsymbol{y}^T} g_0(\boldsymbol{y}) &= -\Sigma^{-1}\exp-\left(\frac{\boldsymbol{y}^T\Sigma^{-1}\boldsymbol{y}}{2}\right) + \Sigma^{-1}\boldsymbol{y}\boldsymbol{y}^T\Sigma^{-1}\exp-\left(\frac{\boldsymbol{y}^T\Sigma^{-1}\boldsymbol{y}}{2}\right) \end{aligned}$$

Ignoring the constant again,

$$\frac{\partial}{\partial \boldsymbol{y}} g_{\boldsymbol{Y}}(\boldsymbol{y}) = -\Sigma^{-1}\int(\boldsymbol{y}-\boldsymbol{\mu})\exp-\frac{1}{2}\left((\boldsymbol{y}-\boldsymbol{\mu})^T\Sigma^{-1}(\boldsymbol{y}-\boldsymbol{\mu})\right)\pi(\boldsymbol{\mu})\,d\boldsymbol{\mu}$$

Also,

$$\begin{aligned} \frac{\partial^2}{\partial \boldsymbol{y}\partial \boldsymbol{y}^T} g_{\boldsymbol{Y}}(\boldsymbol{y}) &= \Sigma^{-1}\int(\boldsymbol{y}-\boldsymbol{\mu})(\boldsymbol{y}-\boldsymbol{\mu})^T\Sigma^{-1}\exp-\frac{1}{2}\left((\boldsymbol{y}-\boldsymbol{\mu})^T\Sigma^{-1}(\boldsymbol{y}-\boldsymbol{\mu})\right)\pi(\boldsymbol{\mu})\,d\boldsymbol{\mu} \\ &- \Sigma^{-1}\int\exp-\frac{1}{2}\left((\boldsymbol{y}-\boldsymbol{\mu})^T\Sigma^{-1}(\boldsymbol{y}-\boldsymbol{\mu})\right)\pi(\boldsymbol{\mu})\,d\boldsymbol{\mu} \end{aligned}$$

Hence,

$$\begin{aligned} \frac{g'_{\boldsymbol{Y}}(\boldsymbol{y})}{g_{\boldsymbol{Y}}(\boldsymbol{y})} &= \Sigma^{-1} E(\boldsymbol{\mu}|\boldsymbol{y}) - \boldsymbol{y} \\ E(\boldsymbol{\mu}|\boldsymbol{y}) &= \boldsymbol{y} + \Sigma l'(\boldsymbol{y}) \end{aligned} \tag{6.18}$$

Now for the posterior variance

$$\begin{aligned} Var(\boldsymbol{\mu}|\boldsymbol{y}) &= E\left[(\boldsymbol{\mu} - E(\boldsymbol{\mu}|\boldsymbol{y}))(\boldsymbol{\mu} - E(\boldsymbol{\mu}|\boldsymbol{y}))^T\right]|\boldsymbol{y} \\ &= E\left[(\boldsymbol{\mu} - \boldsymbol{y} - \Sigma l')(\boldsymbol{\mu} - \boldsymbol{y} - \Sigma l')^T\right]|\boldsymbol{y} \\ &= E(\boldsymbol{\mu} - \boldsymbol{y})(\boldsymbol{\mu} - \boldsymbol{y})^T - \Sigma(l'(\boldsymbol{y}))(l'(\boldsymbol{y}))^T \Sigma \end{aligned}$$

Substituting, we get

$$\begin{aligned} \Sigma\left(\frac{1}{g_{\boldsymbol{Y}}(\boldsymbol{y})}\frac{\partial^2}{\partial \boldsymbol{y}\partial \boldsymbol{y}^T} g_{\boldsymbol{Y}}(\boldsymbol{y})\right)\Sigma &= -\Sigma + E\left((\boldsymbol{\mu} - \boldsymbol{y})(\boldsymbol{\mu} - \boldsymbol{y})^T|\boldsymbol{y}\right) \\ &= -\Sigma + \; Var(\boldsymbol{\mu}|\boldsymbol{y}) + \Sigma(l'(\boldsymbol{y}))(l'(\boldsymbol{y}))^T \Sigma \end{aligned}$$

Hence, the posterior variance is

$$\begin{aligned} Var(\boldsymbol{\mu}|\boldsymbol{y}) &= \Sigma + \Sigma\left(\frac{1}{g_{\boldsymbol{Y}}(\boldsymbol{y})}\frac{\partial^2}{\partial \boldsymbol{y}\partial \boldsymbol{y}^T} g_{\boldsymbol{Y}}(\boldsymbol{y}) - (l'(\boldsymbol{y}))(l'(\boldsymbol{y}))^T\right)\Sigma \\ &= \Sigma + \Sigma l''(\boldsymbol{y})\Sigma \end{aligned} \tag{6.19}$$

6.6. The Saddlepoint Approximation and Notions of Information

Saddlepoint approximations originated with the paper of Daniels [1954]. The paper by Good [1957] describes saddlepoint methods for the multinomial distribution. Davidson and Hinkley [1988] describe their use in resampling methods. Recent expository papers are those of Goutis and Casella [1999]; Reid [1988].

Saddlepoint methods provide a very good approximation to the density of the mean of a distribution. Suppose we have a random sample of size n from density $f_X(x)$ which we may embed into an exponential family:

$$\pi_\phi(x) = \exp\{\phi x - K(\phi)\} f_X(x) \tag{6.20}$$

Then the density of $\bar{X}$ is given by

$$f_{\bar{X}}(\bar{x}) = \exp\left[n\left\{K(\phi) - \phi\bar{x}\right\}\right] f_{\bar{X}}(\bar{x};\phi)$$

We proceed by deriving an Edgeworth expansion Field and Ronchetti [1990] for the density $f_{\bar{X}}(\bar{x};\phi)$. Under this expansion,

$$E_{\phi}\bar{X} = K'(\phi), Var_{\phi}\bar{X} = K''(\phi)$$

In order to ensure that the approximation is not unbounded in the tails, we may choose for each value $\bar{x}$ to evaluate the expansion at the mean $\hat{\phi} = \hat{\phi}(\bar{x})$. This yields the saddlepoint approximation from Daniels [1954, 1980] for the probability density function of the mean $\bar{X}$:

$$f_{\bar{X}}(\bar{x}) \sim \sqrt{\frac{n}{2\pi K''\left(\hat{\phi}\right)}} \exp\left[n\left(K\left(\hat{\phi}\right) - \hat{\phi}\bar{x}\right)\right]$$

where K is the cumulant, $K''\left(\hat{\phi}\right)$ is the variance of a single x, and $\hat{\phi}$ is the solution to $K'\left(\hat{\phi}\right) = \bar{x}$.

The multivariate version of the saddlepoint approximation is given by

$$f_{\bar{X}}(\bar{x}) \quad \sim \quad (2\pi)^{-k/2}\left(\frac{n}{\left|K''\left(\hat{\phi}\right)\right|}\right)^{1/2} \quad \exp\left[n\left(K\left(\hat{\phi}\right) - \hat{\phi}'\bar{x}\right)\right]$$

where the saddlepoint $\hat{\phi}$ is the solution to $K'\left(\hat{\phi}\right) = \bar{x}$ and the $k \times k$ matrix

$$K''\left(\hat{\phi}\right) = \left(\frac{\partial^2 K(\phi)}{\partial\phi_i\partial\phi_j}\right)$$

has determinant $\left|K''\left(\hat{\phi}\right)\right|$.

Example 6.5. Suppose $X_1, \ldots, X_n$ is a random sample with a density from an exponential family

$$\pi_{\phi}(x) = \exp\left\{\theta^T a(x) - \psi(\theta) - d(x)\right\} \tag{6.21}$$

where the vector $A(x) = (A_1(x), \ldots, A_k(x))^T$ is the minimal sufficient statistic.

The density of $A = \sum A(X_i)$ is

$$f_T(a;\theta) = \exp\left\{\theta^T a(x) - n\psi(\theta) - h(a)\right\} \tag{6.22}$$

We may approximation $\exp\{-h(a)\}$ using the saddlepoint approximation. Specifically, we have

$$\begin{aligned} \exp\{-h(a)\} &= g_T(a;\theta)\exp\left\{\phi^T a - n\psi(\phi)\right\} \\ &\sim (2\pi)^{-k/2}\left(\left|n\psi''\left(\hat{\phi}\right)\right|\right)^{-1/2}\exp\left[\left(n\psi\left(\hat{\phi}\right) - \hat{\phi}' a\right)\right] \end{aligned} \quad (6.23)$$

where the saddlepoint equation is $n\psi'(\phi) = t$. The saddlepoint occurs at the maximum likelihood estimate $\hat{\phi} = \hat{\theta}$. Hence, we have

$$\begin{aligned} f_T(a;\theta) &= \exp\left\{\theta^T a - n\psi(\theta) - h(a)\right\} \\ &\sim (2\pi)^{-k/2}\left(\left|n\psi''\left(\hat{\theta}\right)\right|\right)^{-1/2}\exp\left\{\left(\theta - \hat{\theta}\right)^T a - n\psi(\theta) + n\psi\left(\hat{\theta}\right)\right\} \end{aligned} \quad (6.24)$$

It is of interest to relate (6.24) to notions of information. Various notions of information have been defined in the literature beginning with Fisher (1925) in connection with estimation:

$$I_F(\theta) = \int\left(-\frac{\partial^2}{\partial\theta^2}\ln f(x|\theta)\right) f(x|\theta)\,dx$$

Shannon (1948) defined

$$I_S(\theta) = \int \ln f(x|\theta)\, f(x|\theta)\,dx$$

for communication theory, whereas Kullback (1959) defined information to distinguish between two models indexed by θ_1, θ_2:

$$I_K(\theta_1, \theta_2) = \int \ln\frac{f(x|\theta_1)}{f(x|\theta_2)} f(x|\theta_1)\,dx$$

Fraser [1965] introduced a notion of information to deal principally with location problems where the density is given as $f(x-\theta)$

$$I_{Fr}(\theta|\theta_0) = \int \ln f(x-\theta)\, f(x-\theta_0)\,dx$$

Bentoumi et al. [2019] exploited the latter in connection with length-biased survival data. To highlight the differences between these various notions, suppose that $X \sim$

$N(\theta, \sigma^2)$. Then,

$$\begin{aligned} I_F(\theta) &= \frac{1}{\sigma^2} \\ I_S(\theta) &= -\frac{1}{2} + \ln\left(\frac{1}{\sqrt{2\pi}\sigma}\right) \\ I_K(\theta_1, \theta_2) &= \frac{(\theta_1 - \theta_2)^2}{2\sigma^2} \\ I_{Fr}(\theta|\theta_0) &= -\frac{1}{2} + \ln\left(\frac{1}{\sqrt{2\pi}\sigma}\right) - \frac{(\theta_0 - \theta)^2}{2\sigma^2} \end{aligned}$$

Restricting attention to the exponential family

$$\pi_\theta(y) = e^{\theta y - K(\theta) - d(y)}, \theta \in \Theta, y \in \mathcal{Y}$$

we may define the distance between two densities as the deviance:

$$\begin{aligned} D(\theta_1, \theta_2) &= 2E_{\theta_1}\left\{\log \frac{\pi_{\theta_1}(y)}{\pi_{\theta_2}(y)}\right\} \\ &= 2E_{\theta_1}\{(\theta_1 - \theta_2)y - K(\theta_1) + K(\theta_2)\} \\ &= 2\{(\theta_1 - \theta_2)\hat{\mu} - K(\theta_1) + K(\theta_2)\} \end{aligned} \tag{6.25}$$

Now,

$$\begin{aligned} \frac{\pi_\theta(y)}{\pi_{\hat{\theta}}(y)} &= e^{\left(\theta - \hat{\theta}\right)y - K(\theta) + K\left(\hat{\theta}\right)} \\ &= e^{\left(\theta - \hat{\theta}\right)\hat{\mu} - K(\theta) + K\left(\hat{\theta}\right)} \\ &= e^{-\frac{D\left(\hat{\theta}, \theta\right)}{2}} \end{aligned} \tag{6.26}$$

where we set $\theta_1 = \hat{\theta}$ and $\theta_2 = \theta$.

Hence,

$$\pi_\theta(y) = \pi_{\hat{\theta}}(y)\, e^{-\frac{D\left(\hat{\theta}, \theta\right)}{2}}$$

The saddlepoint approximation to the density of $\bar{Y}$ is

$$\pi_\theta(\bar{y}) \cong \sqrt{\frac{n}{2\pi K''\left(\hat{\theta}\right)}} \exp -\frac{n}{2V_\theta}(\bar{y} - \theta)^2$$

Choosing $\theta = \hat{\theta} = \bar{y}$, we see that

$$\pi_{\hat{\theta}}\left(\bar{y}\right) = \sqrt{\frac{n}{2\pi K''\left(\hat{\theta}\right)}}$$

so that the approximation becomes

$$\pi_{\theta}\left(\bar{y}\right) = \pi_{\hat{\theta}}\left(\bar{y}\right) e^{-\frac{nD\left(\hat{\theta},\theta\right)}{2}} \tag{6.27}$$

which coincides with (6.24).

7. Counting Data Analysis

Counting data is often presented as counts or frequencies in the form of tables which in general though not always, have no particular ordering. Contingency tables permit one to see how two categorical variables are related. Some everyday examples of categorical data may be found in E-commerce, online dating, customer service organizations, surveys and questionnaires, and personality tests. Such data are frequently collected from surveys or questionnaires where respondents often include their gender, opinions, preferences, etc. A good reference for work on categorical data are the books by Agresti [2013] and Fienberg [1980] for the analysis of contingency tables.

An early example of categorical data was in the context of genetics when Gregor Mendel in 1865 predicted the results of cross-fertilization of seeds of pea plants. He observed data that showed that round and wrinkled seed occurred in the proportion 3:1. Similarly, yellow- and green-colored pea occurred in the proportion 3:1. The round yellow variety was dominant. He predicted that if these characteristics were independent of one another, then the combinations round yellow, round green, wrinkled yellow, and wrinkled green would occur in the ratio 9:3:3:1. In an actual experiment, he observed the counts in Table 7.1 which supported his predictions. In what follows, we shall first consider modeling for categorical responses. The principal distributions which model categorical data are the binomial, the Poisson, and the multinomial. In cases where the variance in a Poisson distribution exceeds the mean, the negative binomial is used. The framework through which categorical data can be best analyzed is the generalized linear model. Specifically, this approach consists of three components: a random component Y which describes the response variable, a systematic component which specifies the explanatory variables in a regression, and a link function which indicates the relationship of the mean $\mu = E[Y]$ to the systematic component. To illustrate, suppose that $Y_1, \ldots, Y_n$ represent n independent observations from an exponential family of distributions:

$$\pi_{\theta_i}(y_i) = b(y_i)\, e^{Q(\theta_i)y_i - K(\theta_i)} \tag{7.1}$$

Table 7.1.: Mendel's cross-fertilization experiment

	Yellow	Green
Round	315	108
Wrinkled	101	32

M. Alvo, *Statistical Inference and Machine Learning for Big Data*, Springer Series in the Data Sciences, https://doi.org/10.1007/978-3-031-06784-6_7

Suppose as well that the systematic component is given by

$$\eta_i = \sum \beta_j x_{ij} \tag{7.2}$$

or in matrix notation

$$\eta = X\beta \tag{7.3}$$

and the link function is

$$g(\mu_i) = \eta_i. \tag{7.4}$$

We shall consider some examples where these components are defined.

Example 7.1. (Logit Model) Suppose that Y has a Bernoulli distribution with parameter π:

$$\begin{aligned} P(Y = y) &= \pi^y (1-\pi)^{1-y}; y = 0, 1 \\ &= e^{y \log\left(\frac{\pi}{1-\pi}\right) + \log(1-\pi_i)} \end{aligned}$$

The parameter $\log\left(\frac{\pi}{1-\pi}\right)$ is called the logit or log odds of response 1. The mean is

$$\pi = E[Y]$$

Example 7.2. (Logistic Regression) Consider the Bernoulli model and suppose that we have a vector of explanatory variables $\boldsymbol{x} = (x_1, \ldots, x_p)$. Let

$$P(Y = 1) = \pi(\boldsymbol{x})$$

Let $Y = 1$ and suppose that we model the logit as a linear function:

$$\log\left(\frac{\pi(x)}{1-\pi(x)}\right) = \alpha + \beta x$$

Then the logistic regression model with parameters α, β is given by

$$\pi(x) = \frac{\exp(\alpha + \beta x)}{1 + \exp(\alpha + \beta x)}$$

In order to see where the logistic model arises, suppose that a random variable X has the logistic distribution whose cumulative distribution function is given by

$$F(x) = \frac{\exp(x)}{1 + \exp(x)}, -\infty < x < \infty$$

Suppose as well that we have the Bernoulli variable

$$Y = \begin{cases} 1 & X \leq \alpha + \beta x \\ 0 & X > \alpha + \beta x \end{cases}$$

Then,

$$P(Y = 1) = P(X \leq \alpha + \beta x) = \frac{\exp(\alpha + \beta x)}{1 + \exp(\alpha + \beta x)}$$

We shall consider various applications of logistic regression in Part III when discussing machine learning.

Example 7.3. (Poisson Model for Count Data) Suppose that Y has a Poisson distribution with parameter equal to the mean μ:

$$\begin{aligned} P(Y = y) &= \frac{\mu^y}{y!} e^{-\mu}; y = 0, 1, \ldots \\ &= \tfrac{1}{y!} \quad e^{y \log(\mu) - \mu} \end{aligned}$$

The link function given by $Q(\mu) = \log \mu$ is in general modeled in terms of explanatory variables.

Example 7.4. (Probit Model) Suppose that instead of modeling in terms of the logistic distribution, we have that X has a standard normal distribution. Then, assuming once again

$$Y = \begin{cases} 1 & X \leq \alpha + \beta x \\ 0 & X > \alpha + \beta x \end{cases}$$

we have

$$P(Y = 1) = P(X \leq \alpha + \beta x) = \Phi(\alpha + \beta x) \tag{7.5}$$

We call (7.5) the probit model.

Example 7.5. (Over-Dispersion Poisson Model) For Poisson data, the mean and variance are equal. However, it sometimes happens that the variance is much greater than the mean. In this case, we say we have an over-dispersed model. It is customary then to consider a negative binomial distribution for Y which is given by

$$\pi_{k,\mu}(y) = \frac{\Gamma(y + k)}{\Gamma(k)\Gamma(y + 1)} \left(1 - \frac{k}{\mu + k}\right)^y \left(\frac{k}{\mu + k}\right)^k, y = 0, 1, \ldots$$

where μ, k are parameters. We calculate that

$$E[Y] = \mu, Var[Y] = \mu + \frac{\mu^2}{k}$$

so that when $k \to \infty$, we have a Poisson distribution.

7.1. Inference for Generalized Linear Models

The general linear model may be extended to take into account a dispersion parameter. Specifically, we may define the random component through the generalized density

$$\pi_{\theta_i,\phi}(y_i) = \exp\left\{\frac{[y_i\theta_i - b(\theta_i)]}{a(\phi)} + c(y_i, \phi)\right\} \tag{7.6}$$

with systematic component (7.2), link function (7.4), dispersion parameter ϕ, and natural parameter θ_i. Let

$$L = \sum_{i=1}^{n} L_i$$

where

$$L_i = \log \pi_{\theta_i,\phi}(y_i)$$

The maximum likelihood estimation of the parameters $\{\theta_i, \phi\}$ proceeds formally in the usual way by solving the differential equations:

$$\begin{aligned}\frac{\partial L_i}{\partial \theta_i} &= \frac{[y_i - b'(\theta_i)]}{a(\phi)} \\ \frac{\partial^2 L_i}{\partial \theta_i^2} &= -\frac{b''(\theta_i)}{a(\phi)}\end{aligned}$$

It follows that

$$\begin{aligned}E[Y_i] &= b'(\theta_i) \\ Var[Y_i] &= b''(\theta_i)\, a(\phi)\end{aligned}$$

The maximum likelihood estimation of the parameters $\{\beta_j\}$ from (7.2) are

$$\begin{aligned}\sum_{i=1}^{n} \frac{\partial L_i}{\partial \beta_j} &= \sum_{i=1}^{n} \frac{\partial L_i}{\partial \theta_i}\frac{\partial \theta_i}{\partial \mu_i}\frac{\partial \mu_i}{\partial \eta_i}\frac{\partial \eta_i}{\partial \beta_j} \\ &= \sum_{i=1}^{n} \frac{[y_i - \mu_i]}{Var[Y_i]}\frac{\partial \mu_i}{\partial \eta_i} = 0 \end{aligned} \tag{7.7}$$

Similarly, for the asymptotic covariance estimation, we have

$$\begin{aligned} E\left[-\frac{\partial^2 L(\beta)}{\partial\beta_k \partial\beta_j}\right] &= \sum_{i=1}^{n} E\left[-\frac{\partial^2 L_i(\beta)}{\partial\beta_k \partial\beta_j}\right] \\ &= \sum_{i=1}^{n} E\left[\frac{\partial L_i(\beta)}{\partial\beta_k}\frac{\partial L_i(\beta)}{\partial\beta_j}\right] \\ &= \sum_{i=1}^{n}\left[\frac{x_{ik}x_{ij}}{Var[Y_i]}\left(\frac{\partial\mu_i}{\partial\eta_i}\right)^2\right] \end{aligned}$$

In matrix notation, this yields

$$Cov\left[\hat{\beta}\right] = \left(X'\hat{W}X\right)^{-1}$$

where

$$\hat{W} = diag\left(\frac{\left(\frac{\partial\mu_i}{\partial\eta_i}\right)^2}{Var[Y_i]}\right)$$

is evaluated at $\hat{\beta}$. Further details are given in Agresti [2013, Chapter 4] and regularity conditions are in Cox and Hinkley [1974, Section 4.8]. In general, the maximum likelihood equations are solved numerically. We consider two specific examples.

Example 7.6. Let $n_i Y_i$ have a binomial distribution with mean $E[Y_i] = \pi_i$. Then,

$$Var[Y_i] = \frac{\pi_i(1-\pi_i)}{n_i}$$

The Eqs. (7.7) become

$$\sum_{i=1}^{n}\frac{[y_i-\mu_i]}{Var[Y_i]}\frac{\partial\mu_i}{\partial\eta_i} = \sum n_i[y_i-\pi_i]x_{ij} = 0$$

with

$$\pi_i = F\left(\sum \beta_j x_{ij}\right)$$

and F is the standard logistic distribution.

Example 7.7. Let Y_i have a Poisson distribution with mean $E[Y_i] = \mu_i$. The link function is given by

$$\eta_i = \log\mu_i.$$

and the likelihood equations (7.7) become

$$\sum[y_i-\mu_i]x_{ij} = 0$$

The maximum likelihood equations in (7.7) are nonlinear in the parameters. Consequently, solutions are numerical in nature and require the use of computer algorithms. Inference for the GLM is based on the likelihood function wherein the saturated model is tested against a null hypothesis model. The saturated model provides a separate parameter for each observation. Let $L(\hat{\mu}, y)$ denote the maximum log likelihood under the model, and let $L(y, y)$denote the maximum log likelihood under the saturated model. The test statistic is then based on

$$G^2 = -2\left[L(\hat{\mu}, y) - L(y, y)\right]$$

7.2. Inference for Contingency Tables

In a different direction, cell counts may be modeled using log-linear models in contingency tables. The latter can incorporate interactions and various levels of correlations. As an example, consider an $I \times J$ contingency table with independent observed cell counts n_{ij} assumed to have a Poisson distribution with expected cell frequencies $\mu_{ij} = n\pi_{ij}$. It is common practice to model the logarithm of the expected frequencies. Adopting the same notation as in Agresti [2013], the most complex log-linear model for which it is possible to estimate all the parameters is called the saturated model:

$$\log \mu_{ij} = \lambda + \lambda_i^X + \lambda_j^Y + \lambda_{ij}^{XY} \tag{7.8}$$

where λ is the grand mean and λ_i^X, λ_j^Y represent the differential effects of the ith row and jth column, respectively. Also, we may have conditions on the differential effects:

$$\sum_i \lambda_i^X = \sum_i \lambda_j^Y = 0, \sum_i \lambda_i^X = \sum_j \lambda_j^Y = 0.$$

Under the independence model

$$\mu_{ij} = n\pi_{i+}\pi_{+j} \tag{7.9}$$

where

$$\pi_{i+} = \sum_j \pi_{ij}, \pi_{+j} = \sum_i \pi_{ij}$$

and hence

$$\log \mu_{ij} = \lambda + \lambda_i^X + \lambda_j^Y \tag{7.10}$$

From (7.10),

$$\begin{aligned} \mu_{ij} &= e^{\lambda} e^{\lambda_i^X + \lambda_j^Y} \\ &= e^{\lambda} \pi_{i+} \pi_{+j} \end{aligned}$$

Table 7.2.: Online and in class course

	Online	In class
Pass	70	49
Fail	20	11

so that we may identify $e^{\lambda} = n$, and hence,

$$\lambda = \frac{\sum_i \sum_j \log \mu_{ij}}{IJ} = \log n$$

$$\sum_i \lambda_i^X = \sum_j \lambda_j^Y = 0.$$

Example 7.8. Consider the two-way table of data in Table 7.2 for different sections of the same course taught both online and in class. The test of independence shows $\chi_1^2 = 1.07, p\text{-}value = 0.30$.

There is a Poisson-multinomial connection. Specifically, let $\{\pi_{ij}\}$ be the set of joint probabilities:

$$\pi_{ij} = P\left(X = i, T = j\right)$$

If the n_{ij} are independent Poisson random variables with mean μ_{ij}, then conditional on $\sum \sum n_{ij} = n$,

$$\{n_{ij}\} \sim Multinomial\left(n, \{\pi_{ij}\}\right)$$

with

$$\pi_{ij} = \frac{\mu_{ij}}{\sum \sum \mu_{ij}}$$

We may compute the log odds ratio under the saturated model which compares levels (i, i') of X with levels (j, j') of Y:

$$\begin{aligned} \theta &= \log \left[\left(\frac{\mu_{ij}}{\mu_{ij'}} \right) / \left(\frac{\mu_{i'j}}{\mu_{i'j'}} \right) \right] \\ &= \lambda_{ij}^{XY} + \lambda_{i'j'}^{XY} - \lambda_{ij'}^{XY} - \lambda_{i'j}^{XY} \end{aligned}$$

We see θ only involves the interaction terms. Under independence, $\lambda_{ij}^{XY} = 0$ for all i, j and the odds ratio is equal to 1.

Example 7.9. Consider the data in the following table which records the number of deaths from a certain disease

	Treatment	Control	
Death	$a = 13$	$b = 23$	$a + b = 36$
No death	$c = 150$	$d = 125$	$c + d = 275$
	$a + c = 163$	$c + d = 148$	$N = 311$

The odds ratio is estimated to be $OR = \frac{(a/c)}{(b/d)}$. A value of OR<1 favors the treatment. The log odds is estimated to be

$$\begin{aligned}\hat{\theta} &= \log(OR) \\ &= \log\frac{(a+0.5)(d+0.5)}{(b+0.5)(c+0.5)} = -0.32\end{aligned}$$

The variance

$$\begin{aligned}Var(OR) &= \frac{1}{a+0.5} + \frac{1}{b+0.5} + \frac{1}{c+0.5} + \frac{1}{d+0.5} \\ &= 0.08\end{aligned}$$

A test of independence in $I \times J$ tables may be conducted using the likelihood ratio which compares the independence model (7.9) with the saturated model (7.8). We may express this in terms of a hypothesis testing problem:

$$H_0 \; : \; \lambda_{ij}^{XY} = 0$$

$$H_1 \; ; \; \lambda_{ij}^{XY} \neq 0$$

The likelihood function under Poisson or multinomial sampling is given by

$$L = \begin{cases} \prod_{i,j} e^{-\mu_{ij}} \frac{\mu_{ij}^{n_{ij}}}{n_{ij}!} & \textit{Poisson sampling} \\ n! \prod_{i,j} \frac{\pi_{ij}^{n_{ij}}}{n_{ij}} & \textit{Multinomial sampling} \end{cases}$$

This leads to the likelihood ratio test statistic

$$G^2 = 2\sum_{i,j} n_{ij} \log\left(\frac{n_{ij}}{\left(\frac{n_{i+}n_{+j}}{n}\right)}\right) \tag{7.11}$$

or to the asymptotically equivalent χ^2 goodness-of-fit test

$$\chi_c^2 = \sum_{i,j} \frac{\left(n_{ij} - \frac{n_{i+}n_{+j}}{n}\right)^2}{\left(\frac{n_{i+}n_{+j}}{n}\right)} \tag{7.12}$$

where $n_{i+} = \sum_j n_{ij}, n_{+j} = \sum_i n_{ij}$. For an $I \times J$ table, $c = (I-1)(J-1)$. Fienberg [1980, p.20] discusses a continuity correction to this statistic. Agresti [2013, p 590] shows that the two statistics are asymptotically equivalent. The important summary value of these tests is contained in the p-value. It should be pointed out that these tests are only reliable when used with large samples.

Example 7.10. Higher-order contingency tables can also be analyzed. For a three-way table, the saturated model takes the form

$$\log \mu_{ijk} = \lambda + \lambda_i^X + \lambda_j^Y + \lambda_k^Z + \lambda_{ij}^{XY} + \lambda_{ik}^{XZ} + \lambda_{jk}^{YZ} + \lambda_{ijk}^{XYZ} \tag{7.13}$$

Various hypotheses can be tested in (7.13). These are generally conducted using a likelihood ratio test which tests the model against the alternative saturated model. Computationally, it is convenient to make use of the software package *R*.

For small samples, there is the Fisher's exact test which is used as a general rule of thumb when at least one expected value is lower than 5 [Agresti, 2013, p.91–93]. We describe this test in the following example for two-way tables using the obvious notation $n_{11} + n_{12} = n_{1+}$ etc.

	n_{11}	n_{12}	n_{1+}
	n_{21}	n_{22}	n_{2+}
	n_{+1}	n_{+2}	n_{++}

Example 7.11. Suppose that we have independent multinomial distributions within rows. The row totals are fixed and the joint distribution of the n_{ij} is the product of the multinomials

$$\prod \left(\frac{n_{i+}!}{\prod_j n_{ij}!} \prod p_{ij}^{n_{ij}} \right) = \frac{\left(\prod_i n_{i+}!\right)\left(\prod_j p_{+j}^{n_{+j}}\right)}{\prod_i \prod_j n_{ij}!}$$

conditional on both the row and column marginals in the two-way table. This distribution depends on the parameters p_{+j}. To eliminate this dependence on these nuisance parameters, Fisher proposed to condition on their sufficient statistics the n_{+j}. The resulting conditional distribution is the hypergeometric:

$$\begin{aligned} \frac{\left(\prod_i n_{i+}!\right)\left(\prod_j n_{+j}!\right)}{n_{++}! \prod_i \prod_j n_{ij}!} &= \frac{(n_{1+}!n_{2+}!)(n_{+1}!n_{+2}!)}{n_{++}!n_{11}!n_{12}!n_{21}!n_{22}!} \\ &= \frac{\binom{n_{1+}}{n_{11}}\binom{n_{2+}}{n_{+1}-n_{11}}}{\binom{n_{++}}{n_{+1}}} \end{aligned}$$

Consequently, the only variable is seen to be n_{11}. The tea story described in Agresti [2013, p.91] recalls a colleague of Fisher's who claimed that when drinking tea she could distinguish between whether milk or tea was added to the cup first.

Notes 1. Tests for independence which reject the null hypothesis are often followed up by a study of the residuals in order to determine the degree of association. See Van der Vaart [2000, chapter 17].

2. See Bishop and Fienberg [2007] for related work on discrete multivariate analysis.

3. See Good [1976] for related work on Dirichlet distributions and their mixtures to contingency tables

7.3. Two-Way Ordered Classifications

A public opinion survey was carried out in early 1999 in Hong Kong by the Social Science Research Center of the University of Hong Kong through telephone interviews. It was of interest to determine whether the age of the respondents is related to the level of satisfaction of the Policy Address of the Chief Executive of the Hong Kong Special Administrative Region. The response is an ordinal variable with the seven options being 1, very satisfied; 2, satisfied; 3, neutral; 4, unsatisfied; 5, very unsatisfied; 6, not sure; and 7, refuse to answer. Options 6 and 7 will be classified as missing (nonresponse). The age at last birthday of the respondents was recorded when available. The following table shows the frequencies of female respondents classified by their responses and age groups.

We observe that 29.3% of respondents did not answer either one or both questions. One option for analyzing the data in the table is to treat it as a contingency table and to test for independence. Dropping the "missing" category from both age and response as well as level "1" for response which contains no entries, we obtain for the chi-square test statistic $\chi^2_{12} = 15.806$ with a p-value of 0.20.

An alternative option is to think of the age and the response as ordered variables whose values are scores that can be converted to ranks. The ranking thus produced may contain ties and may be partially incomplete. The objective then is to develop a nonparametric test of independence for such rankings. In Table 7.3, we have an example of a two-way ordinal classification or contingency table with ordered categories. Such data is not well analyzed by the usual chi-square test for independence. Missing values will usually be discarded automatically and the ordinal nature will also be ignored. It is difficult to provide a correlation-type statistic to indicate the direction of influence

Table 7.3.: Data from the public opinion survey

	Age group						
Response	18 – 24	25 – 34	35 – 44	45 – 54	> 54	Missing	Subtotal
1	0	0	0	0	0	0	0
2	10	12	17	16	10	5	70
3	16	29	37	21	3	7	113
4	8	9	10	7	1	3	38
5	1	1	3	1	0	0	6
Missing	10	14	19	10	14	6	73
Subtotal	45	65	86	55	28	21	300

if the two classification criteria are not independent. When the ordinal nature is taken into account, some measures of association are available and are discussed in Agresti [2013]. See also Kendall and Stuart [1979, p.593]. Here, we consider a Spearman-type correlation coefficient as an alternative to the above measures with the advantage that it utilizes all information contained in the incomplete data set.

In this context, it is natural to consider rank-based measures of correlation such as those due to Spearman and Kendall which have been used either as measures of association between two factors or as tests of independence between two random variables. Lehmann [1975] considered these test statistics when some observations are tied. Alvo and Cabilio [1993, 1995a] generalized these test statistics when the data contain missing observations but no ties. Alvo et al. [2002] considered a test of independence when the data contain both missing observations and ties under the hypothesis that the pattern of missing observations is fixed. We reconsider the problem under the hypothesis that the pattern of missing observations is random.

Consider two complete rankings of t objects $\mu = (\mu(1), \mu(2), \ldots, \mu(t))^T$ and $\nu = (\nu(1), \nu(2), \ldots, \nu(t))^T$ which can be viewed as permutations of the integers $(1, 2, \ldots, t)$. The Spearman rank correlation ρ_S Spearman [1904] between the rankings can be expressed as

$$\rho_S(\mu, \nu) = \frac{A_S(\mu, \nu)}{c_S}.$$

where

$$c_S = \frac{t(t^2-1)}{12}, \quad A_S(\mu, \nu) = \sum_{i=1}^{t} \left(\mu(i) - \frac{t+1}{2}\right)\left(\nu(i) - \frac{t+1}{2}\right).$$

In a totally analogous manner, one may define the Kendall rank correlation:

$$\rho_K(\mu, \nu) = \frac{A_K(\mu, \nu)}{c_K}.$$

where

$$c_K = \frac{t(t-1)}{2}, \quad A_K(\mu, \nu) = \sum_{i<j} \{sgn\,(\mu(i) - \mu(j))\,sgn\,(\nu(i) - \nu(j))\}$$

and

$$sgn(x) = \begin{cases} 1 & x > 0 \\ -1 & x < 0. \end{cases}$$

In the sequel, we shall make use of the notion of ties in ranking data. Alvo and Cabilio [1995a] defined the notion of compatibility in order to deal with missing data in the form of ranks. We may extend this notion to deal with tied rankings as well. We proceed with the three definitions.

Definition 7.1. A complete ranking of t objects is said to be compatible with an incomplete ranking of a subset of k of these objects, $2 \leq k \leq t$ if every pair of the specified k objects is given the same relative ranking in both rankings.

Definition 7.2. A tied ordering of t objects is a partition into e sets, $1 \leq e \leq t$, each containing g_i objects, $g_1+g_2+\cdots+g_e = t$, so that the g_i objects in each set share the rank $\sum_{j=1}^{i-1} g_j + g_i(g_i+1)/2, 1 \leq i \leq e$. Such a tie pattern is denoted by $\delta = (g_1, g_2, \ldots, g_e)$.

Definition 7.3. A complete ranking of t objects is said to be compatible with a tied ranking of these objects with tie pattern $\delta = (g_1, g_2, \ldots, g_e)$ if every pair of objects which receive distinct ranks is given the same relative ranking in both rankings.

The definition of compatibility when there are both ties and missing values is then a blend of the two previous definitions. We shall denote an incomplete ranking of k out of t objects with tie pattern δ by $\mu^* = (\mu^*(1), \mu^*(2), \ldots, \mu^*(k))^T$. At times, this k-vector is written as a t-vector in which the missing ranks are denoted by the symbol '_'. In the presence of ties, the concept of compatibility suggests that $\mu^*(i)$ is defined as the mid rank of all the items tied with item i. We shall denote by $C_\delta(\mu^*)$ the class of complete rankings compatible with ranking μ^*.

Example. The observations $X_1 = X_2 < X_3$ and X_4 missing have $t = 4$, $k = 3$, $e = 2$, $g_1 = 2$, $g_2 = 1$, and the incomplete tied ranking written as a 4-vector becomes $\mu^* = (1.5, 1.5, 3, \_)^T$. There are $2!1! = 2$ incomplete rankings compatible with $(1.5, 1.5, 3, \_)^T$; namely, $(1, 2, 3, -)^T$ and $(2, 1, 3-)^T$. The $4!/3! = 4$ complete rankings compatible to $(1,2,3,\_)^T$ are $(1,2,3,4)^T$, $(1,2,4,3)^T$, $(1,3,4,2)^T$, and $(2,3,4,1)^T$, while the 4 complete rankings compatible to $(2,1,3,\_)^T$ are $(2,1,3,4)^T$, $(2,1,4,3)^T$, $(3,1,4,2)^T$, and $(3,2,4,1)^T$. The $4!2!1!/3! = 8$ complete rankings are compatible with $(1.5, 1.5, 3, \_)^T$.

For two incomplete rankings with tie patterns, we have generalized versions of the Spearman and Kendall test statistics $A_S^*(\mu^*, \nu^*)$ and $A_K^*(\mu^*, \nu^*)$, respectively:

$$A_S^*(\mu^*, \nu^*) = \frac{(t+1)^2}{(k_1+1)(k_2+1)} \sum_{j=1}^{t} \delta(j) \left[\mu^*(j) - \frac{k_1+1}{2}\right] \left[\nu^*(j) - \frac{k_2+1}{2}\right]$$

where $\delta(j) = 1$ if both rankings of item j are not missing, or 0 otherwise and

$$A_K^*(\mu^*, \nu^*) = \sum_{i<j} a_1(i,j) a_2(i,j).$$

The $a_1(i,j)$ are the scores for ranking 1 given by

$$a_1(i,j) = \begin{cases} 0 & \text{in case of a tie} \\ 1 - \frac{2\mu^*(i)}{k_1+1} & \delta(i) = 1, \delta(j) = 0 \\ \frac{2\mu^*(j)}{k_1+1} - 1 & \delta(i) = 0, \delta(j) = 1 \\ sgn\,(\mu^*(i) - \mu^*(j)) & \delta(i) = 1, \delta(j) = 1. \end{cases}$$

with the scores for ranking 2 defined similarly.

We may now define two hypotheses: H_1 and H_2 for testing independence between two random variables. For both hypotheses, we assume that a number of ranked observations in μ^* and ν^* are fixed as are the tie patterns. In addition, for hypothesis H_1, the pattern of missing observations is fixed, whereas under H_2, the patterns of missing observations are randomly selected from the set of all possible patterns. Therefore, under both hypotheses, the elements in the class of complete compatible rankings $C_{\delta_1}(\mu^*)$ are equally likely and are independent of those in $C_{\delta_2}(\nu^*)$. Hence, it is easily shown that under both hypotheses, the measures $A^*(\mu^*, \nu^*)$ for both Spearman and Kendall are conditional expectations given the classes of complete compatible rankings:

$$A^*(\mu^*, \nu^*) = E\left[A(\mu, \nu)|C_{\delta_1}(\mu^*), C_{\delta_2}(\nu^*)\right].$$

As noted in Alvo and Cabilio [1995a], this remark along with the fact that

$$E\left(A_K - \frac{4}{t}A_S\right)^2 = O\left(t^2\right)$$

implies that the generalized test statistics are asymptotically equivalent as $t \to \infty$. Consequently, in what follows, we shall be concerned only with the Spearman case. The asymptotic distribution of $A_S^*(\mu^*, \nu^*)$ under the hypothesis H_1 was derived in Alvo et al. [2002]. In the next section, we obtain the asymptotic distribution under the hypothesis H_2.

Without loss in generality, assume $k_1 \leq k_2$ and relabel the items in such a way that in ranking 2, the first k_2 objects are the one ranked and similarly tied items can be arranged arbitrarily among themselves in any sequence accordingly. Hence, the missing items can be placed at the end and the new rankings $\nu^*(j)$ appear in natural order in ranking 2. We let o_j be the label of the jth item ranked in ranking 1, and k^* be the number of items ranked in ranking 1 among the k_2 ranked in ranking 2. Define

$$o_j^* = \begin{cases} \nu^*(o_j) & \text{if } 1 \leq j \leq k^* \\ \dfrac{k_2+1}{2} & \text{if } k^*+1 \leq j \leq k_1, \end{cases}$$

and $\bar{o} = \sum_{j=1}^{k_1} \frac{o_j^*}{k_1}$.

For the analyses which follow, we shall focus on the similarity measure A_S^*. Define the function $a_{k_1}(u)$ by

$$k_1 a_{k_1}(u) = \begin{cases} \dfrac{1}{2}(g_{11}+1) & \text{if } 0 < u \leq \dfrac{g_{11}}{k_1} \\ g_{11} + \dfrac{1}{2}(g_{12}+1) & \text{if } \dfrac{g_{11}}{k_1} < u \leq \dfrac{g_{11}+g_{12}}{k_1} \\ \vdots & \vdots \\ (k_1 - g_{1,e_1}) + \dfrac{1}{2}(g_{1,e_1}+1) & \text{if } \dfrac{k_1 - g_{1,e_1}}{k_1} < u \leq 1. \end{cases} \tag{7.14}$$

Rewrite A_S^* as

$$\begin{aligned} A_S^* &= \frac{(t+1)^2}{(k_1+1)(k_2+1)} \sum_{j=1}^{k_1} \left[o_j^* - \frac{k_2+1}{2} \right] \left[\mu^*(o_j) - \frac{k_1+1}{2} \right] \\ &= \frac{(t+1)^2 \, k_1}{(k_1+1)(k_2+1)} \sum_{j=1}^{k_1} \left[o_j^* - \bar{o} \right] \frac{\mu^*(o_j)}{k_1}. \end{aligned} \tag{7.15}$$

The following theorem was demonstrated in Alvo et al. [2002].

Theorem 7.1. *Assume that*

1. $k^* \to \infty$ (hence $k_1 \to \infty$, $k_2 \to \infty$ and $t \to \infty$) with $k^*/t \to \lambda > 0$ (and hence $\lambda \leq \frac{k_1}{t} < 1$ and $\lambda \leq \frac{k_2}{t} < 1$);
2. $\max\limits_{j=1,\ldots,e_1} \frac{g_{1j}}{k^*}$ is bounded away from 1;
3. $\max\limits_{j=1,\ldots,e_2} \frac{g_{2j}}{k^*}$ is bounded away from 1.

Then under H_1, the distribution of A_S^* is asymptotically normal as $k^* \to \infty$. The expected value of A_S^* is zero and the exact variance of A_S^* is

$$Var(A_S^*) = \left[\frac{(t+1)^2 \, k_1}{(k_1+1)(k_2+1)} \right] \frac{\sum_{j=1}^{k_1} \left(o_j^* - \bar{o} \right)^2}{12} \left\{ 1 - \frac{\sum_{j=1}^{e_1} \left[g_{1j}^3 - g_{1j} \right]}{k_1^3 - k_1} \right\}.$$

We now consider the asymptotic distribution of A_S^* under hypothesis H_2.

Theorem 7.2. *Let H_2 represent the null hypothesis under which the pattern of missing observations is randomly distributed so that the t-tuples of integers $1, \ldots, k_i$ and $t - k_i$ zeros are equally likely for $i = 1, 2$. Let $k_1 \to \infty$ (and hence $k_2 \to \infty$, $t \to \infty$) with $k_i/t \to \lambda_i > 0$. Then under H_2, A_S^* is asymptotically normal with mean 0 and variance*

$$Var(A_S^*) = \frac{(t+1)^4}{144(t-1)} \prod_{i=1}^{2} \left\{ \frac{k_i(k_i-1)}{k_i+1} \left[1 - \frac{\sum_{j=1}^{e_i} \left(g_{ij}^3 - g_{ij} \right)}{k_i^3 - k_i} \right] \right\}.$$

We shall consider the cases with the variable age either ungrouped or grouped as in Table 7.3. We have missing and tied observations for both variables in the ungrouped and grouped cases, but the problem of ties is less severe in the ungrouped case. Four test statistics all based on the Spearman statistic A_S^* are used to test the hypothesis H_1 of independence between age and the level of satisfaction. When discarding all the observations with at least one missing variable, the test statistic A_S^* reduces to the modified Spearman rank statistic. The first two tests which discard all the observations with at least one missing variable are tests of independence for the grouped and ungrouped data, respectively. The last two tests utilize all the information from the survey for the grouped and ungrouped data, respectively. The values of the test statistics for a two-sided test, the p-value, and the corresponding Spearman-type correlation are tabulated in Table 7.4. From the results shown in Table 7.4, it is seen that the variables have a moderate negative association, indicating that younger people tend to be less satisfied with the Policy Address. However, the correlation coefficients may not be directly comparable as they depend on the normalizing constant γ (see Lemma 1) which can be very different with different missing and tie patterns. Instead, one may use the p-value which is a standardized measurement and can be compared. It can be seen that the p-values of the four tests do not give consensus results. At the 5% significance level, the tests based on the reduced sample reject H_1, but the tests based on the complete sample accept H_1. This indicates that the proposed test provides a simple way of handling the missing values in a fair manner such that the information carried in the partially missing observations is also utilized. The rank correlation is helpful in providing an insight into the structure of the data. Similar results can be obtained when H_2 is considered.

Table 7.4.: Results of the analyses

Test	A_S^*	Standardized statistic	p-value	Correlation
1	−107165.00	−2.2039	0.0276	−0.3148
2	−110140.00	−2.1993	0.0278	−0.3061
3	−210504.34	−1.8812	0.0600	−0.4193
4	−212370.93	−1.8473	0.0648	−0.3661

Rank tests are widely applicable in many contexts. The proposed rank test for independence is a generalization of the usual Spearman/Kendall rank correlation based on a natural extension of the concept of distance between two incomplete rankings in order to include data consisting of both missing observations and ties. The test is simple and easily applicable. The test statistic reduces to the modified Spearman/Kendall rank statistic when there are ties but no missing values; it reduces to the test proposed by Alvo and Cabilio [1995a] when there are missing values but no ties. Moreover, it further reduces to the traditional Spearman/Kendall rank test when the data are complete.

7.4. Survival Analysis

The time until the occurrence of an event such as death or failure of a component constitutes the subject of survival analysis. In psychological or medical studies, for example, subjects are observed until the event of interest occurs or the study ends. Survival data is characterized by a number of features. Firstly, the variable measure, time, is always nonnegative. Secondly, there may be censoring on the "right," on the "left," or both. In the case of the former, it may be caused by the fact that the study ends before the event occurs. If on the left, it may be that the event occurred before the start of the study.

Table 7.5.: Survival distributions

	$f(t)$	$S(t)$	$\lambda(t)$
Exponential	$\rho\exp -\rho t$	$\exp -\rho t$	ρ
Weibull	$\kappa\rho(\rho t)^{\kappa-1}\exp-(\rho t)^{\kappa}$	$\exp-(\rho t)^{\kappa}$	$\kappa\rho(\rho t)^{\kappa-1}$
Log logistic	$\kappa\rho(\rho t)^{\kappa-1}[1+(\rho t)^{\kappa}]^{-2}$	$[1+(\rho t)^{\kappa}]^{-1}$	$\kappa\rho(\rho t)^{\kappa-1}[1+(\rho t)^{\kappa}]^{-1}$

Let T represent the time of occurrence of an event and suppose it has density $f(t)$. The probability of the event $(T > t)$ is called the survival function:

$$S(t) = P(T > t).$$

It follows that $S(0) = 1, S(\infty) = 0$. Associated with the survival function is the hazard function defined to be

$$\begin{aligned}\lambda(t) &= \lim_{dt\to 0}\frac{P(t < T < t + dt)}{S(t)\,dt} \\ &= \frac{f(t)}{S(t)} = -\frac{S'(t)}{S(t)}\end{aligned}$$

The hazard function or rate represents the probability that an event will occur in the interval $[t, t + dt)$ given that it has not occurred up to time t divided by the width of the interval. We note that the hazard function is characterized by only two conditions:

(i) $\lambda(t) \geq 0$ and (ii) $\int_0^{\infty}\lambda(t)\,dt = -\int_0^{\infty}\frac{d}{dt}\log S(t)\,dt = \infty$. Consequently, the hazard function may be increasing and decreasing or exhibit a "bathtub" shape as in common for certain electronic component.

Letting $\Lambda(t) = -\log S(t)$, we see

$$\Lambda(t) = \int_0^t \lambda(x)\,dx$$

which may be viewed as the cumulative hazard function.

Some examples of density functions describing survival data are given in Cox and Oakes [1984, chapter 3] and are listed in Table 7.5.

As an example for the baseline hazard, suppose that we have a Weibull distribution with hazard function given by

$$\lambda(t) = \kappa\rho(\rho t)^{\kappa-1}, \kappa > 0, \rho > 0$$

When $\kappa = 1$, we have an exponential distribution with constant time dependence. The exponential distribution expresses a "lack of memory" property in the sense that the conditional distribution of $T - t_0$ given that $T > t_0$ is the same as the unconditional distribution of T.

Suppose now that $t_1, t_2, \ldots, t_n$ represent the times when subjects or units are observed and let d_i be the indicator function if "death" occurs at time t_i.Then the likelihood function is given by

$$L = \prod_{i=1}^{n} (\lambda(t_i))^{d_i} S(t_i)$$

The likelihood function may then be used to estimate parameters of the density.

Example 7.12. In the simple case of the exponential,

$$L = \rho^{\sum d_i} \exp -\rho \sum_i t_i$$

and the maximum likelihood estimator of ρ is $\hat{\rho} = \frac{\sum d_i}{\sum_i t_i}$.

Similarly, one may determine estimates of the parameters for the Weibull distribution. An interesting test for $\kappa = 1$ is described in Cox and Oakes [1984, page 43].

7.4.1. Kaplan-Meier Estimator

The Kaplan-Meier estimator of the survival function, also known as the product limit estimator, is an important example of a nonparametric estimator. It has been used in medical studies to estimate the fraction of patients living for a certain amount of time after treatment. The estimator of the survival function is

$$\hat{S}(t) = \prod_{\{i:t_i \leq t\}} \left(1 - \frac{d_i}{n_i}\right)$$

where d_i is the number of events (e.g., deaths) that happened at time and n_i is the individuals known to have survived (have not yet had an event or been censored) up to time t_i. This estimator may be derived as the maximum likelihood estimator as follows.

The survival rate is

$$S(t) = \prod_{\{i:t_i \leq t\}} (1 - h_i)$$

where the discrete hazard rate h_i is the probability of an individual with an event at time t_i. The likelihood function is then

$$L = \prod_{j=1}^{i} h_j^{d_i} (1 - h_j)^{n_j - d_j}$$

It follows that the maximum likelihood estimator of h_j is $\hat{h}_j = \frac{d_i}{n_i}$.

Efron [1988] proposed the use of logistic regression to estimate h_j. He defined the logistic parameter

$$\lambda_j = \log \frac{h_j}{1 - h_j}$$

and fit the model

$$\lambda_j = x_i^T \beta$$

x_i represents a known covariate vector and β is a vector parameter. As an example, he chose

$$x_i^T = \left(1, t, t^2, t^3\right)$$

When no truncation or censoring occurs, the Kaplan-Meier curve is the complement of the empirical distribution function. In medical statistics, a typical application might involve grouping patients into categories, for instance, those with Gene A profile and those with Gene B profile. The Kaplan-Meier estimator is useful only when the predictor variable is categorical (e.g., treatment A vs treatment B; males vs females) and not when we have quantitative predictors such as gene expression, weight, or age. An alternative method makes use of the Cox proportional hazards regression analysis, which works both for quantitative predictor variables and for categorical variables. Furthermore, the Cox regression model extends survival analysis methods to assess simultaneously the effect of several risk factors on survival time.

Example 7.13. We consider an application to lung data available from the survival R package. Survival time in days in 168 patients with advanced lung cancer from the North Central Cancer Treatment Group was modeled as a linear function of the auxiliary variables below. The censoring status variable was equal to 1 if censored and 2 if the patient is dead. There were 121 events.

Variable	Description
Age	In years
Sex	Male = 1; female = 2
ECOG performance score	0=asymptomatic
	1= symptomatic but completely ambulatory
	2= in bed <50% of the day
	3= in bed > 50% of the day but not bedbound
	4 = bedbound
ph.karno	Karnofsky performance score (bad=0-good=100)
pat.karno	Karnofsky performance score as rated by patient
meal.cal	Calories consumed at meals wt.loss
Weight loss	Last 6 months

It can be seen from Table 7.6 that sex, ph.ecog, and ph.karno are significant in the model. In Fig. 7.1, we indicate the plot of the survival function overall and by gender.

Figure 7.1.: Lung data analysis

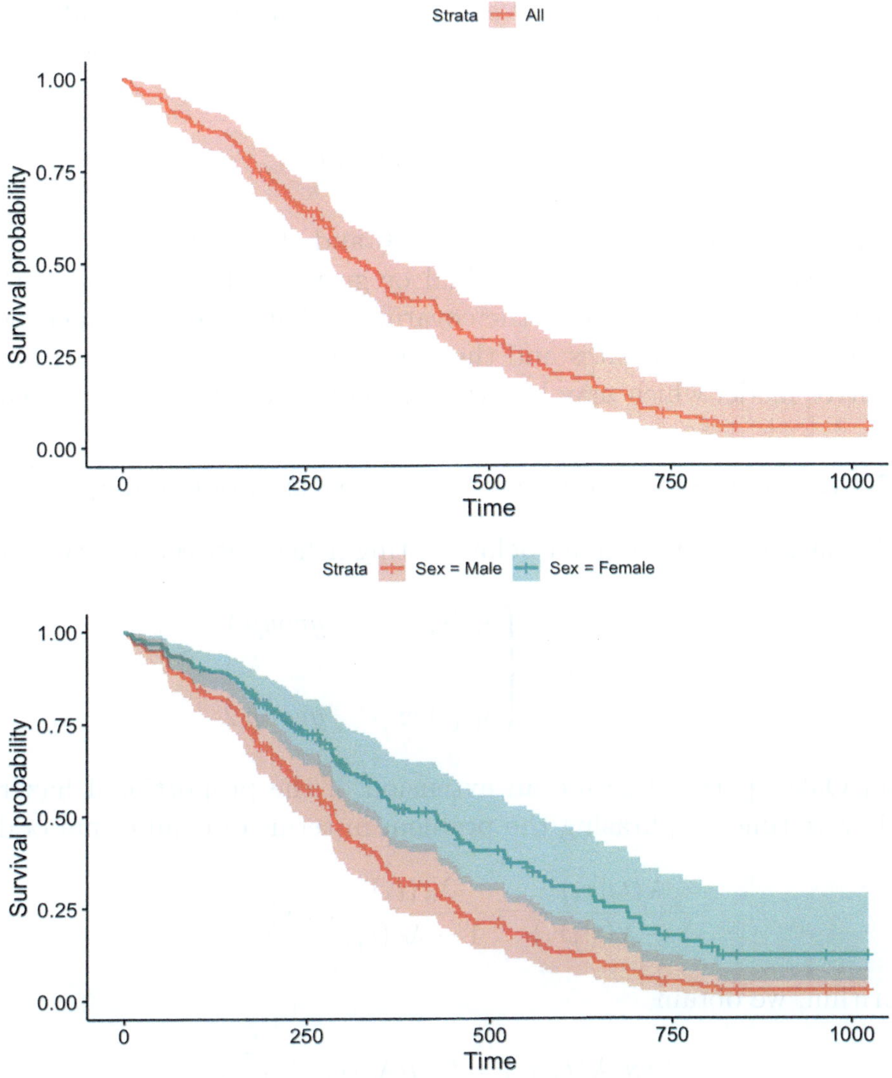

Table 7.6.: Lung data model

Variable	coef	exp(coef)	se(coef)	z	P(>|z|)
Age	1.065e − 02	1.011e+0	1.61e − 02	0.917	0.35906
Sex	−5.509e − 01	5.765e − 01	2.008e − 01	−2.743	0.00609
ph.ecog	7.342e − 01	2.084e + 0	2.233e − 01	3.288	0.00101
ph.karno	2.246e − 02	1.023e+0	1.124e − 02	1.998	0.04574
pat.karno	−1.242e − 02	9.877e − 01	8.8054e − 03	−1.542	0.12316
meal.cal	3.329e − 05	1.000e + 0	2.595 − 04	0.128	0.89791
wt.loss	−1.433e − 02	9.858e − 01	7.771e − 03	−1.844	0.06518

7.4.2. Modeling Survival Data

Turning attention to modeling, we may consider the continuous survival data model due to Cox (1972) described in Cox and Oakes [1984] which focuses on the hazard function. Specifically, the proportional hazard model is given by

$$\lambda_i(t|x_i) = \lambda_0(t)\exp\left(x_i^T\beta\right) \tag{7.16}$$

where $\lambda_0(t)$ is the baseline hazard when $x_i = 0$ and the term $\exp\left(x_i^T\beta\right)$ for vector x_i represents the relative risk for an individual or group i. In this model, the effects of time are separated from those due to the covariates. This model can incorporate a time dependence on both the covariates and the parameters β. Taking logarithm, we obtain a simple additive model which may be fitted using the usual regression techniques.

Example 7.14. Suppose that we have a two-sample situation and $x_i = \begin{cases} 0 & group\,1 \\ 1 & group\,2 \end{cases}$

Then the hazard function given below distinguishes between the two groups:

$$\lambda_i(t|x_i) = \begin{cases} \lambda_0(t) & group\,1 \\ \\ \lambda_0(t)\exp^{\beta} & group\,2 \end{cases}$$

Cox and Oakes [1984] describe an extension of the proportional hazard model to deal with discrete time by phrasing the problem in terms of conditional odds:

$$\frac{\lambda(t_j|x_i)}{1-\lambda(t|_j x_i)} = \frac{\lambda_0(t_j)}{1-\lambda_0(t_j)}\exp\left(x_i^T\beta\right)$$

Taking logarithm, we obtain

$$logit\lambda(t_j|x_i) = logit\lambda_0(t_j) + x_i^T\beta$$

This is the logit model of the conditional probability of dying at time t_j given survival up to that time. A useful link to using R to fit survival data is http://www.sthda.com/english/wiki/cox-proportional-hazards-model. For an application of copula methods to survival data, see Takeshi and Yi-Hau [2018]

8. Time Series Methods

As the name implies, time series data consist of data indexed by time. Examples include the following:

1. Financial time series: exchange rates, closing day Dow Jones averages, stock prices, interest rates, and export numbers
2. Meteorological time series: temperatures and precipitation per hour/day/week/month
3. Physical time series: sunspot numbers and measurements of an experiment
4. Biological time series: genetic values of progeny
5. Medical time series: patient data and clinical test data
6. Ecological time series: pollutants (CO2, ozone, etc.)
7. Demographic time series: population size and monthly income

Time series data may exhibit trend, seasonal variation, and periodicity, either separately or in combination. In addition, consecutive observations may be correlated. As a result, any analysis of time series data must address each of these components. Traditionally, there have been two approaches: either a time domain analysis or a spectrum analysis. In what follows, we provide a bird's-eye view of time series methods. Readers interested in pursuing the subject in depth are referred to the excellent texts by Box et al. [1994]; Brockwell and Davis [1987]; Kendall et al. [1983]. We begin with a description of the classical methods for time series analysis and then consider state space modeling.

8.1. Classical Methods of Analysis

Intuitively, an ordered series of data exhibits a trend if the values are generally either increasing or decreasing. One may test for trend using nonparametric tests as described in Chap. 5. If a trend exists, it is then important to remove it before further analysis. In the simplest situation if no other factors exist, a trend may be modeled as a polynomial function using regression methods. Hence, if the data is represented as

$$X_t = \mu_t + \varepsilon_t, t = 1, \ldots, n$$

M. Alvo, *Statistical Inference and Machine Learning for Big Data*, Springer Series in the Data Sciences, https://doi.org/10.1007/978-3-031-06784-6_8

where $\{\varepsilon_t\}$ represents the random error, we may model the mean μ_t by a low-dimensional polynomial in t. In more complex situations, a moving average may be fitted to the data to smooth out the "noise" from one arising from consecutive observations. Hence, the mean at time t becomes a weighted estimate

$$\hat{\mu}_t = \sum_{i=-\infty}^{\infty} w_i X_{t+i}$$

where the weights w_i are usually chosen in such a way as to ensure a smooth trend. Alternatively, a trend may be removed as opposed to modeled by successive forward differences of the series

$$\begin{aligned} \Delta \mu_t &= X_{t+1} - X_t \\ \Delta^2 \mu_t &= X_{t+2} - 2X_{t+1} + X_t \end{aligned}$$

or by successive backward differences

$$\begin{aligned} \nabla \mu_t &= X_t - X_{t-1} \\ \nabla^2 \mu_t &= X_t - 2X_{t-1} + X_{t-2} \end{aligned}$$

When the series contain a seasonal component of period τ, we may compute the differences

$$\nabla_\tau \mu_t = X_t - X_{t-\tau}$$

following which we remove the trend as indicated before. The analysis proceeds on the series of differences. For further details on removing seasonal trends, see Kendall et al. [1983, Chapter 46]. A time series from which trend and seasonality have been removed will likely exhibit only oscillations about a constant. The Ergodic theorem due to Birkhoff [1931] and Khinchin [1932] states that under certain conditions, the time average of a series is equal to the spatial average, a result which permits one to estimate the average of a term of the series from the time average of a single realization.

Definition 8.1. A time series $\{X_t, t\epsilon T\}$ is said to be stationary if

$$\begin{aligned} E[X_t] &= \mu \\ Var[X_t] &= \sigma^2 < \infty \\ Cov[X_t, X_{t+k}] &= \gamma_k \end{aligned}$$

A stationary stochastic process is in "stochastic equilibrium"; i.e., sections of a sample path "look alike" or fluctuations are purely random. For a nonstationary stochastic process, either the mean level is not constant or the average size of the fluctuations is not constant.

We refer to γ_k as the autocovariance and to $\rho_k = \frac{\gamma_k}{\sigma^2}$ as the autocorrelation. The sample autocovariance function for a sample of size n is estimated by

$$\hat{\gamma}_h = \frac{\sum_{i=1}^{n-h} (x_{i+h} - \bar{x})(x_i - \bar{x})}{n}, 0 \leq h < n$$

where $\bar{x}$ is the sample mean. The choice of n in the denominator ensures that the estimate of the autovariance matrix is nonnegative definite. It is common to plot the autocorrelation function, ACF, as a function of the lags.

An example of an autoregressive time series is the Markoff scheme defined as

$$X_t = -\alpha X_{t-1} + \varepsilon_t$$

It may be shown that

$$Cov\left[X_t, X_{t-1}\right] = \rho\sigma^2$$

Another example due to Yule is given by

$$X_t = -\alpha_1 X_{t-1} - \alpha_2 X_{t-2} + \varepsilon_t$$

from which we obtain

$$\begin{aligned} \rho_1 &= -\frac{\alpha_1}{1+\alpha_2} \\ \rho_2 &= -\alpha_2 + \frac{\alpha_1^2}{1+\alpha_2} \end{aligned}$$

A general linear autoregressive process may be expressed as

$$\sum_{i=1}^{h} \alpha_i X_{t-i} = \varepsilon_t$$

from which we obtain the Yule-Walker equations

$$\rho_k + \sum_{i=1}^{h} \alpha_i \rho_{k-i} = 0, k > 0$$

where $\rho_k = \rho_{-k}$ and $\rho_0 = 1$. Those may be expressed in matrix notation as

$$P\phi = \rho$$

where $\phi = (\alpha_1, \ldots, \alpha_h)^T$, $\rho = (\rho_1, \ldots, \rho_h)^T$, and the matrix P is given by

$$P = \begin{pmatrix} 1 & \rho_1 & \rho_2 & \cdots & \rho_{h-1} \\ \rho_1 & 1 & \rho_1 & \cdots & \rho_{h-2} \\ \rho_2 & \rho_1 & 1 & \cdots & \rho_{h-3} \\ \vdots & \vdots & & \cdots & \vdots \\ \rho_{h-1} & \rho_{h-2} & \rho_{h-3} & \cdots & 1 \end{pmatrix}$$

Solving for ϕ, we have

$$\phi = P^{-1}\rho$$

The order of an autoregressive process is generally not known in advance but may be identified by using the partial autocorrelation function, denoted PACF. Let ϕ_{kj} denote the jth coefficient in an autoregressive scheme of order k so that the Yule-Walker equations may be rewritten as

$$\rho_j + \sum_{i=1}^{k} \phi_{ki}\rho_{j-i} = 0, j = 1, 2, \ldots, k$$

Intuitively, the partial autocorrelation function at lag k is equal to the partial regression coefficient of X_t on X_{t-k} adjusted for $X_{t-1}, \ldots, X_{t-k+1}$, which can be shown to be

$$\begin{aligned} \phi_{kk} &= \frac{cov\left[X_t - \hat{X}_t, X_{t-k} - \hat{X}_{t-k}\right]}{var\left[X_{t-k} - \hat{X}_{t-k}\right]} \\ &= \frac{\left[\rho_k - \sum_{i=1}^{k-1} \phi_{k-1,i}\rho_{k-i}\right]}{\left[\rho_k - \sum_{i=1}^{k-1} \phi_{k-1,i}\rho_i\right]} \end{aligned}$$

[Box et al., 1994, chapter 3]. An important feature of the partial autocorrelation function for an autoregressive process of order h is that it is equal to 0 for $k > h$. The autocorrelation function on the other hand will either decay geometrically as h increases or follow a damped sine wave.

In addition to the autoregressive process, we may also have correlated error terms. Such a mixed process is labeled an ARMA process defined in Box et al. [1994] in terms of operators.

Definition 8.2. A process $\{X_t, t\epsilon T\}$ is said to be an ARMA (p, q) process if the series is stationary and if for every t it can be expressed as

$$\phi(B) X_t = \theta(B) \varepsilon_t$$

where ϕ and θ are the autoregressive and moving average polynomials, respectively, given by

$$\phi(z) = 1 - \sum_{i=1}^{p} \phi_i z^i$$
$$\theta(z) = 1 + \sum_{i=1}^{q} \theta_i z^i$$

The $\{\varepsilon_t\}$ are $N(0, \sigma^2)$ with autocovariance $\rho_h = \begin{cases} \sigma^2 & h = 0 \\ 0 & h \neq 0 \end{cases}$

For a moving average process of order q, denoted MA(q), the autocorrelation function will be equal to 0 for lags greater than q, whereas the partial autocorrelation function will decay exponentially.

An ARMA (p, q) process for which the trend component has been removed by differencing is labeled an autoregressive integrated moving average process and denoted ARIMA (p, d, q) where d is the order of differences taken.

The estimation of the parameters in a general linear autoregressive process proceeds from the likelihood function. Suppose that

$$X_t = \sum_{i=1}^{p} \phi_i X_{t-i} + \varepsilon_t, t = 1, \ldots, n$$

where the errors are distributed as a random sample from $N(0, \sigma^2)$. Then, following the presentation in Box et al. [1994, Chapter 7, Appendix A7.1 P.258–284], the likelihood function can be shown to be the product

$$\begin{aligned} L\left(\phi, \sigma^2\right) &= \frac{|V|^{-\frac{1}{2}}}{\left(2\pi\sigma^2\right)^{\frac{p}{2}}} \exp -\frac{1}{2\sigma^2} x_0^T V^{-1} x_0 \\ &\times \frac{1}{\left(2\pi\sigma^2\right)^{\frac{n-p}{2}}} \exp -\frac{1}{2\sigma^2} \sum_{t=p+1}^{n} \left(x_t - \sum_{i=1}^{p} \phi_i x_{t-i}\right)^2 \end{aligned}$$

where $\phi = (\phi_1, \ldots, \phi_p)$, $x_0^T = (x_1, \ldots, x_p)$ and $V = \left(\frac{\gamma_{|i-j|}}{\sigma^2}\right)$ and $\gamma_i = cov\left(X_t, X_{t+i}\right)$. The estimation of the parameters ϕ, σ^2 proceeds from a least squares analysis by maximizing the logarithm of the likelihood function. Estimation in a moving average model proceeds in a similar manner.

Another aspect of a time series is that it may contain hidden periodicities. The investigation of periodicities falls in the domain of spectral analysis. In analogy with a Fourier series analysis where the periods are integral divisors of the series length n, we may define

$$
\begin{aligned}
a(\alpha) &= \frac{2}{n}\sum_{t=1}^{n} X_t \cos \lambda t \\
b(\alpha) &= \frac{2}{n}\sum_{t=1}^{n} X_t \sin \lambda t
\end{aligned}
$$

where $\{X_t\}$ is stationary with mean equal to 0. Then, the sample periodogram is the graph of

$$
I(\alpha) = a^2(\alpha) + b^2(\alpha)
$$

against $-\pi \leq \lambda \leq \pi$ which is called the frequency. In a Fourier analysis where the periods are given by $\frac{n}{k}$, the frequency is expressed as $\lambda = \frac{2\pi k}{n}$.

Anderson [1971] discusses the statistical aspects of time series including spectral analysis and provides procedures for testing hypotheses. See also Brockwell and Davis [1987, Chapter 10] or Jenkins and Watts [1968] for additional details. Kendall [1973, chapter 8] presents an elementary development. In order to illustrate the concepts introduced, we consider the example of the Nile data.

Example 8.1. The Nile data set contains the annual flow of the river Nile at Aswan from 1871 to 1970. The data is available in the R data sets package. A plot is shown in Fig. 8.1. The autocorrelation and partial autocorrelation functions are exhibited in Fig. 8.2.

The graph of the data indicates that the data is nonstationary, which can also be seen in Fig. 8.2 from the sample autocorrelation function having a significant value for the first 8 lags.

After taking the difference, the Nile data in Fig. 8.3 appears stationary. The sample autocorrelation and partial autocorrelation functions are plotted to help determine the model in Fig. 8.4. The ACF and PACF plots suggest a single MA term with a significant autocorrelation at lag 1 and the exponential decrease in absolute value of the partial autocorrelations.

The Q-statistic can be obtained using the Box.test function on the residuals of the model for the first 12 and 24 lags. The autocorrelations of the model residuals are also plotted in Fig. 8.5.

Both the tests and the autocorrelation function suggest the residuals are uncorrelated. Using the model, a graph as in Fig. 8.6 can be generated showing the forecasts.

A periodogram can be used to analyze the frequency domain of the differenced series. This was done in De Putter et al. [1998] for the period A.D. 622–1470, and it was discovered there were periodicities. We see from the periodogram in Fig. 8.7 that the peaks occur in the range 0.16–0.50 of frequencies implying approximate cycles of 2 to 3 years.

Figure 8.1.: Nile data

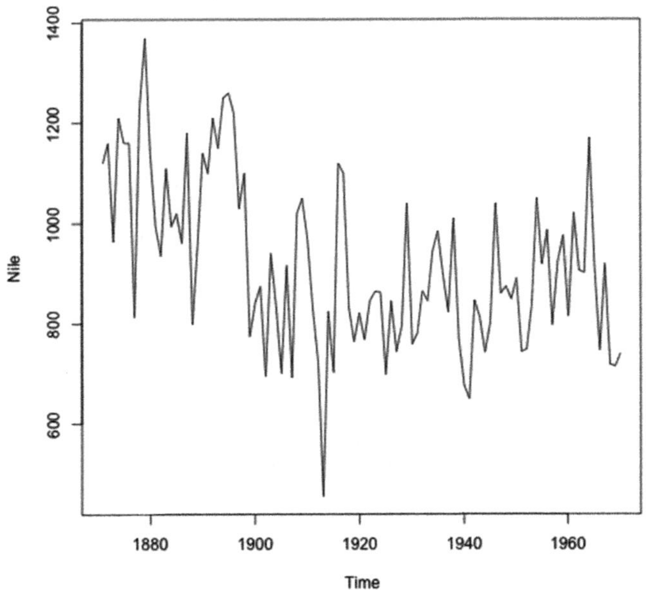

Figure 8.2.: Auto- and partial autocorrelation

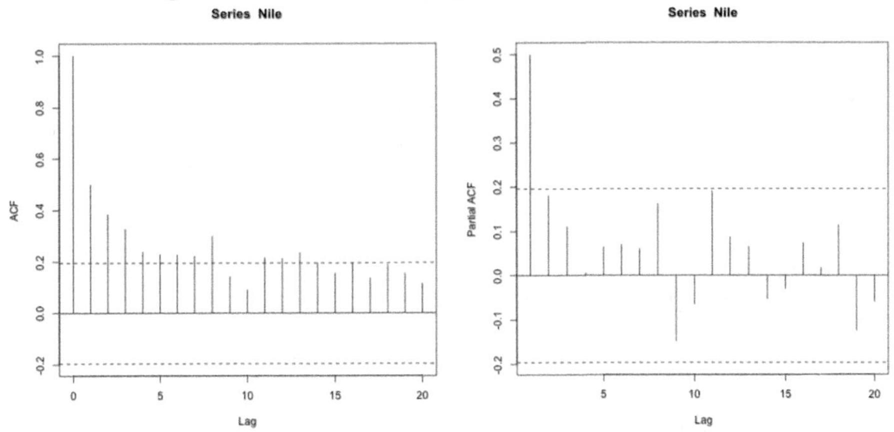

Figure 8.3.: First differences for the Nile data

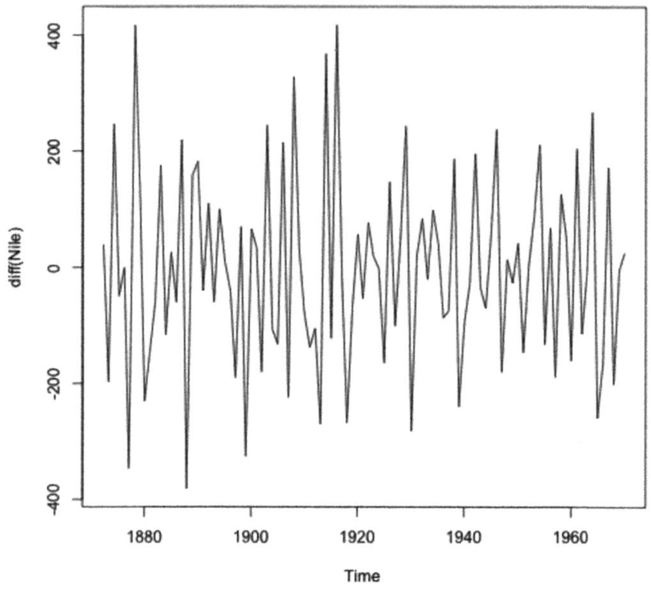

Figure 8.4.: Auto- and partial correlation for differences

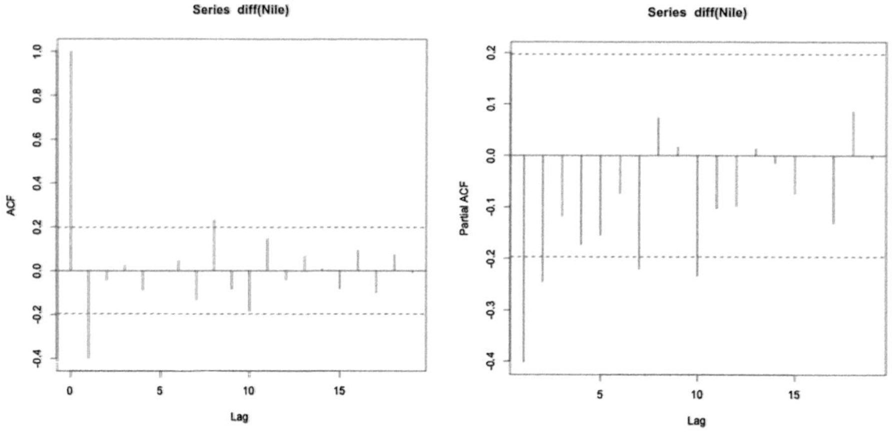


```
##
## Call:
## arima(x = Nile, order = c(0, 1, 1))
##
## Coefficients:
##           ma1
##       -0.7329
## s.e.   0.1143
##
## sigma^2 estimated as 20600:  log likelihood = -632.55,  aic = 1269.09
```

Figure 8.5.: Autocorrelation of model residuals

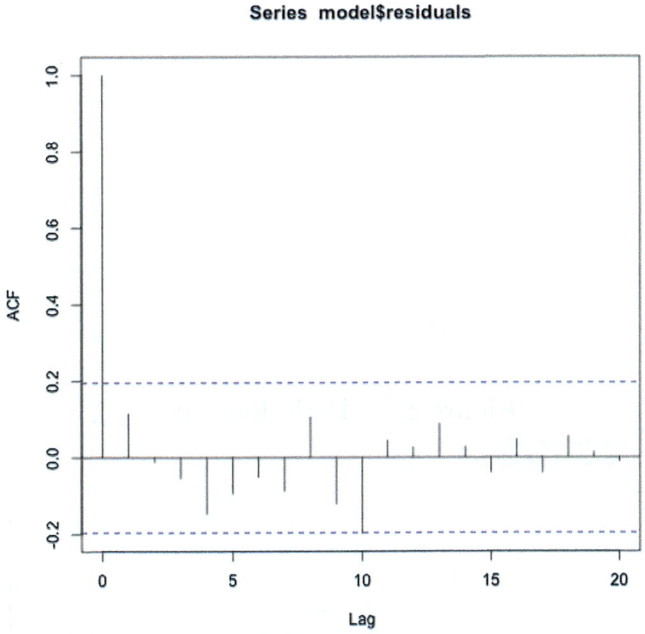


```
##
##  Box-Pierce test
##
## data:  model$residuals
## X-squared = 12.48, df = 12, p-value = 0.408
```

```
##
##  Box-Pierce test
##
## data:  model$residuals
## X-squared = 15.465, df = 24, p-value = 0.9064
```

Figure 8.6.: Forecast for the Nile data

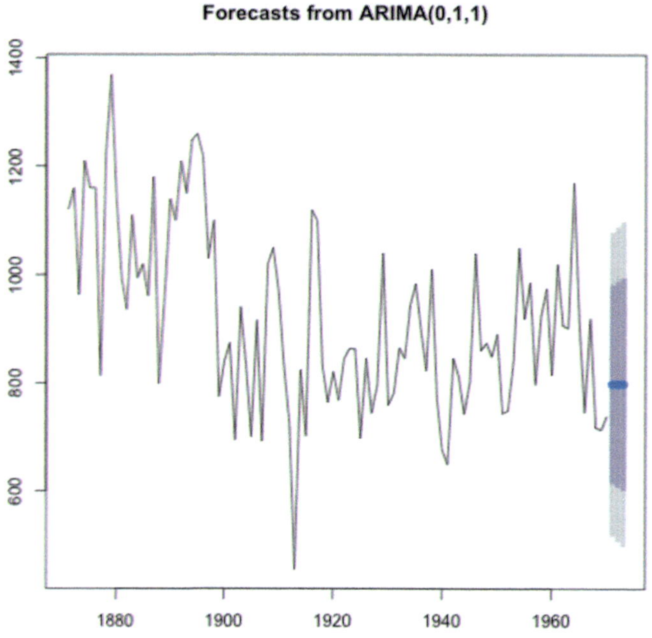

Figure 8.7.: Periodogram

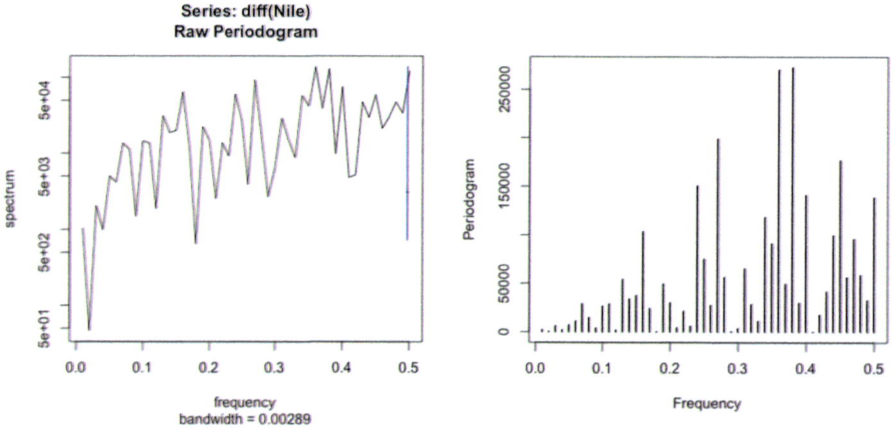

Frequency	0.36	0.38	0.50	0.27	0.40	0.16	0.24	
Period	2.77	2.63	2.00	3.70	2.50	6.25	4.17	

8.2. State Space Modeling

In the state space modeling of a time series, there is a de-coupling of the observations and the state space. This approach has been used for modeling dynamical systems and for forecasting. We begin with some examples drawn from Durbin and Koopman [2012] who provide a thorough treatment of the subject. Partial treatments also appear in several textbooks on time series, for example, [Box et al., 1994, p.163].

Example 8.2. The local level model may be defined from the simple random walk model

$$\begin{aligned} X_t &= \alpha_t + \varepsilon_t, \varepsilon_t \sim \mathcal{N}\left(0, \sigma_\varepsilon^2\right) \\ \alpha_{t+1} &= \alpha_t + \eta_t, \eta_t \sim \mathcal{N}\left(0, \sigma_\eta^2\right) \end{aligned}$$

where $\{\varepsilon_t\}, \{\eta_t\}, \{\alpha_t\}$ are all mutually independent. Here, the states, α_t, are unobserved, whereas the X_t represent the observations, the objective being to draw inference on α from knowledge of X. We may express this model in the form

$$X_t = \alpha_1 + \sum_{i=1}^{t-1} \eta_i + \varepsilon_t, t = 1, \ldots, n$$

from which we may derive expressions for the variance-covariance matrix of the vector $X = (X_1, \ldots, X_n)$.

The local linear trend model may be obtained by adding a trend term as

$$\begin{aligned} X_t &= \alpha_t + \varepsilon_t, \varepsilon_t \sim \mathcal{N}\left(0, \sigma_\varepsilon^2\right) \\ \alpha_{t+1} &= \alpha_t + \nu_t + \eta_t, \eta_t \sim \mathcal{N}\left(0, \sigma_\eta^2\right) \\ \nu_{t+1} &= \nu_t + \zeta_t, \zeta_t \sim \mathcal{N}\left(0, \sigma_\zeta^2\right) \end{aligned}$$

Example 8.3. The autoregressive model is given by

$$X_t = \sum_{i=1}^{p} \phi_i X_{t-i} + \varepsilon_t, t = 1, \ldots, n$$

where $\varepsilon \sim N(0, \sigma^2)$ may be expressed as a state space model as follows. We may write the state space equation

$$\alpha_t = \begin{pmatrix} \phi_p & \phi_{p-1} & \cdots & \cdots & \phi_1 \\ 0 & 0 & 1 & \cdots & \\ \cdots & \cdots & \cdots & \cdots & \cdots \\ 0 & 0 & \cdots & \cdots & 1 \\ 0 & 0 & & 1 & 0 \end{pmatrix} \alpha_{t-1} + \eta_t$$

where $\alpha_t = [X_t, X_{t-1}, \ldots, X_{t-p+1}]^T$ and $\eta_t = [\varepsilon_t, 0, \ldots, 0]^T$. The observation equation is then

$$X_t = [1, 0, \ldots, 0]\, \alpha_t$$

Example 8.4. The general linear state space model may be expressed in terms of an observation equation and a state space equation respectively

$$\begin{aligned} X_t &= Z_t\alpha_t + \varepsilon_t, \varepsilon_t \sim \mathcal{N}(0, H_t) \\ \alpha_{t+1} &= T_t\alpha_t + R_t\eta_t, \eta_t \sim \mathcal{N}(0, Q_t) \end{aligned}$$

$t = 1, \ldots, n$, where X_t is a $p \times 1$ vector of observed values and α_t is an unobserved vector of states. The matrices Z, H, T, Q are assumed known and the error terms are mutually independent.

The Kalman filter approach used for state estimation is optimal provided both the transition and the observation densities are linear and Gaussian. The approach consists of a recursive method requiring an initialization of the parameter values of the initial state density. An easy-to-read development in connection with the local level model is provided in Durbin and Koopman [2012] and illustrated on the Nile data. The parameter estimation is also developed for this model as well as for the general linear state space model.

State space models may also be treated within the context of Bayesian hierarchical models [Murray, 2015]. In that context, there is a set of parameters θ, a state space process α_t, and observations X_t, $t_1 < t_2 < \cdots < t_n$.

The joint density takes the form

$$\begin{aligned} f\left(X_{(t_2:n)}, \alpha_{(t_1:n)}, \theta\right) &= \left[f\left(\alpha_{(t_1:n)}, \theta\right)\right]\left[\left(f\left(X_{(t_2:n)}|\alpha_{(t_1:n)}, \theta\right)\right)\right] \\ &= \left[\pi(\theta) f(\alpha_{t_1}|\theta)\left(\prod_{i=1}^{n-1} f\left(\alpha_{t_{i+1}}|\alpha_{t_i}, \theta\right)\right)\right]\left[\prod_{i=1}^{n} f(X_{t_i}|\alpha_{t_i}, \theta)\right] \end{aligned}$$

where $\prod_{i=1}^{n} f(X_{t_i}|\alpha_{t_i}, \theta)$ represents the likelihood function and $\pi(\theta)$ is the prior density on θ.

The posterior density is then given by

$$\pi\left(\alpha_{(t_1:n)}, \theta|X_{(t_2:n)}\right) = \pi\left(\theta|X_{(t_2:n)}\right) f\left(\alpha_{(t_1:n)}|\theta, X_{(t_2:n)}\right)$$

where the first factor on the right emphasizes the estimation of the parameter θ, while the second factor focuses on state estimation.

In comparing the Box-Jenkins approach with state space models, we see that in the former, trend and seasonal components are eliminated by a process of differencing the initial observations. The resulting series can then be treated as a stationary series. On the other hand, the state space approach proposes a model for each of the individual components.

We conclude by quoting from Durbin and Koopman [2012]: "Other advantages of a state space analysis are (i) its treatment of missing observations; (ii) explanatory variables can be incorporated into the model without difficulty; (iii) associated regression coefficients can be permitted to vary stochastically over time if this seems to be called for in the application; (iv) trading-day adjustments and other calendar variations can be readily taken care of; (v) no extra theory is required for forecasting since all that is needed is to project the Kalman filter forward into the future."

In other works, [Koopman and Mesters, 2017] present an empirical Bayes approach to the state space model, and [Shumway and Stoffer, 2016, p.352] apply state space modeling to analyze influenza data. Also Napier and Lee [2019] developed a Bayesian hierarchical mixture model for geographical clustering based on disease trends.

9. Estimating Equations

Estimating equations have been successfully used to study complex systems when it is difficult to define a likelihood function. This would be the case in non-Gaussian multivariate settings where there is a lack of abundance of distributions. In addition, as pointed out by Jesus and Chandler [2011], there is the difficulty that arises owing to the different measurement scale between the theoretical model and the actual data collected. The former may be aggregated spatially and temporally, whereas the latter may be single measurements collected daily. Estimating equations unify the subjects of least squares, quasi-likelihood, and maximum likelihood. They have also been used to model over-dispersed data. Frequently, the variance of observed data is greater than that expected from simple exponential models. Cox and Hinkley [1974] suggested that the maximum likelihood estimation of regression parameters is not seriously impacted but that is not the case for the standard errors and subsequent tests and confidence intervals. See Desmond [1997] for an excellent review article as well as the development in Van der Vaart [2000, Chapter 5].

The subject of estimating equations had its roots with the work of Godambe [1960] who noted a property of maximum likelihood estimation. Let G be the class of estimating functions $g(x,\theta)$ satisfying the following:

(i) $Eg(X,\theta) = 0$

(ii) $\frac{\partial g(x,\theta)}{\partial \theta}$ exists for all $\theta \in \Theta$ a.e.

(iii) $\int g(x,\theta) f(x,\theta)\, d\mu$ is differentiable under the integral sign

(iv) $\left[E\frac{\partial g(X,\theta)}{\partial \theta}\right]^2 > 0$ for all $\theta \in \Theta$

Let $\boldsymbol{X}_n = (X_1, \ldots, X_n)$ be a random sample from a distribution indexed by a parameter θ. Estimation then proceeds by solving for θ the estimating equation:

$$g_n(\theta; \boldsymbol{X}_n) = 0 \tag{9.1}$$

In this connection, we may define an M-estimator.

Definition 9.1. An M-estimator is one which satisfies a system of equations of the type

$$\frac{1}{n}\sum_{i=1}^{n} \psi_\theta(X_i) = 0$$

where ψ_θ are known vector-valued functions.

M. Alvo, *Statistical Inference and Machine Learning for Big Data*, Springer Series in the Data Sciences, https://doi.org/10.1007/978-3-031-06784-6_9

M-estimators generalize the method of maximum likelihood and can be viewed as finding the zero of an estimating equation. Under certain conditions, M-estimators can be shown to be consistent and to have an asymptotic normal distribution [Van der Vaart, 2000, Chapter 5].

Definition 9.2. The choices of $\psi_\theta(x) = x - \theta$ and $\psi_\theta(x) = |x - \theta|$ lead to estimates of the mean and median, respectively. Under conditions similar to those underlying the Cràmer-Rao inequality, Godambe showed that for all $g \in G$

$$\frac{E_\theta(g^2)}{\left[E_\theta \frac{\partial g(X,\theta)}{\partial \theta}\right]^2} \geq \frac{1}{E_\theta\left[\left(\frac{\partial \log f}{\partial \theta}\right)^2\right]}$$

with equality attained by the score function

$$g^* = \frac{\partial \log f}{\partial \theta}.$$

The latter leads to the usual maximum likelihood estimation. In that case,

$$\frac{E_\theta(g^{*2})}{\left[E_\theta \frac{\partial g^*(X,\theta)}{\partial \theta}\right]^2} \geq \frac{E_\theta(g^2)}{\left[E_\theta \frac{\partial g(X,\theta)}{\partial \theta}\right]^2} \tag{9.2}$$

In another exposition, define Q by

$$\frac{\partial Q}{\partial \mu} = \frac{Y - \mu}{V(\mu)}$$

where $\mu, V(\mu)$ represent the mean and variance of Y. Then it follows that

(i) $E\left(\frac{\partial Q}{\partial \mu}\right) = 0$

(ii) $E\left(\frac{\partial Q}{\partial \mu}\right)^2 = -E\left(\frac{\partial^2 Q}{\partial \mu^2}\right) = \frac{1}{V(\mu)}$.

Property (i) is obvious, whereas to prove (ii), note that

$$\frac{\partial}{\partial \mu}\left(V(\mu)\frac{\partial Q}{\partial \mu}\right) = \left(\frac{\partial}{\partial \mu}V(\mu)\right)\frac{\partial Q}{\partial \mu} + V(\mu)\frac{\partial^2 Q}{\partial \mu^2}$$

Taking expectations and applying (i) yields the result. Properties (i) and (ii) characterize the usual score function for likelihoods and for that reason, Q is called a quasi-likelihood, log-likelihood function or quasi-score function. McCullagh [1983] develops properties of quasi-likelihood functions. Some basic assumptions and accompanying results for estimating equations detailed in Jesus and Chandler [2011] are as follows.

Assumption 1 There exists a sequence of matrices $\{\eta_n\}$independent of θ such as $n \to \infty$

$$\eta_n g_n(\theta; \boldsymbol{X}_n) \to g_l(\theta)$$

uniformly for all θ in probability where

(a) $g_l(\theta) = 0$ has a unique root at $\theta = \theta_0$ and

(b) $g_l(\theta)$ is bounded away from 0 except in a neighborhood of θ_0.

Under Assumption 1, consistency can be demonstrated; that is, the estimating function asymptotically has a unique estimator $\hat{\theta}_n \xrightarrow{P} \theta_0$. The requirement that the convergence be uniform can usually be verified in practice. The next assumption provides conditions under which an estimator has an asymptotic normal distribution.

Assumption 2 Let $G_n(\theta) = \frac{\partial g_n}{\partial \theta}$ be the matrix of partial derivatives and assume the matrices $\{G_n(\theta)\}$ are invertible. Moreover, suppose there exist sequences $(\gamma_n), (\delta_n)$ of invertible matrices independent of θ such that

(a) at $\theta = \theta_0$, the covariance of $\tilde{g}_n(\theta, \boldsymbol{X}_n) = \gamma_n g_n(\theta, \boldsymbol{X}_n)$ exists and converges to a positive definite matrix $\tilde{\Sigma}$

(b) Set $\tilde{G}_n(\theta) = \frac{\partial \tilde{g}_n}{\partial \theta}$ within a neighborhood of θ_0 so that $\tilde{G}_n(\theta)\delta_n$ converges uniformly in probability to an invertible matrix $M(\theta)$ with elements continuous functions of θ.

Theorem 9.1. *Under Assumptions 1 and 2,*

$$E\left[\hat{\theta}_n\right] \longrightarrow \theta_0$$

and the covariance of $\delta_n^{-1}\hat{\theta}_n$ *converges to* $M_0^{-1}\tilde{\Sigma}\left(M_0^{-1}\right)^T$ *where* $M_0 = M(\theta_0)$. *If in addition, the normalized* $\tilde{g}_n(\theta, \boldsymbol{X}_n)$ *converges in distribution to a normal, then so does* $\delta_n^{-1}\hat{\theta}_n$.

Summarizing then, if

$$\Sigma_n = cov\left(g_n(\theta, \boldsymbol{X}_n)\right)$$

and

$$G_0 = E\left[\frac{\partial g_n}{\partial \theta}|\theta = \theta_0\right]$$

exists, then

$$\hat{\theta}_n \xrightarrow{\mathcal{L}} N\left(\theta_0, G_0^{-1}\Sigma_n\left(G_0^{-1}\right)'\right)$$

Example 9.1. Suppose that we have a random sample $X_1, \ldots, X$ with mean μ. Define

$$\begin{aligned} n^{-1} g_n(\theta, \boldsymbol{X}_n) &= n^{-1} \Sigma_{i=1}^n (X_i - \theta) \\ &\longrightarrow \mu - \theta \end{aligned}$$

The convergence is uniform since

$$\left| n^{-1} g_n(\theta, \boldsymbol{X}_n) - (\mu - \theta) \right| = \left| \bar{X}_n - \mu \right|$$

is independent of θ and hence **Assumption** 1 holds.

Suppose as well that var$X_i = \sigma_i^2$ and that $\frac{\sum \sigma_i^2}{n} \to \sigma^2$. Then it can be shown **Assumption** 2 holds with $\gamma_n = \delta_n = 1/\sqrt{n}$. Moreover, $\hat{\theta}_n = \bar{X}_n$ and consequently,

$$\sqrt{n}\left(\bar{X}_n - \mu\right) \xrightarrow{\mathcal{L}} N\left(0, \sigma^2\right)$$

A recent application of estimating equations is provided in the next example.

Example 9.2. Schiopu-Kratina et al. [2021] Independent data $d_i^T = \{S_{ij}, x_{ij}, C_i, j = 1, \ldots,\}$ are collected on subject i, $i = 1, \ldots, n$ which are generated by a distribution indexed by a true parameter $\eta_0 = (\theta_0, \sigma_o^2)^T \in R^{p+1}$. The x_{ij} are random covariates and the occurrence times satisfy

$$0 < S_{i1} < \cdots < S_{ij} < \cdots$$

The jth gap time is

$$Y_{ij} = S_{ij} - S_{i,j-1}, i, j = 1, \ldots$$

Observation stops at a finite time $0 < C_i, i = 1, \ldots$ The basic modeling assumptions are on the conditional mean and variance:

$$E_\theta\left[Y_{ij} | F_{i,j-1}\right] = \mu_{ij}(\theta), Var_\theta\left[Y_{ij} | F_{i,j-1}\right] = \sigma^2 V_{ij}^2(\theta)$$

We may think of θ as a regression parameter and of σ^2 as the over-dispersion parameter.

In a specific application, we consider the menstrual cycles of a sample of Lese women from Zaire whereby we parameterize and analyze the mean length of the menstrual cycle, conditioned on the past cycles, and covariates such as age and the BMI (body mass index). Here we assume

$$
\begin{aligned}
\mu_{i1}(\theta) &= \gamma_0 + \gamma_1 x_{i1} \\
\mu_{ij}(\theta) &= \gamma_0 + \gamma_1 x_{ij} + f_j(\rho)\left\{\sum_{l=1}^{j-1}(Y_{il} - \gamma_0 - \gamma_1 x_{i1})\right\}, j > 1 \\
V_{ij}(\theta) &= |1 + f_j(\rho)|^{1/2}
\end{aligned}
$$

where

$$
\begin{aligned}
x_{ij} &= (B\bar{M}I)_{ij} \\
f_j(\rho) &= \rho\{\rho(j-1) + 1 - \rho\}^{-1}, \rho \neq -(j-2)^{-1}
\end{aligned}
$$

The covariate represents an average of several body mass index measurements taken on individual i at the $(l-1)$th occurrence of the event. This study encapsulates most of the challenges encountered in the analysis of recurrent events: the covariates are time-dependent random variables and there is right censoring. To estimate the parameters, we make use of the estimating equations

$$
\begin{aligned}
g_{n,1}(\theta) &= \sum_{i=1}^{n}\sum_{j=1}^{\infty} f_{ij}(\theta) Z_{ij}(\theta) I\{S_{i,j-1} < C_i\} \qquad (9.3) \\
g_{n,2}(\theta) &= \sum_{i=1}^{n}\sum_{j=1}^{\infty} b_{ij}(\eta)\left\{Z_{ij}^2(\theta) - \sigma^2\right\} I\{S_{i,j-1} < C_i\}
\end{aligned}
$$

where

$$
\begin{aligned}
Z_{ij}(\theta) &= \frac{Y_{ij} - \mu_{ij}(\theta)}{V_{ij}(\theta)} \\
f_{ij}(\theta) &= \frac{\partial \mu_{ij}(\theta)}{\partial \theta} V_{ij}^{-1}(\theta)
\end{aligned}
$$

Under some mild conditions, it is shown that (9.3) are unbiased and as $n \to \infty$, for $k = 1, 2$

$$
n^{-1} g_{n,k}(\eta_0) \to 0, a.s.
$$

In other works, [Liang and Zeger, 1986] proposed a generalized estimating equations (GEE) approach for analyzing discrete longitudinal data. Binder and Patak [1994] applied estimating equations to sampling.

9.1. Composite Likelihood

When either conditional or marginal densities are multiplied together, they form what has been called a composite likelihood which may or may not be a likelihood function in its own right. The approach is appealing as it makes it easier to model dependence in smaller dimensions. Composite likelihood may be viewed as a specific application of estimating equations where the estimating equation involves likelihood functions. Below, we shall describe the important features of composite likelihood based on the excellent review provided by Varin et al. [2011]. We also make use of their notation for ease of reference.

Let Y be an $d-$ dimensional random vector with density $f_Y(y;\theta)$ where θ is a $p-$ dimensional vector parameter taking values in Θ. Let $\{\mathcal{A}_1,\ldots,\mathcal{A}_K\}$ be a set of marginal or conditional events such that the corresponding likelihood function is given by $\mathcal{L}_\mathrm{k}(\theta;\mathrm{y}) = f(y\in\mathcal{A}_\mathrm{k};\theta)$. A composite likelihood is then defined to be the product

$$\mathcal{L}_\mathrm{C} = \prod_{\mathrm{k}=1}^{\mathrm{K}} \left(\mathcal{L}_\mathrm{k}(\theta;\mathrm{y})\right)^{\mathrm{w_k}}$$

where the$\{w_k\}$ are nonnegative weights usually selected to improve efficiency. The composite log-likelihood is

$$cl(\theta;y) = \sum_{k=1}^{K} w_k l_k(\theta;y) \tag{9.4}$$

with

$$l_k(\theta;y) = \log\mathcal{L}_\mathrm{k}(\theta;\mathrm{y})$$

Composite likelihoods have appeared in one of two varieties: composite conditional likelihoods or composite marginal likelihoods. An example of the former is given by

$$\mathcal{L}(\theta;\mathrm{y}) = \prod_{\mathrm{r}=1}^{\mathrm{d}} f_\mathrm{Y}(\mathrm{y_r}|\mathrm{y_{-r}};\theta)$$

where $y_{-r} = (y_1,\ldots,y_{r-1},y_r,\ldots,y_d)$. Examples of the latter include the simplest case of independence where

$$\mathcal{L}_\mathrm{ind}(\theta;\mathrm{y}) = \prod_{\mathrm{r}=1}^{\mathrm{d}} f_\mathrm{Y}(\mathrm{y_r};\theta)$$

In cases where our interest is in parameters involving dependence, we may consider likelihoods based either on the pairs $\{(y_r,y_s)\}$

$$\prod_{r=1}^{d}\prod_{s\neq r}^{d} f_Y(y_r,y_s;\theta_r)$$

or on their pairwise differences $\{(y_r - y_s)\}$

$$\mathcal{L}_{\text{diff}}(\theta; y) = \prod_{r=1}^{d-1} \prod_{s=r+1}^{d} f(y_r - y_s; \theta)$$

Define the following quantities:

$$\begin{aligned} u(\theta; y) &= \frac{\partial}{\partial \theta} cl(\theta; y) && (9.5) \\ H(\theta) &= E_\theta \left\{ -\frac{\partial}{\partial \theta} u(\theta; Y) \right\} && (9.6) \\ J(\theta) &= var_\theta \{ u(\theta; Y) \} && (9.7) \\ G(\theta) &= H(\theta) J(\theta)^{-1} H(\theta) && (9.8) \\ I(\theta) &= var_\theta \left\{ \frac{\partial}{\partial \theta} \log f_Y(Y; \theta) \right\} && (9.9) \end{aligned}$$

We have seen that composite likelihood is not necessarily a likelihood function. Bartlett has proposed as a criteria a set of identities satisfied in the situation of a random sample

$$\begin{aligned} E[U(\theta)] &= 0 \\ Var[U(\theta)] &= I(\theta) \end{aligned} \qquad (9.10)$$

where $[U(\theta)] = \sum_{i=1}^{n} \frac{\partial}{\partial \theta} \log f_Y(Y_i; \theta)$ and $I(\theta)$ is the Fisher information matrix. Mykland [1999] addresses the issue of when a function resembles a likelihood.

We identify $u(\theta; y)$ as the score function obtained by differentiation of the estimating equation. Let $\hat{\theta}_{CL}$ denote that value of θ which maximizes (9.4) found by solving the estimating equation:

$$u(\theta; y) = 0$$

The function $H(\theta)$, labeled the Hessian, is known as the sensitivity matrix. When $Y_1, \ldots, Y_n$ are independent and identically distributed $(i.i.d.)$ random variables, $H(\theta) = I(\theta)$. The matrix $G^{-1}(\theta)$, named after Godambe, plays the role of the information matrix in a single observation in the more general case of composite likelihood. It is also known as the sandwich information matrix, and it is computed under the assumption that the model is incorrectly specified. Under the regularity conditions outlined for estimating equations, we have that as $n \to \infty$,

$$\sqrt{n}\left(\hat{\theta}_{CL} - \theta\right) \xrightarrow{\mathcal{L}} N_p\left(0, G^{-1}(\theta)\right)$$

A good source for information on composite likelihood and its applications is Varin et al. [2011]. See also Zhao and Joe [2005] for the development in multivariate data. An application of composite likelihood leading to a new signed rank statistic is given in Sect. 5.5 of this book.

9.2. Empirical Likelihood

In the parametric setting, when the joint distribution of the observations is known up to one or more parameters, likelihood methods in statistics have been effectively used to provide tests of hypotheses and confidence intervals. On the other hand, estimating equation methods have been used in estimation problems when the complete probability distribution is not specified. Boos [1992] discusses the use of score tests in this more general context. Aitchison and Silvey [1958] considered maximum likelihood estimation and hypothesis testing subject to constraints on the parameters. In cases when the likelihood function is miss-specified, serious errors in inference can result. Empirical likelihood is a nonparametric technique that is entirely data driven. Owen [2001] provides a treatment of the subject and discusses its relation to the bootstrap. Qin and Lawless [1995] have extended empirical likelihood methods to deal with constraints on parameters. DiCiccio et al. [1989] have shown that unlike the bootstrap, empirical likelihood is Bartlett-correctable [Van der Vaart, 2000, p.238]), thus yielding second-order approximations. Below we present the basic principles of empirical likelihood and cite an application to ranking problems.

Suppose that we have a random sample from some distribution with mean μ and variance σ^2. In Sect. 3.1, we saw that the usual estimator of the variance $S^2 = \frac{\sum (X_i - \bar{X}_n)^2}{n-1}$ has mean σ^2 and variance $\sigma^4 \left(\frac{2}{n-1} + \left(\frac{\kappa_4}{\kappa_2^2} \right) \frac{1}{n} \right)$. Under the normal theory model, we have that

$$\frac{(n-1)\,S^2}{\sigma^2} \sim \chi^2_{n-1}$$

In circumstances where $\kappa_4 \neq 0$, the construction of confidence intervals based on normal theory would have incorrect coverage probability. Empirical likelihood is an approach which allows one to make use of a likelihood function determined directly from the data itself. The one distinct disadvantage is due to its computational complexity.

Let $X_1, \ldots, X_n$ be a random sample of n d-dimensional vectors from some cumulative distribution function $F(x)$ associated with a $p-$ *dimensional* parameter θ. The nonparametric likelihood is

$$\begin{aligned} L(F) &= \prod_{i=1}^{n} F(X_i) \\ &\equiv \prod_{i=1}^{n} p_i \end{aligned}$$

Let δ_x denote the distribution under which $X = x$ with probability 1. It can be shown that $L(F)$ is maximized by the empirical cumulative distribution function

$$F_n(x) = \frac{1}{n}\sum_{i=1}^{n} I[X_i \leq x]$$

provided $F \neq F_n$ [Owen, 2001, p.8]. This is analogous to the maximum likelihood method in parametric statistics where the likelihood function forms the basis for inference. We may define the nonparametric likelihood ratio

$$R(F) = \frac{L(F)}{L(F_n)} = \prod_{i=1}^{n} np_i$$

and for a parameter $\theta = T(F)$, the profile likelihood when $\theta = \mu$

$$\mathcal{R}(\theta) = \sup\{R(F)\,|T(F) = \theta, F \in \mathcal{F}\} \tag{9.11}$$

$$= \sup\left\{\prod_{i=1}^{n} np_i | p_i \geq 0, \sum p_i = 1, \sum X_i p_i = \mu\right\} \tag{9.12}$$

The confidence region is given by

$$C_{r,n} = \sup\left\{\sum_{i=1}^{n} nX_i |\ \prod_{i=1}^{n} np_i \geq r, p_i \geq 0, \sum p_i = 1\right\} \tag{9.13}$$

Owen has shown that a unique value exists for the right side of (9.11) provided μ is in the convex hull of the observed points $X_1, \ldots, X_n$ .The maximum is attained when

$$p_i = \frac{1}{n\left(1 + \lambda^T (X_i - \mu)\right)}$$

where λ satisfies

$$\sum_{i=1}^{n}\left(\frac{X_i - \mu}{1 + \lambda^T (X_i - \mu)}\right) = 0$$

The function $\mathcal{R}(\theta)$ is maximized with respect to μ at $\mu = \bar{X}_n$. Owen [2001] has shown the following theorem which can be used to obtain the critical value of r.

Theorem 9.2. *Let $X_1, \ldots, X_n$ be a random sample from some cumulative distribution function F_0. Let μ_0 be the common mean and assume a finite variance covariance matrix V of rank $q > 0$. Then $C_{r,n}$ is a convex set and as $n \to \infty$*

$$-2\log\{\mathcal{R}(\mu_0)\} \xrightarrow{\mathcal{L}} \chi_q^2$$

Owen has generalized the empirical likelihood approach to include estimation of general functions g.

Qin and Lawless [1994] established a link between empirical likelihood and estimating equations. Suppose that we have functions $g_j(\theta, x), j = 1, \ldots, r$ for which

$$E[g(\theta, X)] = 0$$

where

$$g(\theta, x) = (g_1(\theta, x), \ldots, g_r(\theta, x))^T$$

The case where $r > p$ allows one to take into account various information about a distribution.

As an example, suppose that $X \sim N(\theta, 1)$. We may set

$$g(\theta, x) = \left(x - \theta, x^2 - m(\theta)\right)^T$$

where $EX^2 = m(\theta) = 1 + \theta^2$. The empirical likelihood approach then consists of redefining the profile likelihood as

$$\mathcal{R}(\theta) = \sup\left\{\mathop{\Pi}_{i=1}^{n} np_i | p_i \geq 0, \sum p_i = 1, \sum g(\theta, X_i) p_i = 0\right\}$$

The maximization of the function

$$\sum \log p_i + \lambda\left(1 - \sum p_i\right) - n\xi^T \sum g(\theta, X_i) p_i$$

yields

$$p_i = \frac{1}{n}\frac{1}{1 + \xi^T g(\theta, X_i)}$$

The Lagrange multiplier ξ satisfies

$$0 = \sum g(\theta, X_i) p_i$$

Under some mild conditions, [Qin and Lawless, 1994] show that the empirical log-likelihood attains its minimum value $\tilde{\theta}$ in the interior of the ball $\|\theta - \theta_0\| \leq n^{-1/3}$ where θ_0 is the true value of θ. Moreover, $\tilde{\theta}, \xi\left(\tilde{\theta}\right)$ satisfy the estimating equations

$$\begin{aligned}
\sum \frac{1}{1 + \xi^T g(\theta, X_i)} g(\theta, X_i) &= 0 \\
\sum \frac{1}{1 + \xi^T g(\theta, X_i)} \left(\frac{\partial}{\partial\theta} g(\theta, X_i)\right)^T \xi &= 0
\end{aligned}$$

As well,

$$\sqrt{n}\left(\tilde{\theta}-\theta_0\right) \underset{d}{\to} N(0,V)$$

where

$$V=\left[E\left(\frac{\partial}{\partial\theta}g\right)^T\left(Egg^T\right)^{-1}E\left(\frac{\partial}{\partial\theta}g\right)\right]^{-1}.$$

We may estimate $E\left(\frac{\partial}{\partial\theta}g\right)$ and $E(gg^T)$ respectively by

$$\frac{1}{n}\sum\frac{\partial}{\partial\theta}g\left(\tilde{\theta},X_i\right),\frac{1}{n}\sum\frac{\partial}{\partial\theta}g\left(\tilde{\theta},X_i\right)g^T\left(\tilde{\theta},X_i\right)$$

In the next two sections, we apply empirical likelihood to the one- and two-sample ranking problems in randomized block designs. In the one-sample case, the test statistic coincides with the Friedman statistic.

9.2.1. Application to One-Sample Ranking Problems

Suppose that Y_{ij} for $0\le j\le r$, $1\le i\le n$ represent the jth response of $k=r+1$ treatments in the ith replication. Let Y_{ij} have a continuous cumulative distribution function F_{ij}. We would like to test the hypothesis

$$H_0: F_{i0}=\cdots=F_{ir}, i=1,\ldots,n \tag{9.14}$$

against the alternative

$$H_1: F_{ij}(x)=F_i(x-\theta_j), 0\le j\le r, i=1,\ldots,n \tag{9.15}$$

Let R_{ij} denote the rank of Y_{ij} among the k responses $\{Y_{i0},Y_{i1},\ldots,Y_{ir}\}$. Since

$$R_{ij}=1+\sum_{t=0}^{r}I_{[Y_{ij}>Y_{it}]}$$

it follows that

$$ER_{ij}=1+\sum_{t=0}^{r}P(Y_{ij}>Y_{it})$$

which clearly depends on $\theta=(\theta_1,\ldots,\theta_k)^T$. We shall further assume that the $k-$dimensional distributions $\{F_i(\cdot,\ldots,\cdot), i=1,\ldots,n\}$ are independent and identical to some cumulative distribution function $F(\cdot,\ldots,\cdot)$. As well, we suppose that F is differentiable and symmetric in its arguments. In that case, it is possible to drop the first subscript and consider the random vector of ranks $\mathbf{R}=(R_0,R_1,\ldots,R_r)^T$. Under the null hypothesis (9.14), all rankings are equally likely in the space $\mathcal{P}=\{\omega_i\}$ of all possible $k!$ permutations, where ω is a column vector permutation of the integers $(1,2,\ldots,k)$,$k=r+1$. It follows that for any components R,R'

$$P(R=i)=\frac{1}{k}, P(R=i,R'=j)=\frac{1}{k(k-1)}, i\neq j$$

and consequently

$$ER=\frac{(k+1)}{2}, VarR=\frac{k^2-1}{12}, Cov(R,R')=-\frac{k+1}{12}$$

The covariance matrix

$$Cov(\mathbf{R})=\begin{cases}\frac{k^2-1}{12} & \text{on the diagonal}\\ \\ -\frac{k+1}{12} & \text{off the diagonal}\end{cases}$$

is a k x k singular matrix of rank $(k-1)$. Let $\mathbf{1}_k$ be a $k-$dimensional vector of ones and let A be the k x $k!$ matrix given by

$$A=\left(\omega_1-\frac{k+1}{2}\mathbf{1}_k,\ldots,\omega-\frac{k+1}{2}\mathbf{1}_k\right).$$

Noting that

$$\sum_{i=1}^{k!}\omega_i=(k-1)!\left[\frac{k(k+1)}{2}\right]\mathbf{1}_k$$

and

$$\mathbf{R}=\sum_{i=1}^{k!}\omega_i I_{[\mathbf{R}=\omega_i]}$$

it follows that

$$\begin{aligned}E\mathbf{R} &= \sum_{i=1}^{k!}\omega_i P(\mathbf{R}=\omega_i)\\ &= \sum_{i=1}^{k!}\omega_i\pi_i\\ &= A\pi+\frac{k+1}{2}\mathbf{1}_k \qquad (9.16)\end{aligned}$$

where $\pi=(\pi_1,\ldots,\pi_{k!})'$.

In view of the fact that the covariance matrix of $\mathbf{R}$ is singular, we shall consider instead the reduced ranking $\mathbf{S}=\{R_1,\ldots,R_r\}$. The covariance matrix of $\mathbf{S}$ is nonsingular of full rank r with

$$Cov(\mathbf{S}) = \begin{cases} \frac{k^2-1}{12} & \text{on the diagonal} \\ \\ -\frac{k+1}{12} & \text{off the diagonal} \end{cases}$$

The inverse takes the form of

$$\begin{aligned} (Cov(\mathbf{S}))^{-1} &= \frac{12}{k(k+1)} \begin{cases} 2 & \text{on the diagonal} \\ \\ 1 & \text{off the diagonal} \end{cases} \\ &= \frac{12}{k(k+1)} (J_r + I_r) \end{aligned}$$

where J_r is an r x r matrix of ones and I_r is the identity matrix of order r.

Set $\mu = E\mathbf{S}$. Let us consider an empirical likelihood approach for testing hypotheses about the mean μ Owen [2001]. Suppose that we observe a random sample of n rankings $\mathbf{S}_1, \ldots, \mathbf{S}_n$ and suppose that $\Psi(Y_{ij}, 0 \leq j \leq r, \mu)$ is a vector-valued function. Let μ_0 be a fixed point of μ for which the variance-covariance matrix of $\Psi(Y_{ij}, 0 \leq j \leq r, \mu_0)$ is finite and has rank $q > 0$. Let w_i be the probability mass placed at $\{Y_{ij}, 0 \leq j \leq r\}$. If μ_0 satisfies $E(\Psi(Y_{ij}, 0 \leq j \leq r, \mu_0)) = 0$, then

$$l_E(\mu_0) = -2 \log \mathcal{R}(\mu_0) \to \chi_q^2 \tag{9.17}$$

where the profile empirical likelihood ratio function is given by

$$\mathcal{R}(\mu) = \max \left\{ \Pi_{i=1}^n n w_i | \sum_{i=1}^n w_i \Psi(Y_{ij}, 0 \leq j \leq r, \mu) = \mathbf{0}, w_i \geq 0, \sum_{i=1}^n w_i = 1 \right\} \tag{9.18}$$

We now choose

$$\Psi(Y_{ij}, 0 \leq j \leq r, \mu) = (\mathbf{S}_i - \mu), q = r, \mu_0 = \frac{k+1}{2} \mathbf{1}_r$$

The results of the maximization adapted to this choice show that an empirical log-likelihood-ratio statistic $l_E(\mu)$ for the mean can be obtained with

$$\hat{w}_i = \frac{1}{n} \frac{1}{\{1 + \mathbf{u}^T (\mathbf{S}_i - \mu_0)\}}, i = 1, \ldots, n \tag{9.19}$$

and $\mathbf{u}$ is a $(r-1)$ x 1 vector of Lagrange multipliers satisfying

$$\sum_{i=1}^n \frac{(\mathbf{S}_i - \mu_0)}{1 + \mathbf{u}^T (\mathbf{S}_i - \mu_0)} = \mathbf{0}.$$

Owen (Theorem 3.2 p.219) has shown that $\|\mathbf{u}\| = O_p\left(n^{-\frac{1}{2}}\right)$, and

$$l_E(\mu_0) = n\left(\bar{\mathbf{S}} - \mu_0\right)^T (Cov(\mathbf{S}))^{-1}\left(\bar{\mathbf{S}} - \mu_0\right) + o_p(1) \tag{9.20}$$

where $\bar{\mathbf{S}} = n^{-1}(\mathbf{S}_1 + \cdots + \mathbf{S}_n)$ and that hence $l_E(\mu_0) \to_d \chi^2_{k-1}$. The empirical likelihood ratio test would then reject the null hypothesis whenever

$$l_E(\mu_0) = -2\sum \log(n\hat{w}_i) > \chi^2_{k-1}(\alpha)$$

where $\chi^2_{k-1}(\alpha)$ is the upper $100(1-\alpha)$ % point of a chi-square distribution with $(k-1)$ degrees of freedom and the $\{\hat{w}_i\}$ are given in (9.19).

As in Owen, a $100(1-\alpha)$ % confidence region may be written as

$$C_{r,n} = \left\{\sum_{i=1}^{n} \hat{w}_i \mathbf{S}_i | \Pi_{i=1}^n n\hat{w}_i \geq r\right\}$$

with $r = \exp\left(-\frac{\chi^2_{k-1}(\alpha)}{2}\right)$.

We may now relate (9.20) to the vector of relative frequencies $\hat{\pi}$. Setting $\bar{R}_0 = n^{-1}\sum_i R_{0i}$, we have

$$\begin{aligned}
\left(\bar{\mathbf{S}} - \mu_0\right)^T (Cov(\mathbf{S}))^{-1}\left(\bar{\mathbf{S}} - \mu_0\right) &= \frac{12}{k(k+1)}\left(\bar{\mathbf{S}} - \mu_0\right)^T (J_r + I_r)\left(\bar{\mathbf{S}} - \mu_0\right) \\
&= \frac{12}{k(k+1)}\left(\bar{\mathbf{S}} - \mu_0\right)^T \left(\mathbf{1}_r\mathbf{1}_r^T + I_r\right)\left(\bar{\mathbf{S}} - \mu_0\right) \\
&= \frac{12}{k(k+1)}\left\{\left(\bar{\mathbf{S}} - \mu_0\right)^T \left(\mathbf{1}_r\mathbf{1}_r^T\right)\left(\bar{\mathbf{S}} - \mu_0\right) + \left(\bar{\mathbf{S}} - \mu_0\right)^T\left(\bar{\mathbf{S}} - \mu_0\right)\right\} \\
&= \frac{12}{k(k+1)}\left\{\left(\bar{R}_0 - \frac{k+1}{2}\right)^2 + \left(\bar{\mathbf{S}} - \mu_0\right)^T\left(\bar{\mathbf{S}} - \mu_0\right)\right\} \\
&= \frac{12}{k(k+1)}\left\{\left(\bar{\mathbf{R}} - \mu_0\right)^T\left(\bar{\mathbf{R}} - \mu_0\right)\right\} \\
&= \frac{12}{k(k+1)}(A\hat{\pi})^T (A\hat{\pi}) \qquad (9.21)
\end{aligned}$$

since $A\mu_0 = \frac{k+1}{2}A\mathbf{1}_r = 0$. We recognize that (11.33) is the usual Friedman statistic. Alvo et al. [1982] and that in view of (9.16)

$$(A\hat{\pi})^T (A\hat{\pi}) = \sum_{i=0}^{k}\left(\bar{R}_i - \frac{(t+1)}{2}\right)^2$$

where $\bar{R}_i$ is the average of the ranks assigned to object i. Hence, the empirical log-likelihood-ratio statistic is asymptotically equivalent in distribution to the Friedman statistic. As such, the Friedman statistic shares, at least asymptotically, many of the properties of the empirical likelihood ratio statistic. Specifically, we can obtain narrower confidence intervals for the mean.

The expression for the power function may be derived from Sen [1968]. Specifically, it was shown that the power function for the Friedman statistic is given by the non-central chi-square with non-centrality parameter

$$\lambda = \frac{nk}{k+1}\left[\int_{-\infty}^{\infty} f(x,x)\,dx\right]^2 \sum_{j=0}^{r} \theta_j^2$$

where $f(\cdot,\cdot)$ is the joint marginal density function derived from F of any pair of elements.

9.2.2. Application to Two-Sample Ranking Problems

Consider now the two-sample problem. Suppose that Y_{lij} for $0 \le j \le r$, $1 \le i \le n_l, l = 1,2$ represent the jth response of $k = r+1$ treatments in the ith replication, lth population. Let Y_{lij} have a continuous cumulative distribution function F_{lij}. We assume that $F_{lij}(x) = F_{lj}(x - \theta_l)$ for all $1 \le i \le n_l$. We would like to test the null hypothesis

$$H_0 : F_{lij}(x) = F_j(x)$$

against the alternative

$$H_1 : F_{lij}(x) = F_j(x - \theta_l)$$

Let R_{lij} denote the rank of Y_{lij} among the k responses $\{Y_{li0}, \ldots, Y_{lik}\}$ and let $\mathbf{S}_{li}$ $= (R_{li1}, \ldots, R_{lik})$ be the ith vector of rankings in the lth population. We shall suppose that we observe a random sample of n_l rankings $\{\mathbf{S}_{li}\}, l = 1,2$. Set $\gamma_l = E\mathbf{S}_{li}, \gamma = \gamma_2 - \gamma_1$ where

$$\gamma_1 = \mu - \frac{n_2}{n_1}\gamma, \gamma_2 = \mu - \frac{n_1}{n_2}\gamma.$$

Under the null hypothesis, $\gamma = \mathbf{0}$. Suppose that $\Psi_l(Y_{lij}, 0 \le j \le r, 1 \le i \le n_l, \mu, \gamma)$ is a vector-valued function. Let γ_0 be a fixed point of γ for which the variance-covariance matrix of $\Psi_l(Y_{lij}, 0 \le j \le r, 1 \le i \le n_l, \mu, \gamma)$ is finite and has rank $q > 0$. Let w_{li} be the probability mass placed at

$$\Psi_l(Y_{lij}, 0 \le j \le r, 1 \le i \le n_l, \mu, \gamma).$$

If γ_0 satisfies

$$E(\Psi_l(Y_{lij}, 0 \le j \le r, 1 \le i \le n_l, \mu, \gamma_0)) = 0,$$

then we shall show that

$$l_E(\mu,\gamma_0) = -2\log\mathcal{R}(\mu,\gamma_0) \to \chi_q^2$$

where the profile empirical likelihood ratio function is given by

$$\mathcal{R}(\mu,\gamma) = \max\left\{\Pi_{l=1}^2\Pi_{i=1}^{n_l} n_l w_{li} \middle| \sum_{l=1}^{2}\sum_{i=1}^{n_l} w_{li}\Psi_l\left(Y_{lij}, 0\le j\le r, 1\le i\le n_l,\mu,\gamma\right) = \mathbf{0}, w_{ij}\ge 0, \sum_{i=1}^{n_l} w_{ij} = 1\right\}$$

The maximization is solved by using Lagrange multipliers. The Lagrangian is

$$L = \sum_l\sum_i \log w_{li} - \sum_{l=1}^{2}\nu_l\left(\sum_{i=1}^{n_l} w_{li} - 1\right) - \sum_l n_l t_l \sum_i w_{li}\Psi_l\left(Y_{lij}, 0\le j\le r, 1\le i\le n_l,\mu,\gamma\right)$$

where $\{\nu_l, t_l\}$ are the Lagrange multipliers. Setting

$$\Psi_l\left(Y_{lij}, 0\le j\le r, 1\le i\le n_l,\mu,\gamma\right) = \left(\mathbf{S}_{li} - \gamma_l\right), q = r,$$

we see that

$$\hat{w}_{li} = \frac{1}{n_l}\frac{1}{\left\{1+\mathbf{u}_l^T\left(\mathbf{S}_{li}-\gamma_l\right)\right\}}, i = 1,\ldots,n_l, l = 1,2$$

and $\mathbf{u}_l$ is a $(r-1)$ x 1 vector of Lagrange multipliers satisfying

$$\sum_{i=1}^{n_l}\frac{\left(\mathbf{S}_{li}-\gamma_l\right)}{1+\mathbf{t}_l^T\left(\mathbf{S}_{li}-\gamma_l\right)} = \mathbf{0}, l = 1,2.$$

Owen [2001, Theorem 3.2 p.219] has shown that $\|\mathbf{u}_l\| = O_p\left(n_l^{-\frac{1}{2}}\right)$, for each $l = 1,2$. Moreover, it can be seen that under the null hypothesis

$$l_E(\mu) = \sum_{l=1}^{2} n_l\left(\bar{\mathbf{S}}_l - \mu\right)'\left(Cov(\mathbf{S})\right)^{-1}\left(\bar{\mathbf{S}}_l - \mu\right) + o_p(1)$$

where $\bar{\mathbf{S}}_l = n_l^{-1}\sum_{i=1}^{n_l}\mathbf{S}_{li}$. Estimating μ by

$$\hat{\mu} = \frac{n_1\bar{\mathbf{S}}_1 + n_2\bar{\mathbf{S}}_2}{n_1+n_2}$$

we have that

$$l_E(\hat{\mu}) = \frac{n_1 n_2}{n_1+n_2}\left(\bar{\mathbf{S}}_1 - \bar{\mathbf{S}}_2\right)^T\left(Cov(\mathbf{S})\right)^{-1}\left(\bar{\mathbf{S}}_1 - \bar{\mathbf{S}}_2\right) + o_p(1)$$

from which it follows that $l_E(\hat{\mu}) \to_d \chi_{k-1}^2$ *as* $n\to\infty$, with $n_l/n \to \lambda_l$, $l = 1,2$.

We have seen that the empirical likelihood approach does not require the formulation of parameters. This makes it more robust than the usual parametric likelihood function. Chen et al. [2009] discuss how large the vector size can be in relation to the data. The two-sample ranking problem has also been discussed in Alvo and Yu [2014]. Tang and Leng [2010] address the issue of whether or not the empirical likelihood method can be adapted to variable selection. We do not pursue these topics in this book.

10. Symbolic Data Analysis

10.1. Introduction

Big data nowadays are available in the Internet, in businesses, and in government databases, among others, all the while challenging the capacity of computers. The purpose of symbolic data analysis (SDA) consists of analyzing, visualizing, classifying, and reducing the information contained in the symbolic data matrix. In this paradigm, data comes in several forms. In particular, it may take the form of classes that can be described by intervals, histograms, sets of categories, etc. For example, rather than specifying a single number for systolic blood pressure, one may specify an interval which then incorporates the uncertainty about the measurement. This is analogous to constructing confidence intervals in inferential statistics. The new type of data obtained is first used to build the symbolic data table where the rows are classes and the variables appearing as columns can take symbolic values. In that format, it can then be transformed into a structured data table with paired symbolic-valued variables. Among the common data analysis methods, principal component analysis (PCA) is widely used to discover and to visualize the main structure of a multidimensional data set. The proposed methods for analyzing symbolic data have been described in several papers [Diday, 2000, 2008, 2016]. A software is available at https://www.ceremade.dauphine.fr/~touati/SODAS/. The analyses which follow were performed in Liu [2019].

10.2. Some Examples

To illustrate the basic concepts and data types, suppose we have a classical data set with information on films. Table 10.1 shows the classical data set. Table 10.2 exhibits the data aggregated by directors.

Here, we call film names from the classical data set as *individuals* and call director A and director B *symbolic objects*. In Table 10.2, *filming year* is an interval-valued variable, which is described by a range of values. *Filming location 1* is a multi-nominal variable shown as a set of values. The value can be numerical or categorical in real situations. *Filming location 2* is a modal variable, which is multi-nominal data with corresponding probabilities. Hence, London appears as a location 50% of the time.

In a second example, suppose we wish to record the daily activities of a consumer. If Y represents time spent watching television per day, then we may record it as a discrete (probability) distribution:

$$Y(k) = ((0, 0.5), (1, 0.4), (2, 0.05), (3, 0.05))$$

M. Alvo, *Statistical Inference and Machine Learning for Big Data*, Springer Series in the Data Sciences, https://doi.org/10.1007/978-3-031-06784-6_10

Film name	Director	Filming year	Filming location
Film1	A	1990	London
Film2	A	1998	Vancouver
Film3	B	1995	Toronto
Film4	B	2001	Ottawa
Film5	B	1998	Toronto
Film6	B	2006	Vancouver

Table 10.1.: Synthetic classical film data

meaning television is not watched with probability 0.50, only 1 h of television is watched with probability 0.40, and 2 or 3 h are watched with probability 0.05 for each.

A further example, in analyzing groups of individuals, if k denotes the country "Belgium," the linguistic variable Y "official language" can be specified either by a set of languages: Y (k) = {Dutch, French, German} or by a frequency distribution: Y (k) = ((Dutch, 0.50), (French, 0.4), (German, 0.10)) which means that 50% of all Belgian people live in towns where Dutch is the official language, etc. (All non–specified languages are supposed to have the frequency zero). Other types of symbolic data include probabilistic data, fuzzy data, measurements with tolerance intervals, etc.

The Symbolic Data Analysis Software, SODAS for short, is a tool used to process symbolic data. It consists of various modules for data processing and symbolic analysis. The software works on symbolic data tables which means that prior to conducting any analyses, the classical data must be converted into symbolic data. This can be done by the built-in module *SB2SO*. While aggregating, the first column from the original data set will be recognized as the unique key of individuals, whereas the second and subsequent columns will be defined as the symbolic objects. Table 10.3 is an example of a symbolic data table in SODAS.

In the main interface of SODAS, we can conduct various analyses on the symbolic data set, including basic descriptive statistics, discriminant analysis, clustering and constructing a decision tree, etc. We can create analysis pipelines by adding a new *chaining*. We will use some data sets from the SODAS official website to illustrate the calculation of descriptive statistics and their realization in SODAS. One of the examples is film data, which is aggregated by directors. The data set contains five interval variables and four modal variables. Another example is the traveling data aggregated by service type which consists of four interval variables, four categorical variables, and five modal variables.

10.3. Interval Data

10.3.1. Frequency

Interval data may occur in many different situations. It may be used to describe ranges of variable values, such as income of individuals in a company, the weight range and size of horses in a competition, or imprecise data arising from repeated measures or

Director	Filming year	Filming location 1	Filming location 2
A	[1990:1998]	[London, Vancouver]	[London (0.5), Vancouver] (0.5)
B	[1995:2006]	[Toronto, Vancouver, Ottawa]	[Toronto (0.5),Vancouver (0.25),Ottawa (0.25)]

Table 10.2.: Synthetic symbolic film data

Table 10.3.: Symbolic table

	Genre	Nombre_Oscars	Acteur_Principa	Bénéfice(K)	Continent_Tourn	Pays	Nombre_Films	Nombre_Oscars_r
Kubrick	Scien (0.50), Film (0.50)	[1.00 : 4.00]	Keir (0.50), Kirk (0.50)	[78000.00 : 180200.00]	Europ (0.50), Amér (0.50)	USA (1.00)	[13.00 : 13.00]	[1.00 : 1.00]
Mctiernan	Suspe (1.00)	[1.00 : 1.00]	Sean (1.00)	[196200.00 : 196200.00]	Amér (1.00)	USA (1.00)	[11.00 : 11.00]	[0.00 : 0.00]
Cameron	Scien (0.25), Fanta (0.50), Drame (0.25)	[1.00 : 11.00]	Ed Ha (0.25), Sigou (0.25), Arnol (0.25), Leona (0.25)	[38000.00 : 1635000.00]	Europ (0.25), Amér (0.75)	Canad (1.00)	[12.00 : 12.00]	[3.00 : 3.00]
Coppola	Fanta (0.33), Drame (0.33), Polic (0.33)	[2.00 : 3.00]	Marlo (0.67), Gary (0.33)	[45000.00 : 262000.00]	Amér (0.67), Asie (0.33)	USA (1.00)	[22.00 : 22.00]	[5.00 : 5.00]
Howard	Drame (0.33), Film (0.33), Comé (0.33)	[1.00 : 2.00]	Tom H (0.33), Russe (0.33), Jim C (0.33)	[268000.00 : 277000.00]	Amér (1.00)	USA (1.00)	[15.00 : 15.00]	[1.00 : 1.00]
Fleming	Drame (0.50), Comé (0.50)	[2.00 : 8.00]	Vivie (0.50), Judy (0.50)	[1800.00 : 388000.00]	Amér (1.00)	USA (1.00)	[39.00 : 39.00]	[1.00 : 1.00]
Burton	Fanta (0.25), Actio (0.25), Comé (0.25), Comé (0.25)	[1.00 : 2.00]	Micha (0.50), Johnn (0.50)	[-6000.00 : 378200.00]	Europ (0.50), Amér (0.50)	USA (1.00)	[16.00 : 16.00]	[0.00 : 0.00]
Gibson	Drame (1.00)	[5.00 : 5.00]	Mel G (1.00)	[130600.00 : 130600.00]	Europ (1.00)	USA (1.00)	[2.00 : 2.00]	[2.00 : 2.00]
Welles	Drame (1.00)	[1.00 : 1.00]	Josep (1.00)	[9200.00 : 9200.00]	Amér (1.00)	USA (1.00)	[20.00 : 20.00]	[1.00 : 1.00]
Spielberg	Fanta (0.75), Film (0.25)	[2.00 : 5.00]	Tom H (0.25), Dee W (0.25), Sam N (0.25), Richa (0.25)	[317700.00 : 850000.00]	Europ (0.25), Amér (0.50), Asie (0.25)	USA (1.00)	[24.00 : 24.00]	[3.00 : 3.00]
Zemeckis	Fanta (0.25), Comé (0.25), Comé (0.25), Anima (0.25)	[1.00 : 6.00]	Tom H (0.25), Bruce (0.25), Bob H (0.25), Micha (0.25)	[94000.00 : 618800.00]	Amér (1.00)	USA (1.00)	[12.00 : 12.00]	[1.00 : 1.00]
Scott	Film (0.50), Avent (0.50)	[1.00 : 5.00]	Russe (0.50), Susan (0.50)	[73500.00 : 307000.00]	Europ (0.50), Amér (0.50)	Angle (1.00)	[13.00 : 13.00]	[0.00 : 0.00]
Allen	Comé (1.00)	[3.00 : 3.00]	Caine (1.00)	[73600.00 : 73600.00]	Amér (1.00)	USA (1.00)	[37.00 : 37.00]	[3.00 : 3.00]
Ford	Avent (0.50), Weste (0.50)	[1.00 : 2.00]	Jon H (0.50), John (0.50)	[1469.00 : 1600.00]	Amér (1.00)	USA (1.00)	[136.00 : 136.00]	[4.00 : 4.00]
Emmerich	Scien (1.00)	[1.00 : 1.00]	Bill (1.00)	[722900.00 : 722900.00]	Europ (1.00)	Allem (1.00)	[10.00 : 10.00]	[0.00 : 0.00]
Stone	Drame (0.50), Film (0.25), Film (0.25)	[1.00 : 4.00]	Kevin (0.25), Tom C (0.25), Tom B (0.25), Micha (0.25)	[71000.00 : 161400.00]	Amér (0.75), Asie (0.25)	USA (1.00)	[20.00 : 20.00]	[3.00 : 3.00]
Lucas	Scien (1.00)	[6.00 : 6.00]	Mark (1.00)	[502700.00 : 502700.00]	Europ (1.00)	USA (1.00)	[15.00 : 15.00]	[1.00 : 1.00]
Minghella	Drame (1.00)	[9.00 : 9.00]	Ralph (1.00)	[185000.00 : 185000.00]	Afriq (1.00)	Angle (1.00)	[8.00 : 8.00]	[1.00 : 1.00]
Demme	Polic (1.00)	[5.00 : 5.00]	Jody (1.00)	[238000.00 : 238000.00]	Amér (1.00)	USA (1.00)	[20.00 : 20.00]	[1.00 : 1.00]

confidence interval estimation. Interval-valued data may also arise from the aggregation of a typically large data set into one of more manageable size, when real values are generalized by intervals. In this section, we illustrate the calculation of frequency for interval data as described by Billard and Diday [2003].

Suppose we have an interval variable whose observation value for object u is in the form of the interval $Z(u) = [a_u, b_u]$, for $u \in E = \{1, \ldots, m\}$. Then the frequency p of a class I_g is given by

$$p = \frac{1}{m} \sum_{u \in E} \frac{\|Z(u) \bigcap I_g\|}{\|Z(u)\|} \tag{10.1}$$

where $Z(u)$ is the length of the interval, and $\|Z(u) \bigcap I_g\|$ is the part of the interval covered by the classI_g.

To illustrate, consider the interval-valued variable *Year (Année)* from the cinema data. We divide the range of years 1937–2001 into 10 classes. There are a total of 25 symbolic objects in the data set. Table 10.4 indicates the items involved in Class1, Class4, and Class9.

First, we calculate the frequency of Class 1[1937,1943.4]. The "length" of Ford[1937, 1939] is 1939–1937 and the part covered by Class1 is 1939–1937. Hence, we can have the percentage of covered length as$\frac{1939-1937}{1939-1937}$. Fleming, Welles, and Hitchcock is single valued, and all are covered by Class 1, so their percentage count is 1. Therefore, the frequency of Class 1 is by calculation

$$p\,(Class1) = \frac{1}{25} \times \left(\frac{1939 - 1937}{1939 - 1937} + 1 + 1 + 1 \right) = 0.16.$$

For Class 4[1956.2,1962.6], the related director Kubrick only produced a movie from 1960 to 1968, so only part of the year is covered by Class 4. We can obtain the frequency as follows:

$$p\,(Class4) = \frac{1}{25} \times \frac{1962.6 - 1960}{1968 - 1960} = 0.013.$$

Similarly, we calculate the frequency for Class 9:

$$p\,(Class9) = 0.2105.$$

10.3.2. Sample Mean and Sample Variance

The sample mean $\overline{Z}$ and sample variance S^2 for interval data $\{(a_u, b_u)\}$ are defined to be respectively

$$\overline{Z} = \frac{1}{m} \sum_{u \in E} (b_u + a_u)/2$$

$$S^2 = \frac{1}{3m} \sum_{u \in E} \left(b_u^2 + a_u b_u + a_u^2 \right) - \frac{1}{4m^2} \left[\sum_{u \in E} (b_u + a_u) \right]^2$$

Class range	Relevant symbolic objects
Class1[1937,1943]	Ford[1937,1939],Fleming[1939:1939], Welles[1941,1941],Hitchcock[1940,1940]
Class4[1956,1962]	Kubrick[1960,1968]
Class9[1988,1994]	Demme[1990,1990],Stone[1986,1991], Scott[1991,2000],Spielberg[1977,1998],Zemeckies[1985,1994] Burton[1988,1999], Coppola[1972–1992],Mctiernan[1990,1990], Cameron[1986,1997]

Table 10.4.: Film data—relevant symbolic objects of classes

10.3.3. Realization In SODAS

The module *View* in SODAS allows one to visualize each symbolic object by a star-shaped graph. When we click the axes on the graph, we are allowed to see the frequency or range of the variable. The 2D and 3D graphs from the *film data* for a single symbolic object Director Cameron are exhibited in Figs. 10.1 and 10.2. It can be seen that director Cameron made movies from 1987 to 1997, half of which are fantasy films and one quarter each are in science fiction and drama. He earned three Oscars and one of his films has received 11 Oscars. In addition, one of his films earned a profit of 1.635 billion dollars which is the maximum of his total 12 films.

Figure 10.1.: Film data—2D graph of symbolic object Cameron

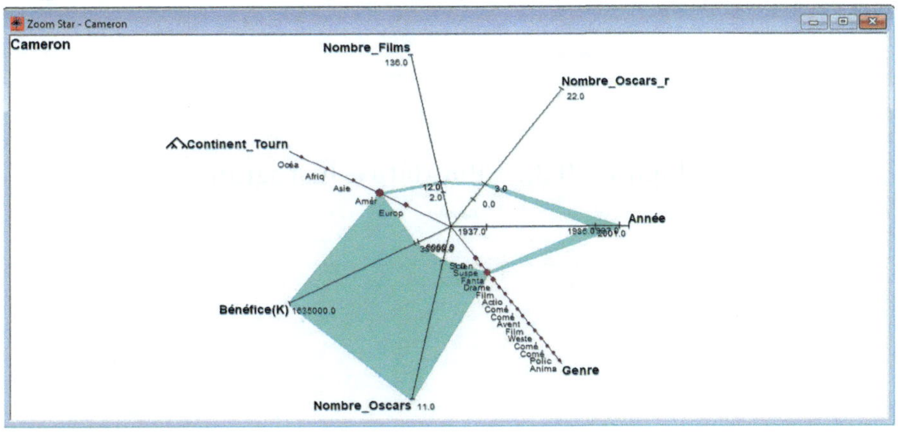

Figure 10.3 is the histogram of variable "Année." It can be drawn by module DSTAT in SODAS.

The numerical mean and standard deviation for interval-valued variables can be obtained from the *Numerical and Symbolic Characteristics* in *the DSTAT* module. For example, in Fig. 10.4, we see the mean value for *profit (benefice)* is \$ 239,717.3 and the standard deviation is \$ 208,934.

In addition, we can draw a biplot in *the DSTAT* module. The plot presents two interval variables as a rectangle in the plane, where each side of the rectangle is the range of the axis variable on the symbolic object.

10.4. Multi-nominal Data

10.4.1. Frequency

To illustrate how to construct the histogram of multi-nominal data, consider the variable *price* from *travel data*. There are 18 symbolic objects in total. The related data for frequencies calculation are gathered in Table 10.5. The frequency of $\xi_k \in Z$ is

$$f(\xi_k) = \frac{\sum_{u \in E} CTR(\xi_k)}{m}$$

Figure 10.2.: Film data—3D graph of symbolic object Cameron with distribution of film type

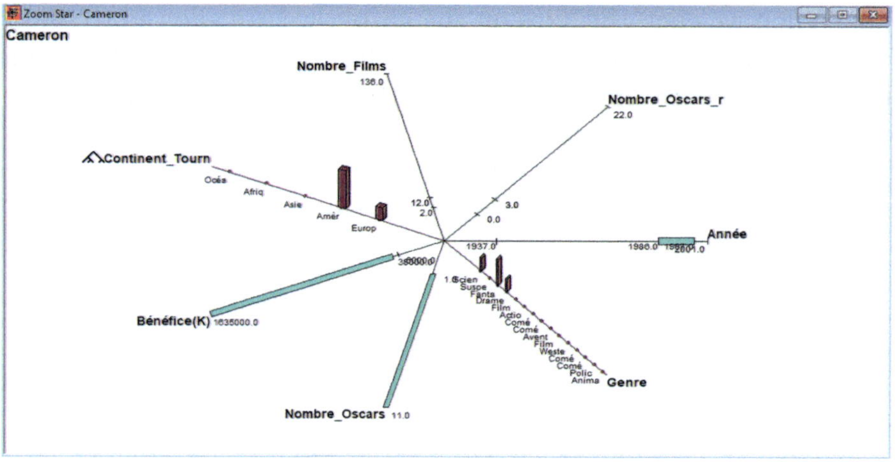

Figure 10.3.: Film data—histogram

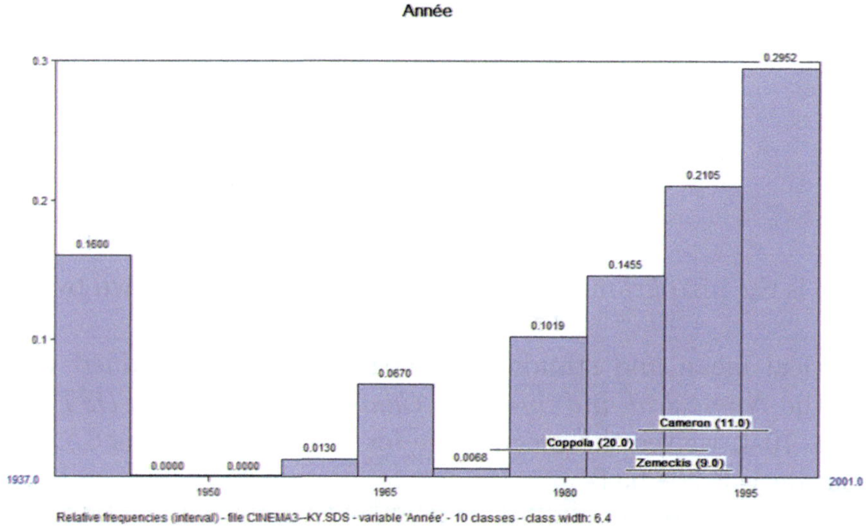

where $CTR(\xi_k) = \frac{C_u}{N_u}$, C_u is the number of occurrence of ξ_k in a cell and N_u is number of all items in a cell.

Here, the prices are not numbers. The value of CTR for restaurant in France is 1, whereas for poolside bar in the USA is $\frac{1}{2}$, so we get the frequency below:

$$f\,(Price - 85) = \frac{1}{18} \times \left(1 + \frac{1}{2}\right) = 0.8333$$

Similarly, we have

$$f\,(Price - 170) = \frac{1}{18} \times 1 = 0.0.0556$$

Figure 10.4.: Film data—mean and standard deviation

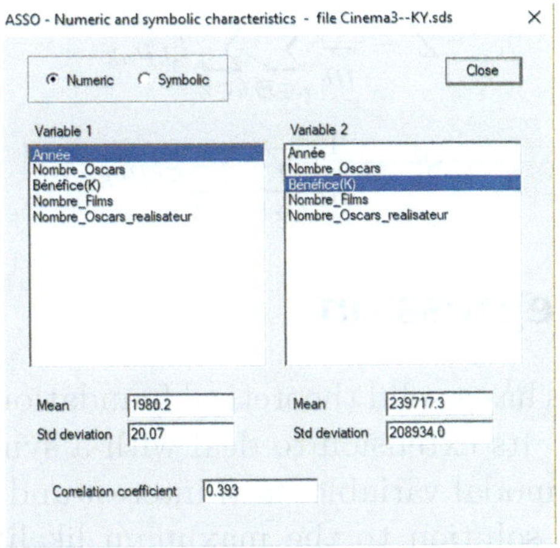

Price	Relevant data
Price—85	Restaurant in France (85), poolside bar in the USA (80,85)
Price—170	Hotel room in France (170), poolside bar in the USA (80,85)
Price—195	Hotel room in the USA (192,195)
Price—120	Bungalow in France (120), poolside bar in France (120)

Table 10.5.: Traveling data—relevant data for price classes

$$f\left(Price - 195\right) = \frac{1}{18} \times \frac{0.5}{18} = 0.278$$

$$f\left(Price - 120\right) = \frac{1}{18} \times (1+1) = 0.1111.$$

Some definitions for the computation of the sample mean and sample variance follow.

Definition 10.1. The sample mean and sample variance for quantitative multi-valued variable are defined to be

$$\overline{Z} = \sum_{\xi_k \in Z} \xi_k f\left(\xi_k\right)$$

$$S_Z^2 = \sum_{\xi_k \in Z} \left[\xi_k - \overline{Z}\right] f\left(\xi_k\right)$$

where $f\left(\xi_k\right)$ is the related frequency.

Suppose p_{uk} is the probability of ξ_k in object u. The frequency for modal data is defined to be

$$f\left(\xi_k\right) = \frac{1}{m} \sum_{u \in E} p_{uk}\left(\xi_k\right)$$

Definition 10.2. The sample mean and variance for quantitative modal variable are defined to be

$$\overline{Z} = \frac{1}{m} \sum_{u \in E} \sum_{k \in Z} \xi_k p_{uk}$$

$$S^2 = \frac{1}{m} \sum_{u \in E} \sum_{k \in Z} \xi_k^2 p_{uk}$$

10.5. Symbolic Regression

Classical regression analysis has a solid theoretical foundation which works well on single-valued numerical variables. Its extension to deal with a symbolic data set consisting of multi-nominal, interval, or modal variables is of interest and is the object of this section. In classical regression, the solution to the maximum likelihood optimization problem yields the estimate $\hat{\beta}$ given by

$$\hat{\beta} = \left(X^T X\right)^{-1} X^T Y$$

where X is the design matrix containing the predictor or independent variables and Y is the vector of the dependent variable. When $p = 1$, we have

$$\hat{\beta}_1 = \frac{\sum_{i=1}^{n} \left(X_i - \overline{X}\right)\left(Y_i - \overline{Y}\right)}{\sum_{i=1}^{n} \left(X_i - \overline{X}\right)^2} = Cov\left(Y, X\right)/Var\left(X\right)$$

$$\hat{\beta}_0 = \overline{Y} - \hat{\beta}_1 \overline{X}$$

The symbolic regression methodology is based heavily on classical regression theory, and the calculation of the variance and covariance is the key. Here, we only introduce symbolic regression for both interval and modal data.

10.5.1. Symbolic Regression for Interval Data

Suppose we have an interval-valued data set where the dependent variables are $Y = [a, b]$, and the independent variables are $X = (X_1, \ldots, X_p)$ with$X_j = [c_j, d_j], j = 1, \ldots, p$. For $p = 1$, we can estimate the coefficients$\hat{\beta}_1$ and $\hat{\beta}_0$ as below:

$$\hat{\beta}_1 = Cov\left(Y, X\right)/S_x^2 \tag{10.2}$$

$$\hat{\beta}_0 = \overline{Y} - \hat{\beta}_1 \overline{X} \tag{10.3}$$

where $\overline{Y} = \sum_{u=1}^{m} \left(a_u + b_u\right)/2m$, $\overline{X} = \sum_{u=1}^{m} \left(c_u + d_u\right)/2m$ and S_x^2 can be estimated as before. The $Cov\left(Y, X\right)$ term is the empirical covariance between two interval-valued variables.

10.5.2. Symbolic Regression for Modal Data

The idea of symbolic regression for modal data is to treat each modality of a modal-valued variable as a single variable. For simplicity of notation, we assume $p = 1$ to explain the methodology. Suppose we have a data set with a single-valued numerical-dependent variable Y and a modal-valued variable $X = \{X_1, \ldots, X_4\}$. The modality is ξ_k and p_{uk} is the corresponding probability of ξ_k for object u. If $X_1 = [\xi_1 (p_{11}), \xi_2 (p_{12}), \xi_3 (p_{13}), \xi_4 (p_{14})]$, it will be interpreted as shown in Table 10.6.

Variable	ξ_1	ξ_2	ξ_3	ξ_4
Value	p_{11}	p_{12}	p_{13}	p_{14}

Table 10.6.: Interpret modal variable into several variables

Thus, the modal variable can be transformed into classical variables so that classical regression can be conducted on the new data. What needs to be mentioned is, even though there are k modalities, we only need $(k-1)$ new variables for regression.

10.5.3. Symbolic Regression in SODAS

In SODAS, the option *SREG* in *the Discriminant & Regression* module can be used to implement symbolic regression. In the parameter interface, we are asked to choose a dependent variable and the predictor variables. Also, we can modify the tests of hypothesis and the confidence level in the parameter interface as shown in Fig. 10.5.

Figure 10.5.: Symbolic regression (SREG) parameter interface

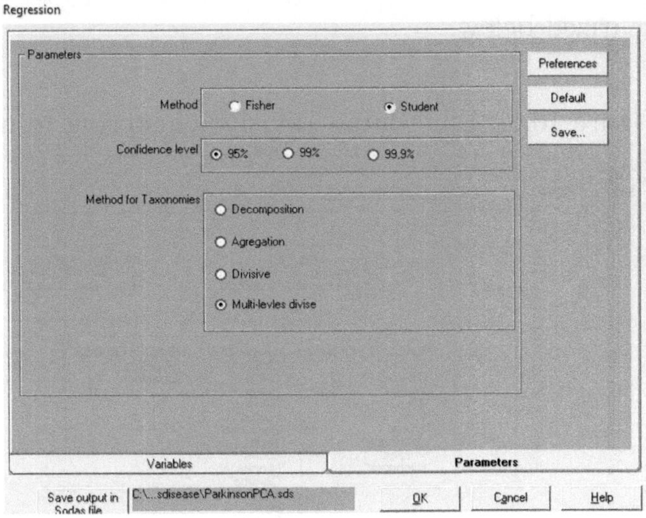

10.6. Cluster Analysis

Classical clustering methods are based on the pairwise distance between observations. The object is to minimize the sum of the distances inside each class and to maximize the distance between classes. The idea is similar in symbolic clustering. Only the way to measure the distance is different. For interval variables, there are several measures such as the Gowda-Diday dissimilarity measure, the Ichino-Yaguchi dissimilarity measure, and the generalized (weighted) Minkowski distance.

A very common method of clustering is that of the top-down divisive clustering process in which the entire data set is treated initially as a whole class and then subdivided into two classes at the first level. The data set can be further subdivided into as many classes as we want by adding more levels. Suppose the collection of all the classes is P, and $P = (C_1, C_2, \ldots, C_k)$. Define $Q(C_k)$ to be a measurement of the quality of cluster C_k, given by

$$Q(C_k) = \frac{1}{2n_k} \sum_{u_i \in C_k} \sum_{u_j \in C_k} d^2(w_i, w_j) \tag{10.4}$$

where n_k is the number of symbolic objects in C_k and $d(w_i, w_j)$ is the distance between objects w_i and w_j. At each level, the original cluster C will be divided into two subclusters C_1 and C_2. The subclusters are chosen by minimizing $W(C)$ and maximizing $\Delta(Q)$:

$$W(C) = Q(C_1) + Q(C_2) \tag{10.5}$$

$$\Delta(Q) = |Q(C) - Q(C_1) - Q(C_2)| \tag{10.6}$$

The cutoff value S is the center of the chosen interval for interval variables and the median value for modal data. With this background, the steps in the divisive clustering process can be summarized below.

Figure 10.6.: Film data—divisive clustering result

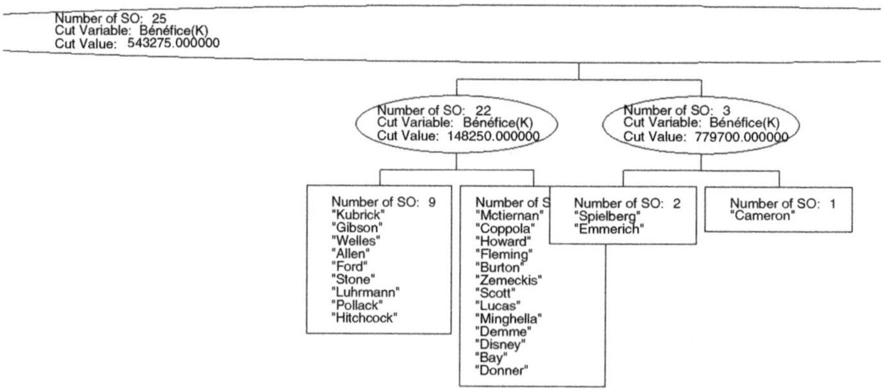

Step 1: Place all objects in one cluster C.

Step 2: Divide successively cluster C into smaller subclusters (C1,C2) according to the within-clusters inertia criterion.

Step 2.1: For each variable y, find the cutoff S which minimizes $W(C) = Q(C_1) + Q(C_2)$.

Step 2.2: Choose the variable y and the cutoff S which optimizes $W(C)$.

Step 3: Split the cluster C (C1,C2) to maximize $\Delta(Q)$.

Step 4: Each division is carried out using a single variable and by separating objects possessing some specified values of this variable, from those lacking them.

The clustering module in SODAS includes as options the methods of divisive clustering, dynamic clustering, and pyramidal clustering. We can choose an appropriate method from the collection base as needed.

To illustrate the methodology, we performed the divisive clustering analysis on the film data set. Suppose we are interested in distinguishing the directors, and we specifically consider the benefit and the number of Oscar awards. After adding method *DIV* in the chaining, we can decide how to normalize the data and how many clusters to have. Here, we chose the inverse of the dispersion as the normalization method and decided to separate the symbolic objects into four groups. The clustering result and the dividing criterion are shown in Fig. 10.6.

We could see only the variable "benefit" is used for the clustering. The entire data set is divided into two clusters by the cutoff value benefit 543,275, and then each of these two clusters is further subdivided into two classes with the cutoff values 148,250 and 779,700, respectively. The result can be explained as the benefit level. For example, the directors in cluster 3 made the most benefits exceeding on average 779,700.

10.7. Factor Analysis

To initiate a factor analysis, it is common to first perform a principal component analysis (PCA) which can be conducted by the *SPCA* module in SODAS. The goal of PCA is to determine a set of linear combinations of variables that can best account for most of the total variance. These linear combinations are ordered by their variances. Usually, the first few components account for the majority of the total variance. An important part of PCA consists of interpreting the physical meaning of the components. Factor analysis attempts to explain the data in terms of linear combinations of a few latent "factors."

There are three optional methods from which to choose, including symbolic object principal component analysis (SO-PCA), range-transformation principal component analysis (RT-PCA), and mixed-strategy principal component analysis (mixed-PCA). *SO-PCA* performs regular principal component analysis on the midpoints of interval data. The midpoints are standardized by a variance calculation based on the vertices matrix, which would be computationally more costly than the pure midpoint method. RT-PCA method performs a classical principal component analysis on the maximum vertices after being centered on the minimum vertices, which is proposed after SO-PCA to simplify the computation. Mixed-PCA method performs a principal component analysis on the recorded midpoints obtained by projecting the midpoints of the observations onto the factorial axes found from the RT-PCA method.

10.8. Factorial Discriminant Analysis

The *SFDA* module performs factorial discriminant analysis on symbolic data in order to classify the objects into different classes on the basis of the given predictive variables. In the parameter selection interface, we can select one *class identifier variable* and some *explanatory variables.* The *class identifier variable* has to be categorical, while the *explanatory variables* can be nominal, multi-nominal, multi-nominal with associated probability (modal), intervals, and real number variables. The goal of factorial discriminant analysis for symbolic objects is to define a geometrical classification rule to best separate the classes. The method will perform a numerical transformation on the symbolic data and then output a symbolic interpretation. All the modal data, multi-nominal data, and interval data are first transformed to a vertices matrix and then the data can be analyzed by regular FDA.

Figure 10.7.: Parkinson's disease—classical data table

Parkinsons

ID	age	test_time	motor_UPDRS	total_UPDRS	Jitter(Absolu	Jitter:RAP	Jitter:PPQ5	Jitter:DDP
2907	73	186.88	39.511	50.418	0.00002202	0.00146	0.0017	0.00438
2928	73	186.88	39.511	50.418	0.00003308	0.00252	0.00263	0.00756
2826	73	186.88	39.511	50.417	0.0000626	0.00374	0.00423	0.01123
2886	73	186.88	39.511	50.418	0.00005859	0.00439	0.00424	0.01318
2867	73	186.88	39.511	50.418	0.00004376	0.00316	0.00293	0.00947
2846	73	186.88	39.511	50.418	0.00003965	0.00238	0.00328	0.00714
5603	68	105.49	37.957	45.957	0.00003878	0.00319	0.00299	0.00956
5576	68	105.49	37.957	45.957	0.00010379	0.00335	0.00528	0.01005
5714	68	105.5	37.957	45.957	0.00003708	0.00291	0.00313	0.00873
5631	68	105.49	37.957	45.957	0.00005428	0.00527	0.00483	0.01582
5686	68	105.5	37.957	45.957	0.0000353	0.00373	0.00322	0.01118
5658	68	105.5	37.957	45.957	0.00004248	0.00275	0.00324	0.00825
5079	56	175.82	37.664	48.53	0.00002093	0.00244	0.00213	0.00733
5033	56	175.82	37.664	48.53	0.00003659	0.00399	0.00388	0.01198
5126	56	175.82	37.664	48.53	0.00000919	0.00107	0.00117	0.00321
5056	56	175.82	37.664	48.53	0.00003002	0.00341	0.00344	0.01024
5102	56	175.82	37.664	48.53	0.00001401	0.00147	0.00166	0.00442
5010	56	175.82	37.664	48.53	0.000033	0.00399	0.00376	0.01197
5605	68	112.56	37.659	45.727	0.00004183	0.00322	0.00354	0.00966

10.9. Application to Parkinson's Disease

We applied symbolic data analysis on the Parkinson's disease data downloaded from Kaggle.com. It contains all the indicators of the unified Parkinson's Disease Rating Scale (*UPDRS*) as well as other measurements of symptoms. *UPDRS* is a scale that reflects the severity of Parkinson's disease. The larger the number, the more severe the disease. The data set also includes variables measuring speaking ability, such as noise-to-harmonics ratio (*NHR*), harmonics-to-noise ratio (*HNR*), recurrence period density entropy (*RPDE*), pitch period entropy (*PPE*), detrended fluctuation analysis (*DFA*), voice frequency instability (*Jitter*), and voice amplitude instability (*shimmer*). There are 5874 rows and 19 quantitative variables in the table. We are interested in the overall UPDRS score, and as a result, we decided to use the *Total_UPRDS* as a symbolic object. Specifically, for the regression analysis on the aggregated Parkinson data, the variable *SCALE,* which was a transformation of *Total_UPDRS* was the dependent variable. The other measurements served as predictive variables. The object of the regression was to be able to predict the severity of the condition of a patient on the basis of just a few measurements.

Figure 10.8.: Import data—Parkinson's disease

DB2SO

Step 1. Create a new data source

Click on "File" >>> click on "New" >>> click on "Machine Data Source" >>> Click on "New" >>> Choose "User Data Source" >>> Choose "Microsoft Access Driver(*.mdb)" >>> Click "Next" >>> Click "Finish" >>> Enter a name (eg."Access MS") for the access approach in box "Data Source Name>>> Click "OK" >>> Check there is a new data source appearing in "Machine Data Source"

Step 2. Import the data

Click on "File" >>> click on "New" >>> click on "Machine Data Source" >>> Click on "Access MS" >>> click "OK" >>> Select the database file (with ".mdb" suffix) >>> Click "OK" >>> click on "Write Query" >>> write a SQL query to select the data required (eg. "select * from Parkinson_Query1; ") >>> Click "OK"

Step 3. Add a variable to the symbolic set (Optional)

Click on "Modify" >>> choose "add single-valued variable" or "add one set-valued multiple variable" " >>> click on "Write Query" >>> write a SQL query to select the variable wish to add (eg. "select * from UPRS; ") >>> Click "OK"

Step 4. Export symbolic data table

Click on "File" >>> click on "Export" >>> enter a title for the symbolic data set in the box "Title" >>> Click on "Select after box "File" >>> Enter the file name and choose "Sodas File (*.sds)" at the box "Save as type" >>> Click "Save"

10.9.1. Data Processing

We may create a new table, labeled Parkinson's in Microsoft Access, and then right-click on it to import the CSV data we download. We can add a unique key *ID* to the data set. Fig. 10.7 exhibits the resulting data in Microsoft Access.

Some preparation is needed before importing the data to SODAS. We need to round off the variable *total_UPDRS* and transform it from a numerical variable to a string variable so that SODAS can read it as symbolic objects. Besides, we multiply 10,000 with the variable *jitter (absolute)* to reduce the number of decimal places, so the value can be processed properly by SODAS. Then, we create a *view* based on the original table to construct a numerical target variable *SCALE*, which can be used as the dependent variable in regression analysis. We then have one table *Parkinson's* and a view of *UPRS*. After all the preparation is done, we save the Access file as a *.mdb* file, a database format which can be read by DB2SO (Database to Symbolic Object) module of SODAS.

The transformation from classical data to symbolic data can be carried out by the module *DB2SO*. To open the DB2SO module, we click on *Sodas File* in the main interface of SODAS software, then we choose *Import*, and then click on *Import with DB2SO*. In the new pop-up window, we can import our classical data and convert it to symbolic data. The specific operation for this is described in Fig. 10.8, where step 1 is only needed for the first time to import data from the Microsoft Access database. Fig. 10.9 shows the aggregated data in SODAS.

Figure 10.9.: Parkinson's disease—aggregated data in SODAS

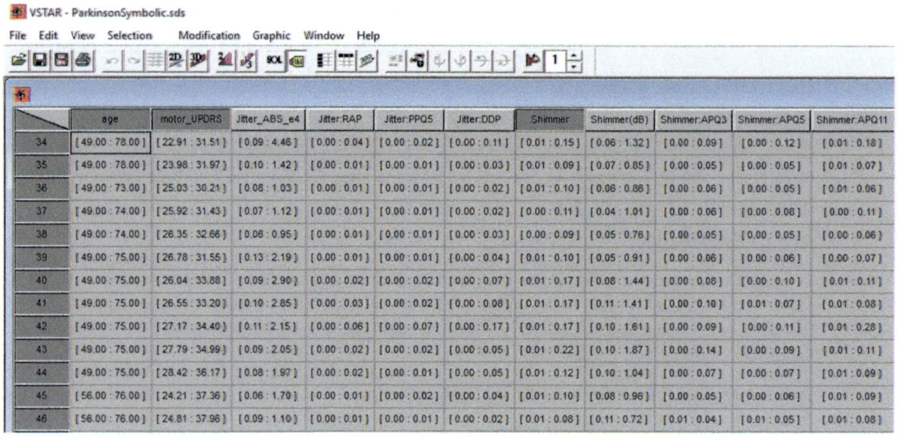

	age	motor_UPDRS	Jitter_ABS_e4	Jitter:RAP	Jitter:PPQ5	Jitter:DDP	Shimmer	Shimmer(dB)	Shimmer:APQ3	Shimmer:APQ5	Shimmer:APQ11
34	[49.00 : 78.00]	[22.91 : 31.51]	[0.09 : 4.46]	[0.00 : 0.04]	[0.00 : 0.02]	[0.00 : 0.11]	[0.01 : 0.15]	[0.06 : 1.32]	[0.00 : 0.09]	[0.00 : 0.12]	[0.01 : 0.18]
35	[49.00 : 78.00]	[23.98 : 31.97]	[0.10 : 1.42]	[0.00 : 0.01]	[0.00 : 0.01]	[0.00 : 0.03]	[0.01 : 0.09]	[0.07 : 0.85]	[0.00 : 0.05]	[0.00 : 0.05]	[0.01 : 0.07]
36	[49.00 : 73.00]	[25.03 : 30.21]	[0.08 : 1.03]	[0.00 : 0.01]	[0.00 : 0.01]	[0.00 : 0.02]	[0.01 : 0.10]	[0.06 : 0.86]	[0.00 : 0.06]	[0.00 : 0.05]	[0.01 : 0.06]
37	[49.00 : 74.00]	[25.92 : 31.43]	[0.07 : 1.12]	[0.00 : 0.01]	[0.00 : 0.01]	[0.00 : 0.02]	[0.00 : 0.11]	[0.04 : 1.01]	[0.00 : 0.06]	[0.00 : 0.08]	[0.00 : 0.11]
38	[49.00 : 74.00]	[26.35 : 32.66]	[0.06 : 0.95]	[0.00 : 0.01]	[0.00 : 0.01]	[0.00 : 0.03]	[0.00 : 0.09]	[0.05 : 0.76]	[0.00 : 0.05]	[0.00 : 0.05]	[0.00 : 0.06]
39	[49.00 : 75.00]	[26.78 : 31.55]	[0.13 : 2.19]	[0.00 : 0.01]	[0.00 : 0.01]	[0.00 : 0.04]	[0.01 : 0.10]	[0.05 : 0.91]	[0.00 : 0.06]	[0.00 : 0.06]	[0.00 : 0.07]
40	[49.00 : 75.00]	[26.04 : 33.88]	[0.09 : 2.90]	[0.00 : 0.02]	[0.00 : 0.02]	[0.00 : 0.07]	[0.01 : 0.17]	[0.08 : 1.44]	[0.00 : 0.08]	[0.00 : 0.10]	[0.01 : 0.11]
41	[49.00 : 75.00]	[26.55 : 33.20]	[0.10 : 2.85]	[0.00 : 0.03]	[0.00 : 0.02]	[0.00 : 0.08]	[0.01 : 0.17]	[0.11 : 1.41]	[0.00 : 0.10]	[0.01 : 0.07]	[0.01 : 0.08]
42	[49.00 : 75.00]	[27.17 : 34.40]	[0.11 : 2.15]	[0.00 : 0.06]	[0.00 : 0.07]	[0.00 : 0.17]	[0.01 : 0.17]	[0.10 : 1.61]	[0.00 : 0.09]	[0.00 : 0.11]	[0.01 : 0.28]
43	[49.00 : 75.00]	[27.79 : 34.99]	[0.09 : 2.05]	[0.00 : 0.02]	[0.00 : 0.02]	[0.00 : 0.05]	[0.01 : 0.22]	[0.10 : 1.87]	[0.00 : 0.14]	[0.00 : 0.09]	[0.01 : 0.11]
44	[49.00 : 75.00]	[28.42 : 36.17]	[0.08 : 1.97]	[0.00 : 0.02]	[0.00 : 0.01]	[0.00 : 0.05]	[0.01 : 0.12]	[0.10 : 1.04]	[0.00 : 0.07]	[0.00 : 0.07]	[0.01 : 0.09]
45	[56.00 : 76.00]	[24.21 : 37.36]	[0.06 : 1.70]	[0.00 : 0.01]	[0.00 : 0.02]	[0.00 : 0.04]	[0.01 : 0.10]	[0.08 : 0.96]	[0.00 : 0.05]	[0.00 : 0.06]	[0.01 : 0.09]
46	[56.00 : 76.00]	[24.81 : 37.96]	[0.09 : 1.10]	[0.00 : 0.01]	[0.00 : 0.01]	[0.00 : 0.02]	[0.01 : 0.08]	[0.11 : 0.72]	[0.01 : 0.04]	[0.01 : 0.05]	[0.01 : 0.08]

Figure 10.10.: Parkinson's disease data—star plot for a single object

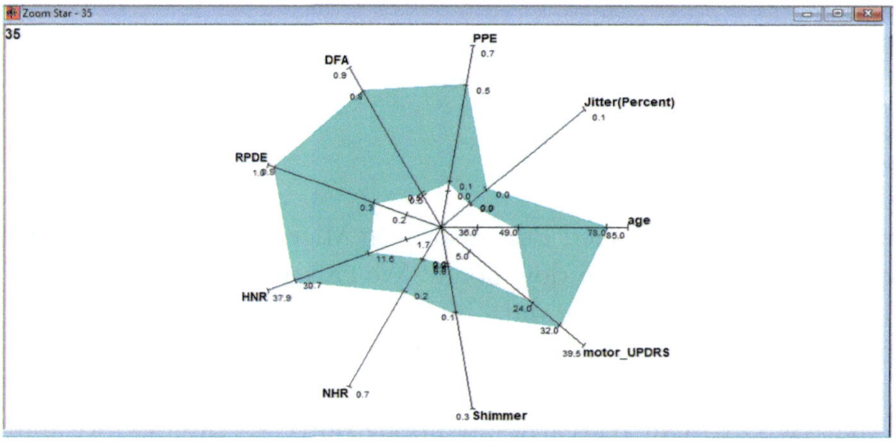

10.9.2. Result Analysis

10.9.2.1. Viewer

The Viewer function is a visualization tool that allows further exploration of the data set. Fig. 10.10 is the star plot representing the distribution of a single symbolic object, and the example given here is the scale 35. We can see the object in different dimensions. Fig. 10.11 is a 2D comparison graph for scales 35 and 45. We observe that scale 35 tends to involve higher DFA and PRDE, and it also has a larger range for age.

10.9.2.2. Descriptive Statistics

The histograms of *age*, *total_UPDRS*, *Shimmer*, *Jittering*, *RPDE*, and *PPE* drawn by the *DSTAT* module are presented in Fig. 10.12. The age group with the highest frequency is 70–75, whereas the group with the lowest frequency is 80–85. Measurements Shimmering and Jittering have a clear decreasing trend along with increasing value, while motor_UPDRS has no obvious pattern.

Figure 10.11.: Parkinson's disease data star plot comparison

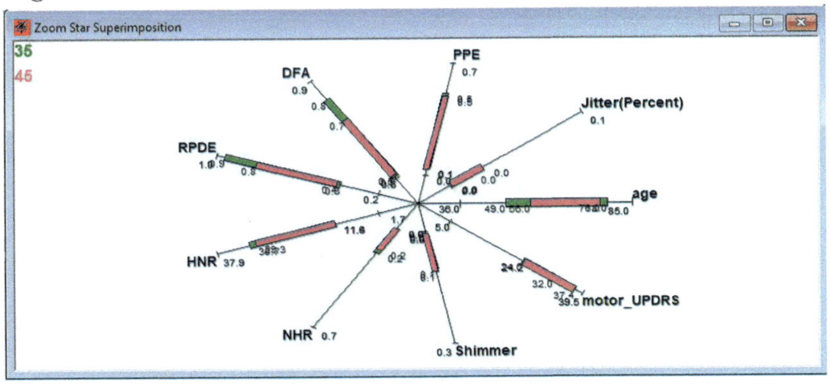

Figure 10.12.: Parkinson's disease histogram for interval data

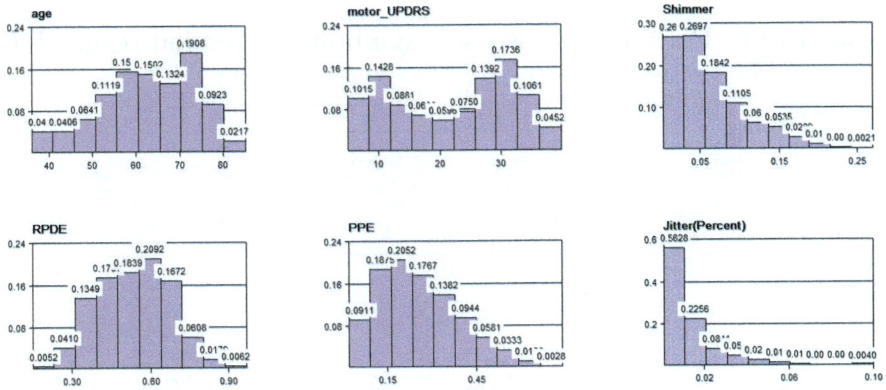

10.9.2.3. Symbolic Regression Analysis

We conducted a symbolic regression analysis on Parkinson's disease data using the *SREG* module. The dependent variable we chose is *SCALE*, a single-valued numerical variable, and all the other variables are used as predictive variables. We obtain the results shown in Figs. 10.13 and 10.14.

In the output of the symbolic regression method, the F-test is used to select the significant variables, and the T-test is used to conduct individual tests. We can see as a result that only two variables, *motor_UPDRS* and *Age*, are selected with 95% confidence, and the predictive equation to estimate the severeness of Parkinson's disease is

$$SCALE = 1.38 \times motor\_UPDRS + 0.29 \times age - 15.67 \tag{10.7}$$

10.9.2.4. Symbolic Clustering

We performed symbolic clustering using the Divisive Classification *DIV* module. The symbolic objects are divided into four groups as shown below by all the measurements. The dividing criteria is exhibited in Fig. 10.15 and described in Table 10.7.

Figure 10.13.: Parkinson's disease—symbolic regression result F-test

Predictive power of the variables (we look for the good variables with the F-test) :

Variables	Variances Ratio F	Quantile f(0.95)	Fisher test F>f?	R2=determination coefficient	Quality
age	52.24154	4.171 00	accepted	0.526408	***
motor_UPDRS	1377.58	4.171 00	accepted	0.9670078	*****
Jitter_ABS_e4	6.540623e-002	4.171 00	refused	1.389688e-003	
Jitter:RAP	0.1755051	4.171 00	refused	3.720259e-003	
Jitter:PPQ5	1.989456e-002	4.171 00	refused	4.231093e-004	
Jitter:DDP	0.1755641	4.171 00	refused	3.721506e-003	
Shimmer	0.9090673	4.171 00	refused	1.897485e-002	
Shimmer(dB)	0.6894545	4.171 00	refused	1.445717e-002	
Shimmer:APQ3	1.603423	4.171 00	refused	3.298993e-002	
Shimmer:APQ5	1.114288	4.171 00	refused	2.31592e-002	
Shimmer:APQ11	1.35787e-002	4.171 00	refused	2.888251e-004	
Shimmer:DDA	1.603389	4.171 00	refused	3.298924e-002	
NHR	2.016349e-002	4.171 00	refused	4.288265e-004	
HNR	7.839637	4.171 00	accepted	0.1429557	*
RPDE	28.9186	4.171 00	accepted	0.3809159	**
DFA	3.635638	4.171 00	refused	7.179998e-002	
PPE	6.249979	4.171 00	accepted	0.1173705	*
Jitter(Percent)	8.898543e-002	4.171 00	refused	1.889729e-003	

Figure 10.14.: Parkinson's disease—symbolic regression result T-test

The Linear regression with the 'successful' variables gives us the following equation :

Dependent variable	Ordered explanatory variables	Weight	Standard deviation si	ti = Weight/si	Student test |ti|<=t(95%)
SCALE	motor_UPDRS	1.381985	4.534155e-002	30.47944	
	age	0.2898031	5.546609e-002	5.22487	
	HNR	0.2467026	0.1587519	1.554014	weight=0
	PPE	-3.628784	7.060846	-0.5139305	weight=0
	RPDE	-11.6468	7.481693	-1.556706	weight=0
	Constant	-15.66818	6.708349	-2.335623	

*****statistics*****

Determination coefficient R2=0.981087

AnaVar	Degree DLE, DLR	Sum of Squares SSE, SSR	Mean Square CME=SSE/DLE, CMR	F=Fisher test F=CME/CMR	Quantiles of the Fisher-Snedecor function f(99,9%)	f(99%)	f(95%)
regression	5	9614.65	1922.93	446.1067	5.534 00	3.699 00	2.534 00
residuals	43	185.3502	4.310471		F>f accepted	F>f accepted	F>f accepted
total	48	9800. 00	1927.24				

****

We see that the severity scales of Parkinson's disease have been separated into four classes. Further looking at the cluster rules, *motor_UPDRS* $\leq$ 14.1296 is the criteria used to distinguish *Cluster 1* from the other classes.

10.9.2.5. Principal Component Analysis

The *SO-SPCA* result shown in Fig. 10.16 indicates that the first two principal components represent more than 80% of the total variance.

Fig. 10.17 reveals the correlation between the original variable and the factors. We see that factor 1 is highly correlated with HNR and least correlated with *motor_UPDRS*, while factor 2 heavily relies on *age*, *motor_UPDRS*, *RPDE*, and *PPE*.

Figure 10.15.: Parkinson's disease—clusters

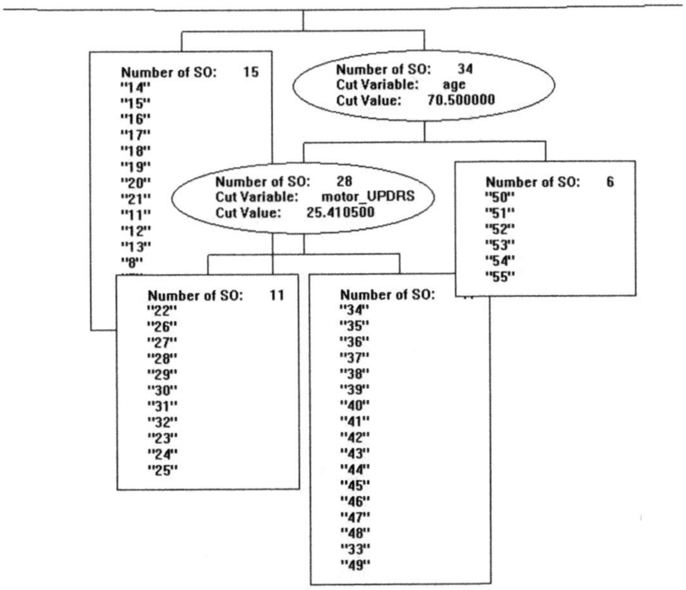

Figure 10.16.: Parkinson's disease—principal component analysis

```
SO-PCA

         Eigenvalues Explained Cumulated                  Histogram
                                        0--------25%--------50%-------75%-------100%
Ev.1       2.46394   56.36613  56.36613 |**********************
Ev.2       1.09137   24.96675  81.33289 |*********
Ev.3       0.34400    7.86959  89.20248 |***
Ev.4       0.15057    3.44447  92.64695 |*
Ev.5       0.08764    2.00497  94.65192 |
Ev.6       0.07214    1.65037  96.30228 |
Ev.7       0.05069    1.15954  97.46182 |
Ev.8       0.04337    0.99211  98.45393 |
Ev.9       0.02292    0.52444  98.97838 |
```

Figure 10.17.: Parkinson's disease—principal component analysis correlations

Correlations between variables and factors (18 vars,9 fact)=

Var.	Factor 1	Factor 2	Factor 3	Factor 4	Factor 5	Factor 6	Factor 7	Factor 8
age	0.66862	0.91222	0.75494	0.75591	0.80014	0.78845	0.80153	0.80792
motor_UPDR	0.26245	0.81440	0.29537	0.35737	0.31797	0.32683	0.32565	0.32935
Jitter_ABS	0.57922	0.70087	0.82339	0.80431	0.83427	0.86500	0.84373	0.83000
Jitter:RAP	0.48214	0.56958	0.80275	0.74556	0.78132	0.77856	0.76231	0.74971
Jitter:PPQ	0.43803	0.54082	0.76385	0.70472	0.74517	0.72696	0.72171	0.70598
Jitter:DDP	0.48217	0.56960	0.80276	0.74557	0.78134	0.77857	0.76232	0.74972
Shimmer	0.62260	0.66923	0.83809	0.86297	0.86892	0.87755	0.88226	0.88616
Shimmer(dB	0.63029	0.68228	0.84707	0.87149	0.87885	0.88521	0.89131	0.89531
Shimmer:AP	0.62368	0.65227	0.81239	0.84770	0.85033	0.86560	0.86618	0.87465
Shimmer:AP	0.60697	0.65652	0.83395	0.84833	0.86401	0.86610	0.87453	0.87161
Shimmer:AP	0.58517	0.67009	0.81742	0.83894	0.85346	0.83720	0.85159	0.84159
Shimmer:DD	0.62368	0.65227	0.81239	0.84770	0.85033	0.86560	0.86618	0.87465
NHR	0.45139	0.56837	0.74811	0.70423	0.73424	0.74116	0.73702	0.71755
HNR	0.92534	0.68367	0.89907	0.92670	0.92826	0.93231	0.93100	0.93136
RPDE	0.75927	0.86385	0.86606	0.90457	0.90004	0.91250	0.91980	0.92093
DFA	0.73800	0.73119	0.87617	0.91427	0.92505	0.92216	0.94193	0.94189
PPE	0.70866	0.82604	0.89706	0.91270	0.92399	0.92727	0.93714	0.93728
Jitter(Per	0.47953	0.57146	0.80117	0.74529	0.78167	0.77435	0.76193	0.74950

Groups	Rules
Cluster 1	[$motor\_UPDRS \leq 14.129600$] is true
Cluster 2	[$motor\_UPDRS \leq 14.129600$] is false and [$motor\_UPDRS \leq 25.410500$] is true and [$age \leq 70.500000$] is true
Cluster 3	[$motor\_UPDRS \leq 14.129600$] is false and [$age \leq 70.500000$] is false
Cluster 4	[$motor\_UPDRS \leq 14.129600$] is false and [$motor\_UPDRS \leq 25.410500$] is false and [$age \leq 70.500000$] is true

Table 10.7.: Parkinson's Disease data–clustering rules

Figure 10.18.: Parkinson's disease—classical regression result

```
Coefficients:
                    Estimate Std. Error t value Pr(>|t|)
(Intercept)        6.775e-01  1.093e+00   0.620 0.535411
Xage               6.749e-02  5.232e-03  12.899  < 2e-16 ***
Xtest_time         2.433e-03  8.130e-04   2.993 0.002775 **
Xmotor_UPDRS       1.230e+00  5.760e-03 213.592  < 2e-16 ***
XJitter.Absolute.  3.141e+04  3.139e+03  10.008  < 2e-16 ***
XJitter.RAP        3.232e+03  1.587e+04   0.204 0.838620
XJitter.PPQ5       1.583e+02  6.383e+01   2.480 0.013166 *
XJitter.DDP       -1.009e+03  5.290e+03  -0.191 0.848707
XShimmer          -3.991e+01  2.196e+01  -1.817 0.069199 .
XShimmer.dB.      -3.504e-01  1.641e+00  -0.214 0.830880
XShimmer.APQ3     -1.396e+04  1.594e+04  -0.876 0.381114
XShimmer.APQ5      8.656e+01  1.876e+01   4.615 4.01e-06 ***
XShimmer.APQ11    -3.048e+01  8.426e+00  -3.618 0.000300 ***
XShimmer.DDA       4.654e+03  5.313e+03   0.876 0.381122
XNHR              -8.490e+00  2.064e+00  -4.114 3.94e-05 ***
XHNR              -7.579e-02  2.339e-02  -3.240 0.001201 **
XRPDE              4.132e+00  6.147e-01   6.722 1.96e-11 ***
XDFA              -2.856e+00  7.947e-01  -3.593 0.000329 ***
XPPE              -4.623e+00  9.958e-01  -4.643 3.51e-06 ***
XJitter.Percent.  -2.999e+02  7.229e+01  -4.148 3.40e-05 ***
---
Signif. codes:  0 '***' 0.001 '**' 0.01 '*' 0.05 '.' 0.1 ' ' 1

Residual standard error: 3.302 on 5855 degrees of freedom
Multiple R-squared:  0.9051,    Adjusted R-squared:  0.9048
F-statistic:  2939 on 19 and 5855 DF,  p-value: < 2.2e-16
```

10.9.3. Comparison with Classical Method

To compare the symbolic analysis results with the classical method, we conducted a traditional regression analysis on the classical data using the software R. We obtained the results shown in Fig. 10.18.

It can be seen that more variables passed the T-test in the classical regression than in symbolic regression. In symbolic regression, only the variables age and motor_UPRDS passed the T-test. However, the two variables selected by symbolic regression are also the most significant variables in classical regression because of the highest t values. In this study, the symbolic method successfully identified the most significant variables, age and motor_UPDRS. The comparison reveals that we should not expect the performance to be as precise as in classical analysis, but it can help identify the most important variables. When we conduct symbolic analysis, the data have been aggregated by specific rules, so the loss of information is inevitable. However, when we have to deal with a large data set, the symbolic method is a great tool to save time and capture important information.

10.10. Application to Cardiovascular Disease Analysis

The cardiovascular disease data set can be obtained from the website Kaggle. The data set contains information on 70,000 patients and their medical examination results. There are 12 variables in total, where the first 11 variables are predictors, and the 12th variable *cardio* is the target variable. Descriptions of the variables are displayed in Table 10.8.

As we are interested in the age of patients, so we will aggregate the data by age. Since the interval value is easily affected by extreme values while aggregating, we decide to transform the variable into categorical groups. This step will transform the corresponding variable into a modal variable, keeping more information than interval variable for a large data set.

Variable	Description	Value
Age	Age	Days
Height	Height	cm
Weight	Weight	kg
Gender	Gender	1:women,2:men
ap_hi	Systolic blood pressure	Integer
ap_lo	Diastolic blood pressure	Integer
Cholesterol	Cholesterol	1: normal, 2: above normal 3: well above normal
Gluc	Glucose	1: normal, 2: above normal 3: well above normal
Smoke	Smoke	1:Yes, 0:No
Alco	Alcohol intake	1:Yes, 0:No
Active	Physical activity	1:Yes, 0:No
Cardio	Presence of cardiovascular disease	1:Yes, 0:No

Table 10.8.: Cardiovascular disease data—variable description

Figure 10.26.: Cardiovascular disease data for classical regression

cholesterol_above_normal	gluc_normal	gluc_above_normal	smoke	alco	active	Cardio
0	1	0	0	0	1	0
0	1	0	0	0	1	1
0	1	0	0	0	0	0
0	1	0	0	0	1	0
0	1	0	0	0	0	0
1	0	1	0	0	0	1
0	1	0	0	0	1	1
0	0	0	0	0	1	1
0	1	0	0	0	1	0
0	1	0	0	0	0	0
0	1	0	0	0	1	1
0	1	0	0	0	1	0
0	1	0	0	0	0	0
0	1	0	0	0	1	0
0	1	0	1	1	1	0
0	1	0	0	0	0	0
0	1	0	0	0	0	1
0	0	0	0	0	1	0
0	1	0	0	0	0	0
0	1	0	1	0	1	1
0	1	0	0	0	1	0
0	1	0	0	0	1	1
1	0	1	0	0	1	1

Figure 10.27.: Cardiovascular disease data—classical regression result

```
Coefficients:
                          Estimate Std. Error z value Pr(>|z|)
(Intercept)               1.557412   1.545110   1.008 0.313473
gender_W                 -0.117032   0.020307  -5.763 8.26e-09 ***
hight_under_150           0.281552   1.544436   0.182 0.855346
hight_150_160            -0.096558   1.543515  -0.063 0.950119
hight_160_170            -0.315411   1.543463  -0.204 0.838078
hight_170_180            -0.500730   1.543506  -0.324 0.745627
hight_180_190            -0.694240   1.544110  -0.450 0.652995
hight_190_200            -0.570717   1.557753  -0.366 0.714088
weight_under_40           0.764558   0.305150   2.506 0.012227 *
weight_40_50             -0.169754   0.075193  -2.258 0.023973 *
weight_50_60             -0.092697   0.035077  -2.643 0.008225 **
weight_60_70             -0.005052   0.027493  -0.184 0.854197
weight_70_80              0.079119   0.026915   2.940 0.003287 **
weight_80_90              0.111066   0.029013   3.828 0.000129 ***
ap_hi_70_90              -0.916116   0.252546  -3.628 0.000286 ***
ap_hi_90_110             -1.372560   0.078061 -17.583  < 2e-16 ***
ap_hi_110_130            -0.875518   0.066961 -13.075  < 2e-16 ***
ap_hi_130_150            -0.452920   0.066712  -6.789 1.13e-11 ***
ap_hi_150_170            -0.265268   0.067340  -3.939 8.17e-05 ***
ap_lo_40_60               0.079705   0.205080   0.389 0.697533
ap_lo_60_80              -0.062660   0.043660  -1.435 0.151236
ap_lo_80_100              0.181178   0.038198   4.743 2.10e-06 ***
cholesterol_normal       -0.637013   0.029741 -21.419  < 2e-16 ***
cholesterol_above_normal -0.558020   0.035335 -15.792  < 2e-16 ***
gluc_normal              -0.217417   0.034637  -6.277 3.45e-10 ***
gluc_above_normal        -0.134133   0.044924  -2.986 0.002828 **
smoke                    -0.225191   0.031598  -7.127 1.03e-12 ***
alco                     -0.160140   0.038211  -4.191 2.78e-05 ***
active                   -0.033925   0.019896  -1.705 0.088163 .
---
Signif. codes:  0 '***' 0.001 '**' 0.01 '*' 0.05 '.' 0.1 ' ' 1
```

10.10.2. Comparison with the Classical Method

To compare the symbolic analysis result with the classical method, we conduct logistic regression on the original cardiovascular disease using R software. To apply regression on categorical data, we need to transform the categorical data into several binary variables. For example, the variable *gluc* has three values, including 1 (normal), 2 (above normal), and 3 (well above normal), so *gluc* is converted to two dummy variables: *gluc* normal (1, 0) and *gluc* above normal (1, 0). Since these two dummy variables can reflect the third modality, we do not need to create a new variable for it. The data are shown in Fig. 10.26.

We show in Fig. 10.27 the results from the R program. There are several important variables identified, such as *gender*, *ap_hi*, *cholesterol*, and *gluc*. Comparing with symbolic regression, more variables are identified to be significant.

Figure 10.28.: Regression for cardiovascular data aggregated by height

Predictive power of the variables (we look for the good variables with the F-test) :

Variables	Variances Ratio F	Quantile f(0.95)	Fisher test F>f?	R2=determination coefficient	Quality
weight	5.147279	4.171 00	accepted	4.759509e-002	*
ap_hi	14.4454	4.171 00	accepted	0.1229967	*
ap_lo	28.12873	4.171 00	accepted	0.2145123	**
cholesterol	1.676732	3.316 00	refused	3.183059e-002	
gluc	1.713964	3.316 00	refused	3.251443e-002	
smoke	2.540033	4.171 00	refused	2.406701e-002	
alco	2.364809	4.171 00	refused	2.244401e-002	
active	22.89854	4.171 00	accepted	0.1818809	*

The Linear regression with the 'successful' variables gives us the following equation :

Dependent variable	Ordered explanatory variables	Weight	Standard deviation si	ti = Weight/si	Student test |ti|<=t(95%)
Cardio_PER	1(active)	0.4332323	0.1062691	4.076749	
	ap_lo	1.270102e-002	3.745171e-003	3.391306	
	ap_hi	3.451618e-003	2.291912e-003	1.505999	weight=0
	weight	5.292758e-004	9.353299e-004	0.5658707	weight=0
	Constant	-1.328884	0.2980726	-4.458258	

The difference may be caused by the information loss during the aggregating process. Therefore, it is reasonable to infer that a different aggregating method may lead to different outcomes. To verify the hypothesis, we tried to aggregate the data by height and then conduct regression analysis. The results in Fig. 10.28 implied that the diastolic blood pressure and the active level are important factors of cardiovascular disease. Comparing with the regression results aggregated by ages, this result emphasizes different factors of cardiovascular disease. Combining both results, we can have a more comprehensive understanding that glucose, diastolic blood pressure, and the active level are closely related to cardiovascular diseases.

The results underline that the symbolic method may lose some information during aggregation. It is nonetheless able to detect the important variables and it is applicable on large data sets. It also reminds us that a different way of aggregation may lead to different results in symbolic data. Therefore, to gain a complete understanding of the data, it is perhaps best to analyze the data by different choices of aggregation.

Part III.

Machine Learning for Big Data

11. Tools for Machine Learning

11.1. Regression Models

In this section we shall describe one of the cornerstones of modern machine learning, namely, the *regression model.* It was first introduced over 100 years ago as a statistical method for discovering the existence and extent of the relationships between a *response* variable and several *explanatory* variables also called *predictors.* Since then, the regression model has become very well known and it is the subject of many books and research papers. Despite its simplicity, the regression model is quite effective and still widely used by statisticians, machine learning engineers, and other researchers. One of several examples is a recent paper [Hsu and Yeh, 2018] in which the authors employ a regression model to predict house prices based on several *features* (see Fig. 11.1). Numerous questions are to be answered in such a research. Is there a relationship at all? If there is a relationship, how strong is it? Is it linear? What is the contribution of each predictor? Can our model be generalized to make predictions for new data? Answers to these and many other questions fall under the scope of *regression analysis.* We will go into some detail when discussing regression models. Readers familiar with simple regression may wish to go directly to Sect. 11.4 on regression in machine learning.

A standard setup for building a regression model is as follows. During the observation stage, the data is recorded in pairs:

$$(\mathbf{x}_1, y_1), (\mathbf{x}_2, y_2), \ldots, (\mathbf{x}_m, y_m),$$

where each $\mathbf{x}_i = (x_1, x_2, \ldots, x_n)$ for $i = 1, 2, \ldots, m$ is the observed set of values of *an explanatory* (or *predictor*) variable $\boldsymbol{X} \in \mathbb{R}^n$ and $y_1, y_2, \ldots, y_m$ are the observed values of the *response* variable Y. It is assumed that m observations of Y are independent of each other. We treat Y as a random variable and assume a probability distribution for it depending on the type of values that Y takes. A regression model relates the conditional expected value $\mu(x) = \mathbf{E}[Y|x]$ given the predictors to a linear combination of the predictors $x_1, x_2, \ldots, x_n$ using a *link* function $g(\cdot)$:

$$g(\mu(x)) = \beta_0 + \beta_1 x_1 + \cdots + \beta_n x_n. \tag{11.1}$$

There are several commonly used types of link functions. The simplest - is the *identity link function*

$$\mu(x) = \beta_0 + \beta_1 x_1 + \cdots + \beta_n x_n, \tag{11.2}$$

M. Alvo, *Statistical Inference and Machine Learning for Big Data*, Springer Series in the Data Sciences, https://doi.org/10.1007/978-3-031-06784-6_11

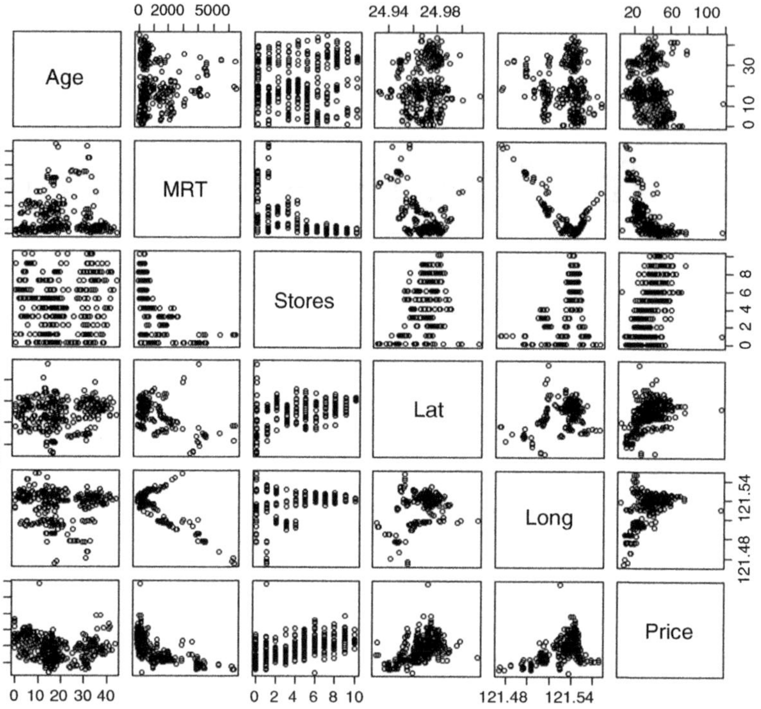

Figure 11.1.: A *real estate valuation* data set includes information about *age, distance to the nearest metro station, number of convenience stores within the walking distance, geographic location,* and *price of a unit area* for a number of houses from Xindian District, New Taipei City, Taiwan

which is used if Y is a *quantitative* variable, i.e., $y_i \in \mathbb{R}$ for $i = 1, 2, \ldots, m$. However, if Y is a *qualitative* variable, i.e., $y_i \in \{1, 2, \ldots, K\}$ for $i = 1, 2, \ldots, m$, another link function is needed. For a *binary* response Y with $\mathbf{E}[Y] = \theta$, the *logit link function* is used:

$$\log\left(\frac{\theta}{1-\theta}\right) = \beta_0 + \beta_1 x_1 + \cdots + \beta_n x_n. \tag{11.3}$$

The most general form of a regression model can be given by

$$Y = h(\boldsymbol{X}) + \varepsilon, \tag{11.4}$$

where $h(\cdot) = g^{-1}(\cdot)$ and ε is a random error. We interpret $h(\boldsymbol{X})$ as the expectation of Y given $\boldsymbol{X}$:

$$h(\boldsymbol{X}) = \mathbf{E}\left[Y = y \mid \boldsymbol{X} = (x_1, x_2, \ldots, x_n)\right] \tag{11.5}$$

The choice of a link function determines the applicability of a regression model to different research tasks. We shall focus our attention on *quantitative* and *classification* models, based on the nature of the response variable. A quantitative model is usually considered if the response variable takes on numeric values $y \in \mathbb{R}$, for example, the market value of a house, a person's annual income, or the closing value of a stock. On the other hand, if the response variable is *qualitative*, i.e., takes its values in one of K

categories $y \in \{1, 2, 3, \dots, K\}$, which are called *classes*, a classification model is used. Some examples of qualitative variables are the brand of a product (Apple, Microsoft, Dell, etc.), a person's education level (high school diploma, college, university, etc.), or the stock price movement direction (up, down).

Quantitative model is realized by choosing the *identity link function* (11.2), which gives

$$h(\boldsymbol{X}) = \beta_0 + \beta_1 X_1 + \cdots + \beta_n X_n. \tag{11.6}$$

Values $\beta_0, \beta_1, \cdots \beta_n$ are called the *regression coefficients* or *weights.* Note that the linear regression model only requires linearity in terms of the regression coefficients; thus, it can be easily adapted to problems with nonlinear relationship between predictors and the response variable.

Classification model is based on the *logit link function* (11.3), which gives

$$h(\boldsymbol{X}) = \frac{e^{\beta_0 + \beta_1 X_1 + \cdots + \beta_n X_n}}{1 + e^{\beta_0 + \beta_1 X_1 + \cdots + \beta_n X_n}}. \tag{11.7}$$

In its basic form, the logistic regression model can only be applied to *binary* classification problems ($K = 2$). We shall later discuss computational techniques that make it suitable for tasks that involve classification into $K > 2$ categories. Explanatory variables in each of the above models could be qualitative or quantitative. However, qualitative predictors must be *properly coded* before they can be included in the regression model.

11.2. Simple Linear Regression

Our starting point is a regression method for investigation of the relationship between a response variable $Y \in \mathbb{R}$ and a single predictor $X \in \mathbb{R}$. This is a special case of a general model (11.4) with the number of features in each observation. The name *simple linear regression* speaks for itself. This is the basic, simplest regression model with a well-established theoretical background that is easy to interpret in practical examples. We assume that the relationship between X and Y is approximately linear and consider the quantitative model with the identity link function, that is,

$$h(X) = \beta_0 + \beta_1 X. \tag{11.8}$$

The coefficients β_0 and β_1 represent the *intercept* and the *slope* of the linear correspondence (11.8). It is usually assumed that $\mathbf{E}[\varepsilon] = 0$ and $\mathbf{Var}[\varepsilon] = \sigma^2$. This gives a simple linear regression model

$$Y = \beta_0 + \beta_1 X + \varepsilon. \tag{11.9}$$

Since the regression coefficients are usually unknown, estimates $\hat{\beta}_0$ and $\hat{\beta}_1$ must be computed using the data set. Once this is done, an estimate $\hat{y}$ of the price can be produced for a newly observed value of $X = x$

$$\hat{y} = \hat{\beta}_0 + \hat{\beta}_1 x. \tag{11.10}$$

We now need to find a way of computing the estimates $\hat{\beta}_0$ and $\hat{\beta}_1$. For a simple linear regression, this is typically achieved by the *least squares method.* Recall that there are m observations in our data set. The data is recorded in pairs:

$$(x_1, y_1), (x_2, y_2), \ldots, (x_m, y_m).$$

We want to have $y_i \approx \hat{\beta}_0 + \hat{\beta}_1 x_i$ for $i = 1, 2, \ldots, m$. Hence, our goal is to estimate $\hat{\beta}_0$ and $\hat{\beta}_1$ so that

$$\epsilon_i = y_i - (\hat{\beta}_0 + \hat{\beta}_1 x_i)$$

is as small as possible for all $i = 1, 2, \ldots, m$. The empirical error term ϵ_i is the estimation for unobserved true error ε_i. Since we cannot observe ε_i, the true regression variance σ^2 is also unknown.

11.2.1. Least Squares Method

We assume independent ε_i's with $\mathbf{E}[\varepsilon_i] = 0$ and $\mathbf{Var}[\varepsilon_i] = \sigma^2$ for all $i = 1, 2, \ldots, m$. Applying (11.9) to each element (x_i, y_i) of the random sample and summing over $i = 1, 2, \ldots, m$, we observe that

$$\sum_{i=1}^{m} y_i = m\beta_0 + \beta_1 \sum_{i=1}^{m} x_i + \sum_{i=1}^{m} \varepsilon_i.$$

Now divide by m and recall the assumptions on ε_i's to obtain

$$\bar{y} = \beta_0 + \beta_1 \bar{x},$$

where $\bar{x} = \frac{1}{m}\sum_{i=1}^{m} x_i$, and $\bar{y} = \frac{1}{m}\sum_{i=1}^{m} y_i$. We see that the point $(\bar{x}, \bar{y})$ satisfies Eq. (11.10). Applying (11.10) to each pair of observations, define the *residual variance* as

$$\mathbf{S}_{\mathbf{R}}^{\mathbf{2}} = \frac{1}{m}\sum_{i=1}^{m}(y_i - \hat{y}_i)^2. \tag{11.11}$$

The essence of the least squares approach is to find values for $\hat{\beta}_0$ and $\hat{\beta}_1$ that would minimize

$$\mathbf{S}_{\mathbf{R}}^{\mathbf{2}} = \frac{1}{m}\sum_{i=1}^{m}\Big(y_i - (\hat{\beta}_0 + \hat{\beta}_1 x_i)\Big)^2 = \frac{1}{m}\sum_{i=1}^{m}\epsilon_i^2$$

The solution for this problem is obtained from the system of differential equations

$$\frac{\partial}{\partial \beta_0}\sum_{i=1}^{m}\Big(y_i - (\beta_0 + \beta_1 x_i)\Big)^2 = 0,$$

$$\frac{\partial}{\partial \beta_1} \sum_{i=1}^{m} \left(y_i - (\beta_0 + \beta_1 x_i) \right)^2 = 0.$$

It can be shown that the solutions are

$$\hat{\beta}_1 = \frac{\sum_{i=1}^{m} (x_i - \bar{x})(y_i - \bar{y})}{\sum_{i=1}^{m} (x_i - \bar{x})^2}, \tag{11.12}$$

$$\hat{\beta}_0 = \bar{y} - \hat{\beta}_1 \bar{x}. \tag{11.13}$$

Parameters $\hat{\beta}_1$ and $\hat{\beta}_0$ are unbiased estimators of the true regression parameters with the following distributions (under the normality assumptions on ε_i's mentioned above)

$$\hat{\beta}_1 \sim N\left(\beta_1, \frac{\sigma^2}{\sum_{i=1}^{m} (x_i - \bar{x})^2}\right); \quad \hat{\beta}_0 \sim N\left(\beta_0, \sigma^2 \left(\frac{1}{m} + \frac{\bar{x}^2}{\sum_{i=1}^{m} (x_i - \bar{x})^2}\right)\right) \tag{11.14}$$

The regression residuals ϵ_i allow us to estimate σ^2 with

$$S^2 = \frac{1}{m-2} \sum_{i=1}^{m} (y_i - \hat{y}_i)^2 = \frac{1}{m-2} \sum_{i=1}^{m} \left(y_i - (\hat{\beta}_0 + \hat{\beta}_1 x_i) \right)^2. \tag{11.15}$$

The quantity S is called the *residual standard error* and it is an unbiased estimator of σ.

Example 11.2.1. Suppose, from the real estate valuation data set, we wish to estimate how the distance to the nearest metro station affects the price of a unit area. Data in Table 11.1 shows rows 21–40 for variables of interest.

According to (11.9), we can model the relationship in our data set as

$$\text{Price} = \beta_0 + \beta_1 * \text{Distance} + \varepsilon$$

Since β_0 and β_1 are unknown, we shall estimate the approximations $\hat{\beta}_0$ and $\hat{\beta}_1$ from data and build our fitted regression line

$$\widehat{\text{Price}} = \hat{\beta}_0 + \hat{\beta}_1 * \text{Distance}$$

We use (11.12) and (11.13) to compute estimates for least squares coefficients β_0 and β_1

$$\hat{\beta}_0 = 45.87, \quad \hat{\beta}_1 = -0.007.$$

An illustration of the least squares line in relation to data points is given in Fig. 11.2. Squared errors ε_i, $i = 1, \ldots, 20$ are represented by the gray vertical line segments connecting the data points (in red) to the regression line (blue line). Coefficients $\hat{\beta}_0$, $\hat{\beta}_1$ were estimated to minimize the sum of squared errors $\sum_{i=1}^{m} \epsilon_i^2$. According to this illustration,

Distance (m)	Price ($/unit^2)	Distance (m)	Price ($/unit^2)
2275.8770	29.3	4519.6900	22.1
279.1726	51.6	769.4034	25.0
1360.1390	24.6	488.5727	34.2
279.1726	47.9	323.6550	49.3
480.6977	38.8	205.3670	55.1
1487.8680	27.0	4079.4180	27.3
383.8624	56.2	1935.0090	22.9
276.4490	33.6	1360.1390	25.3
557.4780	47.0	577.9615	47.7
451.2438	57.1	289.3248	46.2

Table 11.1.: Rows 21–40 from *real estate valuation* data set. Only the data for *distance to the nearest metro station* and *price of a unit area* are displayed

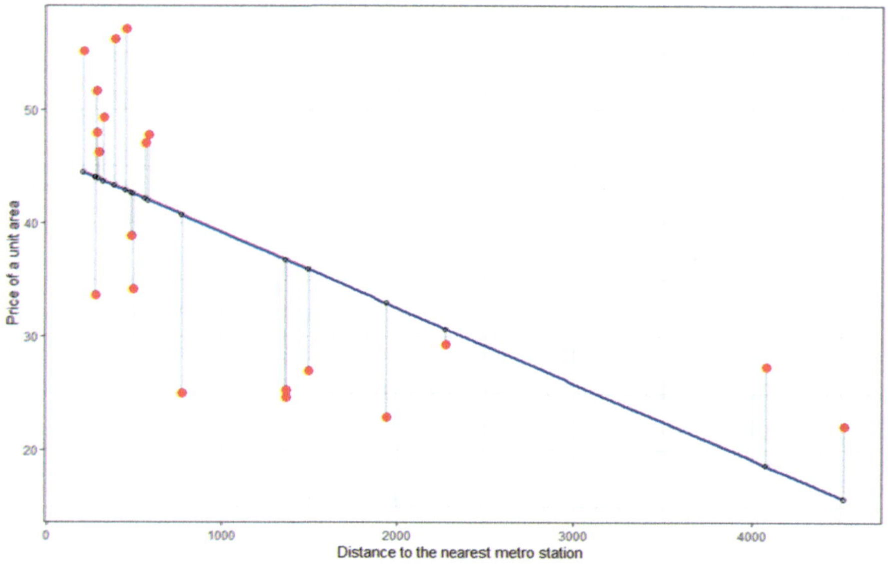

Figure 11.2.: For rows 21–40 of the *real estate valuation* data set, the least squares fit for the regression of *distance to the nearest metro station* onto *price of a unit area* shown. The fit is found by minimizing the sum of squared errors represented by gray line segments

every additional 1000 meter distance to the nearest metro station causes an average drop of $7 in the price of a unit area.

11.2.2. Statistical Inference on Regression Coefficients

We have previously assumed that there is a linear relationship between X and Y. In order to test hypotheses about the coefficients in the relationship, we shall first define an important characteristic of the regression model. Let

$$\mathbf{S_T^2} = \frac{1}{m}\sum_{i=1}^{m}(y_i - \bar{y})^2 \tag{11.16}$$

be the total variance for the regression response y_i. We can partition this expression as

$$\mathbf{S_T^2} = \frac{1}{m}\sum_{i=1}^{m}(y_i - \bar{y})^2 = \frac{1}{m}\sum_{i=1}^{m}(y_i - \hat{y}_i)^2 + \frac{1}{m}\sum_{i=1}^{m}(\hat{y}_i - \bar{y})^2 = \mathbf{S_R^2} + \mathbf{S_M^2}, \tag{11.17}$$

where $\mathbf{S_R^2}$ is the residual variance (11.11) and $\mathbf{S_M^2}$ is the *model variance.* In other words, $\mathbf{S_M^2}$ is the portion of the total variance $\mathbf{S_T^2}$ that is explained by the model, and $\mathbf{S_R^2}$ is the portion of the total variance which cannot be explained by the model.

Assuming that $\varepsilon_i \sim N(0, \sigma^2)$ i.i.d., $i = 1, 2, \ldots, m$, we can test the hypothesis

$$H_0 : \beta_1 = 0 \quad \text{vs} \quad H_1 : \beta_1 \neq 0.$$

If we fail to reject H_0, then our model reduces to $Y = \beta_0 + \varepsilon$, which means that there is no association between X and Y. Under H_0 we have the following distributions:

$$\mathbf{S_T^2} \sim \sigma^2\chi^2_{m-1}; \qquad \mathbf{S_R^2} \sim \sigma^2\chi^2_{m-2}; \qquad \mathbf{S_M^2} \sim \sigma^2\chi^2_1.$$

The test statistic for a one-sided, upper-tailed F test is given by

$$F = \frac{(m-2)\mathbf{S_M^2}}{\mathbf{S_R^2}} \sim F_{1,m-2}.$$

Once we have rejected H_0, another question is in order.

How well do our estimates (11.13) and (11.12) approximate the true (unobserved) regression parameters of the model (11.9)? Recall that $\hat{\beta}_1$ follows a normal distribution with mean β_1 and variance $\dfrac{\sigma^2}{\sum_{i=1}^{m}(x_i - \bar{x})^2}$, where σ^2 is the (unknown) variance of ε. We can estimate σ^2 with S^2 using (11.15). Then, a two-sided test statistic for testing

$$H_0 : \beta_1 = \beta_{10} \quad \text{vs} \quad H_1 : \beta_1 \neq \beta_{10}$$

is given by

$$t = \frac{(\hat{\beta}_1 - \beta_{10})\sqrt{\sum_{i=1}^{m}(x_i - \bar{x})^2}}{S} \sim t_{m-2}.$$

It follows Student's t distribution with $m-2$ degrees of freedom. This gives a $(1-\alpha)100\%$ confidence interval

$$\left[\hat{\beta}_1 - t_{\alpha/2,m-2}\frac{S}{\sqrt{\sum_{i=1}^{m}(x_i - \bar{x})^2}},\ \hat{\beta}_1 + t_{\alpha/2,m-2}\frac{S}{\sqrt{\sum_{i=1}^{m}(x_i - \bar{x})^2}}\right].$$

Similarly, for $\hat{\beta}_0 \sim N\left(\beta_0,\ \sigma^2\left(\frac{1}{m}+\frac{\bar{x}^2}{\sum_{i=1}^m (x_i-\bar{x})^2}\right)\right)$ we can test

$$H_0 : \beta_0 = \beta_{00} \qquad \text{vs} \qquad H_1 : \beta_0 \neq \beta_{00}.$$

The test statistic is given by

$$t = \frac{\hat{\beta}_0 - \beta_{00}}{S\sqrt{\frac{1}{m}+\frac{\bar{x}^2}{\sum_{i=1}^m (x_i-\bar{x})^2}}} \sim t_{m-2}.$$

A $(1-\alpha)100\%$ confidence interval is

$$\left[\hat{\beta}_0 - t_{\alpha/2,m-2} S\sqrt{\frac{1}{m}+\frac{\bar{x}^2}{\sum_{i=1}^m (x_i-\bar{x})^2}},\ \hat{\beta}_0 + t_{\alpha/2,m-2} S\sqrt{\frac{1}{m}+\frac{\bar{x}^2}{\sum_{i=1}^m (x_i-\bar{x})^2}}\right].$$

We can use the residual standard error S as a measurement of regression fit. It gives the average amount of deviation of the response from the true regression line. Another useful measure of how well the regression model fits the data is

$$R^2 = 1 - \frac{\mathbf{S_R^2}}{\mathbf{S_T^2}} = 1 - \frac{\sum_{i=1}^m (y_i - \hat{y}_i)^2}{\sum_{i=1}^m (y_i - \bar{y})^2}. \tag{11.18}$$

Note that $0 < R^2 < 1$, and it gives the proportion of total variation in data explained by our regression model.

11.2.3. Verifying the Assumptions on the Error Terms

We assumed in the previous section that $\varepsilon_i \sim N(0, \sigma^2)$ i.i.d., $i = 1, 2, \ldots, m$. We now demonstrate a technique that allows to check the correctness of such an assumption. First note that for residual terms $\epsilon_i = y_i - \hat{y}_i$ we have

$$\mathbf{E}[\epsilon_i] = \bar{y} - \frac{1}{m}\sum_{i=1}^m \hat{y}_i = \bar{y} - \frac{1}{m}\sum_{i=1}^m (\hat{\beta}_0 + \hat{\beta}_1 x_i) = \bar{y} - \frac{1}{m}\sum_{i=1}^m (\bar{y} - \hat{\beta}_1\bar{x} + \hat{\beta}_1 x_i) = 0. \tag{11.19}$$

Also

$$\begin{aligned}\mathbf{Var}[\epsilon_i] &= \mathbf{Var}[y_i] + \mathbf{Var}[\hat{y}_i] - 2\mathbf{Cov}[y_i, \hat{y}_i] \\ &= \sigma^2 + \sigma^2\left(\frac{1}{m} + \frac{(x_i - \bar{x})^2}{\sum_{i=1}^m (x_i - \bar{x})^2}\right) - 2\mathbf{Cov}[y_i, \bar{y} - \hat{\beta}_1\bar{x} + \hat{\beta}_1 x_i].\end{aligned} \tag{11.20}$$

For the covariance term, we have

$$\begin{aligned}\mathbf{Cov}[y_i, \bar{y} - \hat{\beta}_1\bar{x} + \hat{\beta}_1 x_i] &= \mathbf{Cov}[y_i, \bar{y}] + (x_i - \bar{x})\mathbf{Cov}[y_i, \hat{\beta}_1] \\ &= \frac{\sigma^2}{m} + \frac{(x_i - \bar{x})}{\sum_{i=1}^m (x_i - \bar{x})^2}\mathbf{Cov}\left[y_i, \sum_{i=1}^m (x_i - \bar{x})(y_i - \bar{y})\right] \\ &= \frac{\sigma^2}{m} + \frac{(x_i - \bar{x})}{\sum_{i=1}^m (x_i - \bar{x})^2}\mathbf{Cov}\left[y_i, \sum_{i=1}^m (x_i - \bar{x})y_i\right] \\ &= \frac{\sigma^2}{m} + \sigma^2\frac{(x_i - \bar{x})^2}{\sum_{i=1}^m (x_i - \bar{x})^2} \\ &= \sigma^2\left(\frac{1}{m} + \frac{(x_i - \bar{x})^2}{\sum_{i=1}^m (x_i - \bar{x})^2}\right)\end{aligned} . \tag{11.21}$$

Combining (11.20) and (11.21), we obtain

$$\mathbf{Var}[\epsilon_i] = \sigma^2\left(1 - \left(\frac{1}{m} + \frac{(x_i - \bar{x})^2}{\sum_{i=1}^m (x_i - \bar{x})^2}\right)\right). \tag{11.22}$$

Recall that σ^2 in (11.22) can be estimated with S^2 using (11.15). If the normality assumption on the error terms is correct, the standardized residuals

$$\frac{\epsilon_i}{S\left(1 - \left(\frac{1}{m} + \frac{(x_i - \bar{x})^2}{\sum_{i=1}^m (x_i - \bar{x})^2}\right)\right)}$$

should be very similar to the normal standard random variable which means that their respective quantiles should also be similar to one another. This can be checked by plotting the quantiles of the standardized regression residual terms against the normal quantiles using the quantile-quantile technique. If the normality assumption is correct, plotted points will follow a straight line very closely.

Example 11.2.2. We now demonstrate the application of the theory using our real-life data set. In Example 11.2.1, we used a small fraction of data from the real estate valuation data set to estimate regression coefficients. Let's use all 414 observations on the distance to the nearest metro station as predictors and the price of a unit area as response variables to build a simple linear regression model. Recall that we used

$$\text{Price} = \beta_0 + \beta_1 * \text{Distance} + \varepsilon$$

to model the relationship between predictor and response variables. The approximations $\hat{\beta}_0$ and $\hat{\beta}_1$ of true regression parameters can be produced from the data applying the least squares principle. The regression fit line is given by

$$\widehat{\text{Price}} = \hat{\beta}_0 + \hat{\beta}_1 * \text{Distance}.$$

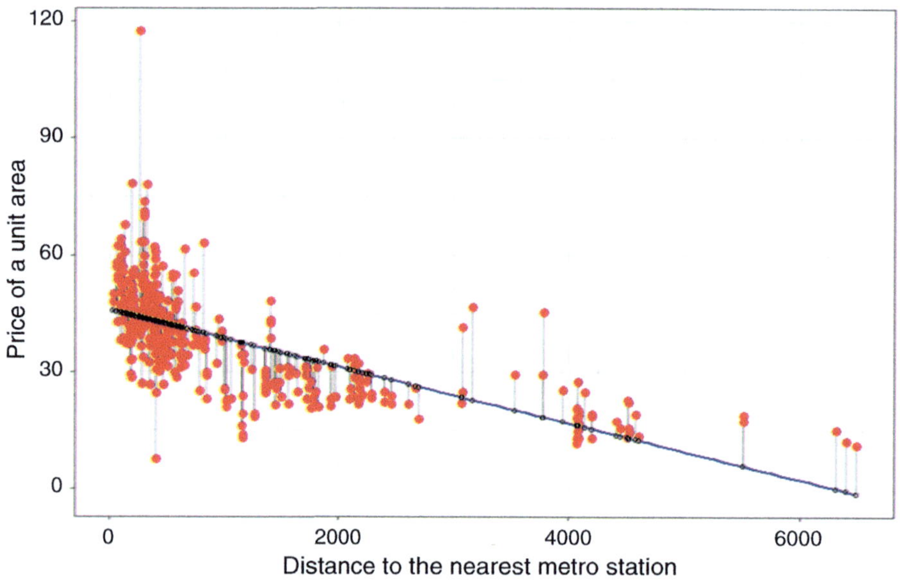

Figure 11.3.: For the *real estate valuation* data set, the least squares fit for the regression of *distance to the nearest metro station* onto *price of a unit area* shown. The fit is found by minimizing the sum of squared errors represented by the gray line segments

Figure 11.3 displays the least squares line and Table 11.2 gives estimates of the regression coefficients. Calculated *p-values* indicate the significance of both parameters. There is a strong evidence in favor of the existence of significant relationship between predictor and response variable.

Coefficients:	Estimate	Standard error	t-value	p-value $(P(>\|t\|))$
$\hat{\beta}_0$	45.8514271	0.6526105	70.26	$< 2 * 10^{-16}$
$\hat{\beta}_1$	−0.0072621	0.0003925	−18.50	$< 2 * 10^{-16}$

Table 11.2.: For the *real estate valuation* data set, coefficients of the least squares model for the regression of *distance to the nearest metro station* onto *price of a unit area* shown

Additionally, we compute (Table 11.3) the residual standard error s and the R^2 coefficient, which demonstrate that as a whole, our simple regression model only explains about 45% of the data variability. In order to better understand the value of $s = 10.07$, the *coefficient of variation* can be computed as follows:

$$\mathrm{CV} = \frac{s}{\bar{y}} * 100\% = 26.51\%. \tag{11.23}$$

The quantity in (11.23) gives a numeric representation of the spread of noise around the regression line relative to the average value of response variable.

s	10.07	df: 412	
R^2	0.4538		
F-statistic	342.2	df: 1, 412	p-value: $< 2.2 * 10^{-16}$

Table 11.3.: For the *real estate valuation* data set, more assessments of the least squares model for the regression of *distance to the nearest metro station* onto *price of a unit area*

Although the computed F-statistic and its corresponding p-value in Table 11.3 suggest the existence of a relationship between the distance to the nearest metro station and price of a unit area in our data, our overall conclusion is that a simple linear regression model is most likely not the best choice for this data set. This conclusion is due to the fact that the calculated R^2 coefficient is low since our proposed model does not explain much of the true variability in the data. We must consider including more explanatory variables to build a model that better explains variability in real estate valuation data set.

A *residual analysis* may be performed to check if the normality assumption on the error terms was correct. First, build a scatter plot of regression residuals ϵ_i against recorded values of distance to the nearest metro station x_i (see Fig. 11.4) to see if there is a fairly even distribution of residuals around zero. This indicates that the estimated least squares line provides a reasonable regression fit.

Furthermore, we build a normal plot (Fig. 11.5) to ensure that the quantiles of the standardized regression residual terms plotted against the quantiles of the standard normal random variable closely follow a straight line. This confirms the correctness of the normality assumption on ϵ_i's.

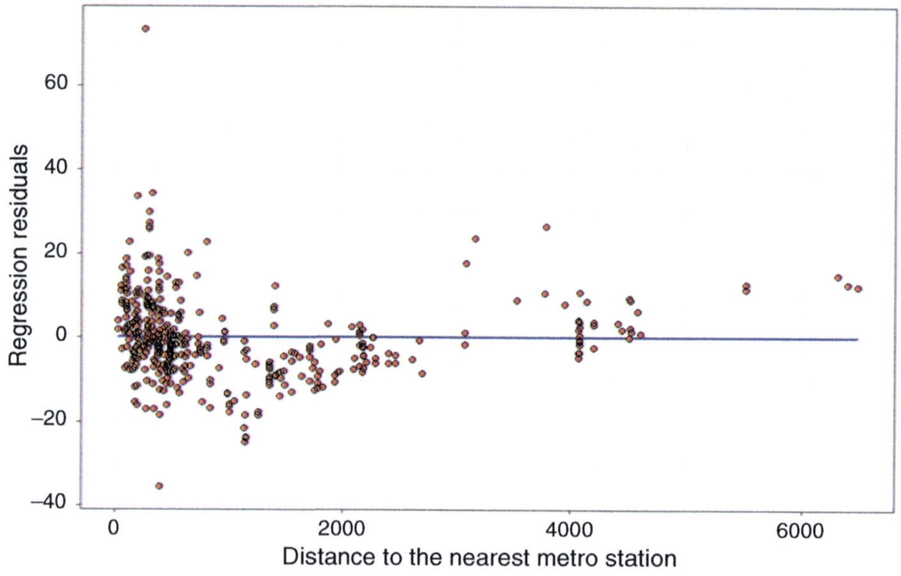

Figure 11.4.: For the *real estate valuation* data set, residual terms are plotted against the distance values for the least squares fit regression of *distance to the nearest metro station* onto *price of a unit area*

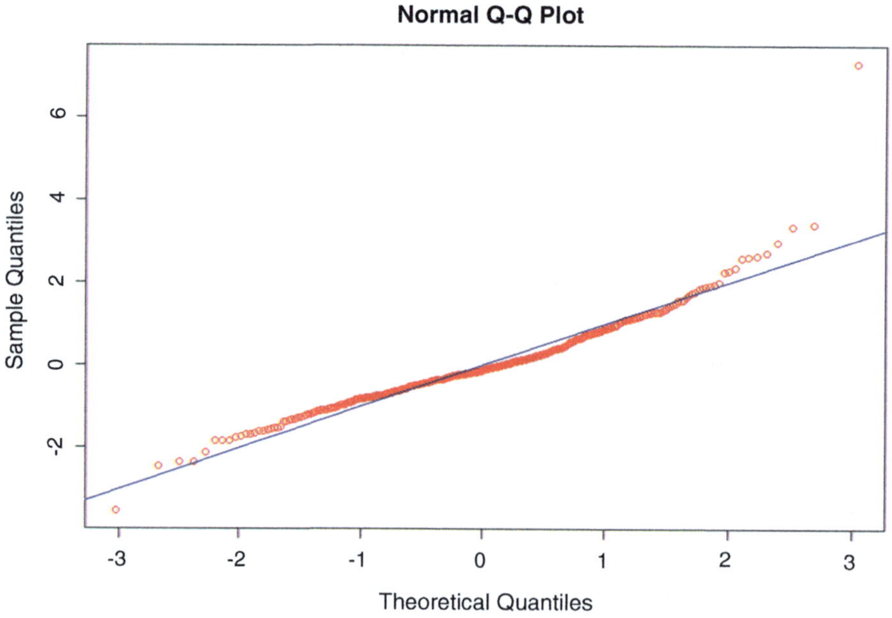

Figure 11.5.: For the *real estate valuation* data set, the *normal plot* of quantiles for the standardized regression residual terms against the normal quantiles is shown. A reference line (in blue) helps to see that the normality assumption on error term was correct for this model

11.3. Multiple Linear Regression

As we have seen in the previous section, a simple linear regression model can be a useful tool for investigating relationships between a response variable and a single predictor variable. In fact, it often serves as a starting point for many research tasks. Clearly, in practice, it is common to have more than one predictor variable available to a researcher in order to draw inference on the response variable. For example, in the *real estate valuation* data set, we have five predictors, namely, *age of a house, distance to the nearest metro station, number of convenience stores within the walking distance, and geographic location (latitude and longitude)* that can be investigated in order to establish their relationship to the price of a unit area. Many more predictors can be identified for such research. In fact, it is not uncommon nowadays to have thousands or even millions of predictors. In this dynamic environment, one must have robust methods for identifying the most important explanatory variables and building a regression model that explains a large portion of the existing relationships between predictor variables and the response variable.

11.3.1. Multiple Linear Regression Model

As shown by Eq. (11.4), the multiple linear regression is given by

$$Y = \beta_0 + \beta_1 X_1 + \beta_2 X_2 + \cdots + \beta_n X_n + \varepsilon, \tag{11.24}$$

where predictor variables form a vector of features $(X_1, X_2, \ldots, X_n)$, and regression parameters form a vector of weights $\boldsymbol{\beta}^\top = (\beta_0, \beta_1, \ldots, \beta_n)$. Note that a feature vector $(X_1, X_2, \ldots, X_n)$ constitutes a *single observation or an experiment* with n data points being recorded. Writing a feature vector as

$$\boldsymbol{X}^\top = (1, X_1, X_2, \ldots, X_n) \tag{11.25}$$

a vector form of (11.24) can be written as

$$Y = \boldsymbol{\beta}^\top \boldsymbol{X} + \varepsilon. \tag{11.26}$$

As before, the random error term ε is assumed to have $\mathbf{E}[\varepsilon] = 0$ and $\mathbf{Var}[\varepsilon] = \sigma^2$. Similar to the simple regression model, each random variable X_i $(i = 1, 2, \ldots, n)$ represents i-th predictor, and each regression parameter β_i $(i \neq 0)$ gives a measure of association between predictor X_i and the response variable Y. The intercept β_0 represents a *null model*, i.e., the case when there is no association between Y and any of the X_i's. Geometrically, each regression parameter can be viewed as a slope of a hyperplane in the corresponding dimension. As such, β_i provides a quantification of an effect on Y of one unit change in X_i, given that all other explanatory variables X_j $(j \neq i)$ remain unchanged.

11.3.2. Normal Equations

In order to estimate regression coefficients, one can use the least squares method introduced for simple linear regression. Recall that we have defined the residual variance $\mathbf{S}_{\mathbf{R}}^2$ as the sum of terms $(y_i - \hat{y}_i)^2$ averaged over m observations and estimated regression coefficients that minimize $\mathbf{S}_{\mathbf{R}}^2$. In a similar fashion, using matrix notation, we can now write

$$\mathbf{S}_{\mathbf{R}}^2 = \frac{(\mathbf{y} - \boldsymbol{M}_{\mathbf{X}}\boldsymbol{\beta})^\top(\mathbf{y} - \boldsymbol{M}_{\mathbf{X}}\boldsymbol{\beta})}{m}, \tag{11.27}$$

where $\mathbf{y}^\top = (y_1, y_2, \dots, y_m)$ is a vector of values of the response variable Y and $\boldsymbol{M}_{\mathbf{X}}$ is $m \times n+1$ matrix, called the *design matrix*, which consists of m observations with n features recorded for each observation. Recall that using the form (11.25) we obtain $n+1$ elements in each row of $\boldsymbol{M}_{\mathbf{X}}$. Finding estimates $\hat{\boldsymbol{\beta}}^\top = (\hat{\beta}_0, \hat{\beta}_1, \dots, \hat{\beta}_n)$ that minimize (11.27) involves solving a system of differential equations:

$$\frac{\partial}{\partial \boldsymbol{\beta}}\left((\mathbf{y} - \boldsymbol{M}_{\mathbf{X}}\boldsymbol{\beta})^\top(\mathbf{y} - \boldsymbol{M}_{\mathbf{X}}\hat{\boldsymbol{\beta}})\right) = 0. \tag{11.28}$$

Under the independence assumption for column vectors of the product $\boldsymbol{M}_{\mathbf{X}}^\top \boldsymbol{M}_{\mathbf{X}}$, the solutions are given by

$$\hat{\boldsymbol{\beta}} = \left(\boldsymbol{M}_{\mathbf{X}}^\top \boldsymbol{M}_{\mathbf{X}}\right)^{-1} \boldsymbol{M}_{\mathbf{X}}^\top \mathbf{y}. \tag{11.29}$$

We label the system of Eqs. (11.28), whose solutions are given by (11.29) the *normal equations*. This gives a relatively simple and mathematically elegant way of optimization for a multiple regression model. It is however not always applicable.

Some Applicability Issues: An important condition

$$m > n + 1 \tag{11.30}$$

must be imposed in order to apply the least squares method to multiple regression. A seemingly unlikely situation when the inequality in (11.30) is not satisfied is in fact quite common in modern-day industrial-level research situations when one has more variables that represent features than observations to estimate regression coefficients. Facing such complication, an educated statistician is still able to conduct research using a number of techniques. For example, a *forward selection, backward selection*, or *principal component analysis* (PCA) can be employed.

Another applicability concern for optimization based on solving the normal equations is its computational complexity. In order to give solutions for (11.28) in the form (11.29), we must compute the inverse of a matrix $\boldsymbol{M}_{\mathbf{X}}^\top \boldsymbol{M}_{\mathbf{X}} \in \mathbb{R}^{n+1 \times n+1}$. This task has *computational complexity* of $\mathcal{O}(n^3)$ and can be effectively executed by many statistical software packages for a moderate value of n. However, for large values of n finding the inverse matrix is a challenging process even for modern computers and can take substantial time to complete. In practice, if the number of features n is greater than $100,000$, other optimization techniques can be used such as the *gradient descent method.*

11.3.3. Statistical Inference on Regression Coefficients for Multiple Regression

The vector $\hat{\boldsymbol{\beta}}$ given by (11.29) is an unbiased estimator of the vector $\boldsymbol{\beta}^\top = (\beta_0, \beta_1, \ldots, \beta_n)$ of regression coefficients, i.e.,

$$\mathbf{E}[\hat{\boldsymbol{\beta}}] = \boldsymbol{\beta}. \tag{11.31}$$

One can show that the variance of $\hat{\boldsymbol{\beta}}$ is given by

$$\mathbf{Var}[\hat{\boldsymbol{\beta}}] = \left(\boldsymbol{M}_{\mathbf{X}}^\top \boldsymbol{M}_{\mathbf{X}}\right)^{-1} \sigma^2, \tag{11.32}$$

where $\sigma^2 = \mathbf{Var}[\varepsilon]$ is the variance of the true (unobserved) error of multiple linear regression. The unbiased estimator of σ^2 is given by

$$S^2 = \frac{\boldsymbol{\epsilon}^\top \boldsymbol{\epsilon}}{m-n-1} = \frac{1}{m-n-1} \sum_{i=1}^{m} (y_i - \hat{y}_i)^2, \tag{11.33}$$

where $\boldsymbol{\epsilon}^\top = (\epsilon_1, \epsilon_2, \ldots, \epsilon_m)$ is the vector of observed error terms $\epsilon_i = (y_i - \hat{y}_i)$ in m experiments.

As in the case with simple linear regression, we wish to answer an important question. Is our assumption about the existence of a relationship between the response variable Y and the explanatory variables $X_1, X_2, \ldots, X_n$ reasonable? Using quantities defined in (11.16) and (11.17), we can give an answer to such a question through hypothesis testing. Assuming $\varepsilon \sim N(0, \sigma^2 \boldsymbol{I}_{n+1})$, we wish to test

$$H_0 : \beta_1 = \beta_2 = \cdots = \beta_n = 0 \quad \text{vs} \quad H_1 : \beta_j \neq 0 \text{ for some } j \in \{1, 2, \ldots, n\}.$$

The test statistic is given by

$$F = \frac{(m-n-1)\mathbf{S}_{\mathbf{M}}^2}{n\mathbf{S}_{\mathbf{R}}^2} \sim F_{n,m-n-1}.$$

and we reject whenever F is large. Recall that under the correct assumption of linearity for a regression model, we have

$$\mathbf{E}\left[\frac{m}{m-n-1}\mathbf{S}_{\mathbf{R}}^2\right] = \mathbf{E}\left[s^2\right] = \sigma^2.$$

Note that if the linearity assumption is correct and H_1 is true, we have

$$\mathbf{E}\left[\frac{m}{n}\mathbf{S}_{\mathbf{M}}^2\right] > \sigma^2$$

and, consequently the F-statistic takes values greater than 1. For a large number of observations m, values of F statistic slightly larger than 1 may provide enough evidence against H_0. On the other hand, a much larger value of the F-statistic is needed to justify H_1 when m is small. We emphasize that rejecting H_0 in this test would mean the existence of approximately linear relationship between at least one of predictors and the response.

Once established that at least one of the predictors $X_1, X_2, \ldots, X_n$ is related to the response variable Y, we may wish to test for an entire vector of regression parameters. In other words, we want to check if $\boldsymbol{\beta} = \boldsymbol{\beta_0}$, where $\boldsymbol{\beta}^\top = (\beta_0, \beta_1, \ldots, \beta_n)$ is a vector of regression parameters and $\boldsymbol{\beta_0}^\top = (\beta_{00}, \beta_{10}, \ldots, \beta_{n0})$ is a constant vector. The hypothesis in vector notation is given as follows:

$$H_0 : \boldsymbol{\beta} = \boldsymbol{\beta_0} \quad \text{vs} \quad H_1 : \boldsymbol{\beta} \neq \boldsymbol{\beta_0}.$$

Under H_0, the test statistic has an F-distribution with degrees of freedom n and $m-n-1$. The matrix form for the F-statistic is

$$F = \frac{(m-n-1)(\hat{\boldsymbol{\beta}} - \boldsymbol{\beta_0})^\top \left(\boldsymbol{M}_{\mathbf{X}}^\top \boldsymbol{M}_{\mathbf{X}}\right)^{-1} (\hat{\boldsymbol{\beta}} - \boldsymbol{\beta_0})}{n\boldsymbol{\epsilon}^\top \boldsymbol{\epsilon}} \sim F_{n,m-n-1},$$

where $\boldsymbol{\epsilon}^\top \boldsymbol{\epsilon} = \sum_{i=1}^m (y_i - \hat{y}_i)^2$ as in (11.33). For a chosen significance level α, we reject H_0 if

$$F < F_{1-\alpha/2;n,m-n-1} \text{ or } F > F_{\alpha/2;n,m-n-1},$$

As mentioned before, an important condition $m > n+1$ must be satisfied in order to use the F-statistic. Otherwise, we may use experimental methods for selection of important predictors. One approach is to try all possible subsets of the set of explanatory variables $X_1, X_2, \ldots, X_n$ (including the empty set) and choose the one with the best fit. Unfortunately, this approach is not always practical, since the number of all possible subsets is 2^n, which is very large even for relatively small number of features n. Most software packages available today limit n to about 40.

11.3.4. Model Fit Evaluation

As in the case with simple linear regression, one can use the residual standard error s (11.33) as a measurement of the multiple linear regression model fit. Smaller values of s indicate a better model fit. We have also introduced the R^2 coefficient as a measure of how well regression model fits the data. Larger values of R^2 will indicate a better model fit. Note that calculating R^2 in its vector form

$$R^2 = 1 - \frac{(\mathbf{y} - \hat{\mathbf{y}})^\top (\mathbf{y} - \hat{\mathbf{y}})}{(\mathbf{y} - \bar{y})^\top (\mathbf{y} - \bar{y})} \tag{11.34}$$

is much more computationally efficient for large numbers of features (n) and observations (m). In the case of multiple regression model, our usual measures of model fit serve a double purpose. In addition to providing the information on how well a particular model fits the data, they give us a method for comparison of different models. In general, given two models, we shall prefer the one with lower value of s and higher value of R^2. Observe that adding predictors to a multiple regression model will always increase the value of

R^2. However, in some cases the difference in corresponding values of R^2 is so low that we may choose a (simpler) model with less predictors given that it satisfies an established threshold for the model performance.

Once the multiple regression model with the best fit is selected, we can verify its performance by computing confidence intervals for the response variable. Assuming a normal distribution $Y \sim N(\mu, \sigma^2)$, for large m (which is usually the case in *big data* research), we can compute a $(1-\alpha)100\%$ confidence interval for μ as follows:

$$\left[\bar{\hat{y}} - z_{\alpha/2}\frac{\sigma}{\sqrt{m}},\ \bar{\hat{y}} + z_{\alpha/2}\frac{\sigma}{\sqrt{m}}\right], \tag{11.35}$$

where $\mu = \boldsymbol{\beta}^\top \mathbf{X}$ by the linearity assumption (see (11.2) for identity link function), $\bar{\hat{y}} = \frac{1}{m}\sum_{i=1}^{m} \hat{\boldsymbol{\beta}}^\top \mathbf{X}_i$, and $\sigma^2 = \mathbf{Var}[\varepsilon]$ is estimated by S^2 (11.33).

Example 11.3.1. We now apply the theory to our real-life data set to see if a multiple regression model provides a better insight into relationship between explanatory variables and the response variable. Recall that the real estate valuation data set has five predictors, namely, *the age of a house* (Age), *distance to the nearest metro station* (MRT), *number of convenience stores within the walking distance* (Stores), *geographical latitude* (Lat), and *geographical longitude* (Long). The response variable in this data set is *the price of a square unit area* (Price). The multiple regression model that includes all predictors is

$$\text{Price} = \beta_0 + \beta_1 * \text{Age} + \beta_2 * \text{MRT} + \beta_3 * \text{Stores} + \beta_4 * \text{Lat} + \beta_5 * \text{Long} + \varepsilon. \tag{11.36}$$

Since the true regression parameters and the error term are unknown, we approximate using the least squares method and compute a vector

$$\hat{\boldsymbol{\beta}}^\top = (\hat{\beta}_0, \hat{\beta}_1, \hat{\beta}_2, \hat{\beta}_3, \hat{\beta}_4, \hat{\beta}_5)$$

from the normal Eqs. (11.28) and the formula for their solutions (11.29). After obtaining the regression fit equation

$$\widehat{\text{Price}} = \hat{\beta}_0 + \hat{\beta}_1 * \text{Age} + \hat{\beta}_2 * \text{MRT} + \hat{\beta}_3 * \text{Stores} + \hat{\beta}_4 * \text{Lat} + \hat{\beta}_5 * \text{Long} \tag{11.37}$$

we can estimate the error terms ϵ_i, $i = 1, \ldots, 414$. This in turn allows for the computation of $s^2 = 80.3712$ (see formula (11.33)) and the vector of standard errors for each regression coefficient using (11.31) and (11.32). We then evaluate the significance of each regression coefficient using the corresponding p-value. Table 11.4 contains the results of the described procedure. Observe that only four of the six regression coefficients provide evidence of significance of the relationships between the corresponding predictors and the response. At the same time, large p-values for $\hat{\beta}_5$ and $\hat{\beta}_0$ indicate that there is not enough evidence in favor of significant relationships between the price and the geographical longitude or the intercept for any reasonable value of the significance level.

Coefficient	Estimate	Standard error	t-value	p-value $(P(>\|t\|))$
$\hat{\beta}_0$	-4946	6211.0	-0.796	0.426
$\hat{\beta}_1$	-0.269	0.0390	-6.896	$2.04 * 10^{-11}$
$\hat{\beta}_2$	-0.004	0.0007	-5.888	$8.17 * 10^{-9}$
$\hat{\beta}_3$	1.1630	0.1902	6.114	$2.27 * 10^{-9}$
$\hat{\beta}_4$	237.80	44.950	5.290	$2.0 * 10^{-7}$
$\hat{\beta}_5$	-7.805	49.150	-0.159	0.874

Table 11.4.: For the *real estate valuation* data set, coefficients of the least squares fit for the multiple regression model with 5 predictors shown

We additionally compute R^2 to get an idea on the proportion of the true variability in data explained by our model. Analyzing the data in Table 11.5, we conclude that there is evidence of a significant relationship between some of the predictors and the response variable, despite the fact that our model as a whole only explains about 57% of the true data variability.

s	8.965	df: 408	
R^2	0.5712		
F-statistic	108.7	df: 5, 408	p-value: $< 2.2 * 10^{-16}$

Table 11.5.: For the *real estate valuation* data set, more assessments of the least squares fit for the multiple regression model with 5 predictors

Combining our observations, we decide to remove one explanatory variable, namely, geographical longitude. Our new multiple regression model is now

$$\text{Price} = \beta_0 + \beta_1 * \text{Age} + \beta_2 * \text{MRT} + \beta_3 * \text{Stores} + \beta_4 * \text{Lat} + \varepsilon. \tag{11.38}$$

Coefficient	Estimate	Standard error	t-value	p-value $(P(>\|t\|))$
$\hat{\beta}_0$	-5916	1113.0	-5.317	$1.74 * 10^{-7}$
$\hat{\beta}_1$	-0.269	0.0390	-6.903	$1.95 * 10^{-11}$
$\hat{\beta}_2$	-0.004	0.0005	-8.473	$4.37 * 10^{-16}$
$\hat{\beta}_3$	1.1650	0.1897	6.141	$1.94 * 10^{-9}$
$\hat{\beta}_4$	238.60	44.560	5.355	$1.43 * 10^{-7}$

Table 11.6.: For the *real estate valuation* data set, coefficients of the least squares fit for the multiple regression model with 4 predictors, excluding the *geographical longitude*

s	8.954	df: 409	
R^2	0.5711		
F-statistic	136.2	df: 4, 409	p-value: $< 2.2 * 10^{-16}$

Table 11.7.: For the *real estate valuation* data set, more assessments of the least squares fit for the multiple regression model with 4 predictors, excluding the *geographical longitude*

Tables 11.6 and 11.7 contain the corresponding calculations for the model parameters and performance evaluation. One can see that there is no significant difference in performance of these two models. Observe that such a moderate decrease in corresponding values of R^2 coefficient tells us that including the geographical longitude into the model was not a rational approach, since it makes almost no difference in terms of explaining the variability in data. It is also worth noting that the significance level of the intercept coefficient $\hat{\beta}_0$ is much higher in the absence of geographical longitude variable. Also, a slightly lower value of the standard error s in model (11.38) will lead to narrower confidence intervals.

While additional analyses can (and should) be done, we conclude this example with a few important notes. First, it is important to realize that for a large number of features n there is a chance that one shall see p-values indicating highly significant relationship between some of the predictors and the response *even if such relationship does not exist.* In fact, for $n > 100$ we are almost guaranteed to see proof of significance for at least one of the predictors. Hence, one should also use the F-statistic to test for existing relationships. Simply relying on individual p-values is misleading. Secondly, it is important to start any linear regression modeling process with a clear understanding that the exact linear relationship is highly unlikely and exact true values of the response variable are very difficult (if at all possible) to predict in quantitative regression models. Hence, confidence intervals must be computed. For example, the 95% confidence interval (11.35) for price in model (11.38) is [37.12, 38.84]. Note that this interval was computed from the least squares fit

$$\widehat{\text{Price}} = -5916 - 0.2695 * \text{Age} - 0.0047 * \text{MRT} + 1.158 * \text{Stores} + 227.1 * \text{Lat} \quad (11.39)$$

and not from the recorded price data itself.

Finally, a very schematic look at the residual analysis can be given as follows. In Fig. 11.6, we plot the regression residuals for model (11.38) against the values of the response variable estimated with the least squares fit (11.39). The green line in Fig. 11.6 suggests that a better fit could be achieved with the piecewise linear regression model. A normal plot (Fig. 11.7) confirms that the normality assumption on ϵ_i's was reasonable.

We conclude this section by noting that there is a Bayesian approach for the analysis of the linear model in Box and Tiao [1973], along with a discussion of hierarchical designs in Chap. 5.

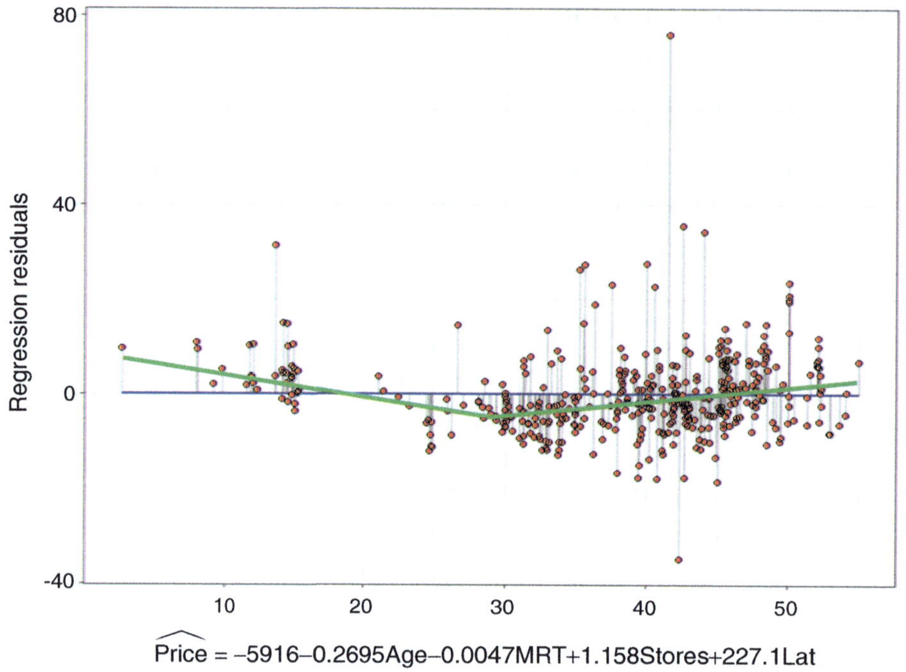

Figure 11.6.: For the *real estate valuation* data set, residual terms are plotted against the fitted values for price of a unit area based on the multiple linear regression model on 4 predictors

11.4. Regression in Machine Learning

We now introduce the *machine Learning* approach to regression modeling. Although all previous theoretical foundations remain unchanged, the main focus is however slightly different. While a statistician is concerned with discovering and explaining the relationship between predictors and the response variable, a machine learning practitioner is mainly interested in the predictive potential of such a relationship. We shall slightly change our notation to adhere to the conventional notation in the machine learning community.

The general experimental setup is the same. Data is recorded in pairs

$$(\mathbf{x}^{(1)}, y^{(1)}), (\mathbf{x}^{(2)}, y^{(2)}), \ldots, (\mathbf{x}^{(m)}, y^{(m)}), \tag{11.40}$$

where each $\mathbf{x}^{(i)} = (x_1^{(i)}, x_2^{(i)}, \ldots, x_n^{(i)}) \in \mathbb{R}^n$, $i = 1, 2, \ldots, m$ is the observed vector of values of *predictor* variable also called *features* and $y^{(1)}, y^{(2)}, \ldots, y^{(m)} \in \mathbb{R}$ are the observed values of the *response* variable Y. The notation $x_j^{(i)}$ refers to the j'th feature in the i'th observation. We predict values of Y with $\hat{y}^{(i)} = h(\mathbf{x}^{(i)})$. Function $h(\cdot)$ is called a *hypothesis* and defined as

$$h(\mathbf{x}^{(i)}) = \mathbf{w}^\top \mathbf{x}^{(i)} + b, \tag{11.41}$$

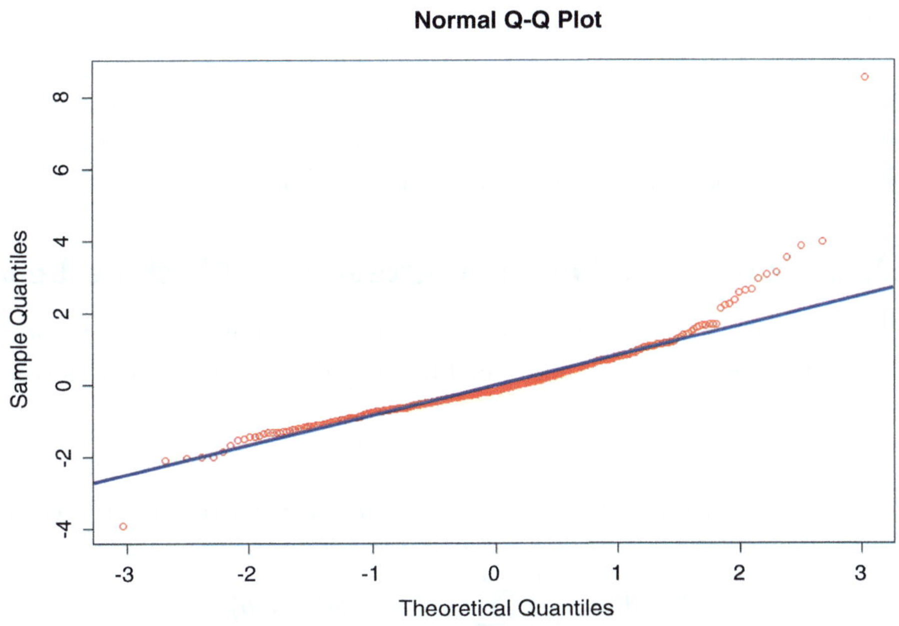

Figure 11.7.: For the *real estate valuation* data set, the *normal plot* of quantiles for the standardized regression residual terms against the normal quantiles shown. A reference line (in blue) helps to see that the normality assumption on error term was correct for this model

where *weights* $\mathbf{w} = (w_1, w_2, \ldots, w_n) \in \mathbb{R}^n$ and intercept $b \in \mathbb{R}$ are model parameters. The interpretation of model parameters is unchanged. Intercept b models the situation of a zero input. If a weight w_j is given a positive value, the corresponding predictor x_j becomes correlated with the response variable in a positive way (i.e., greater values of x_j cause the increase in $\hat{y}$). A negative value of w_j causes a negative correlation between x_j and $\hat{y}$.

By far, the most common measure of the performance of a quantitative regression model is a *loss function:*

$$c_i(\mathbf{w}, b) = y^{(i)} - \hat{y}^{(i)} = y^{(i)} - h(\mathbf{x}^{(i)}) = y^{(i)} - (\mathbf{w}^\top \mathbf{x}^{(i)} + b). \tag{11.42}$$

The expression in (11.42) gives a numerical representation of the empirical error based on a single observation. The performance of the model based on an entire collection of observations (called a *design matrix*) can be evaluated using the *cost function:*

$$\mathcal{C}(\mathbf{w}, b) = \frac{1}{2m} \sum_{i=1}^{m} (y^{(i)} - \hat{y}^{(i)})^2, \tag{11.43}$$

which our reader immediately recognizes as a slightly modified expression for estimating a mean squared residual error. The factor 2 in the denominator is included for computational convenience which will become evident in the sequel.

The goal of a machine learning algorithm is to adjust parameters $\mathbf{w}$, b in a way as to reduce the value of a cost function while gaining experience by observing the change in this value based on inputs from a data design matrix. As described earlier, a statistician typically achieves this goal by applying the least squares method. In machine learning, we often rely on an algorithmic solution for this problem.

11.4.1. Optimization for Linear Regression in Machine Learning

To motivate this approach, we first consider a simple linear regression model with one predictor x in a single observation. The machine learning model for predicting y_i becomes

$$\hat{y}^{(i)} = wx^{(i)} + b, \tag{11.44}$$

where $w, b \in \mathbb{R}$. The cost function $\mathcal{C}(w, b)$ for m observations is still computed as in (11.43)

$$\mathcal{C}(w, b) = \frac{1}{2m} \sum_{i=1}^{m} (y^{(i)} - (wx^{(i)} + b))^2.$$

Looking closely, we notice that it is a quadratic function of arguments w and b. In this particular case, it is called *elliptical paraboloid.* As such, it is a *convex* function with a single *global minimum*—a point that obtains the lowest value of a function on its domain. An illustration of $\mathcal{C}(w, b)$ in its simplest form $\mathcal{C} = w^2 + b^2$ is given in Fig. 11.8. Although the exact shape may be slightly different depending on values $x^{(i)}$ and $y^{(i)}$, the existence of a single global minimum is a common property for all cases.

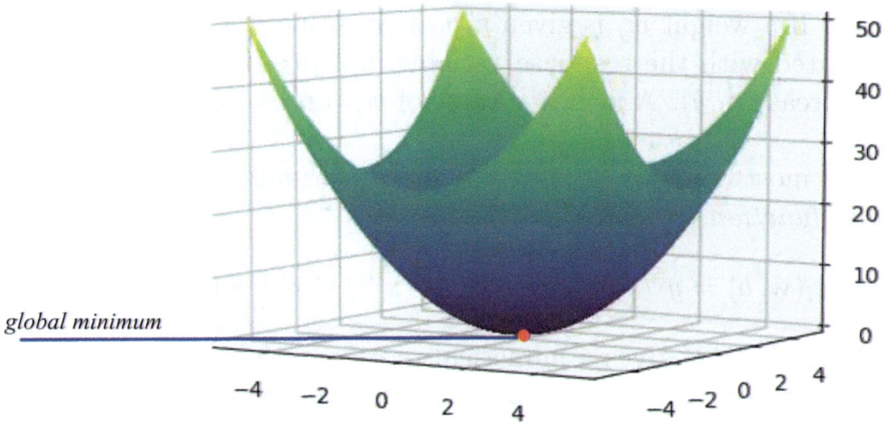

Figure 11.8.: Elliptical paraboloid $\mathcal{C} = w^2 + b^2$ with a single *local minimum*, which is also a *global minimum*

Clearly, there are multiple analytical ways to solve for critical points. One approach is to use partial derivatives of $\mathcal{C}(w, b)$ and find values $w_{\min}, b_{\min}$ such that

$$\frac{\partial \mathcal{C}}{\partial w}(w_{\min}) = 0, \qquad \frac{\partial \mathcal{C}}{\partial b}(b_{\min}) = 0. \tag{11.45}$$

Recall that the vector of all partial derivatives is called *gradient* and is denoted $\nabla \mathcal{C}(w, b)$. Thus, we want to find values $w_{\min}, b_{\min}$ that turn every element of the gradient $\nabla \mathcal{C}(w, b)$ into zero. How does one build an algorithm to solve such a problem?

Example 11.4.1. We consider the simplest possible case by setting $b = 0$ to obtain

$$\mathcal{C}(w) = \frac{1}{2m} \sum_{i=1}^{m} (y^{(i)} - wx^{(i)})^2.$$

We can see that

$$\mathcal{C}(w) = w^2 \frac{1}{2m} \sum_{i=1}^{m} \left(x^{(i)}\right)^2 - 2w \frac{1}{2m} \sum_{i=1}^{m} x^{(i)} y^{(i)} + \frac{1}{2m} \sum_{i=1}^{m} \left(y^{(i)}\right)^2$$

as a function of w takes the form of a parabola. Starting at arbitrary point $\big(w, \mathcal{C}(w)\big)$ of a parabola in Fig. 11.9, we can use the derivative $\mathcal{C}'(w) = \frac{\partial \mathcal{C}}{\partial w}$ to find the direction we want in order to move closer to the local minimum (which in this case is also the global minimum). Recall that the value of a derivative $\mathcal{C}'(w)$ at any given point w gives the slope of a tangent line at this particular point. Note that when our stating point is to

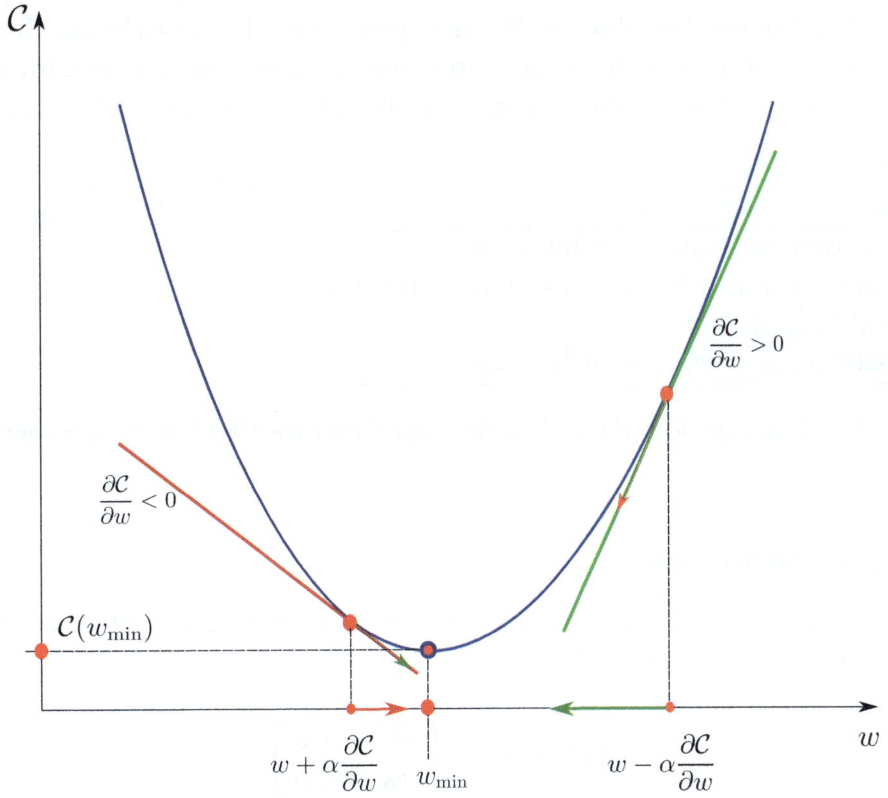

Figure 11.9.: Choosing the direction in order to move closer to the local minimum

the right of the minimum, the slope of a tangent line (green line in Fig. 11.9) is some positive value. On the other hand, if the starting point is to the left of the minimum, the slope of a tangent line (red line in Fig. 11.9) is some negative value. Hence, in order to adjust w so that we get closer to $w_{\min}$ in both cases, we must do the following:

$$w_{next} = w_{prev} - \alpha \mathcal{C}'(w_{prev}), \tag{11.46}$$

where $\alpha \in \mathbb{R}_+$ is called the *learning rate*. The derivative takes a very simple form:

$$\mathcal{C}'(w) = \frac{1}{m} \sum_{i=1}^{m} \left(wx^{(i)} - y^{(i)} \right) x^{(i)}.$$

Applying this adjustment in a recursive manner and choosing the appropriate value of the learning rate, we are guaranteed (*under appropriate assumptions which are met in this case*) to arrive at the local minimum. Note that it is not required that α is the same at each step. In fact, it is a common practice to change the value of α from step to step using various methods. Once we are at the local minimum $\left(w_{\min}, \mathcal{C}(w_{\min})\right)$, the slope of a tangent line becomes zero; hence, we obtain

$$w_{next} = w_{\min} - \alpha \mathcal{C}'(w_{\min}) = w_{\min} - \alpha \cdot 0 = w_{\min},$$

which concludes our search. In practice, it is often acceptable that $0 < |\mathcal{C}'(w)| < \delta$ so that we are almost at the local minimum. We now give a simple algorithm in Table 11.8 that can be followed to find the value of a regression parameter w that minimizes the value of a cost function $\mathcal{C}(w)$ over all training examples $(x^{(1)}, y^{(1)}), (x^{(2)}, y^{(2)}), \ldots, (x^{(m)}, y^{(m)})$.

Algorithm
• choose a random starting point $w = w_{prev}$
• set values for α and δ (small positive numbers)
• **while** $\lvert\mathcal{C}'(w_{next})\rvert > \delta$
• **compute** $w_{next} = w_{prev} - \alpha \mathcal{C}'(w_{prev})$

Table 11.8.: Minimization algorithm for cost function $\mathcal{C}(w)$ using gradient descent

11.4.1.1. Gradient Descent

Generalizing from Example 11.4.1, for a random sample (11.40), let $\mathcal{C}(\mathbf{w}, b)$ be the cost function defined as in (11.43). Let

$$\nabla \mathcal{C}(\mathbf{w}, b) = \left(\frac{\partial \mathcal{C}}{\partial \mathbf{w}}, \frac{\partial \mathcal{C}}{\partial b} \right) \tag{11.47}$$

be the gradient of the cost function $\mathcal{C}(\mathbf{w}, b)$. A *directional derivative* $D_{\mathbf{u}}\mathcal{C}(\mathbf{w}, b)$ in the direction of a unit vector $\mathbf{u} = (u_1, u_2)$ is defined:

$$D_{\mathbf{u}}\mathcal{C}(\mathbf{w}, b) = \mathbf{u}^{\top}\nabla\mathcal{C}(\mathbf{w}, b) = u_1\frac{\partial\mathcal{C}}{\partial\mathbf{w}} + u_2\frac{\partial\mathcal{C}}{\partial b}. \tag{11.48}$$

Recall from calculus:

$$\mathbf{u}^{\top}\nabla\mathcal{C}(\mathbf{w}, b) = \|\mathbf{u}\|_2 \cdot \|\nabla\mathcal{C}(\mathbf{w}, b)\|_2 \cos\theta,$$

where θ is the angle between vectors $\mathbf{u}$ and $\nabla\mathcal{C}(\mathbf{w}, b)$. Since $\mathbf{u}$ is a unit vector, $\|\mathbf{u}\|_2 = 1$ and we obtain

$$D_{\mathbf{u}}\mathcal{C}(\mathbf{w}, b) = \|\nabla\mathcal{C}(\mathbf{w}, b)\|_2 \cos\theta. \tag{11.49}$$

The expression in (11.49) is minimized when $\cos\theta = -1$, i.e., when the gradient $\nabla\mathcal{C}(\mathbf{w}, b)$ and unit vector $\mathbf{u}$ are pointed in the exact opposite directions: $\mathbf{u} = -\nabla\mathcal{C}(\mathbf{w}, b)$. Thus, moving in the direction of the negative gradient, we can decrease the absolute value of the tangent slope. This means that we are moving closer to the local minimum of $\mathcal{C}(\mathbf{w}, b)$ and the value of cost function is decreasing. This method of minimizing the value of cost function is called *gradient descent.*

11.4.1.2. Feature Standardization

Before proceeding to the next example, we describe one useful technique that must be used quite often together with gradient descent method. It is called *feature standardization.* Recall that the method of gradient descent involves summation of $\mathcal{C}'(w)$ over m observations. Since the number of observations may be quite large and computers have a limited number of bits that can be used to store a number, the result of such summation with very large magnitude can be rounded to $-\infty$ or ∞ by the computer. This is known as *overflow* numerical error. Feature standardization is applied to overcome this issue. For a feature x with m observations, standardization is done as follows:

$$x_{\text{std}}^{(i)} = \frac{x^{(i)} - \bar{x}}{\sigma}, \tag{11.50}$$

where

$$\bar{x} = \frac{1}{m}\sum_{i=1}^{m} x^{(i)} \text{ and } \sigma = \sqrt{\frac{1}{m}\sum_{i=1}^{m}(x^{(i)} - \bar{x})^2}.$$

It is also important to note that feature standardization is never to be applied to the values of the response variable.

Example 11.4.2. For the real estate valuation data set, we use a program written in Python to find regression coefficients that correspond to the least squares fit. We choose a constant learning rate $\alpha = 10^{-3}$ and a threshold $\delta = 10^{-6}$ for optimization based on gradient descent. Results are given in Table 11.9. Note that regression coefficients

are different from the ones in Example 11.2.2. This is due to feature standardization that we have applied here. Nevertheless, our model produces equivalent predictions $\hat{y}^{(i)}, i = 1, \ldots, m$, which is confirmed by calculated error terms (Table 11.10).
Figure 11.10 illustrates the list squares line for our model. It is evident that distance from the nearest metro station is displayed on a different scale from the one in Example 11.2.2 and can be now interpreted as the nearness to the average value of the distance from the nearest metro station.

Coefficient	Estimate
Intercept (b)	37.9802
Slope (w)	-9.1655

Table 11.9.: For the *real estate valuation* data set, coefficients of the least squares gradient descent model for the regression of *standardized distance to the nearest metro station* onto *price of a unit area* shown

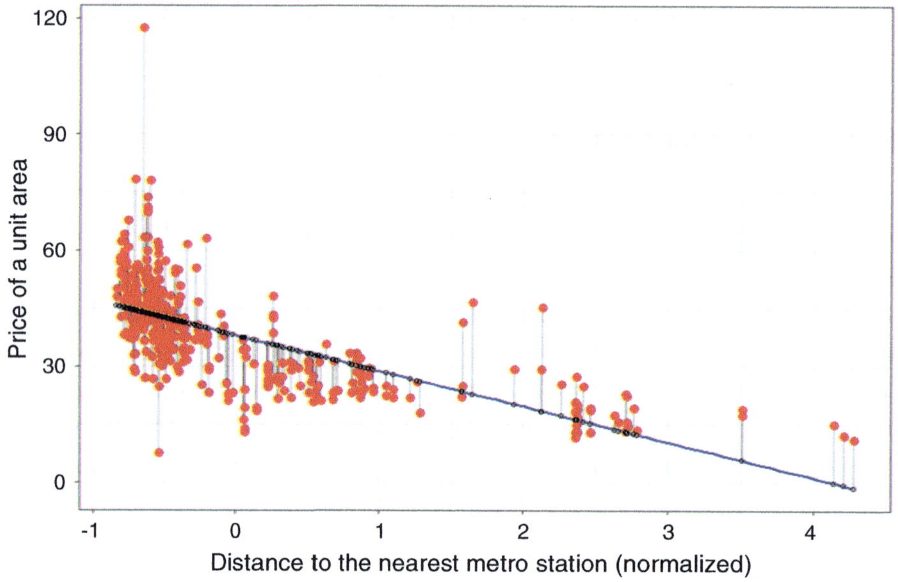

Figure 11.10.: For the *real estate valuation* data set, the least squares fit for the regression of *standardized distance to the nearest metro station* onto *price of a unit area* shown. The fit is found by minimizing the sum of squared errors represented by the gray line segments

We compute the residual standard error s and value of the cost function $\mathcal{C}(w, b)$. It follows from (11.15) and (11.43) that $S = \sqrt{\frac{2m}{m-2} \mathcal{C}(w, b)}$. The coefficient of variation (CV) provides a numeric representation of the spread of noise around the regression line relative to the average value of response variable.

s	10.0685
Cost	50.4429
CV	26.45%

Table 11.10.: For the *real estate valuation* data set, more assessments of the least squares gradient descent model for the regression of *standardized distance to the nearest metro station* onto *price of a unit area*

11.4.1.3. Computing Cost on a Test Set

As mentioned before, a machine learning engineer is interested in the predictive power of a model. Hence, we shall use the value of cost function calculated on a **test set** as a measure of performance for our model. The simplest way of doing this would be randomly splitting our sample (11.40) into two parts, called a *training set* (about 70% of the data) and a *test set* (remaining 30%). We assume that observations in both sets are independent and identically distributed (i.i.d.). Under this assumption, we believe that the expected residual error of our model is the same for a training set as for a test set. A model is allowed to gain experience observing the training set. Regression parameters are produced by minimizing the residual mean squared error on a training set. The value of a mean squared error is then computed using obtained regression coefficients and the data from a test set used to evaluate model performance.

Other configurations are possible. For example, we can divide our collected data set into three subsets: the *training*, *validation* and *test* set, approximately 70%/20%/10% in size proportion. In this setup, as per usual, our model is gaining experience by observing a training set. A validation set is used to assess the model performance by computing the value of the Cost function. Once the best performing model is chosen, it is tested using the data from a test set. Another commonly used technique is a *k-fold cross-validation*. Here, the data set is divided into k approximately equal parts. A model is trained k times, each time $k-1$ sets are used as training sets and one set is used as a test set to evaluate model performance. A test set is changed for each training routine. Residual mean squared error $\mathcal{C}_i$ is computed for $i = 1, 2, \ldots, k$. Finally, we select a model with the lowest error computed on a test set.

Example 11.4.3. Using findings in Example 11.3.1, we demonstrate the simplest approach to performance evaluation for the multiple regression model. A quick look at the data set (Table 11.11) makes it evident that there is a difference in scale for feature measurement. Hence, to avoid computation problems such as overflow, we apply feature standardization (11.50). The result of feature standardization is shown in Table 11.12. Recall that the values of the response variable are not to be standardized. Next, we divide the real estate valuation data set into two parts, the training set (observations 1–300) and the test set (observations 301 – –414). The multiple regression model in matrix form is

$$\hat{\mathbf{y}} = \mathbf{M_x}\mathbf{w} + \boldsymbol{b}, \tag{11.51}$$

Age	MRT	Stores	Lat	Price
32.0	84.88	10	24.98298	37.9
19.5	306.59	9	24.98034	42.2
13.3	561.98	5	24.98746	47.3
13.3	561.98	5	24.98746	54.8
5	390.57	5	24.97937	43.1

Table 11.11.: For the *real estate valuation* data set, the first five measurements shown for the chosen features are *age of a house* (Age), *distance to the nearest metro station* (MRT), *number of convenience stores within the walking distance* (Stores), and *geographical latitude* (Lat)

Age	MRT	Stores	Lat	Price
1.2500	−0.7846	2.0099	1.1152	37.9
0.1453	−0.6113	1.6677	0.9044	42.2
−0.4026	−0.4117	0.2989	1.4728	47.3
−0.4026	−0.4117	0.2989	1.4728	54.8
−1.1362	−0.5457	0.2989	0.82.70	43.1

Table 11.12.: For the *real estate valuation* data set, the first five measurements shown for *standardized features* are *age of a house* (Age), *distance to the nearest metro station* (MRT), *number of convenience stores within the walking distance* (Stores), and *geographical latitude* (Lat)

where $\mathbf{M_x}$ is a design matrix, $\mathbf{w}^\top = (w_1, w_2, w_3, w_4)$ is a vector of regression coefficients (weights), and $\mathbf{b}^\top = (b_1, b_2, b_3, b_4)$ is a bias vector. According to our choice, the training set design matrix has dimensions 300×4, and the test set design matrix has dimensions 114×4. As before, we start by choosing a constant learning rate $\alpha = 10^{-3}$ and a threshold $\delta = 10^{-6}$ for optimization based on gradient descent. The results in Table 11.13 show that it took our algorithm 46,920 steps to reach the set value of the threshold for

$$\|\nabla \mathcal{C}(\mathbf{w}, b)\|_2 = \sqrt{\frac{[\mathbf{M_x}(\hat{\mathbf{y}} - \mathbf{y})]^\top [\mathbf{M_x}(\hat{\mathbf{y}} - \mathbf{y})]}{300}}$$

As per usual, we compute values for the cost function, mean residual squared error, and coefficient of variation (Table 11.14).

A 95% confidence interval for the expected value of the price of a unit area computed from the *training set* data is

$$[36.9901, 39.0732].$$

We now use the least squares fit model

$$\hat{y} = 38.0317 - 3.4315 * \text{Age} - 5.0474 * \text{MRT} + 3.2125 * \text{Stores} + 3.4190 * \text{Lat}$$

to compute predictions $\hat{y}^{(i)}_{\text{test}}, i = 1, 2, \ldots, 114$ for our test set and compare them with

Training set			
Coefficient	Estimate	α	Steps taken
b	38.0317		
w_1	−3.4315		
w_2	−5.0474		
w_3	3.2125		
w_4	3.4190	10^{-3}	46,920

Table 11.13.: For the *real estate valuation* data set (observations 1–300), coefficients of the least squares gradient descent model on the training set for the multiple regression of 4 features onto *price of a unit area* shown

Training set	
s	9.2042
Cost	41.6529
CV	24%

Table 11.14.: For the *real estate valuation* data set (observations 1–300), more assessments of the least squares gradient descent model on the training set for the multiple regression of 4 features onto *price of a unit area*

the actual values $y^{(i)}, i = 1, 2, \ldots, 114$ of response variable. Data in Table 11.15 confirm that our model generalizes well and can be successfully used to predict outputs on a newly observed data with comparable magnitudes of error terms.

Test set	
s	8.511
Cost	34.63
CV	21.88%

Table 11.15.: For the *real estate valuation* data set (observations 301–414), more assessments of the least squares gradient descent model on the test set for the multiple regression of 4 features onto *price of a unit area*

A 95% confidence interval for the expected value of the price of a unit area computed from the *test set* data is

$$[36.4693, 39.5940].$$

Note that the (*predictive*) confidence interval computed on the data previously unseen by the model is, as expected, wider than the one computed on the training set data.

We now wish to test if the selected value of the learning rate was optimal. For that, we choose a set of possible values:

$$\alpha \in [0.0001, 0.0003, 0.001, 0.003, 0.01, 0.03, 0.1, 0.3, 0.6, 0.8, 0.9]$$

Our algorithm computes vectors of parameters $\mathbf{w}_i, \mathbf{b}_i$ (with $i = 1, 2, 3, \ldots, 11$) of the least squares fit for 11 models by observing the training set. Subsequently, the values $\mathcal{C}_i(\mathbf{w}_i, \mathbf{b}_i)$ were computed for each model. As a result, we discover that the initial choice $\alpha = 10^{-3}$ was not optimal.

Test set. Best model.			
b	38.0317		
w_1	−3.4315		
w_2	−5.0474		
w_3	3.2125		
w_4	3.4190	0.8	52

Table 11.16.: For the *real estate valuation* data set (observations 301–414), coefficients of the least squares gradient descent model on the test set for the multiple regression of 4 features onto *price of a unit area* shown

Tables 11.16 and 11.17 show that equivalent value of the cost function (rounded up to 10^{-9}) is achieved with the same regression parameters and $\alpha = 0.8$ in only 52 steps of a gradient descent algorithm. This is a very significant improvement in algorithm running time. To fully appreciate the difference it makes, simply consider an industrial size data set with a few hundred thousands of features and millions of observations, for which our optimization algorithm takes 10 days (240 h) of nonstop computation to find optimal regression parameters with $\alpha = 10^{-3}$. This same algorithm with $\alpha = 0.8$ would only take about 16 min to produce the same regression parameters.

Test set. Best model.	
s	8.511
Cost	34.63
CV	21.88%

Table 11.17.: For the *real estate valuation* data set (observations 301 − −414), more assessments of the least squares gradient descent model on the test set for the multiple regression of 4 features onto *price of a unit area*

Different values of α with the corresponding numbers of steps needed for our algorithm to reach the desired threshold $\delta = 10^{-6}$ are given in Table 11.18. Evidently, the number of steps needed for convergence as a function of the learning rate α achieves its local minimum at $\alpha = 0.8$.

11.5. Classification Models

As it has been mentioned, we often have the response variable Y that takes values in one of K categories (also called classes). For example, Y can denote a person's education level, so we have $y \in \{$high school, college, university, etc., which can be coded

α	0.0001	0.0003	0.001	0.003	0.01	0.03	0.1	0.3	0.6	0.8	0.9
Steps	469266	156417	46920	15635	4685	1557	462	149	71	52	177

Table 11.18.: Values of the learning rate α and corresponding number of steps taken by the gradient descent algorithm to achieve the threshold $\delta = 10^{-6}$ for $\|\nabla \mathcal{C}(\mathbf{w}, b)\|_2$

by assigning an integer value to each category $y \in \{1, 2, 3, \ldots, K\}$ for $2 \leq K \in \mathbb{N}$. In this case, we say that Y is a qualitative variable and use classification methods to predict its value. There exists quite a few classification methods. Some of the most commonly used are logistic regression, K-nearest neighbors, discriminant analysis, and support vector machines (SVM).

Each method has its area of applicability, strong sides, and limitations. For example, the logistic regression is most commonly used when the response variable can take one out of two possible values (in this case it is called a *binary response variable*). Although it is possible to extend the logistic regression to the case with multiple response classes using *one-versus-all* approach, the use of discriminant analysis or support vector machines is more customary for these situations. Logistic regression takes one of the central roles in machine learning, being the base of first deep learning networks such as *feed-forward neural network*. In fact, one of the first neural networks, called the *perceptron* Rosenblatt [1958], in its simplest form is an application of the logistic regression model with computational optimization (e.g., using the gradient descent algorithm described earlier).

11.5.1. Logistic Regression

The experiment setup for this model is usual. The data is recorded in pairs:

$$(\mathbf{x}^{(1)}, y^{(1)}), (\mathbf{x}^{(2)}, y^{(2)}), \ldots, (\mathbf{x}^{(m)}, y^{(m)}),$$

where each $\mathbf{x}^{(i)} = (x_1, x_2, \ldots, x_n)^\top$ for $i = 1, 2, \ldots, m$ is the observed set of values of *predictor* variable $\boldsymbol{X} \in \mathbb{R}^n$ and $y^{(1)}, y^{(2)}, \ldots, y^{(m)}$ are the observed values of the *response* variable Y. It is assumed that m observations of the random variable Y are independent of each other. Here, Y can belong to one of two possible classes. For example, a researcher may want to decide if a randomly chosen patient just admitted to the intensive care unit is having an epileptic seizure or is suffering from other conditions based on the recorded brain activity. It is customary to label the class that represents the condition of interest with "1" and the other class with "0". Hence, we have $y \in \{0, 1\}$ and we use the *logit* link function (11.3). For the feature vector $\mathbf{x}^\top = (x_0, x_1, x_2, \ldots, x_n)$ with $x_0 = 1$ and the parameter vector $\boldsymbol{\beta}^\top = (\beta_0, \beta_1, \beta_2, \ldots, \beta_n)$, we obtain

$$h(\boldsymbol{X}) = \frac{e^{\boldsymbol{\beta}^\top \mathbf{x}}}{1 + e^{\boldsymbol{\beta}^\top \mathbf{x}}}, \tag{11.52}$$

where $h(\boldsymbol{X})$ is the conditional expectation, and, as we assumed, $Y \sim \text{Bernoulli}(\theta)$ takes the form

$$h(\boldsymbol{X}) = \mathbf{Pr}\left[Y = 1 \mid \boldsymbol{X} = \mathbf{x}\right] = \theta. \tag{11.53}$$

Thus, Eq. (11.4) in this case is modeling the probability for Y to be in class "0" or "1" rather than the value of Y. When $h(\boldsymbol{X}) \approx 0.5$ we do not have enough information in $\boldsymbol{X}$ to predict Y. If however $h(\boldsymbol{X}) > 0.5$, we have

$$\mathbf{Pr}\left[Y = 1 \mid \boldsymbol{X} = \mathbf{x}\right] > \mathbf{Pr}\left[Y = 0 \mid \boldsymbol{X} = \mathbf{x}\right],$$

i.e., "1" is more likely to be the correct label. In contrast to the linear regression model, the amount of change in $h(\boldsymbol{X})$ caused by a one unit change in $\boldsymbol{X}$ will depend on the current value of $\boldsymbol{X}$, since the relationship between the predictor variable and $h(\boldsymbol{X})$ is not linear. However, if β_i is positive, then X_i and $h(\boldsymbol{X})$ are positively correlated, e.g., increasing the value of X_i (assuming that all X_j for $j \neq i$ stay unchanged) will cause the increase in the value of $h(\boldsymbol{X})$. If the coefficient β_i is negative, then the correlation between X_i and $h(\boldsymbol{X})$ is negative.

11.5.1.1. Optimization with Maximal Likelihood for Logistic Regression

Since the regression coefficients $\beta_0, \beta_1, \beta_2, \ldots, \beta_n$ are unknown, we would like to obtain the estimates. Earlier we used the least squares approach to estimate the unknown linear regression coefficients. We now introduce a more general approach, which can be applied to variety of models, namely, the *maximal likelihood method.*

The motivation for this method is to find an estimate $\hat{\boldsymbol{\beta}}^\top = (\hat{\beta}_0, \hat{\beta}_1, \hat{\beta}_2, \ldots, \hat{\beta}_n)$ for the unknown parameter vector $\boldsymbol{\beta}$ such that the model (11.52) produces values that are close to the actual observed class labels. In other words, if $y^{(i)} = 1$, we want to obtain the value close to 1 when we integrate $\mathbf{x}^{(i)}$ into Eq. (11.52) with $\hat{\boldsymbol{\beta}}$ estimated by our method. On the other hand, if in i-th experiment we had observed $y^{(i)} = 0$, we would like that (11.52) produces the value close to 0 for the corresponding predictor $\mathbf{x}^{(i)}$ and the estimate $\hat{\boldsymbol{\beta}}$.

For the i-th observation, considering the distribution assumption on Y made earlier and (11.53), we can write:

$$\begin{aligned} p(\mathbf{x};\boldsymbol{\beta}) &= \mathbf{Pr}\left[Y = y \mid \boldsymbol{X} = \mathbf{x}\right] \\ &= (h(\mathbf{x}))^y (1 - h(\mathbf{x}))^{1-y}. \end{aligned} \tag{11.54}$$

The independence assumption on Y renders the *likelihood function*:

$$\mathbf{L}\left(\boldsymbol{\beta}\right) = \prod_{i=1}^{m} p(\mathbf{x}^{(i)};\boldsymbol{\beta}). \tag{11.55}$$

Our goal is to estimate $\hat{\boldsymbol{\beta}}^\top = (\hat{\beta}_0, \hat{\beta}_1, \hat{\beta}_2, \ldots, \hat{\beta}_n)$ to maximize the value of $\mathbf{L}\left(\boldsymbol{\beta}\right)$ for all observed $\mathbf{x}^{(1)}, \ldots, \mathbf{x}^{(m)}$. Since the usual (however not universal) approach to this task is finding the global minima through differentiation, another form of the likelihood function is routinely used. Recalling a nice property of logarithmic function that allows for turning products into sums, we immediately realize its benefit for differentiating the likelihood function in (11.55). Observe that for $\mathbf{L}(\boldsymbol{\beta}) \neq 0$, critical values coincide. Indeed,

$$\frac{d \log \mathbf{L}}{d\boldsymbol{\beta}} = \frac{1}{\mathbf{L}} \cdot \frac{d\mathbf{L}}{d\boldsymbol{\beta}}.$$

The *log-likelihood function* is obtained as follows:

$$\begin{aligned} \boldsymbol{\ell}(\boldsymbol{\beta}) &= \log \mathbf{L}(\boldsymbol{\beta}) \\ &= \log \prod_{i=1}^{m} p(\mathbf{x}^{(i)}; \boldsymbol{\beta}) \\ &= \sum_{i=1}^{m} \log \left(p(\mathbf{x}^{(i)}; \boldsymbol{\beta}) \right), \end{aligned} \tag{11.56}$$

and considering (11.54), we get

$$\boldsymbol{\ell}(\boldsymbol{\beta}) = \sum_{i=1}^{m} y^{(i)} \log \left[h(\mathbf{x}^{(i)}) \right] + (1 - y^{(i)}) \log \left[1 - h(\mathbf{x}^{(i)}) \right], \tag{11.57}$$

which in matrix notation is

$$\boldsymbol{\ell}(\boldsymbol{\beta}) = \mathbf{y}^{\top} \log[h(\boldsymbol{M}_{\mathbf{x}})] + (1 - \mathbf{y})^{\top} \log[1 - h(\boldsymbol{M}_{\mathbf{x}})], \tag{11.58}$$

where $\boldsymbol{M}_{\mathbf{x}} = (\mathbf{x}^{(1)}, \mathbf{x}^{(2)}, \ldots, \mathbf{x}^{(m)})^{\top}$ is the design matrix, and $\mathbf{y}^{\top} = (y^{(1)}, y^{(2)}, \ldots, y^{(m)})$ is the set of observed values of the response variable Y.

An observant reader had surely noticed that the product in (11.55) could be a very small positive number, due to the fact that values in (11.54) represent probabilities. This observation in turn raises a "red flag" for a reader familiar with computational limitations. As it has been mentioned before, computers have a tendency to round up numbers that have large counts of digits. Thus, the value of $\mathbf{L}(\boldsymbol{\beta})$ could be rounded to zero causing *underflow* numerical error. This is another good reason for using the log-likelihood function for numerical optimization.

In conclusion, we would like to mention that estimators obtained with the maximal likelihood method possess several desirable statistical properties. First, if our assumption about the distribution family for the data is correct and the true distribution corresponds to exactly one value of the parameter θ, then, as the number of training examples approaches infinity, the maximal likelihood estimator of a parameter converges to the true value of the parameter. The estimator of θ is then said to have a *consistency property*. Second, the maximal likelihood estimator is asymptotically the most *statistically efficient*, meaning that as the number of training examples increases ($m \to \infty$), the parametric mean squared error (MSE) of a likelihood estimator is minimal among all consistent estimators, as shown by the Cràmer-Rao lower bound.

11.5.1.2. Statistical Inference

Estimating the parameter vector $\boldsymbol{\beta} = (\beta_0, \beta_1, \beta_2, \ldots, \beta_n)^\top$ that maximizes (11.58) is routinely done using simple calculus technique of finding extrema. We must solve a differential equation

$$\nabla \boldsymbol{\ell}(\boldsymbol{\beta}) = \left[\boldsymbol{y} - h(\boldsymbol{M}_{\mathbf{x}})\right]^\top \boldsymbol{M}_{\mathbf{x}} = 0, \tag{11.59}$$

which is equivalent to solving the system of $n+1$ differential equations

$$\frac{\partial \boldsymbol{\ell}}{\partial \beta_k} = \sum_{i=1}^{m} x_k^{(i)} \left(y^{(i)} - h(\mathbf{x}^{(i)})\right) = 0, \quad \text{for } k = 0, 1, 2, \ldots, n. \tag{11.60}$$

Due to the nature of the expectation $h(\boldsymbol{X})$ (see 11.52), equations in (11.60) are not linear and can be solved using a numerical method. One of the most commonly used methods is the *Newton-Raphson method*, which allows one to obtain MLE estimates $\hat{\boldsymbol{\beta}}^\top = (\hat{\beta}_0, \hat{\beta}_1, \hat{\beta}_2, \ldots, \hat{\beta}_n)$ using a recursive approximation.

We would like to know if there is a relationship between any of the explanatory variables $x_1, x_2, \ldots, x_n$ and the response variable y. To answer this question, we may use a general approach of testing if a particular regression coefficient is equal to zero. Hypothesis test has the following setup:

$$H_0 : \beta_k = 0 \qquad \text{vs} \qquad H_1 : \beta_k \neq 0.$$

The null hypothesis corresponds to a reduced model with regression parameters:

$$\boldsymbol{\beta}_R = (\beta_0, \beta_1, \beta_2, \ldots, \beta_{k-1}, \beta_{k+1}, \ldots, \beta_n)^\top,$$

which can be estimated as $\hat{\boldsymbol{\beta}}_R$ from (11.60). We then can use the *likelihood ratio test statistic*:

$$\mathfrak{G}^2 = -2 \log \left[\frac{\mathbf{L}(\hat{\boldsymbol{\beta}})}{\mathbf{L}(\hat{\boldsymbol{\beta}}_R)}\right], \tag{11.61}$$

where $\mathbf{L}(\cdot)$ is given by (11.55). The test statistic $\mathfrak{G}^2$ has the asymptotic χ^2 distribution, i.e., if the number of observations m is large, then approximately $\mathfrak{G}^2 \sim \chi_1^2$. The number of degrees of freedom is 1 since the parameter vector $\boldsymbol{\beta}_R$ has one less element compared to $\boldsymbol{\beta}$. We reject H_0 when $\mathfrak{G}^2 > \chi_1^2(\alpha)$ for a chosen significance level α.

To compute confidence intervals for regression parameters, we must recall and emphasize our assumption that all Y_i are i.i.d. with distributions Bernoulli(θ_i). Under these assumptions, the ML estimators $\hat{\boldsymbol{\beta}}$ follow a large sample normal distribution with covariance matrix given by the inverse of the *Fisher information matrix*. The Fisher information is computed as follows:

$$\mathbf{I}_Y(\boldsymbol{\beta}) = -\mathbf{E}\left[\nabla^2 \boldsymbol{\ell}(\boldsymbol{\beta})\right], \tag{11.62}$$

where the second-order gradient has elements:

$$\frac{\partial^2 \ell}{\partial \beta_k \partial \beta_p} = -\sum_{i=1}^{m} x_k^{(i)} x_p^{(i)} \frac{e^{\boldsymbol{\beta}^\top \mathbf{x}^{(i)}}}{\left(1 + e^{\boldsymbol{\beta}^\top \mathbf{x}^{(i)}}\right)^2}, \quad \text{for } k, p = 0, 1, 2, \ldots, n. \tag{11.63}$$

Denoting the last terms in the sum in the right-hand side of (11.63) as the elements of a diagonal matrix of rank m

$$\mathfrak{W} = \mathbf{diag}\left[\frac{e^{\boldsymbol{\beta}^\top \mathbf{x}^{(1)}}}{\left(1 + e^{\boldsymbol{\beta}^\top \mathbf{x}^{(1)}}\right)^2}, \frac{e^{\boldsymbol{\beta}^\top \mathbf{x}^{(2)}}}{\left(1 + e^{\boldsymbol{\beta}^\top \mathbf{x}^{(2)}}\right)^2}, \ldots, \frac{e^{\boldsymbol{\beta}^\top \mathbf{x}^{(m)}}}{\left(1 + e^{\boldsymbol{\beta}^\top \mathbf{x}^{(m)}}\right)^2}\right], \tag{11.64}$$

we can write the matrix form of the second-order gradient:

$$\nabla^2 \ell(\boldsymbol{\beta}) = -\boldsymbol{M}_{\mathbf{x}}^\top \mathfrak{W} \boldsymbol{M}_{\mathbf{x}}. \tag{11.65}$$

Observe from (11.63) that the expectation on the right-hand side of (11.62) is not a function of Y; hence, the expected information is identical with the observed information and we have for large m

$$\hat{\boldsymbol{\beta}} \sim \mathrm{N}\left(\boldsymbol{\beta},\ \left[\boldsymbol{M}_{\mathbf{x}}^\top \mathfrak{W} \boldsymbol{M}_{\mathbf{x}}\right]^{-1}\right).$$

We may estimate the elements of the diagonal matrix $\mathfrak{W}$ using our parameter estimates $\hat{\boldsymbol{\beta}}$ and obtain the standard error for each $\hat{\beta}_k$ as the square root of the corresponding diagonal element of the matrix $[\boldsymbol{M}_{\mathbf{x}}^\top \widehat{\mathfrak{W}} \boldsymbol{M}_{\mathbf{x}}]^{-1}$

$$\widehat{\mathrm{SE}}_k = \widehat{\mathrm{SE}}(\hat{\beta}_k) = \sqrt{\left[\left[\boldsymbol{M}_{\mathbf{x}}^\top \widehat{\mathfrak{W}} \boldsymbol{M}_{\mathbf{x}}\right]^{-1}\right]_{kk}}, \tag{11.66}$$

where $\widehat{\mathfrak{W}} = \mathfrak{W}(\hat{\boldsymbol{\beta}})$. Finally, a $(1-\alpha)100\%$ confidence interval for β_k is given by

$$\left[\hat{\beta}_k - z_{\alpha/2}\widehat{\mathrm{SE}}_k,\ \hat{\beta}_k + z_{\alpha/2}\widehat{\mathrm{SE}}_k\right]. \tag{11.67}$$

11.5.2. Logistic Regression for Binary Classification in Machine Learning

Logistic regression is used as an effective technique for binary classification problems in machine learning. The criteria of application are slightly different from those practiced by the statisticians. As long as the model has shown satisfactory classification accuracy on the test set (or during k-fold testing), it may be used by the machine learning engineer even if it does not provide an extremely accurate fit to the data and would not be accepted as satisfactory by the statistical standards. In this case, as machine learning practitioners, we view a model as the "best possible" logistic regression model (based on observed data) for the distribution of the response variable. The *Kullback-Leibler divergence* [Kullback, 1968; Kullback and Leibler, 1951] is employed to be the measure of the degree of dissimilarity between the empirical distribution (defined by the data) and the model distribution.

11.5.2.1. Kullback-Leibler Divergence

A short review of Kullback-Leibler divergence may be helpful to our reader. Suppose we use a parametric model $\hat{f}(X|\beta)$ for a random sample $X_1, X_2, \ldots, X_n$ with an unknown pdf f_X. For the sake of preserving generality of all arguments, let us assume that $\hat{f}(X|\beta)$ is the pdf of a continuous distribution.

There is no guarantee that the true pdf f_X is in the model $\hat{f}(X|\beta)$ and that the parameter β is a true parameter associated with f_X. Consequently, measuring the error of our estimator $\hat{\beta}$ as a distance from β (using, e.g., MSE) does not seem to be useful. This is where the Kullback-Leibler divergence proves to be extremely helpful by providing a measure of distance between a pdf represented by our model $\hat{f}(X|\beta)$ and the true pdf f_X of the data.

Definition 11.1. Let f_X be the pdf of a random variable X. Let $\hat{f}$ be the pdf of a continuous distribution. The Kullback-Leibler divergence from $\hat{f}$ to f_X is

$$D_{KL}\left(f_X||\hat{f}\right) = \int f_X(x)\log\left(\frac{f_X(x)}{\hat{f}(x)}\right)dx = \mathbf{E}\left[\log\left(\frac{f_X(X)}{\hat{f}(X)}\right)\right].$$

Observe that D_{KL} has two important properties

1. $D_{KL}\left(f_X||\hat{f}\right) \geq 0$ for any proper pdf's f_X and $\hat{f}$
2. $D_{KL}\left(f_X||\hat{f}\right) = 0$ if and only if $f_X = \hat{f}$.

For (1), note that $g(x) = -\log x$ is a strictly convex function. Then, we can show the non-negativity of D_{KL} using Jensen's inequality:

$$\begin{aligned}
D_{KL}\left(f_X||\hat{f}\right) &= \mathbf{E}\left[\log\left(\frac{f_X(X)}{\hat{f}(X)}\right)\right] = \mathbf{E}\left[-\log\left(\frac{\hat{f}(X)}{f_X(X)}\right)\right] \\
&\geq -\log\mathbf{E}\left[\frac{\hat{f}(X)}{f_X(X)}\right] = -\log\int\frac{\hat{f}(x)}{f_X(x)}f_X(x)dx \\
&= -\log\int f_X(x)dx = 0
\end{aligned}$$

To see (2), first suppose $f_X = \hat{f}$, then

$$D_{KL}\left(f_X||\hat{f}\right) = \mathbf{E}\left[\log\left(\frac{f_X(X)}{\hat{f}(X)}\right)\right] = \mathbf{E}\left[\log(1)\right]. = 0$$

Further, due to the strict convexity of the function $g(x) = -\log x$, the inequality

$$\mathbf{E}\left[-\log\left(\frac{\hat{f}(X)}{f_X(X)}\right)\right] \geq -\log \mathbf{E}\left[\frac{\hat{f}(X)}{f_X(X)}\right]$$

can only become an equality if $\frac{f_X(X)}{\hat{f}(X)}$ is a constant random variable. Hence, the assumption $D_{KL}\left(f_X||\hat{f}\right) = 0$ implies $f_X = \hat{f}$.

For a random sample $X_1, \ldots, X_n$ with pdf f_X and a model $\hat{f}(X|\beta)$, as we have previously discussed, the value $\hat{\beta}$ that maximizes $\boldsymbol{\ell}(\beta)$ (11.56) is the maximal likelihood estimator (MLE) of the true parameter β. The function

$$\frac{1}{n}\boldsymbol{\ell}(\beta) = \frac{1}{n}\sum_{i=1}^{n} \log\left(\hat{f}(X_i|\beta)\right)$$

is also maximized by the MLE $\hat{\beta}$. Moreover, by the weak law of large numbers

$$\frac{1}{n}\sum_{i=1}^{n} \hat{f}(X_i|\beta) \xrightarrow{\Pr} \mathbf{E}_X\left[\log\left(\hat{f}(X_i|\beta)\right)\right].$$

The expectation $\mathbf{E}_X$ can be presented as

$$\begin{aligned}
\mathbf{E}_X\left[\log\left(\hat{f}(X_i|\beta)\right)\right] &= \mathbf{E}_X\left[\log\left(\frac{f_X(X)}{f_X(X)}\hat{f}(X_i|\beta)\right)\right] \\
&= \mathbf{E}_X\left[\log\left(f_X(X)\right) + \log\left(\frac{\hat{f}(X_i|\beta)}{f_X(X)}\right)\right] \\
&= \mathbf{E}_X\left[\log\left(f_X(X)\right) - \log\left(\frac{f_X(X)}{\hat{f}(X_i|\beta)}\right)\right] \\
&= \mathbf{E}_X\left[\log\left(f_X(X)\right)\right] - \mathbf{E}_X\left[\log\left(\frac{f_X(X)}{\hat{f}(X_i|\beta)}\right)\right] \\
&= \mathbf{E}_X\left[\log\left(f_X(X)\right)\right] - D_{KL}\left(f_X(X)||\hat{f}(X_i|\beta)\right).
\end{aligned} \tag{11.68}$$

Since the only term in the last line of (11.68) that depends on β is clearly $D_{KL}\left(f_X(X)||\hat{f}(X_i|\beta)\right)$, the value $\hat{\beta}$ that maximizes $\mathbf{E}_X\left[\log\left(\hat{f}(X_i|\beta)\right)\right]$ must minimize $D_{KL}\left(f_X(X)||\hat{f}(X_i|\beta)\right)$. In plain words, the maximum likelihood estimator $\hat{\beta}$ corresponds to the distribution of the model $\hat{f}(X|\beta)$ that is closest to the true pdf f_X with respect to the Kullback-Leibler (KL) divergence.

In this short review, we have merely "scratched the surface" of a useful concept of the KL divergence and its vast area of applications. An interested reader is recommended to review [Kullback, 1968; Kullback and Leibler, 1951; White, 1982] for a detailed account of the history, argumentation, and proofs.

Example 11.5.1. To demonstrate the logistic regression in action, we shall perform a binary classification on a data set that was created from the sets of electroencephalographic (EEG) time series used by researchers in Andrzejak et al. [2001]: surface EEG recordings from healthy volunteers with eyes closed and eyes open and intracranial EEG recordings from epilepsy patients during the seizure-free interval from within and from outside the seizure-generating area as well as intracranial EEG recordings of epileptic seizures. The original data set used by Andrzejak et al. [2001] consists of 5 different folders, each with 100 files, with each file representing a single subject/person. Each file is a recording of brain activity for 23.6 s. The corresponding time series is sampled into 4097 data points. Each data point is the value of the *EEG* recording at a different point in time. There is a total of 500 individuals, each having 4097 data points for 23.5 s. Every 4097 data points were divided and shuffled into 23 chunks, each chunk contains 178 data points for 1 s, and each data point is the value of the EEG recording at a different point in time. As a result, $23 \times 500 = 11{,}500$ pieces of information (rows) obtained, each row contains 178 data points for 1 s, the last column represents the label $y \in \{1, 2, 3, 4, 5\}$. This modified data set is available at

https://archive.ics.uci.edu/ml/data sets/Epileptic+Seizure+Recognition

We now have the explanatory variables $X_1, X_2, \ldots, X_{178}$, and the response variable is Y in column 179, which contains the category of the 178-dimensional input vector. The categories are described as follows:

1. Recording of seizure activity.
2. The EEG was recorded from the area where the tumor was located.
3. The location region of the brain tumor was identified and the EEG activity was recorded from the healthy brain area.
4. During the recording of the EEG signal, the patient had their eyes closed.
5. During the recording of the EEG signal, the patient had their eyes open.

All subjects falling in classes 2, 3, 4, and 5 did not have epileptic seizure. Only subjects in class 1 had epileptic seizure. We shall use this fact to perform the binary classification, i.e., we shall build a logistic model that will be able to recognize the recording of the seizure activity based on 178 predictors. To do that, we must relabel the response column accordingly. The label for class 1 will stay unchanged, whereas labels 2, 3, 4, and 5 will be replaced with 0 to obtain a binary response $y \in \{0, 1\}$.

We shall employ the gradient descent algorithm given in Table 11.8 to perform optimization with the maximal likelihood. Strictly speaking, we want to estimate parameters $\hat{\boldsymbol{\beta}}$ that minimize $-\boldsymbol{\ell}(\hat{\boldsymbol{\beta}})$, where $\boldsymbol{\ell}(\boldsymbol{\beta})$ is the log-likelihood function defined in (11.56). In order to test our model's predictive potential, after random shuffling, we divide the data set, which consists of 11500 vectors, into two subsets with 7/3 ratio, where 8050 data vectors are used as the training set for our model and the remaining

3450 data vectors will be the test set. We recall that the parameter vector estimates $\hat{\beta}$ are produced based on the data from the training set. The test set is only introduced to our model as a new data to perform the classification, using the parameter vector estimated during the model training phase. As in previous examples, the model parameters α (learning rate) and δ (threshold) are found from a list of possible values by comparing the model performance on the training set. It is clearly impractical to provide all 178 values of the regression parameters. Instead, in Table 11.19 we give the best values of the model parameters together with the corresponding cost function and the average classification accuracy values.

Parameters	
α	1
δ	1/100

Results	
Cost	0.63
Classification accuracy training set	86.5%
Classification accuracy test set	85.1%

Table 11.19.: Logistic regression model parameters and classification accuracy for epileptic seizure recognition data set

In addition to the average classification accuracy (also called the *average precision*), we provide calculations for *precision* and *recall* by activity, which give a better insight into understanding of our results.
"Precision" gives the proportion of correct predictions:

$$\text{Precision} = \frac{\sharp \text{ of } \textbf{Correctly Classified} \text{ Cases}}{\text{Total } \sharp \text{ of } \textbf{Cases Predicted} \text{ for this Class}}$$

"Recall" gives the proportion of actual activities correctly classified as such by our model:

$$\text{Recall} = \frac{\sharp \text{ of } \textbf{Correctly Classified} \text{ Cases}}{\text{Total } \sharp \text{ of } \textbf{True Cases} \text{ in this Class}}$$

In addition, we calculate the F_1 score, which is the harmonic mean of precision and recall:

$$F_1 = 2 * \frac{\text{Precision} * \text{Recall}}{\text{Precision} + \text{Recall}}$$

The F_1 score is commonly used instead of an average since it penalizes extreme values and gives equal weight to both precision and recall. High F_1 score indicates a classification model with optimal balance of precision and recall (Table 11.20). Presenting our results in this form helps in understanding the strong suites and the limitations of our model. In this case, the discrepancy between the average precision and the F_1 score points out a well-known weakness of the logistic regression model in recognizing relatively rare events.

Actual\predicted	Class 0 (no seizure)	Class 1 (seizure)	Recall
Class 0 (no seizure)	2668	92	97%
Class 1 (seizure)	422	268	39%
Precision	86%	74%	$F_1 = 74\%$

Table 11.20.: Logistic regression classification results on the test set for epileptic seizure recognition data set

Note that the total number of true epileptic seizure cases in the training set is 690 out of 3450, which is only 20%.

11.5.3. Logistic Regression with Multiple Response Classes in Machine Learning

The multinomial distribution can be viewed as a generalization of the binomial distribution. In an experiment with $K \in \mathbb{N}$ possible outcomes, if one denotes Y_k the number of occurrences of kth outcome in n trials, then the random vector

$$\boldsymbol{Y}^\top = (Y_1, Y_2, \ldots, Y_k)$$

is a multinomial random vector with PMF:

$$f_{\boldsymbol{Y}}(y_1, y_2, \ldots, y_K) = \begin{cases} \dfrac{n!}{y_1!y_2!\ldots y_K!} p_1^{y_1} p_2^{y_2} \cdots p_K^{y_K} & \text{if } \sum\limits_{k=1}^{K} y_k = n \\ 0 & \text{otherwise} \end{cases},$$

where p_k is the probability of kth outcome and $\sum\limits_{k=1}^{K} p_k = 1$.

For a single experiment, only one element Y_i of a vector $\boldsymbol{Y}$ takes the value 1, while all other elements Y_j $(j \neq i)$ take the value 0. Such a vector is customarily called a *one-hot* vector. The joint PMF is

$$f_{\boldsymbol{Y}}(y_1, y_2, \ldots, y_K) = \begin{cases} p_1^{y_1} p_2^{y_2} \ldots p_K^{y_K} & \text{if } \sum\limits_{k=1}^{K} y_k = 1 \\ 0 & \text{otherwise} \end{cases},$$

where $\sum\limits_{k=1}^{K} p_k = 1$. A random variable with the above PMF is sometimes said to follow the *multinomial distribution*. Observe that in a single random experiment with outcome taking the form of a one-hot vector we have

$$f_{\boldsymbol{Y}}(y_1, y_2, \ldots, y_K) = p_1^{y_1} p_2^{y_2} \cdots p_K^{y_K} = p_1^0 p_2^0 \cdots p_{i-1}^0 p_i^1 p_{i+1}^0 \cdots p_K^0 = p_i. \tag{11.69}$$

In a single experiment, with the observed set of predictors $\mathbf{x}^\top = (x_1, x_2, \dots, x_n)$, the conditional expectation is

$$h_k(\boldsymbol{X}) = \mathbf{Pr}[Y_k = 1 | \boldsymbol{X} = \mathrm{x}] = p_k. \tag{11.70}$$

We model $h_k(\boldsymbol{X})$ $(k = 1, 2, 3, \dots, K)$ as

$$h_k(\boldsymbol{X}, M_\beta) = \frac{e^{\boldsymbol{\beta}_k^\top \mathbf{x}}}{\sum\limits_{j=1}^{K} e^{\boldsymbol{\beta}_j^\top \mathbf{x}}}, \tag{11.71}$$

so that $\boldsymbol{p}^\top = (p_1, p_2, \dots, p_K)$ is a proper probability vector, i.e., $\sum\limits_{k=1}^{K} p_k = 1$. It is worth mentioning that M_β in Eq. (11.71) denotes a parameter matrix:

$$M_\beta = \begin{bmatrix} \beta_{01} & \beta_{02} & \cdots & \beta_{0K} \\ \beta_{11} & \beta_{12} & \cdots & \beta_{1K} \\ \vdots & \vdots & \ddots & \vdots \\ \beta_{n1} & \beta_{n2} & \cdots & \beta_{nK} \end{bmatrix}$$

with each column containing a set of parameters $\boldsymbol{\beta}_k^\top = (\beta_{0k}, \beta_{1k}, \dots, \beta_{nk})$ for k-th outcome of an experiment $(k = 1, 2, 3, \dots, K)$. Recall that we are using an amended feature vector $\mathbf{x}^\top = (1, x_1, x_2, \dots, x_n)$ to enable matrix notation $\boldsymbol{\beta}_k^\top \mathbf{x}$.

The maximal likelihood over m observations can be written as

$$\mathbf{L}(M_\beta) = \prod_{i=1}^{m} p_1^{y_1^{(i)}} p_2^{y_2^{(i)}} \dots p_K^{y_K^{(i)}}, \tag{11.72}$$

where for each $i = 1, 2, 3, \dots, m$, vector $\boldsymbol{y}^{(i)} = (y_1^{(i)}, y_2^{(i)}, \dots, y_K^{(i)})$ is a one-hot vector, i.e., $\sum\limits_{k=1}^{K} y_k^{(i)} = 1$. It can be shown that the log-likelihood takes the form

$$\begin{aligned} \log \mathbf{L}(M_\beta) = \boldsymbol{\ell}(M_\beta) &= \sum_{i=1}^{m} \sum_{j=1}^{K} y_j^{(i)} \log h_j(\mathbf{x}^{(i)}, M_\beta) \\ &= \sum_{i=1}^{m} \sum_{j=1}^{K} \left[y_j^{(i)} \left(\boldsymbol{\beta}_j^\top \mathbf{x}^{(i)} - \log \sum_{k=1}^{K} e^{\boldsymbol{\beta}_k^\top \mathbf{x}^{(i)}} \right) \right]. \end{aligned} \tag{11.73}$$

11.5.4. Regularization for Regression Models in Machine Learning

In this section we shall present two forms of regularization commonly used for regression models, namely, ridge regularization and lasso regularization. These methods make the estimate more regular, by shrinking it toward zero and hence decreasing its variance. It is worth mentioning that this type of regularization is often employed well beyond the linear model setting. Virtually any time we have an estimate defined by optimizing a certain criterion; we can regularize this estimate in a high-dimensional scenario by adding a penalty term. As we shall discuss later, adding a squared penalty term will shrink components of the estimate but not set them to zero; adding a linear penalty term will both shrink and set some estimates to zero exactly. Such practice, of regularizing parameter estimates, is often employed in high-dimensional maximum likelihood settings.

Note however that regularization is not limited to squared and linear penalties, nor is it philosophically limited to the idea of shrinking the components of parameter estimates toward zero. Across various problems, various types of regularization may be appropriate and can be employed.

Let us use a linear regression model to demonstrate several possible issues and suggest how we can resolve those. Recall that we have defined the residual variance $\mathbf{S}^2_{\mathbf{R}}$ as the sum of terms $(y_i - \hat{y}_i)^2$ averaged over m observations and estimated regression coefficients that minimize $\mathbf{S}^2_{\mathbf{R}}$. Using matrix notation, we wrote an Eq. (11.27) as follows:

$$\mathbf{S}^2_{\mathbf{R}} = \frac{(\mathbf{y} - \boldsymbol{M}_{\mathbf{X}}\boldsymbol{\beta})^\top(\mathbf{y} - \boldsymbol{M}_{\mathbf{X}}\boldsymbol{\beta})}{m},$$

where $\mathbf{y}^\top = (y_1, y_2, \ldots, y_m)$ is a vector of values of the response variable Y and $\boldsymbol{M}_{\mathbf{X}}$ is $m \times n + 1$ matrix, called the *design matrix*, which consists of m observations with n features recorded in each observation. Imposing the independence assumption for column vectors of the product $\boldsymbol{M}_{\mathbf{X}}^\top \boldsymbol{M}_{\mathbf{X}}$, we wrote the solutions in Eq. (11.29):

$$\hat{\boldsymbol{\beta}} = \left(\boldsymbol{M}_{\mathbf{X}}^\top \boldsymbol{M}_{\mathbf{X}}\right)^{-1} \boldsymbol{M}_{\mathbf{X}}^\top \mathbf{y}.$$

As noted earlier, when $n + 1 > m$, the matrix $\boldsymbol{M}_{\mathbf{X}}$ cannot have linearly independent columns, and $\boldsymbol{M}_{\mathbf{X}}^\top \boldsymbol{M}_{\mathbf{X}}$ is not invertible. Moreover, even for data with a relationship that is close to linear, several issues arise as a result of the large number of predictors. Indeed, suppose that for a predictor vector $\mathbf{X}$ and response Y we have:

$$y_i = \boldsymbol{\beta}^\top \mathbf{x}_i + \varepsilon_i, \quad i = 1, 2, 3, \ldots, m, \tag{11.74}$$

or, in general:

$$Y = \boldsymbol{\beta}^\top \mathbf{X} + \varepsilon, \tag{11.75}$$

where $\boldsymbol{\beta}$ is a true parameter vector and ε is a random error with $\boldsymbol{E}[\varepsilon] = 0$ and $\boldsymbol{Var}[\varepsilon] = \sigma^2$. Recall that we defined

$$h(\mathbf{X}) = \boldsymbol{E}[Y|\mathbf{X} = \mathbf{x}].$$

We model the relationship in (11.75) with

$$\hat{h}(\mathbf{X}) = \hat{\boldsymbol{\beta}}^\top \mathbf{X}, \tag{11.76}$$

where we choose $\hat{\boldsymbol{\beta}}$ so that $\mathbf{S}^2_{\mathbf{R}}(\hat{\boldsymbol{\beta}})$ is minimized. The expected test error for an arbitrary new feature vector $\mathbf{x}$ can be written as

$$\begin{aligned}
\boldsymbol{E}[\mathfrak{E}_{\text{test}}] &= \boldsymbol{E}\left[\left(Y - \hat{h}(\mathbf{x})\right)^2 \mid \mathbf{X} = \mathbf{x}\right] \\
&= \boldsymbol{E}\left[(Y - h(\mathbf{x}))^2 \mid \mathbf{X} = \mathbf{x}\right] + \boldsymbol{E}\left[\left(h(\mathbf{x}) - \hat{h}(\mathbf{x})\right)^2 \mid \mathbf{X} = \mathbf{x}\right] \\
&= \boldsymbol{E}\left[\varepsilon^2\right] + \boldsymbol{E}\left[\left(h(\mathbf{x}) - \hat{h}(\mathbf{x})\right)^2 \mid \mathbf{X} = \mathbf{x}\right] \\
&= \sigma^2 + \boldsymbol{E}\left[\left(h(\mathbf{x}) - \hat{h}(\mathbf{x})\right)^2\right] \\
&= \sigma^2 + \left(\boldsymbol{E}\left[\hat{h}(\mathbf{x})\right] - h(\mathbf{x})\right)^2 + \boldsymbol{E}\left[\left(\hat{h}(\mathbf{x}) - \boldsymbol{E}[\hat{h}(\mathbf{x})]\right)^2\right] \\
&= \sigma^2 + \left(\mathbf{Bias}[\hat{h}(\mathbf{x})]\right)^2 + \mathbf{Var}[\hat{h}(\mathbf{x})].
\end{aligned} \tag{11.77}$$

The first term on the right-hand side σ^2 is the irreducible error and is the minimal lower bound on the expected test error.

Under the above assumption, the true relationship between $\mathbf{X}$ and Y is close to linear; thus, $\mathbf{Bias}[\hat{h}(\mathbf{x})] \approx 0$. The third term on the right-hand side comes from the variance of the linear regression estimate averaged over the number of observations. It comes entirely from our model and adding features to the model $\hat{h}$ will result in increased $\mathbf{Var}[\hat{h}]$. This may prove problematic if we are dealing with extremely large number of predictors. Therefore, we may try to *shrink small coefficients toward zero*, which potentially introduces some bias but also potentially reduces the variance.

11.5.4.1. Ridge Regression

The ridge coefficients are defined by finding a parameter vector $\hat{\boldsymbol{r}}^\top = (\hat{r}_0, \hat{r}_1, \hat{r}_2, \ldots, \hat{r}_n)$ that minimizes

$$\mathbf{S}^2_{\mathbf{R}}(\hat{\boldsymbol{r}}) + \lambda\, \hat{\boldsymbol{r}}^\top_{\text{Tr}} \hat{\boldsymbol{r}}_{\text{Tr}}, \tag{11.78}$$

where $\hat{\boldsymbol{r}}^\top_{\text{Tr}} = (\hat{r}_1, \hat{r}_2, \ldots, \hat{r}_n)$ is a truncated parameter vector with the intercept coefficient $\hat{r}_0$ excluded. The scalar λ is a *tuning parameter*. The last term on the right-hand side is called a *penalty term*. The idea here is that we are forced to choose smaller values for components of the ridge parameter vector $\hat{\boldsymbol{r}}$ in comparison with the usual least squares estimate $\hat{\boldsymbol{\beta}}$ of the parameter vector $\boldsymbol{\beta}$. It is quite easy to see that

1. When $\lambda = 0$, we get the linear regression estimate;
2. When $\lambda = \infty$ we get $\hat{\boldsymbol{r}} = 0$;
3. For $0 < \lambda < \infty$, we are balancing two ideas: fitting a linear model of y on x and shrinking the coefficients. The bias increases and the variance decreases as λ increases;
4. Ridge regression will produce a different coefficient vector, $\hat{\boldsymbol{r}}_\lambda$, for each value of λ; hence, selecting a good value for λ is critical.

The intercept term $\hat{r}_0$ is not penalized. We do not want to shrink the intercept which is simply a measure of the mean value of the response when $x_1^{(i)} = x_2^{(i)} = x_3^{(i)} = \cdots = x_n^{(i)} = 0$ for all $i = 1, 2, 3, \ldots, m$.

Feature standardization is usually recommended when the predictor variables are not on the same scale. This serves two purposes. First, if the columns of the data matrix $\boldsymbol{M_X\beta}$ have been centered to have mean 0, then

$$\hat{r}_0 = \sum_{i=1}^{m} y_i.$$

Second, if the columns of the data matrix $\boldsymbol{M_X\beta}$ have been scaled to have sample variance 1, then the effect of the penalty term $\lambda \hat{\boldsymbol{r}}_{\mathrm{Tr}}^{\top} \hat{\boldsymbol{r}}_{\mathrm{Tr}}$ is evened out across predictors.

If a group of the true coefficients is exactly zero, the *predictive performance* of ridge regression is the same as with a group of small true coefficients. However, for *interpretation purposes*, ridge regression does not provide as much help as one would hope. This is because it shrinks components of its estimate toward zero but never sets these components to be zero exactly (unless $\lambda = \infty$, in which case all components are zero). So strictly speaking, ridge regression does not perform variable selection. That being said, ridge regression has substantial computational advantages over best subset selection, which requires searching through 2^n models, and consequently, even for moderate values of n, can be computationally infeasible. In contrast, for any fixed value of λ, ridge regression only fits a single model, and the model-fitting procedure can be performed quite quickly. In fact, one can show that the computations required to solve (11.78), simultaneously for all values of λ, are almost identical to those for fitting a model using least squares.

11.5.4.2. Lasso Regression

The *Lasso* (least absolute selection and shrinkage operator) also solves a penalized least squares problem. The lasso parameter vector $\hat{\boldsymbol{l}}^{\top} = (\hat{l}_0, \hat{l}_1, \hat{l}_2, \ldots, \hat{l}_n)$ minimizes

$$\mathbf{S}_{\mathbf{R}}^2(\hat{\boldsymbol{l}}) + \lambda \sum_{i=1}^{m} |l_i|. \tag{11.79}$$

As before, the tuning parameter λ controls the strength of the penalty, and we get $\hat{\boldsymbol{l}} = \hat{\boldsymbol{\beta}}$ when $\lambda = 0$, and $\hat{\boldsymbol{l}} = 0$ when $\lambda = \infty$. For $0 < \lambda < \infty$, we are balancing two ideas: fitting a linear model of y on x and shrinking the coefficients. But the nature of the $\lambda \sum_{i=1}^{m} |l_i|$ penalty term is such that some coefficients are shrunk to zero exactly. This is what makes the lasso substantially different from ridge regression: it is able to perform variable selection in the linear model. As λ increases, more coefficients are set to zero (less variables are selected), and among the nonzero coefficients, more shrinkage is performed. Note that feature standardization has the same effect for the lasso regression as it had in the case of the ridge regression. As in ridge regression, selecting a good value of λ for the lasso is critical and can be done using *cross-validation.*

11.5.4.3. The Choice of Regularization Method

It is clear that the lasso has a major advantage over ridge regression, in that it produces simpler and more interpretable models that involve only a subset of the predictors. However, in general neither ridge regression nor the lasso will universally dominate the other. We can use R^2 defined by (11.34) on the training data to compare performance of regression models with different types of regularization. We generally expect the lasso regression to perform better in a setting where a relatively small number of predictors have substantial coefficients and the remaining predictors have coefficients that are very small or that equal zero. Ridge regression will likely perform better when the response is a function of many predictors, all with coefficients of roughly equal size. However, the number of predictors that are related to the response is never known a priori for real data sets. A technique such as cross-validation can be used in order to determine which approach is better on a particular data set. As with ridge regression, when the least squares estimates have excessively high variance, the lasso solution can provide a reduction in variance at the expense of a small increase in bias and consequently can generate more accurate predictions. Unlike ridge regression, the lasso performs variable selection and hence results in models that are easier to interpret. There are very efficient algorithms for fitting both ridge and lasso models; in both cases the entire coefficient paths can be computed with about the same amount of work as a single least squares fit.

11.5.5. Support Vector Machines (SVM)

11.5.5.1. Introduction

Important tools in optimization theory are the duality theorems. Support vector machines as they have come to be called constitute a tool to classify data points characterized by several features. The separation is made up of a hyperplane, and the support vectors are the data points which are closest to the hyperplane. These are the points which most influence the orientation of the hyperplane and are generally the most difficult to classify.

Suppose we are interested in calculating the distance from the point (x_0, y_0) to the line $L : ax + by = c$. The line has slope $-a/b$.

Hence the line perpendicular to L say L' will have slope b/a. It will have the form:

$$L' : y = a^* + (b/a)\,x$$

The point (x, y) of intersection between the lines L, L' hence satisfies:

$$\begin{cases} y &= a^* + (b/a)\,x \\ y &= (c/b) - (a/b)\,x \end{cases}$$

The solution to these equations yields

$$x = \frac{\left(\frac{c}{b} - a^*\right) ab}{a^2 + b^2}$$

$$y = \frac{a^2 a^* + bc}{a^2 + b^2}$$

We note that if (x_0, y_0) is on the line L':

$$a^* = y_0 - (b/a)\, x_0$$

The square distance from the point (x_0, y_0) to the line $L : ax + by = c$ is then given by

$$(x - x_0)^2 + (y - y_0)^2 = \left(\frac{\left(\frac{c}{b} - a^*\right) ab}{a^2 + b^2} - x_0\right)^2 + \left(\frac{a^2 a^* + bc}{a^2 + b^2} - (a^* + (b/a)\, x_0)\right)^2$$

$$= \frac{(ax_0 + by_0 - c)^2}{a^2 + b^2}.$$

In general then, the distance between two hyperplanes $\mathcal{P}_1$ and $\mathcal{P}_2$ is given by

$$\frac{\left\|\boldsymbol{w}^\top \boldsymbol{x} + b\right\|}{\|\boldsymbol{w}\|}, \tag{11.80}$$

where $\boldsymbol{w}$ is the set of parameters that defines $\mathcal{P}_1$ and $\boldsymbol{x} \in \mathcal{P}_2$.

11.5.5.2. Finding the Optimal Hyperplane

Let $\mathcal{D} = (\mathbf{x}^{(i)}, y^{(i)})$ be a data set with $i = 1, 2, 3, ..., m$. For each i we have: $\mathbf{x}^{(i)} = (x_1^{(i)}, x_2^{(i)}, x_3^{(i)}, ..., x_n^{(i)})^\top$, $y^{(i)} \in \{-1, 1\}$. We can think of each variable $y^{(i)}$ as defining the class of the corresponding vector $\mathbf{x}^{(i)}$. Assuming that our data set is linearly separable, we would like to find a set of parameters that define a hyperplane separating the two classes of elements of $\mathcal{D}$ with the greatest *margin* (i.e., the distance between our hyperplane and the closest data point from each class is maximized. See Fig. 11.11 for illustration of this task in $\mathbb{R}^2$).

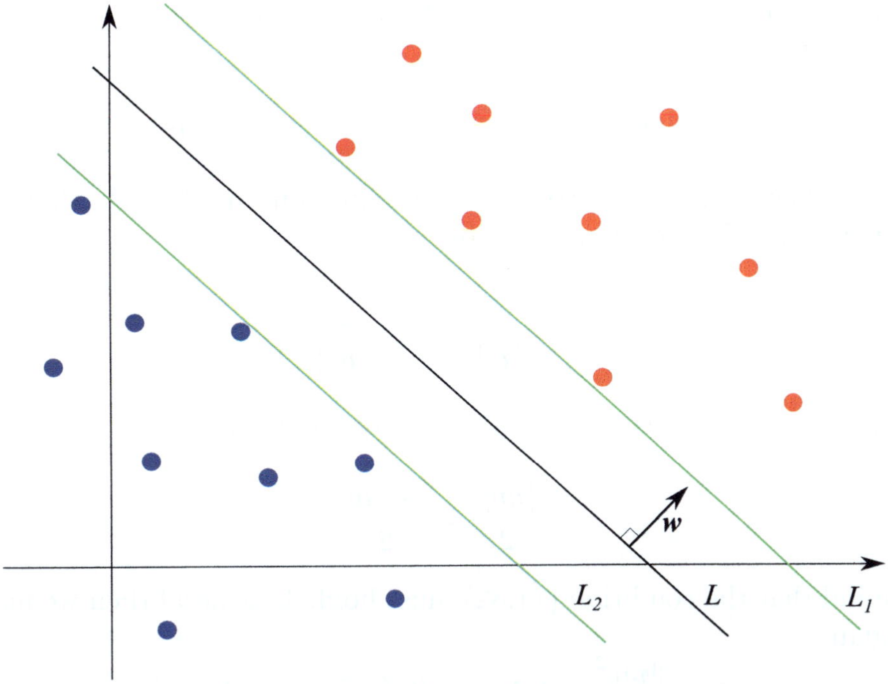

Figure 11.11.: Two sets of points shown are linearly separable. The line L that separates the two sets is a hyperplane in 2D space defined by the vector $\boldsymbol{w}$. Lines L_1 and L_2 that define the margin are parallel to L and at the same distance from it. We aim to find the position of line L, so that the distance between L_1 and L_2 is maximized

Suppose the hyperplane in question is defined by $\mathcal{P}_0 : \boldsymbol{w}^\top \boldsymbol{x} + b = 0$. Let $\mathcal{P}_1$ and $\mathcal{P}_2$ be two hyperplanes that satisfy the following conditions:

- $\mathcal{P}_1$ and $\mathcal{P}_2$ are parallel to $\mathcal{P}_0$,
- $\mathcal{P}_1$ and $\mathcal{P}_2$ are at the same distance from $\mathcal{P}_0$,
- There are no points $\mathbf{x}^{(i)} \in \mathcal{D}$ between $\mathcal{P}_1$ and $\mathcal{P}_2$.

We can define hyperplanes $\mathcal{P}_1$ and $\mathcal{P}_2$ as follows:

$$\begin{aligned}\mathcal{P}_1 &: \boldsymbol{w}^\top \boldsymbol{x} + b = -1,\\ \mathcal{P}_2 &: \boldsymbol{w}^\top \boldsymbol{x} + b = 1.\end{aligned} \tag{11.81}$$

Then, for any $\mathbf{x}^{(i)}$ such that the corresponding $y^{(i)} = 1$, we must have

$$\boldsymbol{w}^\top \mathbf{x}^{(i)} + b \geq 1,$$

and for any $\mathbf{x}^{(i)}$ such that the corresponding $y^{(i)} = -1$, we must have

$$\boldsymbol{w}^\top \mathbf{x}^{(i)} + b \leq -1.$$

Then, for all elements of the data set $\mathcal{D}$, we can derive the following requirement for hyperplanes $\mathcal{P}_1$ and $\mathcal{P}_2$

$$y^{(i)}(\boldsymbol{w}^\top \mathbf{x}^{(i)} + b) \ \geq 1, \quad i = 1, 2, 3, ..., m \tag{11.82}$$

It can be shown, using (11.80) and two equations in (11.81), that the distance between hyperplanes $\mathcal{P}_1$ and $\mathcal{P}_2$ is given by

$$\frac{\left\|\boldsymbol{w}^\top \boldsymbol{x} + b\right\|}{\|\boldsymbol{w}\|} = \frac{2}{\|\boldsymbol{w}\|}. \tag{11.83}$$

To maximize the distance in (11.83), we may minimize

$$\frac{\|\boldsymbol{w}\|^2}{2} = \frac{\boldsymbol{w}^\top \boldsymbol{w}}{2},$$

keeping in mind that the condition (11.82) must hold. In general then we may construct the Lagrangian

$$L = \frac{\|\boldsymbol{w}\|^2}{2} - \sum a^{(i)} \left[y^{(i)} \left(\boldsymbol{w}^\top \mathbf{x}^{(i)} + b\right) - 1\right]$$

and minimize with respect to $(\boldsymbol{w}, b)$. Differentiating yields

$$\frac{\partial L}{\partial \boldsymbol{w}} = \boldsymbol{w} - \sum a^{(i)} y^{(i)} \mathbf{x}^{(i)} = 0 \tag{11.84}$$

$$\frac{\partial L}{\partial b} = \sum a^{(i)} y^{(i)} = 0$$

Hence, the minimum is

$$\frac{\boldsymbol{w}^\top \boldsymbol{w}}{2} - \sum a^{(i)} y^{(i)} \left((\mathbf{x}^{(i)})^\top \sum a^{(i)} y^{(i)} \mathbf{x}^{(i)} + b\right) + \sum a^{(i)} =$$

$$-\frac{1}{2} \sum_i \sum_j a^{(i)} a^{(j)} y^{(i)} y^{(j)} \left((\mathbf{x}^{(i)})^\top \mathbf{x}^{(j)}\right) + \sum a^{(i)} \tag{11.85}$$

We may now appeal to the duality theorems in optimization theory to facilitate the calculations (see Wikipedia). Instead of aiming to determine the $\boldsymbol{w}_i$, the dual problem consists of maximizing (11.85) with respect to $\left\{a^{(i)}\right\}$ such that $\sum a^{(i)} y^{(i)} = 0$, $a^{(i)} \geq 0$. Once the $\left\{a^{(i)}\right\}$ are determined, we may compute $\boldsymbol{w}$ from (11.84). The advantage of the dual formulation is that we only need to compute the inner products $\left((\mathbf{x}^{(i)})^\top \mathbf{x}^{(j)}\right)$. Most of the weights $\boldsymbol{w}_i$ will be equal to 0. Only the support vectors will have nonzero weights.

11.5.5.3. SVM for Nonlinearly Separable Data Sets

In some situations, such as nonlinear classification boundary, we may consider replacing the inner products with a kernel function. Examples are

$$K\left(\mathbf{x}^{(i)}, \mathbf{x}^{(j)}\right) = \left((\mathbf{x}^{(i)})^\top \mathbf{x}^{(j)} + 1\right)^p \tag{11.86}$$

$$K\left(\mathbf{x}^{(i)}, \mathbf{x}^{(j)}\right) = \exp\left(-||\mathbf{x}^{(i)} - \mathbf{x}^{(j)}||/2\sigma^2\right) \tag{11.87}$$

$$K\left(\mathbf{x}^{(i)}, \mathbf{x}^{(j)}\right) = \tanh\left(K\left(\mathbf{x}^{(i)} - \mathbf{x}^{(j)}\right) - \delta\right) \tag{11.88}$$

Kernel functions may be viewed as measures of similarity between the elements in an observed data set. For example, the Gaussian kernel (11.87) will return value close to 1 if the Euclidean distance between $\mathbf{x}^{(i)}$ and $\mathbf{x}^{(j)}$ is close to 0 and a value close to 0 if the Euclidean distance between $\mathbf{x}^{(i)}$ and $\mathbf{x}^{(j)}$ is large. Fig. 11.12 demonstrates the output of Gaussian kernel for a two-dimensional data.

In practice, to determine the SVM classification boundary, we usually use one of many existing automated optimization packages, such as *scikit-learn* for Python.

11.5.5.4. Illustrating SVM

For an example of application of SVM method, we refer our reader to Anguita et al. [2013b], in which researchers conducted experiments on the human activity recognition data set created using smartphones. The data set used for this example is available at **UCI** Machine Learning Repository http://archive.ics.uci.edu/. It contains data from experiments that have been carried out with a group of 30 volunteers. Each person performed six activities wearing a smartphone on the waist. The activities had been labeled as shown in Table 11.21. Researchers captured 3-axial linear acceleration and

Label	1	2	3	4	5	6
Activity	Walking	Walking upstairs	Walking downstairs	Sitting	Standing	Laying

Table 11.21.: Table of activities and the corresponding classes

3-axial angular velocity at a constant rate of 50Hz. The obtained data set has been randomly partitioned into two sets, with 70% being the training data and 30% the test data. The data has been processed and a set of 561 features was produced to include such values as mean, standard deviation, max, min, signal entropy, angle between vectors, median absolute deviation, and so on. All features were normalized and bounded within $[-1, 1]$. The full description for this data set can be reviewed at **UCI** website.

Pairwise plots in Fig. 11.13 show that there is a quite significant overlap in most features with regard to their classification. This suggests that using just some features

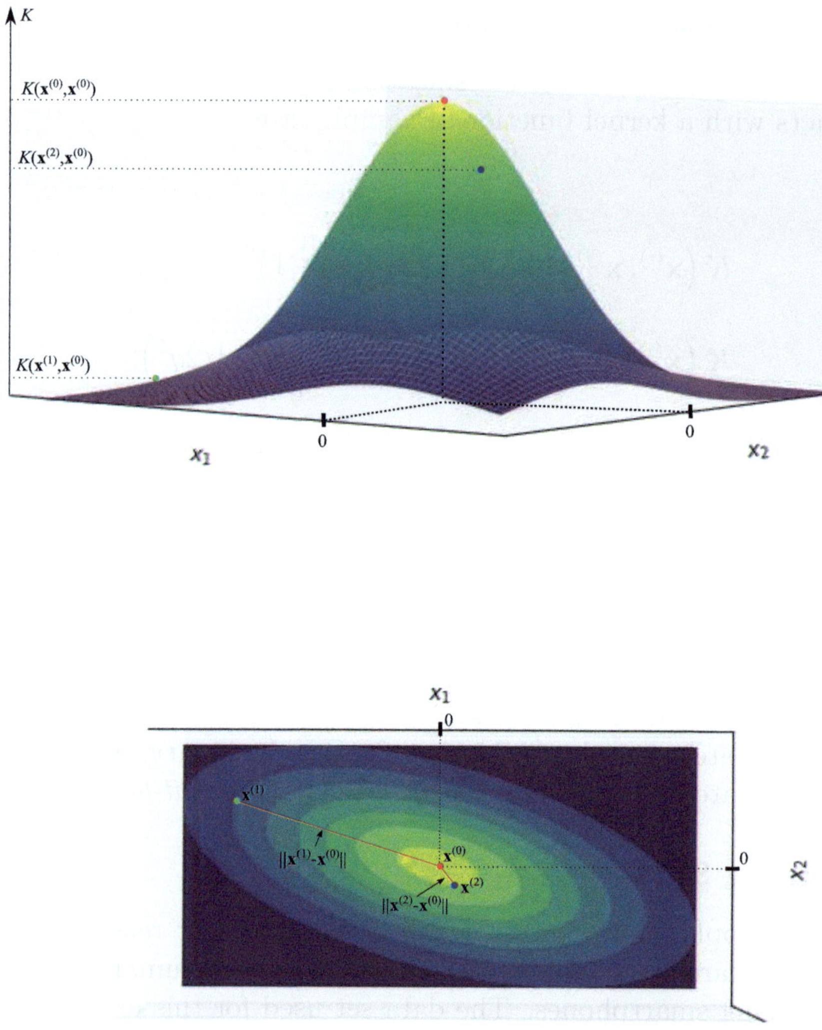

Figure 11.12.: Illustration of the output of Gaussian kernel for two-dimensional inputs $\mathbf{x}^{(i)} = (x_1^{(i)}, x_2^{(i)})^\top$ with the fixed mean value $\mathbf{x}^{(0)} = (0,0)^\top$. Top: $K(\mathbf{x}^{(i)}, \mathbf{x}^{(0)})$ as a function of one two-dimensional variable reaches its maximum when $\mathbf{x}^{(i)} = \mathbf{x}^{(0)} = (0,0)^\top$. Bottom: contour plot illustrates the Euclidean distance $||\mathbf{x}^{(i)} - \mathbf{x}^{(0)}|| = (x_1^{(i)} - 0)^2 + (x_2^{(i)} - 0)^2$. The greater the distance $||\mathbf{x}^{(i)} - \mathbf{x}^{(0)}||$, the smaller the value of the kernel $K(\mathbf{x}^{(i)}, \mathbf{x}^{(0)})$

may not be a feasible option, although may prove effective for reducing the learning time of classification algorithms.

Support vector machine binary classifiers generalized to the multiclass case through a one-vs-all approach were applied. The SVM hyperparameters were selected through a tenfold cross-validation procedure, and Gaussian kernels were used for the experiments. The classification results using the multiclass SVM (MC-SVM) for the six activities

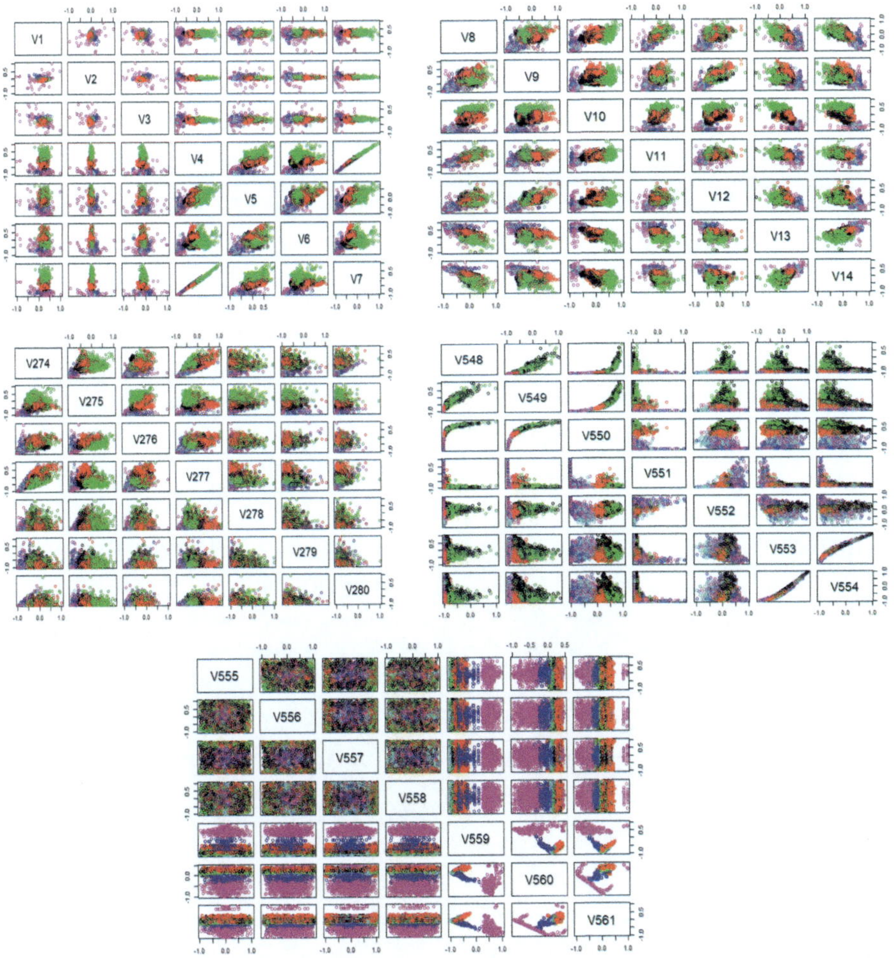

Figure 11.13.: Pairwise plots. Colors chosen according to existing true classification data

of daily living (ADL: walking, walking upstairs, walking downstairs, sitting, standing, laying) show an overall accuracy of 96% for the test data composed of 2947 patterns. Authors compared their results with similar works performed using specialized wearable sensors (such as waist-mounted triaxial accelerometer or chest-mounted accelerometer) and concluded that the use of smartphones, in addition to be more unobtrusive and less invasive than other special purpose solutions (e.g., wearable sensors), is a feasible way to choose for effectively performing HAR.

12. Neural Networks

12.1. Feed-Forward Networks

Deep learning *neural networks* (NN) are playing an increasingly important role in modern machine learning. Deservingly so, as such models provide researchers with versatile methods and powerful tools for supervised learning. Sufficiently extended deep networks can represent quite complex functions. Although not all powerful, in that they cannot replicate sophisticated human thinking, deep networks perform extremely well in most tasks that involve mapping an input vector to an output vector.

Perhaps the most well-known application of neural networks is self-driving vehicles. The experiments in this field had been conducted since the 1920s. However, it was not until the early 2000s that acceptable versions of autonomous vehicles were presented. This is in large measure due to the level of development of neural networks. As a new driver, each of us started by observing other drivers. Equipped with this theoretical knowledge about the most appropriate decisions in many possible situations, one attempts to drive. On each such occasion, in addition to previously observed examples, one takes into account information such as the type of road, the proximity of other vehicles, the presence of an intersection, and so on. All environmental conditions are interconnected into a hierarchical structure. One then makes a decision about the type of action to take or a driving direction to choose. Note that one does not acknowledge publicly all factors that influence the decision made. Moreover, a human driver rarely, if ever, considers a detailed layout of all factors and tries to analyze each factor separately. All information is processed almost simultaneously and stays "hidden" in the back of one's mind. Only the final decision is visible.

For a neural network operating a self-driving vehicle, much like for a human driver, the process begins with *network training* during which a network is allowed to "observe" a human driver operating the vehicle. In a base model, a view of the road and the selected driving direction corresponding to this view is recorded intermittently. The training image is used as an observed data set (predictors). A selected driving direction serves as a response variable. An algorithm is designed to pass the set of predictors through several sets of model parameters. Each set of parameters can be viewed as responsible for different environmental conditions. The observed predictors and all parameter sets are then interconnected into a hierarchical system consisting of several layers. The mathematical operation performed in each layer is in general similar to the one we have seen in logistic regression (see Sect. 11.5.1) with several key differences. First, each parameter set is a matrix rather than a vector. Secondly, after obtaining the output similar to (11.52) using the initial set of predictors and the first parameter set, a model

M. Alvo, *Statistical Inference and Machine Learning for Big Data*, Springer Series in the Data Sciences, https://doi.org/10.1007/978-3-031-06784-6_12

x_1	x_2	x_1 **OR** x_2	x_1 **AND** x_2	x_1 **XOR** x_2	**NOT** x_1
0	0	0	0	0	1
0	1	1	0	1	1
1	0	1	0	1	0
1	1	1	1	0	0

Table 12.1.: Binary logic operations **OR**, **AND**, **XOR, NOT**

utilizes this new data as input for the next parameter set, as in iterated function systems. When the last parameter set is used, we obtain the final model output and compute the cost function.

Note that only the output from the last layer becomes "visible" to the cost function as model output. For this reason, all iterations before the last one are called *hidden layers*. As before, the goal is to determine values of parameters that minimize the cost, hence producing outputs that are maximally close to the observed response (e.g., the driving direction chosen by the human driver).

Trained on many different road types, in many different conditions, the neural network operating a self-driving vehicle is able to perform well in the complicated environment of a busy road with many obstacles.

12.1.1. Motivation

To provide some motivation for using neural networks instead of other classification models, let us consider a simple classification problem such as binary logic operation **OR**. The possible inputs (predictors) are $\mathbf{x} \in \{(0,0),(0,1),(1,0),(1,1)\}$, and the corresponding outputs (values of response) are $y \in \{0,1\}$.

Observe that this classification problem can be easily solved using a binary logistic regression model. Recall that from the Eq. (11.53), one can compute the probability for the response variable Y to be in class "0" or "1", using 0.5 as a "cutoff" value. Consider the function in Eq. (11.52)

$$g(t) \quad = \quad \frac{e^t}{1+e^t}.$$

Note that

$$\begin{aligned} g(t) > 0.5 &\quad \text{for} \quad t > 0, \\ g(t) = 0.5 &\quad \text{for} \quad t = 0, \\ g(t) < 0.5 &\quad \text{for} \quad t < 0. \end{aligned}$$

Hence, for this simplistic problem, an analytical solution can be given in terms of a liner form $t = \boldsymbol{\beta}^\top \mathbf{x}$ for regression coefficients. We must find $\boldsymbol{\beta} = (\beta_0, \beta_1, \beta_2)^\top$ such that

$$\begin{cases} \beta_0 + 0 \times \beta_1 + 0 \times \beta_2 & < 0 \\ \beta_0 + 0 \times \beta_1 + 1 \times \beta_2 & > 0 \\ \beta_0 + 1 \times \beta_1 + 0 \times \beta_2 & > 0 \\ \beta_0 + 1 \times \beta_1 + 1 \times \beta_2 & > 0 \end{cases} \tag{12.1}$$

Clearly, the system of inequalities (12.1) has infinitely many solutions. For instance, $\beta_0 = -1, \beta_1 = \beta_2 = 2$ is a solution. The optimization procedure would surely settle with the parameters of much larger magnitudes to ensure more distinguished classification probabilities but would converge easily nevertheless.

Furthermore, consider a slightly more complicated binary logic operation **XOR**, which can be decomposed as

$$x_1 \textbf{ XOR } x_2 = \Big(x_1 \textbf{ OR } x_2\Big) \textbf{ AND } \Big(\Big(\textbf{NOT } x_1\Big) \textbf{ OR } \Big(\textbf{NOT } x_2\Big)\Big) \tag{12.2}$$

For the same inputs $\big((0,0),(0,1),(1,0),(1,1)\big)$—the corresponding outputs are $(0,1,1,0)$ as shown in Table 12.1. Following the same path as before we may try to use a logistic regression model. Then the solutions for the regression coefficients of such model are given by

$$\begin{cases} \beta_0 + 0 \times \beta_1 + 0 \times \beta_2 & < 0 \\ \beta_0 + 0 \times \beta_1 + 1 \times \beta_2 & > 0 \\ \beta_0 + 1 \times \beta_1 + 0 \times \beta_2 & > 0 \\ \beta_0 + 1 \times \beta_1 + 1 \times \beta_2 & < 0 \end{cases} \tag{12.3}$$

Unfortunately, this system of inequalities has no solutions since the combined inequalities in the second and third line of (12.3) contradict the inequality in the fourth line. Consequently, the logistic regression model described earlier will fail to determine the appropriate estimates for regression coefficients.

Although there exist modifications to the logistic regression model that allow adding feature interactions (in terms of degree greater than 1), these models are inefficient for many real-life data sets that contain a large number of features to begin with. Consider including quadratic terms (an interaction $x_i x_j$) for all possible pairs of a 200-dimensional feature vector $\mathbf{x} = (x_1, ..., x_{200})$. We would be adding 20,000 predictors and estimating a separate parameter for each of these predictors. How about adding cubic or higher degree terms? Thus, it is quite natural to inquire about the existence of a simple computationally based model to solve this seemingly trivial problem which cannot be resolved using a logistic regression model that assumes a linear form for the parameter equation $t = \boldsymbol{\beta}^\top \mathbf{x}$.

Consider a decomposition (12.2) of **XOR** operation. As we can see from Table 12.2, the outputs of the first two operations are used as inputs for the third and final operation.

x_1	x_2	$h_1 = x_1$ **OR** x_2	$h_2 = \big($**NOT** $x_1\big)$ **OR** $\big($**NOT** $x_2\big)$	$y = h_1$ **AND** h_2
0	0	0	1	0
0	1	1	1	1
1	0	1	1	1
1	1	1	0	0

Table 12.2.: Result of a binary **XOR** operation performed in three steps, using a two-level model

Suppose we want to build a model that reflects the approach to evaluate the result of **XOR** operation via decomposition. We shall use three sets of parameters

$$\begin{aligned}\boldsymbol{\beta} &= (\beta_0, \beta_1, \beta_2)^\top, \\ \boldsymbol{\theta} &= (\theta_0, \theta_1, \theta_2)^\top, \\ \boldsymbol{\omega} &= (\omega_0, \omega_1, \omega_2)^\top\end{aligned}$$

to evaluate each of the three parts in formula (12.2) individually, applying the same consideration and solving systems of inequalities seen in (12.3) or (12.1). We have already found from (12.1) that values $\beta_0 = -1, \beta_1 = \beta_2 = 2$ give an acceptable (but not optimal!) coefficient vector for a classification problem x_1 **OR** x_2. It can be checked that values $\theta_0 = 3, \theta_1 = \theta_2 = -2$ and $\omega_0 = -3, \omega_1 = \omega_2 = 2$ provide acceptable (but not optimal!) coefficients for corresponding classification problems $\big($**NOT** $x_1\big)$ **OR** $\big($**NOT** $x_2\big)$ and x_1 **AND** x_2. We shall use a separate logistic regression model for each of the two intermediate steps:

$$\begin{aligned}\boldsymbol{h}_1^{(i)} &= \mathbf{1}_{g\geq 0.5}\left[g(\boldsymbol{\beta}^\top \mathbf{x}^{(i)})\right] = \mathbf{1}_{g\geq 0.5}\left[\frac{e^{\boldsymbol{\beta}^\top \mathbf{x}^{(i)}}}{1+e^{\boldsymbol{\beta}^\top \mathbf{x}^{(i)}}}\right], \\ \boldsymbol{h}_2^{(i)} &= \mathbf{1}_{g\geq 0.5}\left[g(\boldsymbol{\theta}^\top \mathbf{x}^{(i)})\right] = \mathbf{1}_{g\geq 0.5}\left[\frac{e^{\boldsymbol{\theta}^\top \mathbf{x}^{(i)}}}{1+e^{\boldsymbol{\theta}^\top \mathbf{x}^{(i)}}}\right],\end{aligned} \tag{12.4}$$

where $\mathbf{1}_{g\geq 0.5}$ is an indicator function. The elements of resulting vectors $\boldsymbol{h}_1$ and $\boldsymbol{h}_2$ give the values of predictor variables for a logistic regression model used for the final step

$$\hat{\boldsymbol{y}}^{(i)} = \mathbf{1}_{g\geq 0.5}\left[g(\boldsymbol{\omega}^\top \boldsymbol{h}^{(i)})\right] = \mathbf{1}_{g\geq 0.5}\left[\frac{e^{\boldsymbol{\omega}^\top \boldsymbol{h}^{(i)}}}{1+e^{\boldsymbol{\omega}^\top \boldsymbol{h}^{(i)}}}\right]. \tag{12.5}$$

The graph in Fig. 12.1 gives an illustration of the two-level model.

In a real-world application, however, we require that the model produce the final result without explicit reference to the intermediate steps. Such a model in a much greater generality can be built using a neural network. In what follows we shall describe several fundamental architectures of neural networks.

12.1.2. Introduction to Neural Networks

We begin by stating a simplified form of the *Universal Approximation Theorem* (UAT) [Cybenko, 1989], which lays the theoretical foundation for a vast field of applications of this machine learning model.

Theorem 12.1 (UAT). *Let f be a nonconstant, bounded, and monotone-increasing continuous function. Then for any continuous function $\mathfrak{g}(\mathbf{x})$ with $\mathbf{x}^\top = (x_1, x_2, ..., x_n)$, and $\delta > 0$, there exists an integer M, real-valued vectors $\boldsymbol{\alpha}^\top = (\alpha_1, ..., \alpha_M)$, $\mathbf{b}^\top = (b_1, ..., b_M)$, and a real-valued $M \times n$ matrix Ω, such that for all $\mathbf{x}$ that lie in the input space, we have*

$$|\mathfrak{G}(\mathbf{x}) - \mathfrak{g}(\mathbf{x})| < \delta,$$

where

$$\mathfrak{G}(\mathbf{x}) = \boldsymbol{\alpha}^\top f\left(\Omega \mathbf{x} - \mathbf{b}\right) \tag{12.6}$$

is an approximate realization of $\mathfrak{g}(\mathbf{x})$.

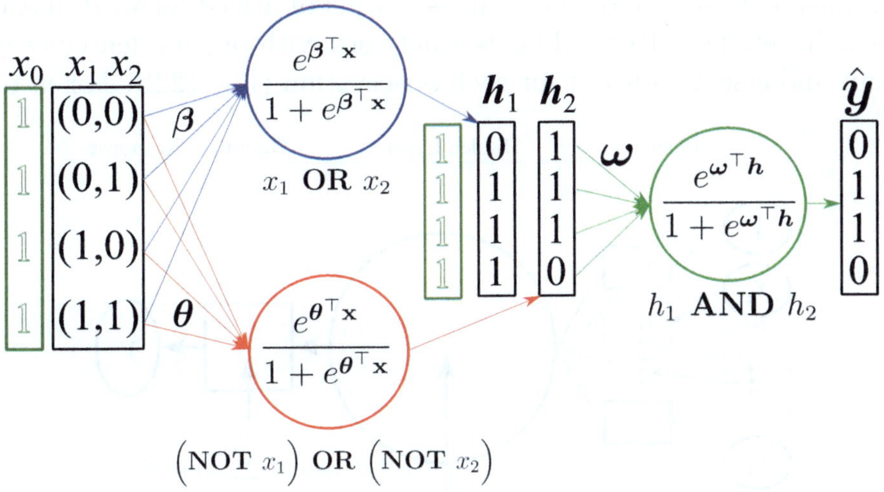

Figure 12.1.: Two-level logistic regression model. The initial observed set $\big((0,0),(0,1),(1,0),(1,1)\big)$ is used as an input for two separate logistic regression models, each having its own regression parameters β and θ. The outputs are composed of an input set for the second-level model with a new parameter vector ω. The final outputs $\hat{\boldsymbol{y}} = (0,1,1,0)$ of the second-level model match the result $(0,1,1,0)$ of **XOR** operation for the initial inputs

This theorem implies that given a sufficient number M of hidden units, a two-layer feed-forward neural network can approximate any continuous function. It is also important to understand that *deep networks* have a hierarchical structure which makes them particularly well adapted to learn the hierarchies of knowledge that seem to be useful in solving real-world problems. For example, in image recognition, one wants to have a system that understands not just individual pixels but also increasingly more complex concepts, such as edges, simple geometric shapes, all the way up to complex, multi-object scenes. It has been discovered that deep networks do a better job than shallow networks at learning such hierarchies of knowledge.

12.1.3. Building a Deep Feed-Forward Network

A model input is a two-dimensional set that contains records of m outcomes of an experiment with n features observed for every single outcome. We shall use an augmented input set

$$\boldsymbol{M}_{\mathbf{x}} = \begin{bmatrix} 1 & x_1^{(1)} & \dots & x_n^{(1)} \\ 1 & x_1^{(2)} & \dots & x_n^{(2)} \\ \dots & \dots & \dots & \dots \\ 1 & x_1^{(m)} & \dots & x_n^{(m)} \end{bmatrix},$$

which is called a design matrix.

As mentioned earlier, one of the simplest forms of a feed-forward network is called the *perceptron* [Rosenblatt, 1958]. This is a network with one hidden layer that consists of a single unit and a scalar output for each observation (Fig. 12.2). For m observations,

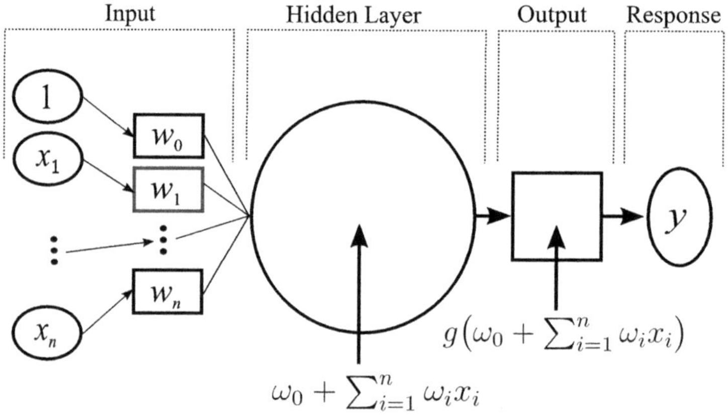

Figure 12.2.: Perceptron on a single observation with n features

a parameter vector $\boldsymbol{\omega}^\top = (\omega_0, \omega_1, ..., \omega_n)$, and an output vector $\hat{\boldsymbol{y}}^\top = (\hat{y}^{(1)}, \hat{y}^{(2)}, ..., \hat{y}^{(m)})$, we have

$$\hat{\boldsymbol{y}} = g(\boldsymbol{M}_{\boldsymbol{X}}\boldsymbol{\omega}), \tag{12.7}$$

where $g(\cdot)$ is called an *activation function.*

One of the simplest activation functions to use in the model above is the *threshold (or step) function*, which can be defined as

$$\operatorname{sgn}(x) = \begin{cases} 1 & x \geq 0 \\ 0 & x < 0 \end{cases} \tag{12.8}$$

The threshold function was in fact one of the first activation functions used in neural networks (Fig. 12.3).

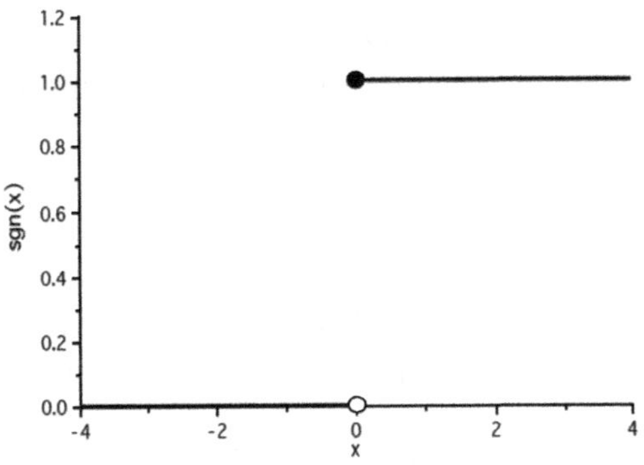

Figure 12.3.: Threshold function sgn(x)

Another popular choice of activation function, which was introduced later, is the *logistic sigmoid function,* (Fig. 12.4).

$$\text{sigmoid}(x) = \frac{1}{1 + e^{-x}}, \tag{12.9}$$

which is an equivalent form of (11.7). It is the sigmoid function that helped G. Cybenko to formulate and prove his UAT [Cybenko, 1989].

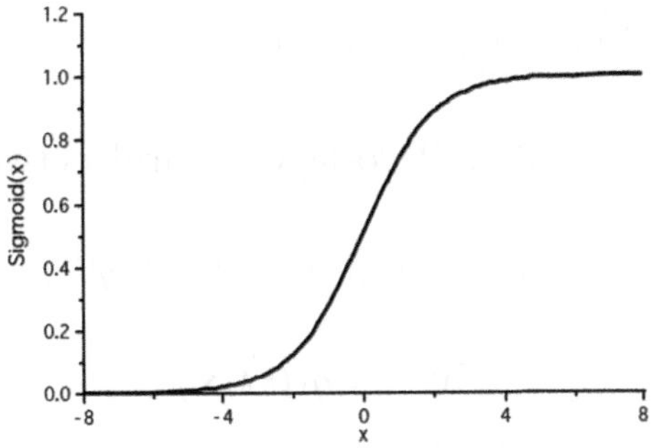

Figure 12.4.: Logistic sigmoid function

We note that a perceptron with the sigmoid activation is a replica of the logistic regression classification model. This observation enables one to consider the application of the same (or very similar) approach to model optimization, or *model training* as it is labeled in the machine learning world. We recall that the goal of the optimization process is to find a set of parameters $\boldsymbol{\omega}^\top = (\omega_0, \omega_1, ..., \omega_n)$ such that the chosen cost function $\mathcal{C}(\boldsymbol{\omega}; \mathbf{x})$ is minimized. In practice, however, we sometimes sacrifice the best solution to the optimization problem as long as our model *generalizes well*, i.e., has an acceptable prediction accuracy on a new data (cf. Sect. 11.5.2.1).

The methods described in Sects. 11.4.1, 11.5.2, 11.5.3 can be applied as model training techniques for perceptron as well as much more complex multilevel neural networks.

Therefore, we begin by describing a general construction of a multilevel feed-forward neural network, also often called, *a deep feed-forward neural network.*

For that, we need to make several simple but important steps. First, consider using a set of weight vectors $\mathbf{\Omega} = (\boldsymbol{\omega}_1, \boldsymbol{\omega}_2, ..., \boldsymbol{\omega}_M)^\top$. Note that the set $\mathbf{\Omega}$ can be viewed as an $M \times (n+1)$ matrix. Then, for a single augmented feature vector $\mathbf{x}$, we obtain

$$\Omega\mathbf{x} = (z_1, z_2, ..., z_M)^\top = \mathbf{z}, \tag{12.10}$$

where $z_i = \omega_{i0} + \sum_{j=1}^{n} \omega_{ij} x_j$. Second, choose an activation function g_1 and consider using the output $\mathbf{h} = g_1(\mathbf{z})$ as a new input vector. Finally, choose another weight vector $\boldsymbol{\omega}^\top = (\omega_1, \omega_2, ..., \omega_M)$ and an activation function g_2 to obtain the output

$$\hat{y} = g_2(\boldsymbol{\omega}^\top \mathbf{h})$$

of a two-layer network (see Fig. 12.5). Following the above routine, for a single observa-

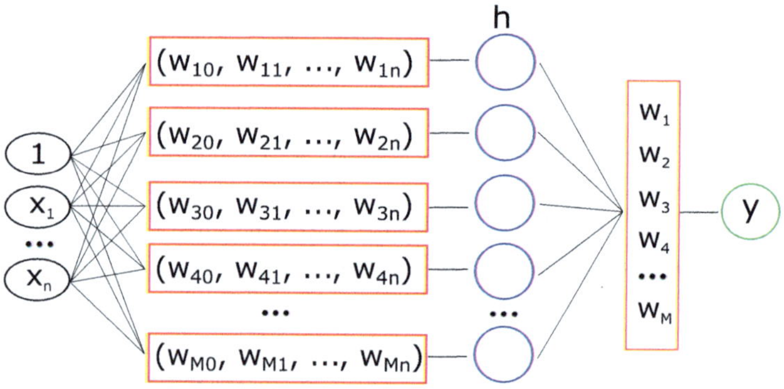

Figure 12.5.: Two-layer neural network

tion $\mathbf{x}$ that was recorded as a set of ordered pairs $(\mathbf{x}^{(i)}, y^{(i)})$, we build a deep feed-forward network as follows:

$$\begin{aligned} \mathbf{h}^{(1)} &= g_1(\mathbf{\Omega}_1 \mathbf{x}) \\ \mathbf{h}^{(i)} &= g_i(\mathbf{\Omega}_i h^{(i-1)}), \quad \text{for } i = 2, 3, \ldots, D \\ \hat{y} &= f(\boldsymbol{\omega}^\top \mathbf{h}^{(D)}) \end{aligned} \tag{12.11}$$

In general, there exists quite a few activation functions $g_i (i = 1, ..., D)$ for learning the parameters in hidden layers. Examples other than the step function (12.8) and the sigmoid function (11.7), are the *hyperbolic tangent* function (12.12) (Fig. 12.6)

$$\tanh(z) = \frac{e^z - e^{-z}}{e^z + e^{-z}}. \tag{12.12}$$

and the *Gaussian function* (12.13) (Fig. 12.7) which may be preferable in some cases

$$\text{Gaussian}(x) = e^{-x^2} \tag{12.13}$$

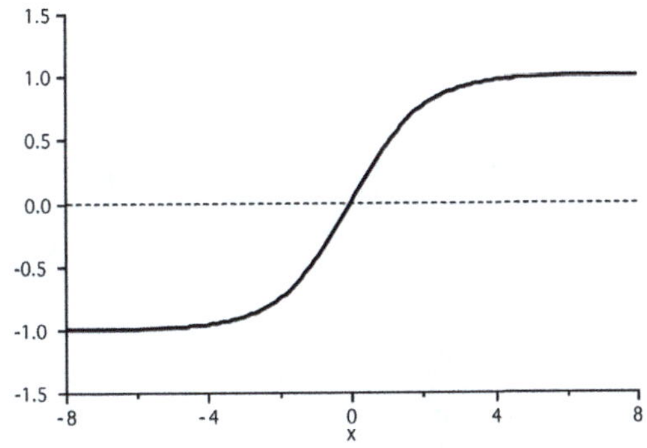

Figure 12.6.: Hyperbolic tangent function $\tanh(x)$

One can use a different activation function for each hidden layer. It is more common,

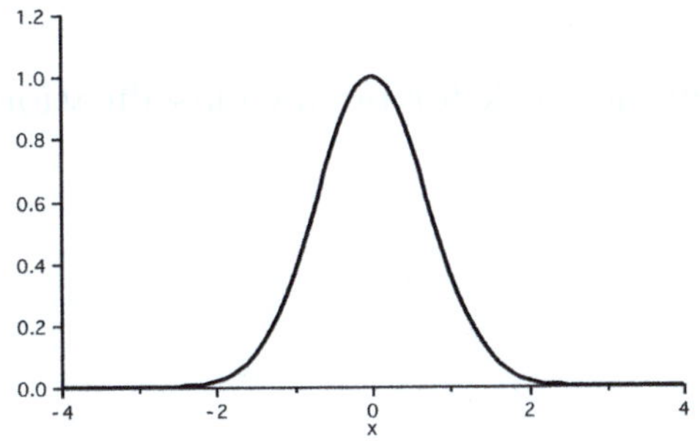

Figure 12.7.: Gaussian function

however, to have the same activation function for all hidden layers except for the output. Here, the choice of activation function depends on the desired result or equivalently on the type of response variable. For binary classification problems, we usually use the sigmoid or the hyperbolic tangent function. A diagram illustrating a deep feed-forward neural network for binary classification and quantitative models is shown in Fig. 12.8. For multiclass classification problems, one may apply the *softmax* function

$$f_i(z) = \frac{e^{z_i}}{\sum_j e^{z_j}}, \tag{12.14}$$

which arises from the multinomial link (11.71). In this case, as previously discussed (11.5.3), an output is the probability vector $(p_1, p_2, ..., p_k)$ with $p_i = P(Y = i|\mathbf{x})$. A graphical representation for the multiclass classification network is given in Fig. 12.9. For the quantitative case $y \in \mathbb{R}$, one can use a linear activation function. Several important questions arise before deciding on the final model structure such as the following:

1. Is there any analysis or preprocessing necessary for the data?

2. How many hidden layers do we need? Note that the number of hidden layers is represented by the parameter D in our model (12.11).

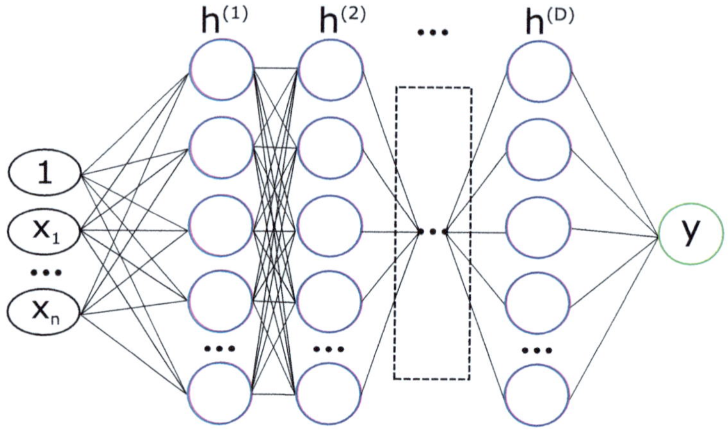

Figure 12.8.: Deep learning network for binary classification and quantitative models

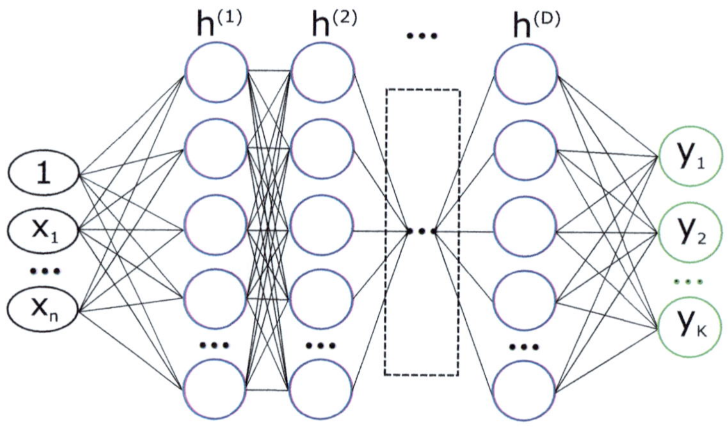

Figure 12.9.: Deep learning network. Each column represents a layer of the network The first column represents the input layer, shown here as a vector with $\mathbf{x}_1, \mathbf{x}_2, \ldots, \mathbf{x}_n$ as features. The last column represents the output layer, where $y_1, y_2, \ldots, y_K$ represent classes. All intermediate columns represent hidden layers for which we do not have explicit outputs. Instead those pass on learned parameters further forward

3. How many elements will be in each hidden layer? This number is given by the parameter M in the simplest model with one hidden layer (12.10). In general, one may choose a different number of hidden units in each layer though in practice it is more common to have the same number of elements in all hidden layers.

4. How does one choose the initial values for the matrices $\mathbf{\Omega}_i$, $i = 1, 2, 3, ..., D$?

5. How does one know when to stop the network training?

We present a few ideas on how to approach such issues. First and foremost, a preliminary data analysis is considered standard practice in modern machine learning. In fact, most data scientists would argue that data analysis, data interpretation, cleaning and imputation account for 60%–70% of their workload. For instance, in the Example 12.1.1, researchers started with only 6 observed explanatory variables. The working model, however, contains 561 features derived from the recorded data. This step contributed significantly to the quality of the result obtained.

Secondly, real-world data often contain errors. For example, a set of predictors may contain noise, i.e., one or more variables with random relation to the response variable. Achieving the perfect prediction accuracy on a training set may then cause our model to perform poorly on new data due to the increased model variance. This is also referred to as *model over fitting*.

Thirdly, as demonstrated in Example 11.4.3, a feature standardization (11.50) is often helpful and sometimes necessary for models based on a gradient descent algorithm.

Regarding the number of hidden layers and hidden units in each layer, Theorem 12.1 states that given a sufficient number M of hidden units, a two-layer feed-forward neural network can approximate any continuous function. Unfortunately there is no precise way to determine the number of hidden units necessary for any particular function. This is not to say we have absolutely no good starting point for any particular problem. It has been determined empirically that a better result could be achieved in many cases by choosing the number of units in a single hidden layer approximately equal to the number of features in a data set (excluding models aimed at dimension reduction and principal component analysis). Unfortunately this means having thousands of hidden units for many real-life data sets. This is of course an inefficient approach given existing computational limitations. We have also learned that increasing the number of hidden layers allows neural networks to learn more complex functions with a decreased number of hidden units in each layer. Moreover, some functions (e.g., logical XOR and NXOR) cannot be adequately represented using a one-layer perceptron model.

Finding an exact solution for the values of weight sets $\mathbf{\Omega}_i$, $i = 1, 2, 3, ..., D$ is not feasible due to the model's complex structure. Therefore, we must rely on numerical algorithms, such as the gradient descent, for the model to learn these parameters. All such algorithms require some initial parameter values. Algorithms like the gradient descent treat all the weights in the same way; hence, using all the same initial values would result in the same learning paths for all hidden units. This invalidates having multiple hidden units in our model. For this reason we use small random numbers (usually sampled from uniform or normal distributions) as the initial values of weights for deep networks.

Note that an algebraic structure of a deep network represents a complex geometric object with many local minima (Fig. 12.10). Hence, the choice of the initial weight values for the gradient descent algorithm actually matters and cannot be completely left to a chance. The optimal initial values are usually determined experimentally by trying several different starting points. Due to several properties of smooth activation functions,

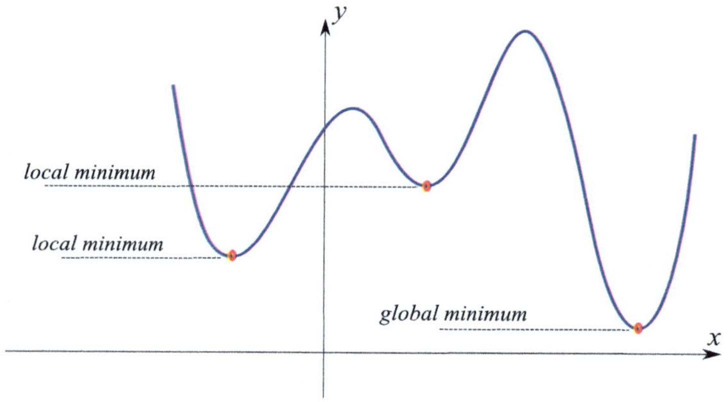

Figure 12.10.: Local and global extrema

a deep learning model can only achieve its actual targets as the *number of steps* $\to \infty$. Thus, we opt to choose an appropriate, problem-specific, threshold δ (11.8) for our cost function $\mathcal{C}(\boldsymbol{\omega}; \mathbf{x})$ and stop the optimization algorithm when this threshold is achieved.

In summary, we would like to build a neural network capable of producing outputs for the input data that were not originally used to design and train our network. Hence, one can distinguish two important phases in the network's life cycle:

1. Learning phase. In this phase the model is trained on an observed data set (training set) with an available set of correct response values. Note that each data point in the abovementioned set refers to the vector of predictors.

2. Generalization phase. In this phase, the set of parameters (weights) produced as a result of the learning process is applied to previously unavailable data set (test set) in order to produce response values. If the model generalizes well, these will be sufficiently close to the true response values of the test set.

12.1.4. Learning in Deep Networks

The same idea as described in Sect. 11.4.1 is applied in a learning process for deep networks. Recall that the goal is to find the set of weights (parameters) that minimize a chosen cost function which evaluates the "level of dissimilarity" (error) between the model outputs and the actual values of a response variable. One of the most effective ways to achieve this is a gradient descent algorithm (11.47). For a set of weights $\boldsymbol{\omega}$, we find the gradient descent weight updates from (11.46)

$$\Delta\boldsymbol{\omega} = \boldsymbol{\omega}_{next} - \boldsymbol{\omega}_{prev} = -\alpha\nabla\mathcal{C}. \tag{12.15}$$

Although only the outputs $\hat{y} = f(\boldsymbol{\omega}^\top \mathbf{h}^{(D)})$ of a final layer are present in the cost function, these final outputs will depend on all weights from all previous hidden layers. Thus, the learning algorithm must be such that it takes into consideration the outputs from all hidden layers. To do that, we must apply weight updates (12.15) to all hidden units in each hidden layer.

12.1.4.1. Quantitative Model

Let us use the cost function

$$\mathcal{C} = \frac{1}{2m} \sum_{i=1}^{m} (y^{(i)} - \hat{y}^{(i)})^2$$

defined in (11.43), with a simple linear output activation function $f(x) = x$ and the sigmoid activation function g (11.7) in the hidden layer for a two-level perceptron ($D = 2$) to demonstrate the computation process of weight updates. For a single augmented feature vector $\mathbf{x} = (1, x_1, x_2, ..., x_n)^\top$, we have

$$\mathbf{z} = \Omega \mathbf{x}, \tag{12.16}$$

with $z_i = \sum_j \omega_{ij} x_j$ $(i = 1, 2, 3, ..., M)$. Further

$$\mathbf{h} = g(\mathbf{z}) = \frac{1}{1 + e^{\mathbf{z}}}, \tag{12.17}$$

and for $\boldsymbol{\omega}^\top = (\omega_1, \omega_2, ..., \omega_M)$

$$z = \boldsymbol{\omega}^\top \mathbf{h}, \tag{12.18}$$

Finally

$$\hat{y} = f(z) = z. \tag{12.19}$$

Now, the cost function for m feature vectors is

$$\mathcal{C}(\boldsymbol{\omega}) = \frac{1}{2m} \sum_{i=1}^{m} (y^{(i)} - \hat{y}^{(i)})^2 = \frac{1}{2m} \sum_{i=1}^{m} (y^{(i)} - z^{(i)})^2. \tag{12.20}$$

The gradient is

$$\nabla \mathcal{C}(\boldsymbol{\omega}) = -\frac{1}{m} \sum_{i=1}^{m} (y^{(i)} - z^{(i)}) \frac{\partial z^{(i)}}{\partial \boldsymbol{\omega}}, \tag{12.21}$$

where for each training example

$$\frac{\partial z}{\partial \boldsymbol{\omega}} = \frac{\partial \boldsymbol{\omega}^\top \mathbf{h}}{\partial \boldsymbol{\omega}} = \mathbf{h}. \tag{12.22}$$

For the weight updates in the hidden layer, we write

$$\nabla \mathcal{C}(\Omega) = -\frac{1}{m}\sum_{i=1}^{m}(y^{(i)} - z^{(i)})\frac{\partial z^{(i)}}{\partial \Omega}. \tag{12.23}$$

Defining the *Jacobian* of $\mathbf{h}$ with respect to $\mathbf{z}$as an $M \times M$ diagonal matrix, with nonzero elements corresponding to $\dfrac{\partial h_i}{\partial z_i}$

$$\mathbf{J}_{\mathbf{z}}(\mathbf{h}) = \mathrm{diag}\Big[g(\mathbf{z})(1 - g(\mathbf{z}))\Big], \tag{12.24}$$

we have for each training example

$$\frac{\partial z}{\partial \Omega} = \mathbf{J}_{\mathbf{z}}(\mathbf{h})\ \boldsymbol{\omega}\ \mathbf{x}^{\top}. \tag{12.25}$$

12.1.4.2. Binary Classification Model

For this model, the output activation function is both the sigmoid function (11.7) and the activation function g in the hidden layer. We shall use a two-level perceptron ($D = 2$) once again. The cost function in this case is the negative maximal likelihood (11.57) averaged over m training examples

$$\mathcal{C} = -\frac{1}{m}\sum_{i=1}^{m} y^{(i)} \log\left[f(z^{(i)})\right] + (1 - y^{(i)})\log\left[1 - f(z^{(i)})\right], \tag{12.26}$$

similar to (11.57). The gradient is similar to (11.59) and gives the weight updates in the output layer

$$\nabla \mathcal{C}(\boldsymbol{\omega}) = -\frac{1}{m}\sum_{i=1}^{m}\left(y^{(i)} - f(z^{(i)})\right)\frac{\partial z^{(i)}}{\partial \boldsymbol{\omega}} \quad . \tag{12.27}$$

The weight updates in the hidden layer are given by (12.23) as in the case of the quantitative model. It is worth noting that for the multiclass classification model with the softmax output function (12.14), we get the same form for weight updates as in (12.23) and (12.27).

12.1.5. Generalization

As a result of the network training process, we want to be able to recreate a (smooth) mapping using an observed imperfect data set. To gain the statistical understanding of such a generalization process, we once again turn to a well-known regression model (11.4). We recall that for a data set recorded in ordered pairs

$$(\mathbf{x}^{(i)}, y^{(i)}),\ i = 1, 2, 3, ..., m, \tag{12.28}$$

the regression is modeled as

$$y^{(i)} = \mathfrak{g}(\mathbf{x}^{(i)}) + \varepsilon^{(i)}, \tag{12.29}$$

where $\mathfrak{g}$ is a true function we want our model to approximate. The random error terms are $\varepsilon^{(i)}$ assumed to be independent with $\mathbf{E}\left[\varepsilon^{(i)}\right] = 0$ and $\mathbf{Var}\left[\varepsilon^{(i)}\right] = \sigma^2$. We interpret $\mathfrak{g}(\mathbf{x}^{(i)})$ as the expectation of Y given $\boldsymbol{X}$

$$\mathbf{E}\left[Y = y \mid \boldsymbol{X} = \mathrm{x}\right] = \mathfrak{g}(\boldsymbol{X}). \tag{12.30}$$

Training our model based on a single observed data set (12.28) and a set of parameters $\boldsymbol{\Omega} = \left\{\omega_{ij}^{(d)}\right\}$ will likely produce an output set $\hat{\boldsymbol{y}} = \mathfrak{G}(\boldsymbol{\Omega}; \mathbf{x})$ that is only an approximation of $\mathfrak{g}$. Since we obviously cannot use all possible sets of observations in order to find the true underlying function $\mathfrak{g}$, our goal is to find the best approximation $\mathfrak{G}(\boldsymbol{\Omega}; \mathbf{x})$ using the data set in hand. Suppose we want to use a well-known cost function

$$\mathcal{C}(\boldsymbol{\Omega}; \mathbf{x}) = \frac{1}{2m} \sum_{i=1}^{m} (y^{(i)} - \hat{y}^{(i)})^2. \tag{12.31}$$

Finding the values of a parameter set $\boldsymbol{\Omega}$ that minimize (12.31) would provide the best fit for an observed data set. This however does not guarantee the best (or even good) approximation to the actual underlying function $\mathfrak{g}$. Hence, we are more interested in minimizing

$$\left|\mathbf{E}\left[Y = y \mid \boldsymbol{X} = \mathrm{x}\right] - \mathfrak{G}(\boldsymbol{\Omega}; \mathbf{x})\right| \tag{12.32}$$

rather than (12.31).

For a single observation $(\mathbf{x}, y)$, with $\mathbf{x} = (x_1, x_2, x_3, ..., x_n)$, consider (with a slight abuse of notation) the squared error estimator

$$\mathfrak{S}^2 = \Big(\mathbf{E}\left[Y = y \mid \mathrm{x}\right] - \mathfrak{G}(\boldsymbol{\Omega}; \mathbf{x})\Big)^2. \tag{12.33}$$

Using the notation from (12.30) and averaging over all possible data sets give

$$\mathbf{E}\Big[\big(\mathfrak{g}(\mathbf{x}) - \mathfrak{G}(\boldsymbol{\Omega}; \mathbf{x})\big)^2\Big] = \Big(\mathbf{E}\big[\mathfrak{G}(\boldsymbol{\Omega}; \mathbf{x})\big] - \mathfrak{g}(\mathbf{x})\Big)^2 + \mathbf{E}\Big[\Big(\mathfrak{G}(\boldsymbol{\Omega}; \mathbf{x}) - \mathbf{E}\big[\mathfrak{G}(\boldsymbol{\Omega}; \mathbf{x})\big]\Big)^2\Big]. \tag{12.34}$$

The first term on the right-hand side of (12.34) represents the squared bias of $\mathfrak{G}$, and the second term is the variance of $\mathfrak{G}$. Thus, we write

$$\mathbf{E}\left[\mathfrak{S}^2\right] = \mathbf{Bias}^2\left[\mathfrak{G}\right] + \mathbf{Var}\left[\mathfrak{G}\right]. \tag{12.35}$$

The bias term $\mathbf{Bias}^2\left[\mathfrak{G}\right]$ gives a measure of the approximation error averaged over all possible observed data sets. The variance term $\mathbf{Var}\left[\mathfrak{G}\right]$ can be viewed as a measure of dependence of the output to the particular data set. Unfortunately, both of these terms cannot be equal to zero at the same time. Suppose we have

$$\mathbf{Var}\left[\mathfrak{G}\right] = 0.$$

This means **constant outputs** of our model for any input data: $\mathfrak{G}(\boldsymbol{\Omega}; \mathbf{x}) = a$. In this case

$$\mathbf{Bias}^2\left[\mathfrak{G}\right] = \left(\mathbf{E}\left[a\right] - \mathfrak{g}(\mathbf{x})\right)^2 = \left(a - \mathfrak{g}(\mathbf{x})\right)^2. \tag{12.36}$$

Unless $\mathfrak{g}(\mathbf{x}) = a$, the expression in (12.36) will be a relatively large number. This is known as *model under-fitting*. Conversely, suppose we achieve a perfect fit of our model on a particular data set, i.e.,

$$\mathbf{E}\left[\mathfrak{S}^2\right] = \mathbf{Bias}^2\left[\mathfrak{G}\right] = 0,$$

or equivalently

$$\mathfrak{G}(\boldsymbol{\Omega}; \mathbf{x}) = \mathfrak{g}(\mathbf{x}) + \epsilon \text{ and } \mathbf{E}\left[\mathfrak{G}(\boldsymbol{\Omega}; \mathbf{x})\right] = \mathfrak{g}(\mathbf{x}),$$

where ϵ is the empirical error on the particular data set with $\mathbf{E}\left[\epsilon\right] = 0$. In this case, the expressions (12.34) and (12.35) give

$$\mathbf{E}\left[\mathfrak{S}^2\right] = \mathbf{Var}\left[\mathfrak{G}\right] = \mathbf{E}\left[\left(\mathfrak{G}(\boldsymbol{\Omega}; \mathbf{x}) - \mathbf{E}\left[\mathfrak{G}(\boldsymbol{\Omega}; \mathbf{x})\right]\right)^2\right] = \mathbf{E}\left[\left(\mathfrak{g}(\mathbf{x}) + \epsilon - \mathfrak{g}(\mathbf{x})\right)^2\right] = \mathbf{E}\left[\epsilon^2\right] = \mathbf{Var}\left[\epsilon\right]. \tag{12.37}$$

Depending on the data set in hand, the expression (12.37) could be equal to a relatively large number. As mentioned earlier, this is referred to as *model over-fitting.*

In order for our model to generalize well, we must compute the optimal values for network weights $\boldsymbol{\Omega} = \left\{\omega_{ij}^{(d)}\right\}$, so that both terms on the right-hand side of (12.35) are as small as possible in order to strike a balance between model under-fitting and over-fitting. We give several general recommendations that are applicable in many cases.

Model over-fitting is commonly caused by using an excessive number of hidden units, prolonged model training process, presence of a very large number of features (high dimension). Thus, it is considered a common practice to try several (usually many) model configurations. One needs to keep in mind that achieving the perfect model fit on a particular training set may cause poor model generalization; hence, stopping the model training "at the right time" is as important as minimizing the model error. Model regularization (11.5.4) is also a very useful tool routinely used to prevent model over-fitting.

On the other hand, having too few hidden units and/or insufficient model training time will likely cause model under-fitting. Thus, finding the optimal number of hidden units and layers that is able to represent an underlying mapping and discover its possible hierarchical structure is of great importance. We shall further describe useful cross-validation techniques that will help in determining optimal stopping points in model training.

12.1.5.1. A Machine Learning Approach to Generalization

Given the extensive modern computing abilities, one does not actually end up calculating the bias and the variance for each model configuration of a deep-learning network. Instead, one uses various combinations of training, validation, and test data sets. We had briefly touched upon this topic in Sect. 11.4.1, where we have discussed several different ways to optimize a regression model. This approach can be generalized to most machine learning models, in particular neural networks. The simplest configuration that only includes the training set, and the test set is the one most commonly used for demonstration purposes or proof-of-concept models. It is hardly ever used in the actual research or industry applications for obvious reasons. Since our data sets are assumed to be random, a single test set could be simply a "lucky draw," and a model that had demonstrated strong results on such test set may still generalize poorly. Thus, when using a single training set and a single test set, we always apply a random shuffle to the entire observed data prior to splitting it into two parts.

A more reliable approach is the K-fold cross-validation or the *leave-one-out* cross-validation. We recall that in K-fold Cross-validation, we randomly divide our data set into K subsets. The training process is performed K times. Each time, we use $K-1$ subsets for training and 1 remaining subset for testing. We then keep the model that produces the best results out of K possible. An average performance on K test sets is usually taken as a final measure of model performance. Note that if several different model configurations must be tested, each of those configurations will have to undergo K rounds of training.

In some cases we set K to be equal to the number of observations m in an entire data set. We then train our model m times, and in each case the size of a training set is equal to 1. Thus, we use $m-1$ observations for training and 1 observation for testing our model, hence, the name of this method—leave-one-out cross-validation.

Cross-validation is computationally costly but is proven to be an effective method. Having such method in our disposal, we can test different model configurations in order to strike a balance between the statistical bias and the variance, i.e., construct a deep learning network with the appropriate number of hidden units and hidden layers that allow for a quick optimization while producing the best generalization results.

Regarding hidden units, there are two most common approaches—*constructive* and *pruning*. The constructive approach consists of adding hidden units to a model that was initially built with a small number of hidden units. It is possible to retrain an entire model when adding units; however, it is more computationally effective to retrain just the part of a model that contains new units.

The pruning approach, on the other hand, is comprised of building an initial model with an excessive number of hidden units and decreasing this number gradually in order to achieve the desired level of performance. Regularization methods discussed in Sect. 11.5.4 are also classified as pruning techniques as they are used to decrease the values of "unwanted" parameters; hence, effectively, "prune those parameters out" of the model.

Another way of pruning is removing certain parameters (weights) with low level of importance in our model. To find such parameters, an *absence cost coefficient* can be defined. Suppose $\mathcal{C}(\boldsymbol{\Omega})$ is the chosen cost function with $\boldsymbol{\Omega} = \left\{\omega_{ij}^{(d)}\right\}$. We can compute

$$\mathcal{C}(\boldsymbol{\Omega} \mid \omega_{sr}^{(k)} = 0)$$

the value of the cost function with certain parameters set to zero. Then the corresponding absence coefficient is

$$\mathcal{A}_{sr}^{(k)} = \mathcal{C}(\boldsymbol{\Omega}) - \mathcal{C}(\boldsymbol{\Omega} \mid \omega_{sr}^{(k)} = 0). \tag{12.38}$$

The value in (12.38) measures the change in the total error caused by removing a certain parameter. We can now compute the values of absence cost coefficients for some (all) weight parameters and remove those for which these values are relatively low. In this way we retain parameters that have the largest impact on the model output.

We can also utilize the training-validation-test set approach to determine the optimal stopping time for the training algorithm. Since deep learning networks are usually constructed with a large number of parameters, sufficient to fit a training set almost perfectly, model over-fitting is quite often an issue. To avoid it effectively, we may stop the training process "early," or rather at the appropriate time. The stopping time can be determined experimentally by computing the value of the cost function on the validation set in certain intervals. The appropriate stopping time is usually chosen so that the total output error is minimal on the validation set. We assume that this is when our model reaches a better generalization, even though this point may not correspond to the minimal total error on the training set.

Example 12.1.1. In this example, we use the basic feed-forward neural network along with a few classification and optimization techniques in application to a public domain data set for human activity recognition using smartphones [Anguita et al., 2013b]. We use the same data set as in Example 11.5.5.4, so that we have a point of reference when assessing our model performance. Let us recall that the data set contains the activities shown in Table 12.3. As described earlier, a feed-forward neural network can be used

Label	1	2	3	4	5	6
Activity	Walking	Walking Upstairs	Walking Downstairs	Sitting	Standing	Laying

Table 12.3.: Table of activities and the corresponding classes

to approximate a mapping $f : X \to Y$, where the variable X takes values in some "feature" space, which can be thought of as $\mathbb{R}^n$, and the variable Y can take values on a set of integers $\{0, 1, \ldots, K\}$. This is referred to as the classification problem. If $K = 1$ i.e., $Y \in \{0, 1\}$—we are dealing with binary classification; if $K \geq 2$—multiclass classification. A feed-forward network takes a feature vector X as its input and learns parameters $\boldsymbol{\omega}$, which give the best approximation $y \approx \hat{f}(x; \boldsymbol{\omega})$. Neural networks are typically constructed by composing many functions $f_0, f_1, f_2, \ldots, f_D$, where f_0 represents the first layer of the network, and f_D—the last, output layer. All layers, except for the input layer and the output layer, are called hidden layers, since we do not have access to the desired output for these layers in observed data. As a rule, we apply the model to a transformed input $g(x)$ instead of X itself, which allows for the representation of nonlinear functions, since $g(\cdot)$ is a nonlinear transformation. In the simplest case, such model is $y = g(\mathbf{x};\boldsymbol{\omega}_1)^\top \boldsymbol{\omega}_2$, where $\boldsymbol{\omega}_1$ is the set of parameters that allow us to learn g, and $\boldsymbol{\omega}_2$ is another set of parameters used to map from $g(\mathbf{x};\boldsymbol{\omega}_1)$ to the output layer.

There are two common approaches for the multiclass classification problem. One consists of solving binary classification problem for each pair of classes and consequently applying one-versus-all approach to choose the most likely outcome. The output function commonly used in this approach is the sigmoid function (11.7), which arises from the binomial logit link (11.3).

Another approach is to find the set of probabilities $p_i(x)$, with the property $\sum_{i=0}^{k} p_i(x) = 1$, which gives a mapping to the output layer by $y = \mathbf{1}[\max_i(p_i)]$. This can be done by using the *softmax* function (11.71) in a form

$$f_i(z) = \frac{e^{z_i}}{\sum_j e^{z_j}} \tag{12.39}$$

In this example we use gradient-based learning model, since the expected training set size can be substantially large. Furthermore, we consider two approaches to optimization problem: deterministic, where we produce calculations based on the entire training set for each step of the gradient descent, and stochastic (mini-batch), where we randomly choose a subset of the training set for each step of gradient descent. We then compare the performance and classification accuracy for chosen methods.

Model

In this example, we choose the $\tanh(\cdot)$ activation function for all hidden layers

$$\tanh(z) = \frac{e^z - e^{-z}}{e^z + e^{-z}}. \tag{12.40}$$

Its derivative is given by

$$\frac{d}{dz}\tanh(z) = 1 - (\tanh(z))^2. \tag{12.41}$$

Further, we shall explore two of the possible choices for the output function f. These choices will dictate the method we use for prediction. Our first attempt was done using the sigmoid function, which led us to a binary classification, i.e., a single output with the conditional probability

$$P(Y = 1|\mathbf{x})$$

We computed such probabilities for each class and then used one-versus-all method, which is essentially assigning to the class with the greatest probability. As a result, we have obtained an average classification accuracy of 93.96% on the training set and 90.13% on the test set.

In an effort to improve our results we used a different approach to classification by choosing the softmax output function (12.14). This approach allows for the output vector

$$p = (p_1, p_2, \ldots, p_6) \text{ with } p_i = P(Y = i|\mathbf{x}),$$

and the property $\sum_{i=1}^{6} p_i(x) = 1$. We used a negative log-likelihood cost function (11.58) for all cases as it has been proven to perform well in gradient-based classification deep learning models. The negative log-likelihood function can be thought of as the cross-entropy between the training data and the model distribution (11.5.2.1). As a result, we have obtained an average prediction accuracy of 98.52% on the training set and 92.94% on the test set. However, the learning time to train our network is close to 30 min, which does not seem very impressive since our training data only has 7352 observations with 561 features in each observation, which cannot be considered a large data set by modern standards.

To reduce the training time for our network while maintaining similar prediction accuracy, we employed the optimization technique known as *stochastic mini-batch gradient descent*. In essence this approach consists of splitting the existing data set into several subsets (*mini-batches*) and computing the gradient updates for each step of gradient descent using a randomly chosen mini-batch, rather than an entire data set. As a result, we obtained a remarkable improvement in training time and a slight improvement in classification accuracy. The training time of our network is now only around 3 min and the average classification accuracy for the test set is 96.57%.

Results

We present detailed results for two most effective classification Methods, namely, the softmax model with the gradient-based learning on an entire training set for each step of the gradient descent algorithm and the stochastic softmax model. In addition to the average precision, we provide calculations for precision and recall by activity, which gives a better insight into understanding of our results (Table 12.4, Table 12.5) .
Precision gives the classification accuracy, i.e., the proportion of correct predictions.

$$\text{Precision} = \frac{\sharp \text{ of } \textbf{Correctly Classified} \text{ Cases}}{\text{Total } \sharp \text{ of } \textbf{Cases Predicted} \text{ for this Class}}$$

Recall gives the proportion of actual activities correctly classified as such by our network.

$$\text{Recall} = \frac{\sharp \text{ of } \textbf{Correctly Classified} \text{ Cases}}{\text{Total } \sharp \text{ of } \textbf{True Cases} \text{ in this Class}}$$

In addition, we calculate the F_1 score, which is the harmonic mean of the precision and the recall.

$$F_1 = 2 * \frac{\text{Precision} * \text{Recall}}{\text{Precision} + \text{Recall}}.$$

The F_1 score is commonly used instead of an average prediction accuracy since it punishes extreme values and gives equal weight to both precision and recall. A high F_1 score indicates a classification model with optimal balance of precision.

Actual \ Predicted	Class 1	Class 2	Class 3	Class 4	Class 5	Class 6	Recall
Class 1	437	28	31	0	0	0	88%
Class 2	7	459	5	0	0	0	97%
Class 3	13	27	380	0	0	0	90%
Class 4	0	1	0	442	48	0	90%
Class 5	0	0	0	23	509	0	96%
Class 6	0	0	0	4	21	512	95%
Precision	96%	89%	91%	94%	88%	100%	$F_1 = 93\%$

Table 12.4.: Feed-forward neural network classification results for softmax classification model with deterministic gradient descent optimization

The classification performance in each class shows that there is a noticeable overlap in miss classification of sitting and standing. As mentioned in Anguita et al. [2013b], the reason for that may be the physical location of the device. An activity with the lowest classification precision of 93% was "standing," which is a nondynamic activity along with "sitting." In fact, both of these activities have relatively low precision and recall, which indicates that some additional measurements or specific features may need to be added in order to better distinguish these nondynamic activities. Further research may be needed in order to improve the classification accuracy.

Actual \ Predicted	Class 1	Class 2	Class 3	Class 4	Class 5	Class 6	Recall
Class 1	488	1	7	0	0	0	98%
Class 2	11	458	2	0	0	0	97%
Class 3	2	8	410	0	0	0	98%
Class 4	0	3	0	447	41	0	91%
Class 5	1	0	0	25	506	0	95%
Class 6	0	0	0	0	0	537	100%
Precision	97%	97%	98%	95%	93%	100%	$F_1 = 97\%$

Table 12.5.: Feed-forward neural network classification results for softmax classification model with stochastic gradient descent optimization

In this example, we have only demonstrated some of the many possible approaches to classification problems. All of these methods were based on the same feed-forward neural network model. We have tested several classification techniques as well as compared different approaches to optimization. As our reader can observe, a slightly different approach often has a significant effect on the final result. Hence, experimentation is encouraged! To support this statement, we suggest a comparison of our (quite simple) model and two published results that were considered to be state-of-the-art at the time of publishing. The results produced by our model outperform those of MC-HF-SVM model used in Anguita et al. [2013a] by almost 8% and slightly more accurate than the

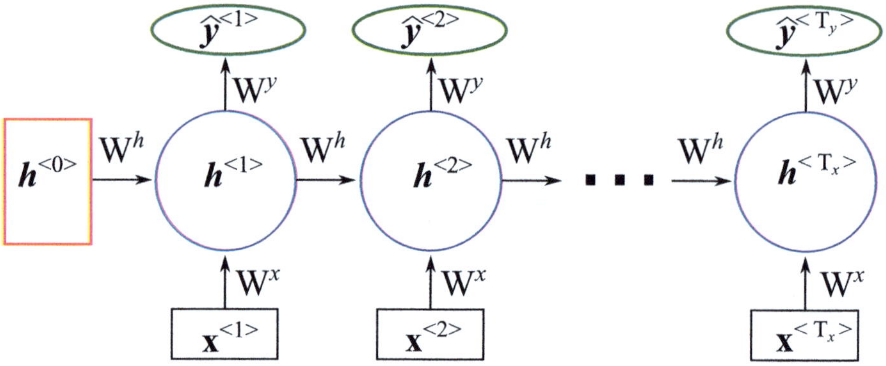

Figure 12.11.: Acyclic graph of a recurrent neural network with the initial value of a hidden layer $\boldsymbol{h}^{\langle 0\rangle}$, and $T_x = T_y$

result of MC-SVM model used in Anguita et al. [2013b]. In addition, our "final" model proves to have balanced precision and recall results and a quite satisfying training time.

12.2. Recurrent Neural Networks

Often we have to deal with sequential data sets $(x^{\langle 1\rangle}, x^{\langle 2\rangle}, ..., x^{\langle t\rangle}, ..., x^{\langle T_x\rangle})$. In such data sets, the value of each element $x^{\langle t+1\rangle}$ depends on all or some of the previous elements $x^{\langle t\rangle}, x^{\langle t-1\rangle}, ..., x^{\langle 1\rangle}$. Although the conventional feed-forward networks can be applied to such data sets (e.g., Weigend and Gershenfeld [1994]) in order to predict the value of $x^{\langle t+1\rangle}$ as a function of the previous t elements in the sequence, there are several drawbacks to this approach. One is the necessity to observe the entire sequence $x^{\langle 1\rangle}, x^{\langle 2\rangle}, ..., x^{\langle t\rangle}$ before a prediction can be made on $x^{\langle t+1\rangle}$. Another extremely important issue is that feed-forward networks learn using a separate parameter for each predictor and hence do not take advantage of an important statistical technique of sharing the parameters between different parts of a model. It seems quite natural that one would greatly benefit if such ability was included in the model, given that our predictors are not just i.i.d. random variables but rather are the elements of a (time) series. This is because parameter sharing enables the generalization to different sequence lengths and allows our model to share the accumulated knowledge between different elements of the sequence. It is worth noting that even though the index t is commonly used to denote a position in time, we mostly view its meaning as a position in the sequence; hence, we assume that $t \in \mathbb{Z}_+$.

12.2.1. Building a Recurrent Neural Network

We now describe one of the simplest architectures of a recurrent neural network, which is similar to a multilevel perceptron (MLP) with the difference that each previous set of hidden unit values used as arguments for the next layer along with the observed value of a feature set and the prediction on the value of the response variable is made at each step.

Let $(\mathbf{x}^{\langle t\rangle}, \boldsymbol{y}^{\langle t\rangle})$ be a data set with $t = 1, 2, 3, ..., T_x$. Here, we assume that $T_x = T_y$. As usual, we call $\mathbf{x}^{\langle t\rangle} = (x_1^{\langle t\rangle}, x_2^{\langle t\rangle}, ..., x_n^{\langle t\rangle})^\top$—a feature vector and $\boldsymbol{y}^{\langle t\rangle} = (y_1^{\langle t\rangle}, y_2^{\langle t\rangle}, ..., y_K^{\langle t\rangle})^\top$—a response vector. We would like to use $(\mathbf{x}^{\langle 1\rangle}, \mathbf{x}^{\langle 2\rangle}, ..., \mathbf{x}^{\langle t\rangle})$ to make a prediction on $\boldsymbol{y}^{\langle t\rangle}$ for all time steps $t = 1, 2, 3, ..., T_x$. This recurrent network can be demonstrated using the *acyclic* graph (shown in Fig. 12.11), which can be formalized as a dynamical system with an added output layer

$$\begin{aligned}
\boldsymbol{z}^{\langle t\rangle} &= W^h \boldsymbol{h}^{\langle t-1\rangle} + W^x \mathbf{x}^{\langle t\rangle} + \boldsymbol{b}^h \\
\boldsymbol{h}^{\langle t\rangle} &= g^{\langle t\rangle}\left(\boldsymbol{z}^{\langle t\rangle}\right) \\
\boldsymbol{\varsigma}^{\langle t\rangle} &= W^y \boldsymbol{h}^{\langle t\rangle} + \boldsymbol{b}^y \\
\hat{\boldsymbol{y}}^{\langle t\rangle} &= f^{\langle t\rangle}\left(\boldsymbol{\varsigma}^{\langle t\rangle}\right),
\end{aligned} \tag{12.42}$$

where $g^{\langle t\rangle}$ is an activation function and $f^{\langle t\rangle}$ is an output function. These functions are commonly chosen to be non-time-dependent (i.e., $g^{\langle t\rangle} = g$ and $f^{\langle t\rangle} = f$ for All $t = 1, 2, 3, ..., T_x$), but variations are possible. In many cases, it is reasonable to use the hyperbolic tangent function (12.12) as an activation function $g^{\langle t\rangle}$. For feed-forward networks, the choice of an output function $f^{\langle t\rangle}$ depends on the nature of the response variable, i.e., sigmoid (12.9) for binary response, or softmax (12.14) for multiclass response. The initial value for $\boldsymbol{h}^{\langle 0\rangle}$ can be set to a null vector or contain small random values. Note that Eqs. (12.42) are in vector form, and element-wise values can be obtained using matrix multiplication rules.

Using the same parameter (weight) matrices W^x, W^h, W^y and the bias parameter vectors $\boldsymbol{b}^h$, $\boldsymbol{b}^y$ for all hidden layers and outputs allows for parameter sharing across the entire data set (time series). This approach allows us to develop a model that can be generalized to various sequence lengths, including those not present in the training set, instead of developing a new model for every possible length of an observed sequence (an impossible approach for most real-world tasks).

An application of such network could be to a word-search algorithm, where inputs $\mathbf{x}^{\langle t\rangle}$ represent words of a sentence (coded as a *one-hot* vector as in Fig. 12.12, for example), and the response $\boldsymbol{y}^{\langle t\rangle}$—is a binary variable that indicates if a certain element of a sequence $(\mathbf{x}^{\langle 1\rangle}, \mathbf{x}^{\langle 2\rangle}, ..., \mathbf{x}^{\langle T_x\rangle})$ is a labeled word (e.g., a name of a person or company). For such application, setting the equal lengths $T_x = T_y$ of sequences $\left(\mathbf{x}^{\langle t\rangle}\right)_{t=1}^{T_x}$ and $\left(\boldsymbol{y}^{\langle t\rangle}\right)_{t=1}^{T_y}$ is very reasonable. However, this is by no means the only possible option, and other network architectures will be described as well. Computing the value of the cost function is fairly straightforward. First, consider a binary response variable $y^{\langle t\rangle} \in \{0, 1\}$. Then, similar to (12.26), for a sequence of length T (which represents a single training example for our RNN) and the sigmoid output, one can use the cost function

$$\mathcal{C}_T = -\sum_{t=1}^{T} \left[y^{\langle t\rangle} \log\left(\hat{y}^{\langle t\rangle}\right) + \left(1 - y^{\langle t\rangle}\right) \log\left(1 - \hat{y}^{\langle t\rangle}\right) \right]. \tag{12.43}$$

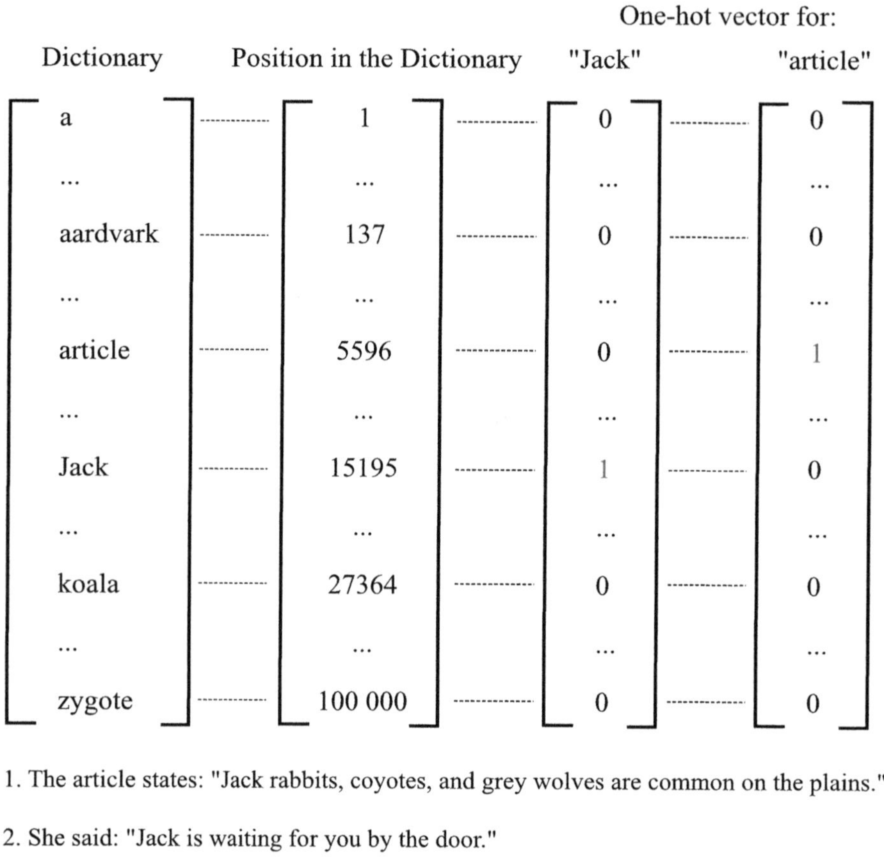

Figure 12.12.: For the above two sentences (training examples), the words “Jack” and “article” are represented as one-hot vectors of length 10^5. To do that, we use the word's position in a 10^5-element Dictionary. An RNN model can be used to recognize “Jack” as a person's name in the second sentence

For a classification problem with $K > 2$ classes, the response variable is a one-hot vector $\boldsymbol{y}^{\langle t\rangle} = (y_1^{\langle t\rangle}, y_2^{\langle t\rangle}, ..., y_K^{\langle t\rangle})$ with $y_k^{\langle t\rangle} = 1$ for some $k \in \{1, 2, 3, ..., K\}$ and $y_j^{\langle t\rangle} = 0$ for all $j \neq k$. For m training examples, referring to Sect. 11.5.3, one can see that maximal likelihood takes the form (11.73). Consequently, for a sequence of length T and the softmax output, the cost function takes the form

$$C_T = -\sum_{t=1}^{T}\sum_{j=1}^{K}\left[y_j^{\langle t\rangle}\left(\zeta_j^{\langle t\rangle} - \log\sum_{k=1}^{K} e^{\zeta_k^{\langle t\rangle}}\right)\right], \tag{12.44}$$

where $\zeta_i^{\langle t\rangle} = b_i^y + \sum_j W_{ij}^y h_j^{\langle t\rangle}$ are the elements of vector $\boldsymbol{\zeta}^{\langle t\rangle}$ from the Eq. (12.42).

12.2.2. Learning in Recurrent Networks

The recursive process of computing gradient descent parameter updates for recurrent networks is often called *back-propagation through time.* We demonstrate the approach for a more general classification problem with $K > 2$ classes. Note that from (12.44), the cost function $\mathcal{C}_T$ can be presented as

$$\mathcal{C}_T = \sum_{t=1}^{T} \mathcal{C}_T^{\langle t \rangle}, \tag{12.45}$$

where for $t = 1, 2, 3, ..., T$

$$\mathcal{C}_T^{\langle t \rangle} = -\sum_{j=1}^{K} \left[y_j^{\langle t \rangle} \left(\varsigma_j^{\langle t \rangle} - \log \sum_{k=1}^{K} e^{\varsigma_k^{\langle t \rangle}} \right) \right]. \tag{12.46}$$

Further, we have for $j = 1, 2, 3, ..., K$ and $t = 1, 2, 3, ..., T$

$$\frac{\partial \mathcal{C}_T}{\partial \varsigma_j^{\langle t \rangle}} = \frac{\partial \mathcal{C}_T}{\partial \mathcal{C}_T^{\langle t \rangle}} \frac{\partial \mathcal{C}_T^{\langle t \rangle}}{\partial \varsigma_j^{\langle t \rangle}} = \frac{\partial \mathcal{C}_T^{\langle t \rangle}}{\partial \varsigma_j^{\langle t \rangle}} = -\left(y_j^{\langle t \rangle} - \frac{e^{\varsigma_j^{\langle t \rangle}} \sum_{j=1}^{K} y_j^{\langle t \rangle}}{\sum_{k=1}^{K} e^{\varsigma_k^{\langle t \rangle}}} \right). \tag{12.47}$$

We can rewrite the last expression in (12.47) as

$$\frac{\partial \mathcal{C}_T}{\partial \varsigma_j^{\langle t \rangle}} = \hat{y}_j^{\langle t \rangle} - \mathbf{1}_{[y_j^{\langle t \rangle}]}, \tag{12.48}$$

where the function $\mathbf{1}_{[y_j^{\langle t \rangle}]}$ returns the j-th element of a vector $\boldsymbol{y}^{\langle t \rangle} = (y_1^{\langle t \rangle}, y_2^{\langle t \rangle}, ..., y_K^{\langle t \rangle})$. Thus, the gradient $\nabla \mathcal{C}_T(\boldsymbol{\varsigma}^{\langle t \rangle})$ is a $K \times 1$ vector with its elements given by (12.48). The gradient with respect to a hidden layer at time T is

$$\nabla \mathcal{C}_T(\boldsymbol{h}^{\langle T \rangle}) = \left[W^y \right]^\top \times \left(\nabla \mathcal{C}_T(\boldsymbol{\varsigma}^{\langle T \rangle}) \right). \tag{12.49}$$

For all other values $t = T-1, T-2, ..., 1$, we have (using the chain rule for differentiation)

$$\nabla \mathcal{C}_T(\boldsymbol{h}^{\langle t \rangle}) = \left[\frac{\partial \boldsymbol{h}^{\langle t+1 \rangle}}{\partial \boldsymbol{h}^{\langle t \rangle}} \right]^\top \times \left(\nabla \mathcal{C}_T(\boldsymbol{h}^{\langle t+1 \rangle}) \right) + \left[\frac{\partial \boldsymbol{\varsigma}^{\langle t \rangle}}{\partial \boldsymbol{h}^{\langle t \rangle}} \right]^\top \times \left(\nabla \mathcal{C}_T(\boldsymbol{\varsigma}^{\langle t \rangle}) \right). \tag{12.50}$$

We can rewrite the right-hand side of (12.50) as

$$\nabla \mathcal{C}_T(\boldsymbol{h}^{\langle t \rangle}) = \left[W^h \right]^\top \times \left(\nabla \mathcal{C}_T(\boldsymbol{h}^{\langle t+1 \rangle}) \right) \times \mathbf{J}_{\boldsymbol{z}}^{\langle t+1 \rangle}(\boldsymbol{h}) + \left[W^y \right]^\top \times \left(\nabla \mathcal{C}_T(\boldsymbol{\varsigma}^{\langle t \rangle}) \right). \tag{12.51}$$

where

$$\mathbf{J}_{\boldsymbol{z}}^{\langle t \rangle}(\boldsymbol{h}) = \operatorname{diag}\left[1 - \left(\boldsymbol{h}^{\langle t \rangle} \right)^2 \right], \tag{12.52}$$

assuming that $g^{\langle t\rangle} = \tanh, \forall t$. We can now write the gradient descent parameter updates

$$\nabla \mathcal{C}_T(W^y) = \sum_{t=1}^{T} \left(\nabla \mathcal{C}_T(\boldsymbol{\varsigma}^{\langle t\rangle})\right) \times \left(\boldsymbol{h}^{\langle t\rangle}\right)^{\top}, \tag{12.53}$$

$$\nabla \mathcal{C}_T(\boldsymbol{b}^y) = \sum_{t=1}^{T} \nabla \mathcal{C}_T(\boldsymbol{\varsigma}^{\langle t\rangle}), \tag{12.54}$$

$$\nabla \mathcal{C}_T(W^h) = \sum_{t=1}^{T} \operatorname{diag}\left[1 - \left(\boldsymbol{h}^{\langle t\rangle}\right)^2\right] \times \left(\nabla \mathcal{C}_T(\boldsymbol{h}^{\langle t\rangle})\right) \times \left(\boldsymbol{h}^{\langle t-1\rangle}\right)^{\top}, \tag{12.55}$$

$$\nabla \mathcal{C}_T(\boldsymbol{b}^h) = \sum_{t=1}^{T} \operatorname{diag}\left[1 - \left(\boldsymbol{h}^{\langle t\rangle}\right)^2\right] \times \left(\nabla \mathcal{C}_T(\boldsymbol{h}^{\langle t\rangle})\right), \tag{12.56}$$

$$\nabla \mathcal{C}_T(W^x) = \sum_{t=1}^{T} \operatorname{diag}\left[1 - \left(\boldsymbol{h}^{\langle t\rangle}\right)^2\right] \times \left(\nabla \mathcal{C}_T(\boldsymbol{h}^{\langle t\rangle})\right) \times \left(\mathbf{x}^{\langle t\rangle}\right)^{\top}. \tag{12.57}$$

Note that there are many different possible models, but the back-propagation algorithm works in the same manner for most of them.

12.2.3. Most Common Design Structures of RNNs

We now describe several basic patterns upon which the majority of contemporary RNNs are built. By the type of input and output data, the classification can be done as follows:

- Recurrent networks that accept a single input and produce a multiple output, i.e., produce $\hat{y}^{\langle t\rangle}$ for every time step (e.g., some text or music generators)
- Recurrent networks that accept a multiple input and produce a single output (e.g., sentiment classifiers, encoders for translator systems)
- Recurrent networks that accept a multiple input and produce a multiple output (e.g., speech recognition, word-search for a person's name)

By the type of connections between units of different (neighboring) Time steps, one could distinguish

- Recurrent networks that have connections between hidden units of different time steps
- Recurrent networks that have connections between output units of a time step to hidden units of the next time step
- Recurrent networks that have connections between hidden units of different time steps and for which the output of a time step serves as an input for the next time step

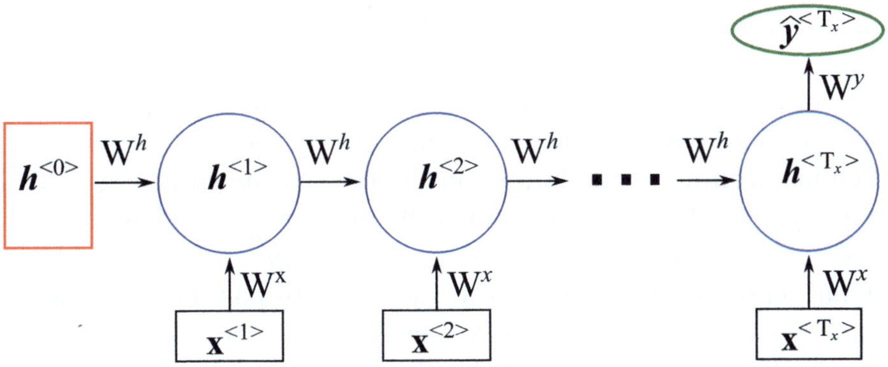

Figure 12.13.: Sentiment classification RNN. Inputs $x^{\langle t \rangle}$ are one-hot vectors coded as shown in Fig. 12.12. A single output $\hat{\boldsymbol{y}}^{\langle T_x \rangle}$ is a softmax probability vector

Combining these patterns one can obtain a recurrent neural network for many areas of applications such as natural language processing, music analysis and generation, sentiment classification, image analysis, DNA sequence analysis, video sequence recognition, word search, and many more. For example, the structure described earlier (Fig. 12.11) is a combination of many-to-many pattern with connections between hidden units at different time steps.

To give an example of an RNN with multiple inputs and a single output, suppose we would like to build a model for a sentiment classification. Such a model would use a user review text as an input and must output a single value (e.g., number of stars) that can be assigned to a product by this review. A construction similar to (12.42) can be used for this purpose. The difference consists in replacing outputs $\hat{\boldsymbol{y}}^{\langle t \rangle} = f^{\langle t \rangle}\left(\zeta^{\langle t \rangle}\right)$ for $t = 1, 2, 3, ..., T_x$ with a single output $\hat{\boldsymbol{y}}^{\langle T_x \rangle} = f^{\langle T_x \rangle}\left(\zeta^{\langle T_x \rangle}\right)$ only for the last time step (see Fig. 12.13 for illustration).
Another application of a network shown in Fig. 12.13 is to summarize a sequence and present a fixed-size output, that in turn can be used as an input for another model. This model architecture is commonly used in translator systems. Such a model can be constructed as a combination of two RNNs. The first one, called an *encoder*, accepts a sentence in one language as its multiple inputs $(\mathbf{x}^{\langle t \rangle}), t = 1, 2, 3, ..., T_x$. It then produces the output $\boldsymbol{\mathcal{Z}}$ in a form of a vector (or a sequence of vectors) as a function of $\boldsymbol{h}^{\langle T_x \rangle}$. The second RNN, called a *decoder*, is using the produced output $\boldsymbol{\mathcal{Z}}$, as shown in Figs. 12.16 or 12.14, to generate the desired output $(\boldsymbol{y}^{\langle t \rangle}), t = 1, 2, 3, ..., T_y$. Note that in this model in general $T_x \neq T_y$. A self-reliant architecture for this type of model was first introduced in Sutskever et al. [2014], where it was called a *sequence-to-sequence architecture*. It was later shown in Bahdanau et al. [2015] that a variable length sequence $\boldsymbol{\mathcal{Z}}$ would produce better results compared to a fixed-size vector. This advancement made it possible for $\boldsymbol{\mathcal{Z}}$ to better represent quite long input sequences in machine translation.

A known difficulty with recurrent networks that have connections between hidden units (of neighboring time steps) is a computationally expensive training process caused by the necessity to perform back-propagation through time for the entire network. To overcome this difficulty, for some models, one can employ a construction that only has connections from the output units to the hidden units only, as shown in Fig. 12.15. Such RNN uses the output $\hat{\boldsymbol{y}}^{\langle t \rangle}$ as the only information for the next time step. The hidden unit $\boldsymbol{h}^{\langle t-1 \rangle}$ is connected to the $\boldsymbol{h}^{\langle t \rangle}$ only via the predictions $\hat{\boldsymbol{y}}^{\langle t-1 \rangle}$. One needs to have a very high-dimensional representation for each $\hat{\boldsymbol{y}}^{\langle t \rangle}$ in order to pass enough useful information through time steps. Consequently, an RNN in Fig. 12.15 can represent a smaller set of functions compared to that in Fig. 12.11. However, the advantage of such a network architecture is that each time step can be trained separately, meaning that the gradient updates for each time step can be computed in a parallel mode with other time steps. This approach is enabled by the availability of the desired output $\boldsymbol{y}^{\langle t \rangle}$ for each time step as a part of the training set $(\mathbf{x}^{\langle t \rangle}, \boldsymbol{y}^{\langle t \rangle})$, thus eliminating the need to compute $\hat{\boldsymbol{y}}^{\langle t \rangle}$ first.

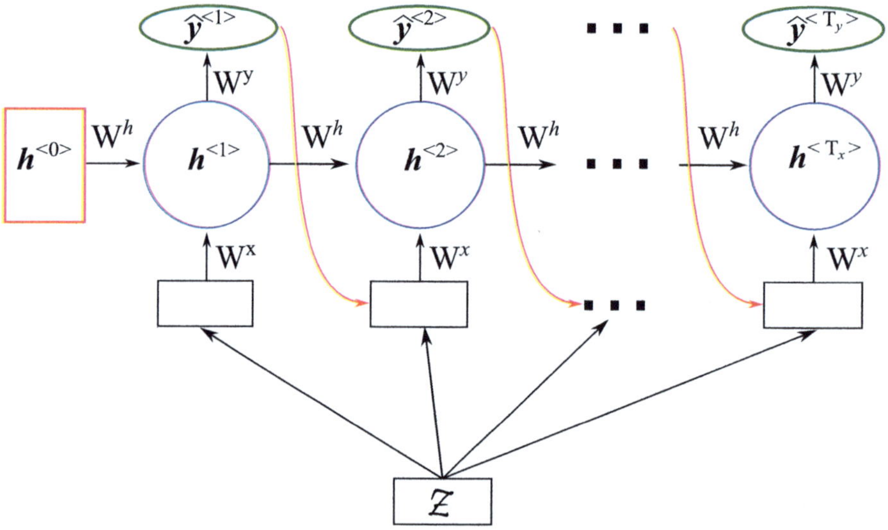

Figure 12.14.: Decoder RNN. Mapping a fixed-length vector (a single input) to input layers at each time step

Using the data forced approach, the network in Fig. 12.15 is passing the actual observed response values $\boldsymbol{y}^{\langle t \rangle}$ (rather than computed $\hat{\boldsymbol{y}}^{\langle t \rangle}$) as inputs for hidden units on the next time step during the training process. A trained network is then deployed on the test set, followed by the real-life data, exactly adhering to the pattern depicted in Fig. 12.15, since the desired output values $\boldsymbol{y}^{\langle t \rangle}$ are either used for control purposes only (test set) or not available (real-life data).

Generating a sequence, such as a musical composition or a text, may be done by combining some of the approaches described above. In this case, we have a single input $\mathbf{x}^{\langle 0 \rangle}$ (which is allowed to be a zero vector), a present connection between hidden units

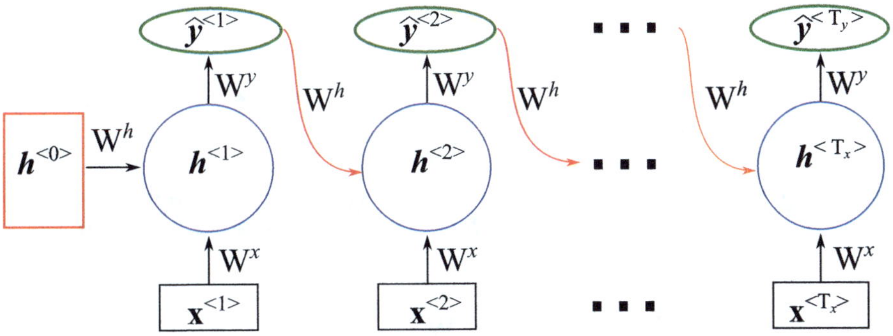

Figure 12.15.: Data forced RNN

$\boldsymbol{h}^{\langle t-1\rangle}$, $\boldsymbol{h}^{\langle t\rangle}$, and in many situations we are passing the output $\hat{\boldsymbol{y}}^{\langle t\rangle}$ directly to the input of the next time step. As in the previous example, such models are trained similarly to data forced RNNs. Namely, we use a large corpus of text in certain language or a music collection of a certain genre as a training set. In addition, a vocabulary is used for vectorization of inputs (as depicted in Fig. 12.12). For a language model, during the training process, we use sentences from the text corpus as training sequences. At each time step, the prediction $\hat{\boldsymbol{y}}^{\langle t\rangle}$ is computed; however, the actual value $\boldsymbol{y}^{\langle t\rangle}$ is used as an input for the next time step. The predictions $\hat{\boldsymbol{y}}^{\langle t\rangle}$ are used in computing the value of a cost function and consequently in computing the gradient updates during the back-propagation through time process. Note that the gradient updates cannot be computed in a parallel mode for such models due to existing connections between hidden units of different time steps. Thus, the back-propagation through time cannot be avoided unlike in data forced RNN as described earlier.

A model trained in such a way can then be deployed as a text generator for language modeling. During the generation process, the outputs $\boldsymbol{y}^{\langle t\rangle}$ are not available; hence, the predicted value $\hat{\boldsymbol{y}}^{\langle t\rangle}$ at each time step is used as an input for the next time step, thus adhering exactly to the pattern in Fig. 12.16.

12.2.4. Deep RNN

In the previous two sections, we introduced an architecture that is only dependent on several separate parameter sets, namely, matrices W^x, W^h, W^y and bias vectors b^h, b^y. These parameters stay unchanged for every input value, hidden state, or output value of our RNN. Thus, all operations described so far correspond to a transformation within a single layer of a multilayer perceptron. A very reasonable question to ask is the following: Would the model benefit from introducing additional layers to its structure? It has been shown (i.e., see Graves et al. [2013]) that the answer is definitely "Yes!" The structure complexity and the computational expense of such models made them only available for the most advance researches in the field for quite some time. However, many modern workflows and packages make these tools available.

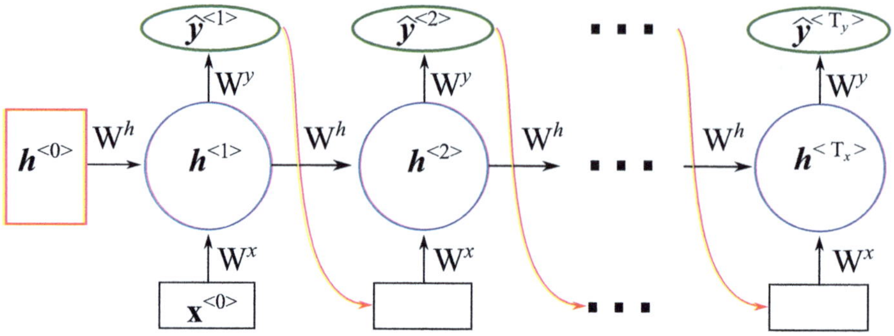

Figure 12.16.: Generator RNN. Input vector $\mathbf{x}^{\langle 0\rangle}$ here could be some initial information, i.e., a music genre or literature style, an output $\boldsymbol{\mathcal{Z}}$ of an encoder RNN, or simply a null vector

We now describe a general structure of a deep recurrent neural network. Let $(\mathbf{x}^{\langle t\rangle}, \boldsymbol{y}^{\langle t\rangle})$ be a data set with $t = 1, 2, 3, ..., T_x$. We shall assume that $T_x = T_y$. Let $(W^{[d]x}), (W^{[d]h})$ be two sets of parameter matrices, and $(\boldsymbol{b}^{[d]h})$—a set of bias parameter vectors for $d = 1, 2, 3, ..., D$. As for a single-layer RNN, we shall require a parameter matrix W^y and a bias parameter vector $\boldsymbol{b}^y$. Activation functions $g^{[d]\langle t\rangle}$ and $f^{\langle t\rangle}$ are defined as before and are as well commonly taken $g^{[d]\langle t\rangle} = g$, $f^{\langle t\rangle} = f$ for all $d = 1, 2, 3, ..., D$ and $t = 1, 2, 3, ..., T_x$. As before, the same parameter sets are used within each hidden layer to enable parameter sharing between different time steps and allow for better generalization for sequences of various lengths. For example, we use $W^{[1]h}, W^{[1]x}, \boldsymbol{b}^{[1]h}$ across all time steps of the first layer. Further, introducing a new set of parameters for the following layer enables our model to represent the output of the previous layer in a more acceptable form, which in turn increases the prediction accuracy. The set of equations for the first hidden layer of a deep RNN is almost identical to what was presented earlier for a single-layer (shallow) RNN, with modification made to the use of the output from hidden units. Namely, for $d = 1$, the equations are

$$\begin{aligned} \boldsymbol{z}^{[1]\langle t\rangle} &= W^{[1]h}\boldsymbol{h}^{[1]\langle t-1\rangle} + W^{[1]x}\mathbf{x}^{\langle t\rangle} + \boldsymbol{b}^{[1]h}, \\ \boldsymbol{h}^{[1]\langle t\rangle} &= g^{[1]\langle t\rangle}\left(\boldsymbol{z}^{[1]\langle t\rangle}\right). \end{aligned} \tag{12.58}$$

The resulting values $\boldsymbol{h}^{[1]\langle t\rangle}$ then serve as inputs for the next layer of deep RNN. This procedure is repeated (i.e., values $\boldsymbol{h}^{[d-1]\langle t\rangle}$ are used as inputs for layer d) until we reach the last hidden layer, at which point the final outputs $\hat{\boldsymbol{y}}^{\langle t\rangle}$ are computed. Hence, for $d = 2, 3, ..., D$, we have

$$\begin{aligned}
\boldsymbol{z}^{[d]\langle t\rangle} &= W^{[d]h}\boldsymbol{h}^{[d]\langle t-1\rangle} + W^{[d]x}\boldsymbol{h}^{[d-1]\langle t\rangle} + \boldsymbol{b}^{[d]h}, \\
\boldsymbol{h}^{[d]\langle t\rangle} &= g^{[d]\langle t\rangle}\left(\boldsymbol{z}^{[d]\langle t\rangle}\right), \\
\boldsymbol{\varsigma}^{\langle t\rangle} &= W^{y}\boldsymbol{h}^{[D]\langle t\rangle} + \boldsymbol{b}^{y}, \\
\hat{\boldsymbol{y}}^{\langle t\rangle} &= f^{\langle t\rangle}\left(\boldsymbol{\varsigma}^{\langle t\rangle}\right).
\end{aligned} \tag{12.59}$$

Fig. 12.17 demonstrates a deep RNN for $D = 3$. As opposed to MLP models, where having a very large number of hidden layers is common, in deep RNN models, large values of D are rarely used in practice. As we have mentioned, even shallow RNNs with connections between hidden states across different time steps are quite computationally expensive due to the necessity of performing the back-propagation through time process. It comes as no surprise that adding more hidden layers with the same number of time steps at each layer would even be a more resource-demanding process as we now have to add a back-propagation through all the layers of deep RNN.

12.2.5. Bidirectional RNN

In many applications, using the information about the "future" proves to be quite beneficial for predicting $\boldsymbol{y}^{\langle t\rangle}$ at each time step t. More formally, we would like to use elements $\mathbf{x}^{\langle t+1\rangle}, \mathbf{x}^{\langle t+2\rangle}, ...$ together with elements $\mathbf{x}^{\langle 1\rangle}, ..., \mathbf{x}^{\langle t-1\rangle}, \mathbf{x}^{\langle t\rangle}$ for computing the approximation $\hat{\boldsymbol{y}}^{\langle t\rangle}$ of $\boldsymbol{y}^{\langle t\rangle}$ at the time step t. Consider, for example, the two sentences in Fig. 12.12.

1. The article states: "Jack rabbits, coyotes, and grey wolves are common on the plains."

2. She said: "Jack is waiting for you by the door."

Clearly, words that follow "Jack" in each sentence present valuable information for a model that recognizes a person's name. Hence, it would be in our best interest if the model built is capable of using this valuable information. However, as it is easy to see, none of the earlier described constructions is suitable. To address this challenge, a model called bidirectional recurrent neural network (BRNN) was introduced in Schuster and Paliwal [1997]. It consists of a combination of two copies of RNN. Both copies are similar to the model shown in Fig. 12.11, with the observed sequence $(\mathbf{x}^{\langle t\rangle})$ used as an input for both copies. An important innovation is that the order of the sequence $(\mathbf{x}^{\langle t\rangle})$ is reversed for one of the copies, and both copies are connected to the same output layer. This way, the output unit $\hat{\boldsymbol{y}}^{\langle t\rangle}$ is computed based on $\mathbf{x}^{\langle t\rangle}$, given both, $\mathbf{x}^{\langle 1\rangle}, ..., \mathbf{x}^{\langle t-1\rangle}$ and $\mathbf{x}^{\langle T_x\rangle}, ..., \mathbf{x}^{\langle t+1\rangle}$ at each time step. One must introduce a separate parameter set for each copy of RNN. We use $W_F^x, W_F^h, W_F^y, \boldsymbol{b}_F^h, \boldsymbol{b}_F^y$ for the copy that accepts a sequence $(\mathbf{x}^{\langle t\rangle})$ in the "direct" order (from 1 to T_x) and $W_R^x, W_R^h, W_R^y, \boldsymbol{b}_R^h, \boldsymbol{b}_R^y$ for the copy that accepts $(\mathbf{x}^{\langle t\rangle})$ in the "reversed" order (from T_x to 1). As before, the same parameter set is used through different time steps of the same hidden layer. No information is transmitted

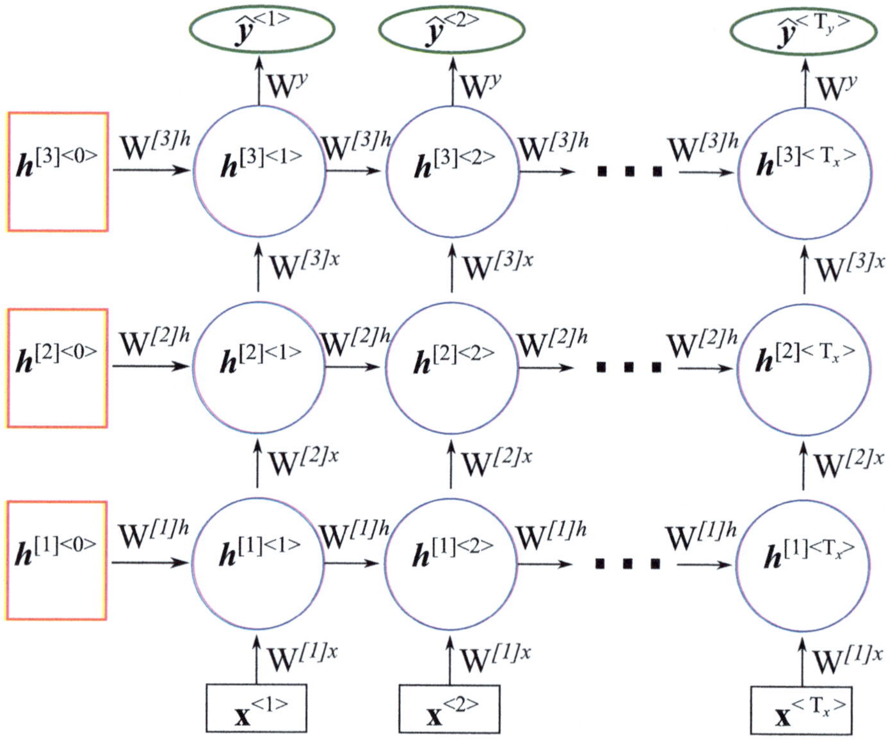

Figure 12.17.: Deep Rnn with three hidden layers

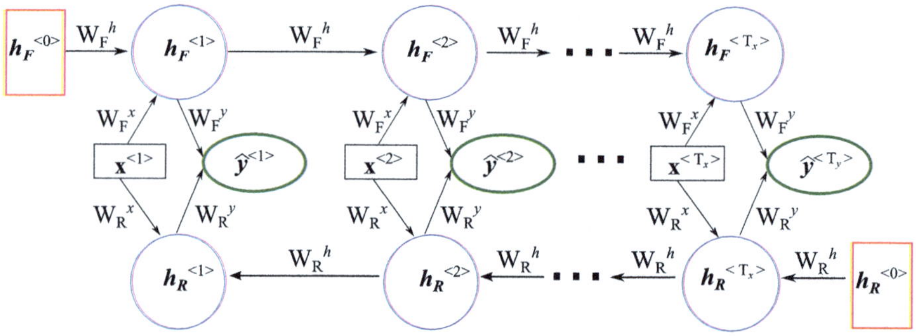

Figure 12.18.: Bidirectional recurrent neural network. In this model, a separate Parameter sets must be used for each of the two RNN copies to transmit information through different time steps

between hidden units of the "direct" and the "reversed" copy.

At each iteration, a model depicted in Fig. 12.18 performs a forward and a backward path. During the forward path:

- BRNN performs computations given by the first two lines of the equation set (12.42) for both the "direct" and the "reversed" RNNs, storing values $\boldsymbol{h}_F^{\langle t\rangle}$ and $\boldsymbol{h}_R^{\langle t\rangle}$ obtained at each time step.
- BRNN computes the values $\hat{\boldsymbol{y}}^{\langle t\rangle}$ simultaneously for all time steps according to the

last two lines of the equation set (12.42) as

$$\begin{aligned}\boldsymbol{\varsigma}^{\langle t\rangle} &= \left(W_F^y \boldsymbol{h}_F^{\langle t\rangle} + \boldsymbol{b}_F^y\right) + \left(W_R^y \boldsymbol{h}_R^{\langle t\rangle} + \boldsymbol{b}_R^y\right) \\ \hat{\boldsymbol{y}}^{\langle t\rangle} &= f^{\langle t\rangle}\left(\boldsymbol{\varsigma}^{\langle t\rangle}\right)\end{aligned} \tag{12.60}$$

During the backward path, the back-propagation through time is performed as follows:

- Compute and save the gradient parameter updates

$$\nabla\mathcal{C}_T(W_F^y), \nabla\mathcal{C}_T(\boldsymbol{b}_F^y), \nabla\mathcal{C}_T(W_R^y), \nabla\mathcal{C}_T(\boldsymbol{b}_R^y)$$

  for all time steps, using (12.53) and (12.54),

- Computing the gradient parameter updates

$$\nabla\mathcal{C}_T(W_F^h), \nabla\mathcal{C}_T(\boldsymbol{b}_F^h), \nabla\mathcal{C}_T(W_R^h), \nabla\mathcal{C}_T(\boldsymbol{b}_R^h), \nabla\mathcal{C}_T(W_F^x), \nabla\mathcal{C}_T(W_R^x)$$

  using formulas (12.55), (12.56), (12.57) and saved values from step one.

The model described above clearly needs an entire input sequence to be available before one can start using such BRNN. Therefore, however powerful, this model may have limited use for tasks where obtaining an information about the "future" is not feasible. An example of such task could be a stock market value prediction. However, in many other tasks, such as protein structure prediction or image processing, input sequences are available beforehand, allowing for the use of BRNNs. In addition, several techniques, such as natural division of speech into phrases and texts into sentences, enable the use of BRNNs leading to significantly improved performance of such models compared to unidirectional RNNs.

12.2.6. Long-Term Dependencies and LSTM RNN

An important advantage of RNNs is their ability to preserve and use the dependencies between the elements of an input sequence $(\mathbf{x}^{\langle t\rangle})$ in order to produce a better prediction. As seen from (12.42), the forward path of an RNN contains a repeated multiplication by the same parameter matrices within hidden units through consecutive time steps. This operation, for an input $\mathbf{x}^{\langle i\rangle}$ at the Time step $i+k$, can be viewed as equivalent to multiplication by the kth power of a parameter matrix $(W^x)^k\mathbf{x}^{\langle i\rangle}$. For a large enough k, the influence of an input $\mathbf{x}^{\langle i\rangle}$ can either become negligible or increase to uncontrollable level (*explode*), making it increasingly difficult to preserve long-term dependencies within the input sequence or leading to instability of the optimization (training) process. This is referred to as *vanishing/exploding gradient problem* [Hochreiter and Schmidhuber, 1997]. The issue mostly arises during the back-propagation through time, since no activation functions are used in this procedure. Gradients explode if the largest eigenvalue of W is larger than 1 and vanish if the largest eigenvalue is smaller than 1 [Hochreiter and Schmidhuber, 1997; Hochreiter et al., 2001]. In the forward pass, a nonlinear activation function is applied at each step, mapping the values to $[-1, 1]$ or $[0, 1]$, thus preventing values from an extreme growth. Due to the linearity of the backwards pass, there is nothing to prevent the derivatives from exploding.

Perhaps the easiest way to demonstrate long-term dependencies on paper is presented by language modeling. Let us consider a simple example.

'After reading a blog post on writing long sentences, when ***she*** *finally understood the difference between left branching and right branching sentences, and how complicated it is to read a sentence where the core is broken up,* ***her*** *writing abilities improved greatly.'*

In such a sentence, in order to correctly predict the word "her," we need to preserve the information on the word "she" as it defines a gender of the person mentioned. However, it is unlikely that a simple RNN will preserve this information, given the fact that there are 26 other words that separate "she" and "her." This means that the forward path of an RNN will have multiplication by a parameter matrix $W$of order 26 between inputs corresponding to these words. Furthermore, the back-propagation through time between these inputs will involve 26 multiplications of Jacobians as a result of differentiation of the cost function. Consequently, longer-term dependencies will be associated either with much smaller or much greater weights in the trained model.

We shall now describe one of the most effective ways to address the issue of preserving long-term dependencies, namely, the long shirt-term memory (LSTM) RNNs which was originally presented in Hochreiter and Schmidhuber [1997] and has been surpassed only several years ago by the transformer RNNs (). The general structure of an LSTM RNN is unchanged and can still be demonstrated by Fig. 12.11. Several important changes are made within each hidden unit in order to create paths through time with derivatives that neither exhibit an extreme growth nor vanish. The changes include using additional parameter sets and equations that allow for passing information through a much longer path.

The basic LSTM *block* has somewhat more complicated structure compared to what was seen before. Figure 12.19 provides an illustration for this basic unit. As in ordinary RNNs, every hidden layer of an LSTM RNN is composed of many such blocks. One of the main elements in an LSTM block is a *block-memory-state* (or block-state) value $B^{\langle t \rangle}$ that has an ability to preserve information over multiple time steps. It has a linear activation function and a reference to the previous block-state $B^{\langle t-1 \rangle}$ (controlled by the *forget gate* $G_f^{\langle t \rangle}$) and the *recall input* $\tilde{R}^{\langle t \rangle}$ (controlled by the *update gate* $G_u^{\langle t \rangle}$).

The forget and the update gates, as all other elements of a block, have their own sets of parameters which are used to compute a linear function of inputs. The sigmoid function (12.9) is then applied in order to keep outputs between 0 and 1. The input $\boldsymbol{h}^{\langle t-1\rangle}$ from the previous block is used in computation of all elements of the current LSTM block. In particular, it helps to compute the value of the *output gate* $G_o^{\langle t\rangle}$, which is identical in structure to other gates, again with its own set of parameters.

Recall that one of the main tasks for each hidden unit of an RNN is to compute the value $\boldsymbol{h}^{\langle t\rangle}$, using $\boldsymbol{h}^{\langle t-1\rangle}, \mathbf{x}^{\langle t\rangle}$ in the first two lines of the equation set (12.42). For an LSTM RNN model, keeping all the connections between hidden units, we replace each hidden unit with an LSTM block. In each such block, we similarly use the values $\boldsymbol{h}^{\langle t-1\rangle}, \mathbf{x}^{\langle t\rangle}$ to compute the value of an additional recall input $\tilde{R}^{\langle t\rangle}$ as well as the three gates, the update gate $G_u^{\langle t\rangle}$, the forget gate $G_f^{\langle t\rangle}$, and the output gate $G_o^{\langle t\rangle}$.

$$\begin{aligned}
\tilde{R}^{\langle t\rangle} &= g_{\tilde{R}}^{\langle t\rangle}\Big(W_{\tilde{R}}^{h}\boldsymbol{h}^{\langle t-1\rangle} + W_{\tilde{R}}^{x}\mathbf{x}^{\langle t\rangle} + \boldsymbol{b}_{\tilde{R}}^{h}\Big), \\
G_u^{\langle t\rangle} &= \sigma\Big(W_u^{h}\boldsymbol{h}^{\langle t-1\rangle} + W_u^{x}\mathbf{x}^{\langle t\rangle} + \boldsymbol{b}_u^{h}\Big), \\
G_f^{\langle t\rangle} &= \sigma\Big(W_f^{h}\boldsymbol{h}^{\langle t-1\rangle} + W_f^{x}\mathbf{x}^{\langle t\rangle} + \boldsymbol{b}_f^{h}\Big), \\
G_o^{\langle t\rangle} &= \sigma\Big(W_o^{h}\boldsymbol{h}^{\langle t-1\rangle} + W_o^{x}\mathbf{x}^{\langle t\rangle} + \boldsymbol{b}_o^{h}\Big).
\end{aligned} \tag{12.61}$$

Note that the activation function $g_{\tilde{R}}^{\langle t\rangle}$ in the first line of (12.61) is not necessarily a sigmoid. In fact using the hyperbolic tangent function (12.12) is a common choice for $g_{\tilde{R}}^{\langle t\rangle}$. The current *block-state* value $B^{\langle t\rangle}$ is then derived combining the current recall unit with the value passed from the previous block of RNN and the current gate values. Below, we denote the element-wise multiplication of vectors or matrices by "$*$".

$$B^{\langle t\rangle} = G_u^{\langle t\rangle} * \tilde{R}^{\langle t\rangle} + G_f^{\langle t\rangle} * B^{\langle t-1\rangle}. \tag{12.62}$$

Finally, the output value $\boldsymbol{h}^{\langle t\rangle}$ for current block is computed using the current block-state value $B^{\langle t\rangle}$ and output gate $G_o^{\langle t\rangle}$

$$\boldsymbol{h}^{\langle t\rangle} = G_o^{\langle t\rangle} * g_o^{\langle t\rangle}(B^{\langle t\rangle}), \tag{12.63}$$

where $g_o^{\langle t\rangle}$ is an activation function, again not necessarily a sigmoid. A predicted output for the response variable is computed at each time step using the last line of Eq. (12.42)

$$\hat{\boldsymbol{y}}^{\langle t\rangle} = f^{\langle t\rangle}\Big(W^{y}\boldsymbol{h}^{\langle t\rangle} + \boldsymbol{b}^{y}\Big). \tag{12.64}$$

Fig. 12.19 demonstrates a single block replacing each hidden unit of an RNN. As seen from an equation set (12.61), each such block at time t has a block-state input $B^{\langle t-1\rangle}$ from the previous time step, in addition to $\boldsymbol{h}^{\langle t-1\rangle}$. Connecting such block for time

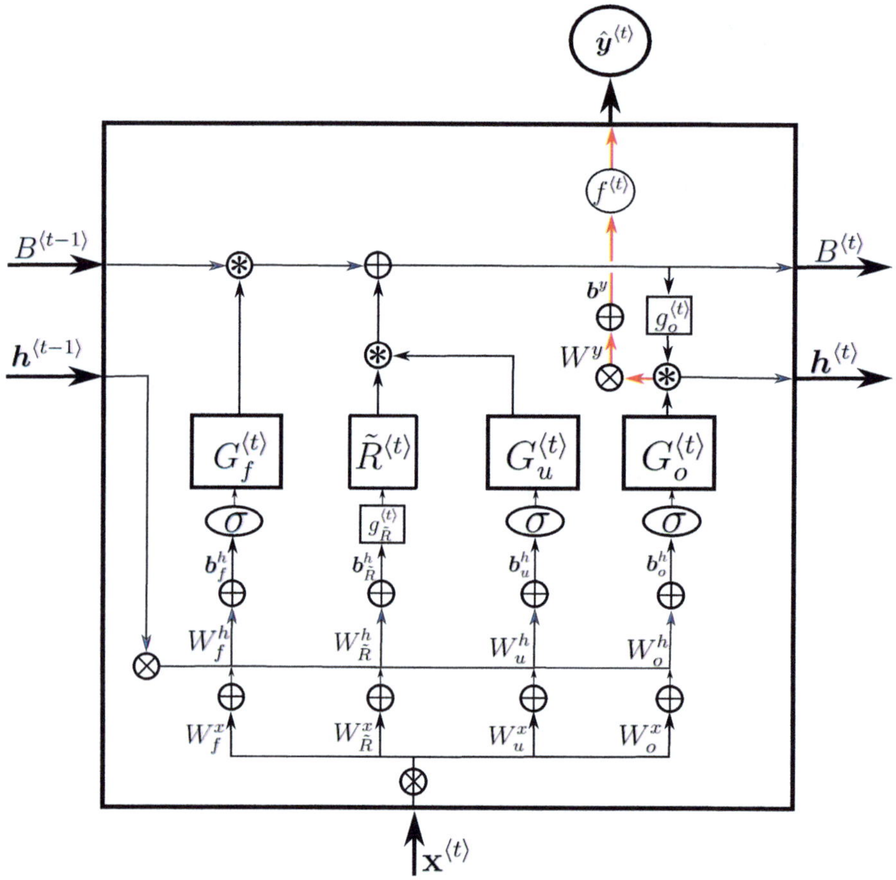

Figure 12.19.: LSTM block replacing each hidden unit of RNN at time t (for $t = 1, 2, 3, ..., T_x$), when building an LSTM RNN

steps $1, 2, 3, ..., T_x$, as shown in Fig. 12.11, one obtains a (shallow) LSTM RNN.

Note that the LSTM blocks shown in Fig. 12.19 can be also used in models depicted by Figs. 12.13, 12.14, 12.15, 12.16, as well as in deep RNNs (Fig. 12.17), bidirectional RNNs (Fig. 12.18), and many other models.

12.2.7. Reduction for Exploding Gradients

As has been mentioned earlier, in addition to vanishing gradients, another existing issue for RNN models is the possibility of a sharp increase of values for some model parameters, referred to as exploding gradients, which leads to model instability or even algorithm failure due to numerical overflow error. Although as damaging as the vanishing gradients problem, the exploding gradient issue is easier to overcome.

To gain some more intuition on exploding gradients, recall that both the forward path and the backward path of an RNN are iterated functions, i.e., they use their own output as an input for the next iteration $x_{t+1} = f(x_t)$. Such functions may produce extremely complicated surfaces after large number of iterations. Consider, for example, a simple quadratic function $f(x) = x(3 - x)$ and see Fig. 12.20 for the illustration of what kind of output it produces after just a few iterations.

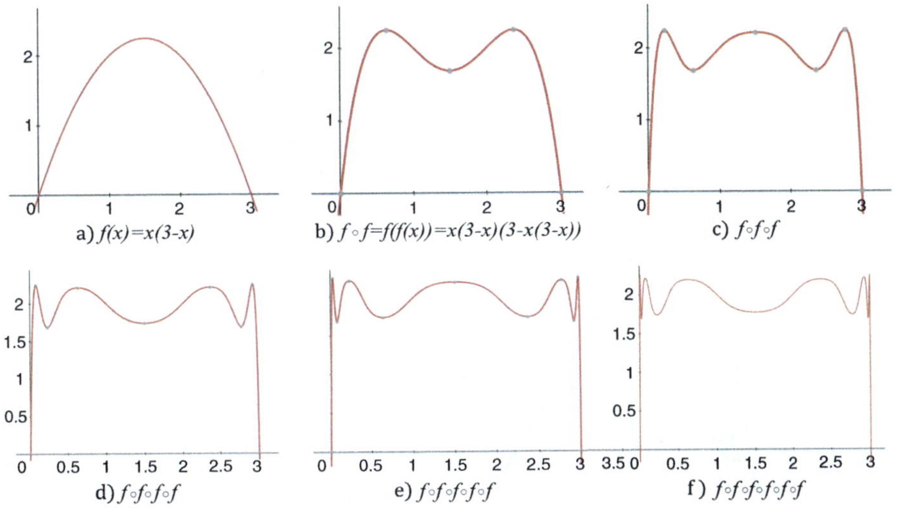

Figure 12.20.: Iterated function. Taking an initial function $f(x) = x(3-x)$, we compute several consecutive compositions, taking the output at a previous step as an input for the next step each time. After just six iterations, the result proves to be a complicated curve

Suppose now we would like to use a gradient descent method to find critical values. Observe the differences in slopes (Fig. 12.21) of tangent lines at different points on an enlarged illustration of curve from Fig. 12.20f. These slopes give a graphical representation of the derivative magnitudes (gradient norms for multivariate case). It is also evident that applying gradient updates of the same magnitude (line segments $1, 2, 3$ in Fig. 12.21), one obtains a very different result. Note that line segments $1, 2,$ and 3 have the same length that represents the magnitude of a gradient update. However, only the line segment 1 connects to the local minimum. Compare that to the line segment 3, which completely misses the entire region of a local minimum. Combining these two observations, one concludes that even with a small learning rate, the gradient obtained as a result of back-propagation through time over a large number of steps (iterations) may give a relatively very large "jump," that will cause missing local minima and destabilizing the network learning process or making it inefficiently long.

A simple yet quite effective approach to this problem was proposed in Pascanu et al. [2013]. In essence it consists of setting up a certain "threshold" $\mathfrak{D}$ for the gradient norm $||\nabla \mathcal{C}||$ and reducing this norm as follows:

$$\text{if } ||\nabla \mathcal{C}|| > \mathfrak{D} \quad \text{then} \quad \nabla \mathcal{C}(W) \leftarrow \frac{\nabla \mathcal{C} \mathfrak{D}}{||\nabla \mathcal{C}||}$$

Because the gradient of weights W and biases $\boldsymbol{b}$ are jointly normalized with a single scaling factor, this method has the advantage of guaranteeing that each step is still in the gradient direction. This method is referred to as the *gradient norm clipping* since it bounds the parameter update vector norm. Such bounded gradient reduces the risk of

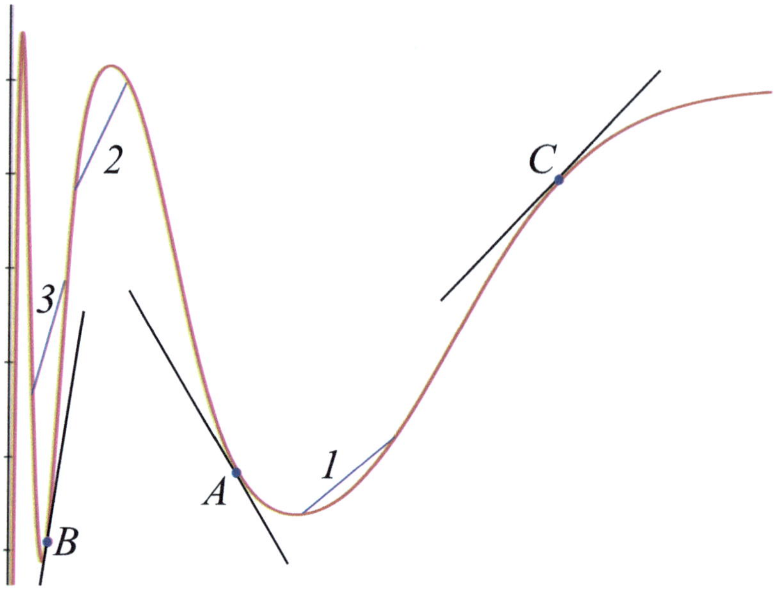

Figure 12.21.: An enlargement of a graph in Fig. 12.20f demonstrates the difference in slopes of tangent lines at different points. The slope of the tangent line at point B is clearly much larger than those at points A and C. These slopes provide a graphical representation of the magnitudes of the derivatives at corresponding points. In addition, the same gradient update applied at different points gives different results. Line segment 1 connects to a local minimum almost perfectly. Line segment 2 misses a local maximum but still within a reasonable distance from it. Line segment 3 completely skips the entire region of a local minimum

making a relatively large, destabilizing step in weight updates even when the gradient explodes.

12.3. Convolution Neural Networks

Vision seems like a very natural task we, as humans, perform every day without really thinking about it. We rarely acknowledge that a multi-element complex system that we are lucky to be equipped with is involved in the process. To get an intuition on why this same process is so difficult for a computer, try to imagine a situation when you are listening to a story told by your friend about their vacation trip to a picturesque place you've never seen before. Without seeing any pictures, it is up to your own imagination, based on the information stored in your memory over your life time, to create a mental image of that place. Notice that other people who are listening to the same story may have a very different image in their minds about the same place, based on their own life experience. Now think what would have happened if your friend told you a name of a flower or an animal you've never seen before. Most definitely, you would need to get some additional description in a form of comparison to other flowers or animals known to you. Or better yet, you would like to see a picture... Instead, your friend is showing you a rectangular sheet of paper filled with numbers...

For a computer, any image is just a numerical array, since computers store information only through digits. Hence, for a computer to learn the similarities or the differences between images, a numerical way of describing those similarities or differences must be presented. There is a vast and active field of research called *computer vision* which deals with developing methods for such numerical representation of images. A noticeable progress in this field has been made due to increasing sophistication of machine learning techniques and the availability of large image collections. Although many applications exist for computer vision models, one that arguably received the most attention and which has brought a lot of the progress in this relatively new field is the task of classifying an image into a set of object categories—*object recognition*.

Recall that the neural networks considered in the prequel were constructed by including layers that computed a linear function followed by a nonlinear activation function. In a very general way, we can denote this as follows:

$$\boldsymbol{h} = g(\Omega \times \mathbf{x}) \tag{12.65}$$

Note that every element in these layers was connected to all elements in neighboring layers leading to a full connection of every element of the input layer to every element of the output layer, as can be seen in Fig. 12.9. For this reason, these are called fully connected layers. Although this model has proven to be quite powerful for many tasks, there are some clear drawbacks.

One drawback is that a network built for a single observation with n predictors in the input variable and k classes in the output variable requires $m \times k$ connections. This may not seem like a large quantity for many data types. However, it grows noticeably faster if the input data is an image. As mentioned before, a computer stores images as numerical data sets. Consider a gray-scale image 1 megapixel in size, that is represented by a 1000×1000 matrix, and we wish to recognize 10 objects in this image. We then need $1000 \times 1000 \times 10 = 10{,}000{,}000$ connections. Keep in mind that a separate computation of a type (12.65) is required for each connection in a network. This makes fully connected networks computationally expensive for processing large images. Moreover, our model must estimate a separate model parameter (weight) for every connection. With such a large number of parameters, in order to reduce the risk of model over-fitting, one must use a very large number of training examples, which again increases the computational expense.

As mentioned before, a fully connected feed-forward network has a separate parameter for each feature, as such they do not share parameters between different parts of a model. In image processing, however, the ability to share parameters is very important as certain structures may be repeatedly seen in different parts of an image. Also, we expect that our model produces the same classification for an image shifted by a few pixels.

In what follows, we discuss the *convolution neural network* (CNN) which is inspired by a mathematical *convolution operator*. These models were first introduced in the early 1980s; yet it is not until 2012 that they've became widely known and truly revolutionized the field of computer vision. The vast majority of papers published in top computer vision conferences use convolution neural networks in some way.

12.3.1. Convolution Operator for Arrays

Here we give a brief introduction of the convolution operator and its properties. For two arrays $\mathbf{x} = (x_0, x_1, x_2, ..., x_n)$ and $\boldsymbol{\omega} = (\omega_0, \omega_1, \omega_2, ..., \omega_m)$, their convolution is an array of length $n + m - 1$ with its elements given by

$$(\mathbf{x} * \omega)_i = \sum_k x_{i-k}\omega_k \tag{12.66}$$

The arrays $\mathbf{x}$ and $\boldsymbol{\omega}$ are usually finite, but it is sometimes convenient to extend the number of elements or treat them as infinite by entering a convention that their elements have values equal to zero everywhere else. This is known as *zero padding.*

For example, consider computing the convolution of arrays $\mathbf{x} = (3, -1, 2)$ and $\boldsymbol{\omega} = (1, 2, 3)$. The result $\mathbf{x} * \boldsymbol{\omega}$ is an array that contains 5 elements which can be computed using (12.67) with zero padding on $\mathbf{x}$, as shown in Fig. 12.22.

$$\begin{aligned}
(\mathbf{x} * \omega)_0 &= x_{0-0}\omega_0 + x_{0-1}\omega_1 + x_{0-2}\omega_2 \\
&= 3 \times 1 + 0 \times 2 + 0 \times 3 = 3 \\
(\mathbf{x} * \omega)_1 &= x_{1-0}\omega_0 + x_{1-1}\omega_1 + x_{1-2}\omega_2 \\
&= (-1) \times 1 + 3 \times 2 + 0 \times 3 = 5 \\
. \quad . &\quad . \\
(\mathbf{x} * \omega)_4 &= x_{4-0}\omega_0 + x_{4-1}\omega_1 + x_{4-2}\omega_2 \\
&= 0 \times 1 + 0 \times 2 + 2 \times 3 = 6
\end{aligned} \tag{12.67}$$

The two-dimensional case is very similar to the one-dimensional case. For two matrices $\mathbf{M}^{\mathbf{x}}$ and Ω, we have

$$[\mathbf{M}^{\mathbf{x}} * \Omega]_{i,j} = \sum_{k,l} \mathbf{M}^{\mathbf{x}}_{i-k,j-l}\Omega_{k,l} \tag{12.68}$$

For example, the convolution of two-dimensional arrays can be done as shown in Fig. 12.23. Using the formula (12.68), and zero padding on $\mathbf{M}^{\mathbf{x}}$ one can, for instance, compute:

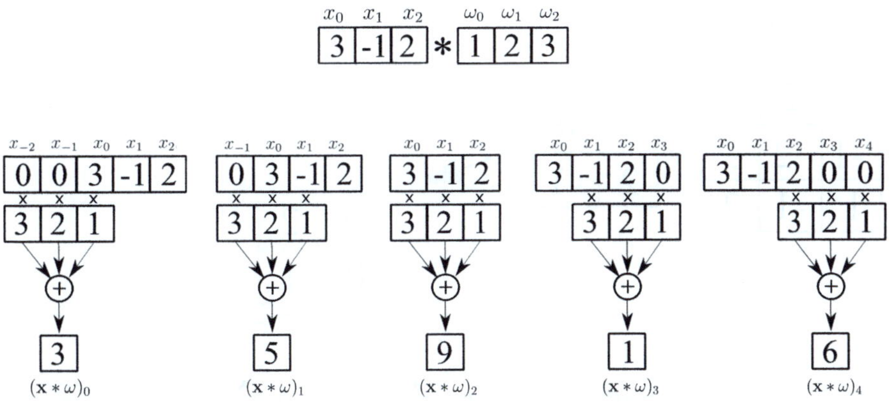

Figure 12.22.: Elements of the convolution $(3, -1, 2) * (1, 2, 3)$. Note that $\boldsymbol{\omega}$ must be reflected horizontally for this graphical representation to illustrate the formula (12.66)

$$
\begin{aligned}
[\mathbf{M^x} * \Omega]_{0,3} &= \mathbf{M}^{\mathbf{x}}_{0-0,3-0}\Omega_{0,0} + \mathbf{M}^{\mathbf{x}}_{0-0,3-1}\Omega_{0,1} + \mathbf{M}^{\mathbf{x}}_{0-1,3-0}\Omega_{1,0} + \mathbf{M}^{\mathbf{x}}_{0-1,3-1}\Omega_{1,1} \\
&= 0 \times 1 + 2 \times 0 + 0 \times 2 + 0 \times (-1) = 0 \\
&\cdots \\
[\mathbf{M^x} * \Omega]_{2,0} &= \mathbf{M}^{\mathbf{x}}_{2-0,0-0}\Omega_{0,0} + \mathbf{M}^{\mathbf{x}}_{2-0,0-1}\Omega_{0,1} + \mathbf{M}^{\mathbf{x}}_{2-1,0-0}\Omega_{1,0} + \mathbf{M}^{\mathbf{x}}_{2-1,0-1}\Omega_{1,1} \\
&= 1 \times 1 + 0 \times 0 + (-1) \times 2 + 0 \times (-1) = -1 \\
&\cdots \\
[\mathbf{M^x} * \Omega]_{3,3} &= \mathbf{M}^{\mathbf{x}}_{3-0,3-0}\Omega_{0,0} + \mathbf{M}^{\mathbf{x}}_{3-0,3-1}\Omega_{0,1} + \mathbf{M}^{\mathbf{x}}_{3-1,3-0}\Omega_{1,0} + \mathbf{M}^{\mathbf{x}}_{3-1,3-1}\Omega_{1,1} \\
&= 0 \times 1 + 0 \times 0 + 0 \times 2 + 2 \times (-1) = -2
\end{aligned}
$$

12.3.1.1. Properties of the Convolution Operator

Let us point out some useful properties of the convolution operator. First, it is commutative and associative.

$$
\begin{aligned}
\boldsymbol{\omega} * \boldsymbol{\beta} &= \boldsymbol{\beta} * \boldsymbol{\omega} \\
(\mathbf{x} * \boldsymbol{\omega}) * \boldsymbol{\beta} &= \mathbf{x} * (\boldsymbol{\omega} * \boldsymbol{\beta})
\end{aligned}
$$

Suppose we wish to add a blur filter Ω^b to the image $\mathbf{M^x}$ and then apply a vertical edge filter Ω^v, represented as

$$(\mathbf{M^x} * \Omega^b) \quad * \quad \Omega^v$$

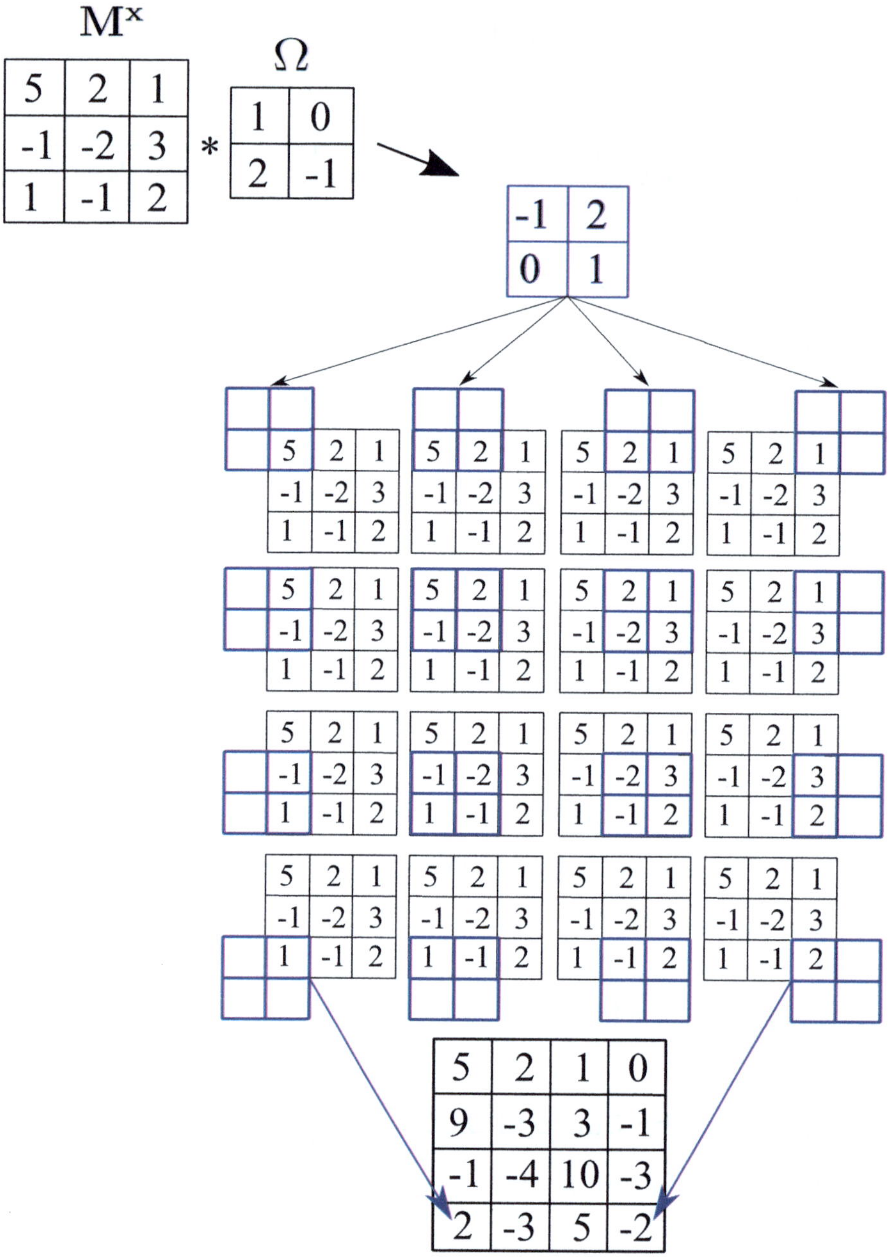

Figure 12.23.: Convolution of two-dimensional arrays. Note that in order for graphical representation to illustrate the formula (12.68), one must perform vertical and horizontal reflection of Ω

By commutativity and associativity, this is equivalent to applying the vertical edge filter and then blurring the result

$$(\mathbf{M}^{\mathrm{x}} * \Omega^{v}) \quad * \quad \Omega^{b}$$

It is also equivalent to convolve the image with a single filter which is obtained by applying the blur to the vertical edge filter

$$\mathbf{M}^{\mathrm{x}} \quad * \quad (\Omega^{v} * \Omega^{b})$$

Another useful property of convolution is that it is linear

$$\begin{aligned}(a\mathbf{x} + b\mathbf{z}) * \boldsymbol{\omega} &= a\mathbf{x} * \boldsymbol{\omega} + b\mathbf{z} * \boldsymbol{\omega} \\ \mathbf{x} * (a\boldsymbol{\omega} + b\boldsymbol{\beta}) &= a\mathbf{x} * \boldsymbol{\omega} + b\mathbf{x} * \boldsymbol{\beta}\end{aligned}$$

Since linear operations are often easier to deal with, it also means that a neural network that consists of convolution computed in sequence can still only compute linear functions. Thus, a nonlinear activation function must be applied in order to deal with most tasks.

A third property of convolution is that it is *equivariant* to translation. Meaning that if the input $\mathbf{x}$ is translated, by some amount, then the output $\mathbf{x}\boldsymbol{\omega}$ is translated by the same amount. This property guarantees that the network's computations behave in a well-defined way for the transformed inputs.

As mentioned above, a sequence of convolutions can only compute a linear function. Convolutions followed by a nonlinear transformation are an even more powerful tool for the detection of a vast amount of image features. We have seen several activation functions before (e.g., (12.8), (12.9), (12.12), (12.13)). An additional very simple but quite often used function is called the *rectified linear unit* function (RELU). It is given by

$$g(z) \quad = \quad \begin{cases} z, & z > 0 \\ 0, & z \leq 0 \end{cases} \tag{12.69}$$

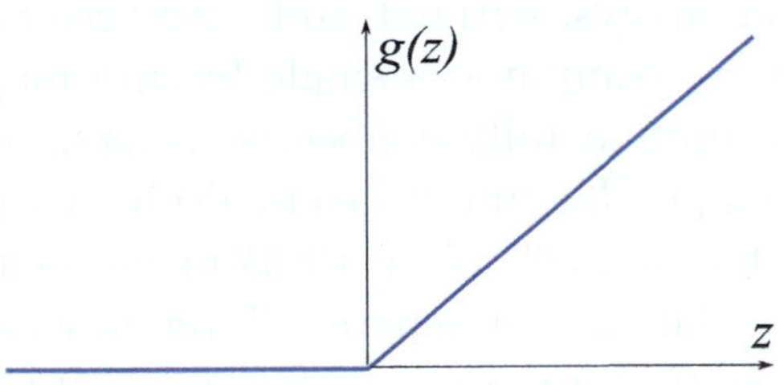

Figure 12.24.: Rectified linear unit (RELU) function

For example, to obtain a map of elements with vertical orientation in an image, one can follow the procedure:

- Compute the convolution with filter for vertical edge detection.
- Apply RELU function.

- Compute the convolution with blur filter.

The resulting output is called a *feature map*, since it can be interpreted as a map of feature strength over parts of an image.

12.3.2. Convolution Layers

A neural network layer that computes convolutions followed by a nonlinear activation function is called a convolution layer. Not uncommonly for machine learning, the operation used in convolution neural networks is not actually a convolution but a closely related operation called *cross-correlation* or *filtering*. For two arrays $\mathbf{x} = (x_0, x_1, x_2, ..., x_n)$ and $\boldsymbol{\omega} = (\omega_0, \omega_1, \omega_2, ..., \omega_m)$, the filtering is defined given by

$$(\mathbf{x} \star \omega)_i = \sum_k x_{i+k}\omega_k. \tag{12.70}$$

The result may have different dimensions, depending on the goals of the researchers. If one chooses to use filtering with no zero padding, the output $\mathbf{x} \star \omega$ has $n - m + 1$ elements, as shown in Fig. 12.25. Note that the filter dimensions are always smaller than the input dimensions. Alternatively, we may wish to have the output $\mathbf{x} \star \omega$ of the same size as an input $\mathbf{x}$. We then apply zero padding of the size $(m-1)/2$ to input $\mathbf{x}$ (on each side). Note that for the latter option, m must be an odd number.

For two-dimensional arrays $\mathbf{M}^{\mathbf{x}}$ and Ω, we have:

$$[\mathbf{M}^{\mathbf{x}} \star \Omega]_{i,j} = \sum_{k,l} \mathbf{M}^{\mathbf{x}}_{i+k,j+l}\Omega_{k,l}. \tag{12.71}$$

We compute dimensions of the output $\mathbf{M}^{\mathbf{x}} \star \Omega$ analogously to the one-dimensional case, applying zero padding if needed. Figs 12.25(Bottom) and 12.26 show examples of filtering for two-dimensional arrays, without and with zero padding correspondingly.

In the above example, we computed a single feature map, but just as we normally use more than one hidden unit in fully connected layers, convolution layers normally compute multiple feature maps. The input layers also consist of multiple feature maps; these could be different color channels of an RGB image or feature maps computed by another convolution layer. There is a separate filter associated with each pair of an input and output feature map. Suppose that for $n \times n$ inputs $\mathbf{M}^{\mathbf{x}}_0, \mathbf{M}^{\mathbf{x}}_1, ..., \mathbf{M}^{\mathbf{x}}_{D-1}$ (for example, $D = 1$ for gray-scale images, or $D = 3$ for RGB images), we wish to compute $\boldsymbol{Z}_0, \boldsymbol{Z}_1, ..., \boldsymbol{Z}_{M-1}$ activations to use in the next layer of our CNN. We then need $m \times m \times D$ filters $\Omega_1, \Omega_2, ..., \Omega_M$, and the activations are computed for $t = 0, 1, 2, ..., M-1$:

$$(\boldsymbol{Z}_{i,j})_t = \sum_{d=0}^{D-1} \sum_{k,l} \mathbf{M}^{\mathbf{x}}_{i+k,j+l,d}(\Omega_{k,l,d})_t \tag{12.72}$$

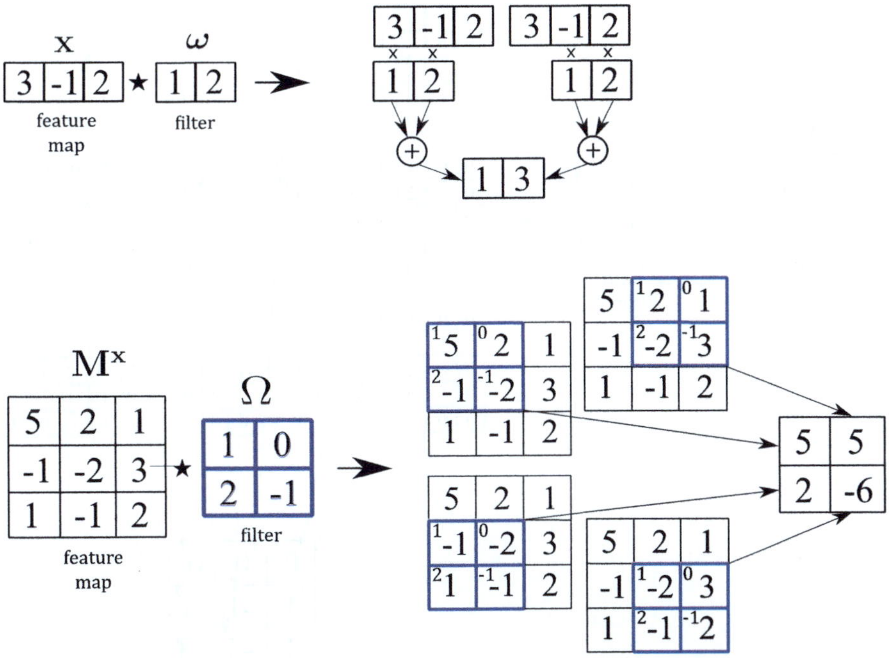

Figure 12.25.: Filtering without zero padding. Top: Example of filtering for one-dimensional arrays using formula (12.70). Filter $\boldsymbol{\omega}$ is applied to the three-element input $\mathbf{x}$. Bottom: Example of filtering for two-dimensional arrays using formula (12.71). Filter Ω applied to the 3×3 input $\mathbf{M}^{\mathbf{x}}$. Both cases were done without zero padding

The activation function is then applied element-wise:

$$\boldsymbol{h}_t = g(\boldsymbol{Z}_t)$$

This procedure is very similar to the one we used for fully connected feed-forward neural networks (e.g., see (12.11)). In fact, setting certain parameters (12.11) to zero and adjusting certain notations, one can obtain a total equivalent. A natural conclusion can be drawn: Any function computed by a convolution layer can be computed by a fully connected layer. An equally natural question is then in order: Why use a more notation-heavy convolution layers if they are not more powerful compared to fully connected layers in terms of representational capacity?

The answer can be given by pointing out two key differences:

- For three-dimensional inputs and filters, we obtain only two-dimensional activations. This means that not every input unit is connected to every output unit. This feature of CNN is referred to as *sparse connectivity.*
- Note that in formula (12.72), the same set of weights (filter) is combined with many different parts of the input array to create the corresponding elements of the output. This ensures that weights are each shared between multiple connections of the network. This feature of CNN is referred to as *weight sharing.*

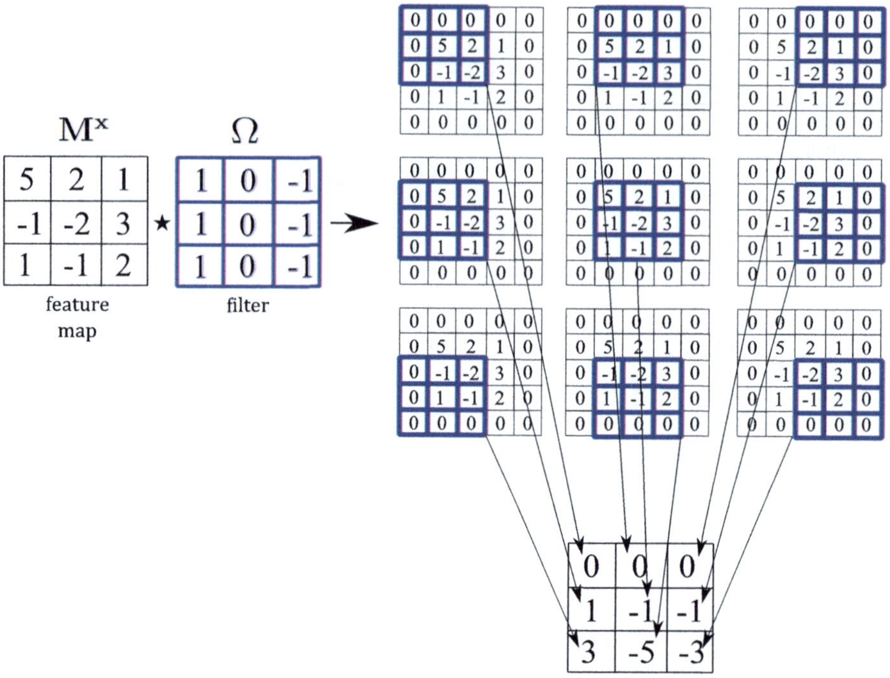

Figure 12.26.: Example of filtering for two-dimensional arrays using formula (12.71). Filter Ω applied to the 3×3 input $\mathbf{M}^{\mathbf{x}}$. In order to have the output dimension the same as the input, zero padding of size $(3-1)/2$ was applied

For example, suppose we wish to use D input inputs (each of size $n \times n$) to compute M outputs (each also of size $n \times n$), and the filters are of size $m \times m$. The number of weights for the convolution layer is

$$m \times m \times D \times M.$$

The number of connections is approximately

$$m \times m \times D \times M \times n \times n.$$

Now, the number of connections created for a fully connected layer with the same set of inputs and outputs is

$$n \times n \times D \times n \times n \times M.$$

This is of course equal to the number of weights that are required to be computed in a fully connected layer. Hence, using the convolution layers allows for the reduction in number of connections by a factor of n^2/m^2 and the number of weights by a factor of n^4/m^2. As mentioned above, the filter dimensions are always smaller than those of the input. Consider using a single 100×100 filter for a feature detection in a 500×500 image, aiming to generate an output of the same size. It is easily computed that using the convolution structure reduces the number of connections in such a model by a factor of 25 and the number of weights by almost a factor of 6.25×10^6!

Another modification to the classical convolution operator, in addition to formulas (12.70) and (12.71), is changing the size of step that a filter takes while "moving through" a feature map. This is referred to as changing the *stride* of the convolution filter. For the chosen stride s, the formula (12.72) becomes

$$(\boldsymbol{Z}_{i,j})_t \quad = \quad \sum_{d=0}^{D-1}\sum_{k,l} \mathbf{M}^{\mathbf{x}}_{s\times i+k,s\times j+l,d}(\Omega_{k,l,d})_t. \tag{12.73}$$

Note that for a filter of linear size m used for a feature detection in an image of linear size n with zero padding of size n_0, the resulting output has linear size

$$\frac{n+2n_0-m}{s}+1, \tag{12.74}$$

which is not necessarily an integer. In this case, a sufficiently large padding size is chosen, and the floor function $\lfloor\cdot\rfloor$ is applied. This function rounds a number down to its integer part.

Due to the modifications described in this section, the resulting linear operator, although still conventionally called convolution, is generally not enjoying the properties of associativity and commutativity of the classical convolution. However, the gain in notation and computational simplicity obtained as a result of such modifications outweighs the loss of those properties.

12.3.3. Pooling Layers

A highly desirable property of the convolution networks is the invariance to small transformations of an input, such as shifting it by a few elements. For example, it is easy to agree that a picture of a vehicle shifted by a few pixels to the left remains a picture of the same vehicle, and we expect our model to support this fact. In order to amplify invariance, another type of convolution layer is used, the *pooling layer.* In image processing, for instance, such layers summarize the feature maps of the previous layer by computing a simple function over small regions of the image. The most common choice for this function is the max$(\cdot)$, in which case the process is called *max pooling.* However, the *average* function could also be used.

For an $n\times n$ inputs $\mathbf{M}^{\mathbf{x}}_0, \mathbf{M}^{\mathbf{x}}_1, ..., \mathbf{M}^{\mathbf{x}}_{D-1}$, in order to compute $\boldsymbol{Z}_0, \boldsymbol{Z}_1, ..., \boldsymbol{Z}_{D-1}$ activations to use in the next layer of our CNN, we use an $m\times m\times D$ filter, so that the max-pooling activations with stride s are computed for $d = 0, 1, 2, ..., D-1$:

$$(\boldsymbol{Z}_{i;j})_d \quad = \quad \max_{0\leq k,l\leq m-1}\left[\mathbf{M}^{\mathbf{x}}_{s\times i+k,s\times j+l,d}\right] \tag{12.75}$$

Note that Ω is absent from the formula (12.75). This is because the "filter" Ω for pooling layers is just a set of hyper-parameters which does not contain any actual model weights. For this reason, the gradient descent or any other chosen optimization algorithm is not applied to pooling layers. Applying formula (12.75) results in the representation being shrunk by a factor of approximately s along each dimension. As before, the exact linear dimensions can be computed using (12.74).

Observe that pooling is increasing the size of the regions of the input called the *receptive fields*, which influence their activations. For instance, consider the network architecture that consists of an input $n \times n$ gray-scale image, followed by the convolution layer, followed by the pooling layer. Let us use a single $m \times m$ filter for each layer and a stride s for the pooling layer. Then each unit in the first convolution layer has a receptive field of size $m \times m$. But the output of the pooling layer guarantees that the next convolution layer has approximately $sm \times sm$ receptive fields, since it does $m \times m$ filtering over a representation which was shrunken by almost a factor of s along each dimension. Hence, pooling allows small filters to account for information over large regions of an image.

12.4. Text Analytics

12.4.1. Introduction

In conclusion of this chapter, we briefly describe the state-of-the-art approach to *natural language processing* (NLP). The *Transformer* model, originally presented in Vaswani et al. [2017] was a major breakthrough in computer-based *natural language understanding* (NLU), which is a subset of NLP. It quickly became an active research field, with dozens of papers published over a relatively short period of time with several improvements proposed. As discussed in Katharopoulos et al. [2020], there exists a relation between transformers and RNNs. In particular, the authors present an improvement to the original transformer model architecture which enables a significantly faster model training process.

Since their introduction, decades ago, LSTM RNNs and CNNs, along with several modifications, became dominant models in computer-based text analytics, or more generally, *sequence transduction*. They far outperform rule-based models and produce truly phenomenal results for several important tasks in modern digitized society, such as language modeling and machine translation, both of which are based on NLU. The main drawback of these sophisticated and complex methods has always been the requirement for significant computing power to perform the network training. Moreover, even provided with such computing power, the model training times were still something to improve on.

Numerous attempts had been made to overcome various difficulties. However, the sequential nature of these models causes difficulty in parallel processing across many training examples, especially for long sequences. In the Vaswani et al. [2017] transformer model, a successful attempt is made at eliminating the model dependence on recurrence and solely relying on attention mechanism [Parikh et al., 2016] to understand the dependence between input and output sequences. This approach allows for greater parallelization across training examples, which in turn reduces model training time needed for achieving state-of-the-art results in machine translation.

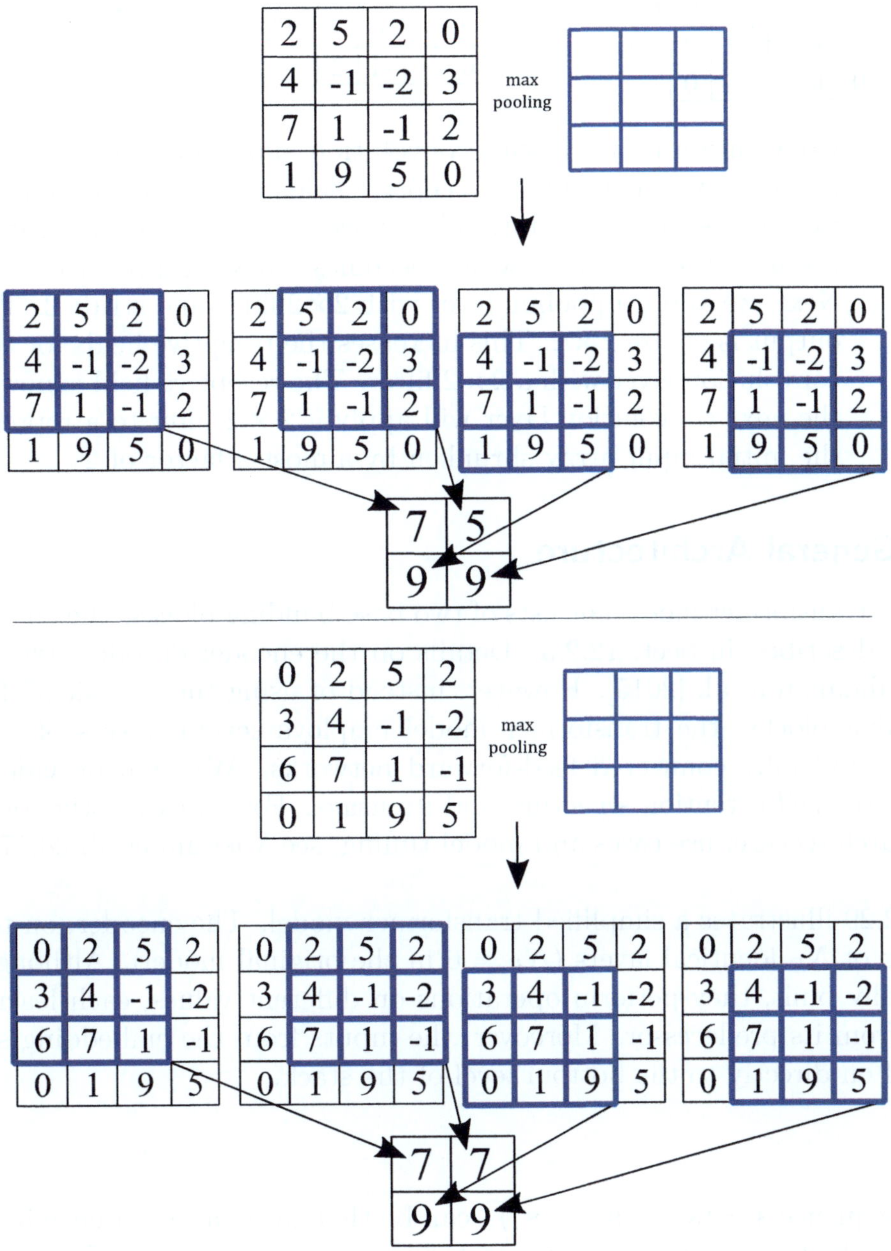

Figure 12.27.: An example of how pooling can provide partial invariance to translations of the input. Top: We apply pooling layer with 3×3 filter and stride 1 to the initial 4×4 input array. Bottom: We shift the initial input array to the right by 1 element and apply pooling with the same filter and stride. Note that three out of four outputs are unchanged since the corresponding maximum values in the input array are still covered by the pooling filter

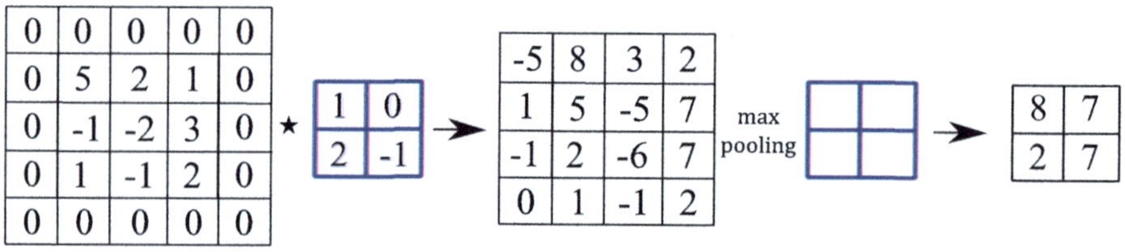

Figure 12.28.: Convolution network with convolution and pooling layers. Observe that pooling layers expand the receptive fields of units in subsequent convolution layers. in a convolution layer, we apply a 2×2 filter with stride 1 to the initial 3×3 array (which becomes a 5×5 array with zero padding). Next, we apply a pooling layer with 2×2 filter and stride 2. The resulting output is of size 2×2. This increases the receptive fields for the following convolution layer with a 2×2 filter. The receptive field is now 4×4, since the next convolution layer will apply a 2×2 filter to a representation of the initial input array shrunken by almost a factor of 2

12.4.2. General Architecture

The original transformer model consists of two base building blocks: the encoder and the decoder, as described in Sect. 12.2.3. Details on the encoder-decoder structure can be found in Bahdanau et al. [2015]. However, instead of using the "classical" RNN within these building blocks, the transformer model employs several layers of *self-attention* mechanisms and fully connected feed-forward networks. We shall provide a brief description of the self-attention structure and its usage. For an exhaustive description of the model architecture, use cases and model tuning; see Vaswani et al. [2017], Rothman [2021].

Fig. 12.29 illustrates a simplified transformer model. The encoder, on top, consists of the stack of N_E identical layers ($N_E = 6$ in the original model). Although all layers are identically built, they perform operations on different values, each learning the information from its predecessor. Moreover, the inputs from the embedding sublayer are only passed on directly to the bottom level of the stack.

Input

The input sequence $\mathbf{s} = (\mathbf{s}_1, \mathbf{s}_2, \mathbf{s}_3, ..., \mathbf{s}_n)^\top$ can be thought of as a sentence in the English language. Each element (word) is represented as a numerical vector of dimension d_{model}, using a learned embedding layer. See, for example, the *word2vec* embedding approach [Mikolov et al., 2013a,b]. The embedding is followed by the *positional encoding,* which allows for adding information about the word position in a sentence to the word embedding. The positional encoding can be done in several ways. The method researchers choose in Vaswani et al. [2017] is the following. Let t be a position of a word $\mathbf{s}_t$ in a sentence $\mathbf{s}$, and let $\mathbf{e}_t = (e_{t,1}, e_{t,2}, e_{t,3}, ..., e_{t,d_{model}})^\top$ be an embedding of $\mathbf{s}_t$. Then, the

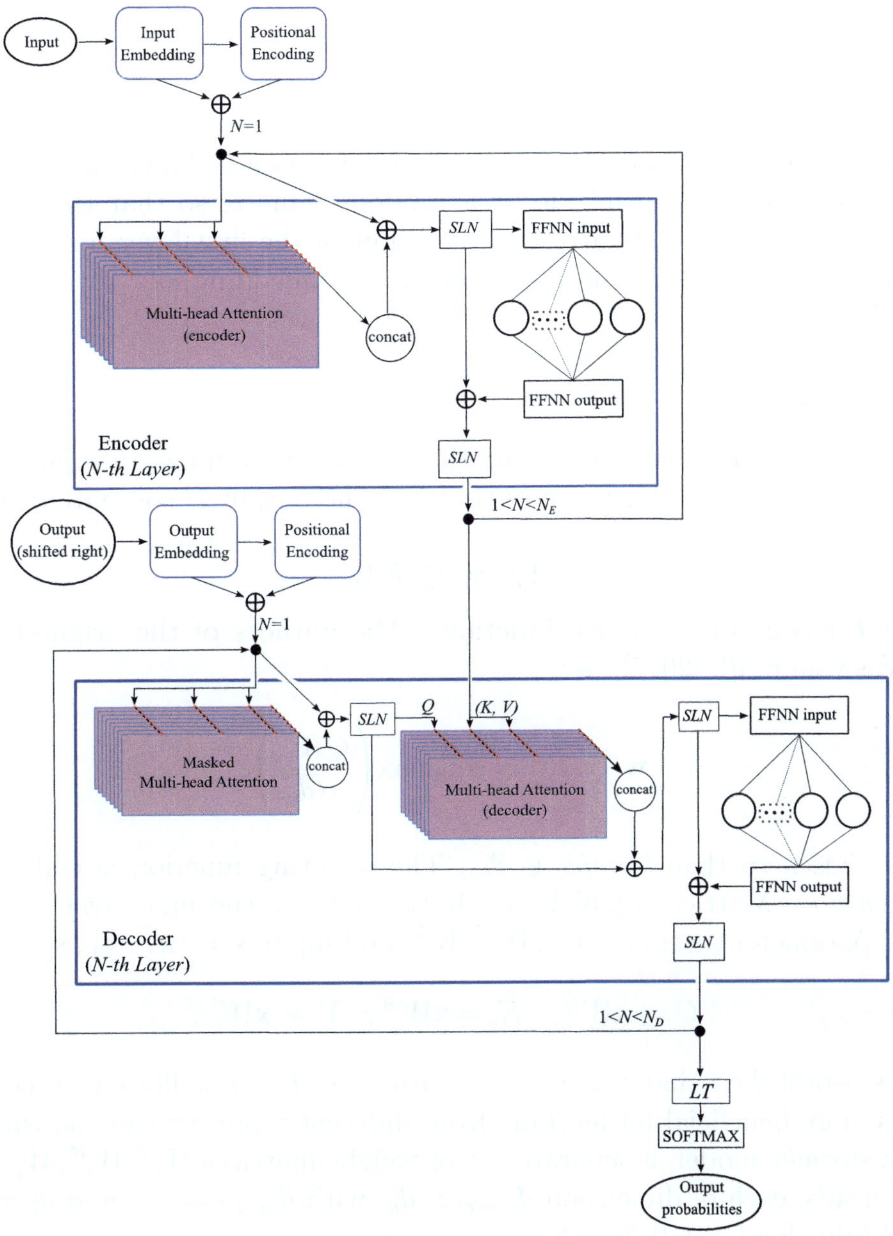

Figure 12.29.: Transformer

positional vector $\mathbf{v}_t = (v_{t,1}, v_{t,2}, v_{t,3}, \dots, v_{t,d_{model}})^\top$ for $\mathbf{s}_t$ is defined element-wise by

$$v_{t,i} = \begin{cases} \sin\left(\dfrac{t}{10000^{2i/d_{model}}}\right), & \text{for } i \text{ - even} \\ \cos\left(\dfrac{t}{10000^{2i/d_{model}}}\right), & \text{for } i \text{ - odd.} \end{cases} \tag{12.76}$$

The positional encoding for $\mathbf{s}_t$ is then

$$\mathbf{x}_t = \mathbf{e}_t + \mathbf{v}_t. \tag{12.77}$$

Note that in order to increase the weight of the word embedding $\mathbf{e}_t$ in the expression (12.77), we may need to multiply it by a random value r, so that the magnitudes of $\mathbf{e}_t r$ and $\mathbf{v}_t$ are comparable. As a result, the input of the first layer for the transformer encoder contains the information about the word embedding and the position of each word $\mathbf{s}_t$ in the sentence $\mathbf{s}$.

Attention Sublayer

In general, the attention function can be thought of as mapping an input query Q and a set of key-value pairs (K, V) to an output A. The output is computed as a weighted sum

$$A = \mathbf{w}(Q, K)V, \tag{12.78}$$

where $\mathbf{w}(Q, K)$ is a compatibility function. The authors of the original transformer model, in Vaswani et al. [2017], use

$$\mathbf{w}(Q, K) = \text{softmax}\left(\frac{QK^\top}{\sqrt{d_k}}\right), \tag{12.79}$$

where d_k is chosen so that $d_{model}/d_k \in \mathbb{Z}_+$. The resulting function is called *scaled dot-product attention.* Matrices Q, K, V are initialized, for the first layer in the encoder stack, using parameter matrices W^Q, W^K, W^V and input $\mathbf{x} = (\mathbf{x}_1^\top, \mathbf{x}_2^\top, \mathbf{x}_3^\top, ..., \mathbf{x}_n^\top)^\top$ as

$$Q = \mathbf{x}W^Q, \quad K = \mathbf{x}W^K, \quad V = \mathbf{x}W^V. \tag{12.80}$$

The authors established that using several *attention heads,* followed by concatenation of results, is more beneficial for learning from different representation subspaces. In the original transformer model, a separate set of weight matrices W_i^Q, W_i^K, W_i^V is used for 8 attention heads, each of dimension $d_{model} \times d_k$, with $d_{model} = 512$ and $d_k = 512/8 = 64$. One then obtains for $i = 1, 2, 3, ..., 8$

$$\begin{aligned} H_i &= \text{softmax}\left(\frac{Q_i K_i^\top}{\sqrt{d_k}}\right) V_i \\ H_M &= \text{concat}(H_1, H_2, H_3, ..., H_8)W^O, \end{aligned} \tag{12.81}$$

where W^O is yet another learned parameter matrix. The described approach is called the *multi-head attention.* This clever construction allows for the output of dimension $n \times d_{model}$ for an arbitrary input sentence length n, and it does not need any modifications for different values of n. As can be seen from Fig. 12.29, the multi-head attention is employed in several sublayers of the transformer. The exact application, however, is slightly different depending on the position in the model.

In the encoder, all three input matrices, Q, K, V are received from the previous layer of the encoder, after the initialization (12.80) in the first layer. Each position in the encoder can attend to all positions in the previous layer of the encoder.

In the decoder, the self-attention mechanism is applied in two separate sublayers. The *masked multi-head attention* sublayer is receiving its inputs entirely from the output sequence. In order to prevent the model from seeing the "future" elements of an output sequence, masking is applied by setting to "$-\infty$" all values of the softmax corresponding to those elements. Hence, each position in the decoder is allowed to attend to all positions in the output sequence up to and including the current position. This ensures the activation of learning and prediction process.

The second multi-head attention sublayer is receiving the query matrix Q from the previous decoder layer, and the key-value pairs (K, V) come from the output of the encoder. Hence, every position in the decoder is allowed to access all positions in the input sequence, similarly to the traditional encoder-decoder attention mechanisms in sequence-to-sequence models (e.g., see Sect. 12.2.3 and Bahdanau et al. [2015] for more details).

Position-Wise Feed-Forward Network Sublayer

A relatively simple fully connected feed-forward neural network follows the multi-head attention in each layer of the encoder and the decoder. The FFNN in the original model is an MLP with one hidden layer, containing 2^{11} hidden units and RELU activation function (12.69). The output layer has the same dimension (d_{model}) as the input and is using the identity output activation function. This sublayer is applied to each position of the model's input separately and identically. Note that the input of FFNN, denoted as *SLN* in Fig. 12.29, is a normalized sum of the multi-head attention sublayer's output and its input.

Sublayer Normalization

The output of every sublayer in the transformer model is not passed directly to the following sublayer. Instead, a normalized sum (*SLN*) of sublayer's output $\mathcal{S}(\mathbf{x})$ and its input $\mathbf{x}$ is used

$$SLN = \mathcal{N}(\mathbf{x} + \mathcal{S}(\mathbf{x})), \tag{12.82}$$

where $\mathcal{N}$ is a normalization function. One of the simplest of such functions is the standardization

$$\mathcal{N}(\mathbf{X}) = \frac{\mathbf{X} - \bar{\mathbf{X}}}{SD[\mathbf{X}]},$$

however, more complex functions with adjustable scaling and bias parameters can be used.

Output

As for the encoder, the decoder consists of the stack of N_D identical layers ($N_D = 6$ in the original model). Each layer is learning the information from its predecessor, and the inputs from the embedding sublayer are only passed on directly to the bottom level of the decoder stack. Each layer of the decoder stack consists of three sublayers. The masked multi-head attention sublayer (starting with the second level of the stack) is getting its inputs from the previous decoder sublayer. Its output query Q, together with the key-value pair (K, V) from the encoder stack, after applying (12.82), is transmitted to the decoder's multi-head attention sublayer. The output is once again normalized with (12.82) and enters the FFNN decoder sublayer. The architecture of the FFNN sublayer in the decoder is identical to the one used in the encoder and is described above. The FFNN output undergoes (yet again) post-layer normalization (12.82) and is transmitted to the next layer in the stack. When all layers of the decoder stack are passed, the output is transmitted to a single linear transformation sublayer denoted as LT in Fig. 12.29. For input $\mathbf{x}$, the outputs of LT sublayer are of the form

$$LT = \mathbf{x}W^L + \boldsymbol{b}, \tag{12.83}$$

where W^L and $\boldsymbol{b}$ are learned parameters, which in the original transformer model Vaswani et al. [2017] are shared between the input/output embedding layers and the linear transformation (12.83). Finally, the SOFTMAX sublayer produces the output probabilities for the predicted output sequence.

Part IV.

Computational Methods for Statistical Inference

13. Bayesian Computation Methods

In this chapter, we suppose that the reader is familiar with a brief description of Markov chains given in Chap. 2. In Bayesian methods, the goal is often to compute the posterior distribution, a task made easier whenever conjugate priors are used as seen in Sect. 3.3. The Bayesian computation of the posterior distribution in most other cases is a challenging task. In what follows, we will describe various ways of computing integrals of the form

$$\frac{\int h(\theta) f(x;\theta) \pi(\theta;\alpha)\, d\theta}{\int f(x;\theta) \pi(\theta;\alpha)\, d\theta}$$

for some function $h(\theta)$. The first approach is to appeal to the Bernstein-von Mises normal distribution approximation to the posterior distribution centered at the mode of θ (see Albert [2009] p.94). This was discussed earlier in Sect. 3.3.

The second approach, labeled rejection sampling, consists of identifying a proposal density say $p(\theta)$ which "resembles" the posterior density with respect to location and spread and which is easier to sample from. The selection of the proposal density in rejection sampling was previously discussed in Chap. 6. We then seek a constant c such that for all θ,

$$\pi(\theta|x_1, \ldots, x_n) \leq cp(\theta)$$

Rejection sampling then proceeds by repeatedly generating a value θ from both $p(\theta)$ and independently a value U from a uniform distribution on the interval $(0, 1)$. We accept θ as coming from the posterior if and only if

$$U \leq \frac{\pi(\theta|x_1, \ldots, x_n)}{cp(\theta)}$$

In the next section, we will present data augmentation methods as well as some of the theory underlying Markov chain Monte Carlo Methods (MCMC) which rely on properties of Markov chains as described in Sect. 2.4.

13.1. Data Augmentation Methods

To motivate the discussion, we consider Rao's genetic linkage model whereby 197 animals are distributed into four categories [Tanner, 1997, chapter 5]. The observed counts, denoted by $x = (x_1, x_2, x_3, x_4)$, are distributed according to a multinomial distribution

M. Alvo, *Statistical Inference and Machine Learning for Big Data*, Springer Series in the Data Sciences, https://doi.org/10.1007/978-3-031-06784-6_13

with probabilities, respectively,

$$\left(\frac{2+\theta}{4}, \frac{1-\theta}{4}, \frac{1-\theta}{4}, \frac{\theta}{4}\right)$$

The posterior distribution of θ given x would then be proportional to

$$\pi\left(\theta|x\right) \sim \left(2+\theta\right)^{x_1}\left(1-\theta\right)^{x_2+x_3}\theta^{x_4}\pi\left(\theta\right)$$

where $\pi\left(\theta\right)$ is the prior on θ. Normally, one would be tempted to choose a conjugate Beta distribution for a prior were it not for the first factor $(2+\theta)$. Suppose now we "augment" the data so as to enable us to split that factor into two with $x_1 = y + (x_1 - y)$ and y attributed to θ. The distribution of y given x_1 and θ is therefore binomial $\left(x_1, \frac{\theta}{2+\theta}\right)$, denoted $f\left(y|x,\theta\right)$ and the "augmented" posterior distribution of θ given the augmented data would then be proportional to

$$\pi\left(\theta|x,y\right) \sim \theta^{y+x_4}\left(1-\theta\right)^{x_2+x_3}\pi\left(\theta\right)$$

In this framework, y is of course random.

In general, the posterior distribution satisfies the relationship

$$\pi\left(\theta|x\right) = \int \pi\left(\theta|x,y\right) f\left(y|x\right) dy \tag{13.1}$$

where $f\left(y|x\right)$ is the predictive density of y given x which may be expressed as

$$f\left(y|x\right) = \int f\left(y|x,\phi\right)\pi\left(\phi|x\right) d\phi \tag{13.2}$$

The algorithm by which we sample from the posterior $\pi\left(\theta|x\right)$ proceeds by the following steps as follows. Suppose that at stage i the posterior distribution is labeled as $\pi_i\left(\theta|x\right)$. Using (13.2),

(a) Sample θ from the current estimate of the posterior say $\pi_i\left(\theta|x\right)$.
(b) Generate y from $f\left(y|x,\phi\right)$ where ϕ is the value of θ in step (a).
(c) Repeat steps (a) and (b) m times to obtain

$$y_1, y_2, \ldots, y_m$$

(d) The estimate of the posterior at stage $i+1$ using (13.1) is

$$\pi_{i+1}\left(\theta|x\right) = \frac{1}{m}\sum_{k=1}^{m}\pi_i\left(\theta|x,y_k\right) \tag{13.3}$$

which is justified by the strong law of large numbers. Once the process has converged for large enough m, we may then sample θ from the posterior in (13.3) by selecting at random one of the $\{\pi\left(\theta|x,y_k\right)\}$ and sampling θ from it.

To make this process more concrete, suppose the prior to begin with, say $\pi(\theta)$, is the uniform on $(0,1)$. Then the augmented posterior is the beta distribution

$$\pi(\theta|x,y)=\frac{\Gamma(\alpha+\beta)}{\Gamma(\alpha)\Gamma(\beta)}\theta^{\alpha-1}(1-\theta)^{\beta-1},0<\theta<1$$

where

$$\alpha=y+x_4+1,\beta=x_2+x_3+1$$

In the posterior step, we may estimate $\pi(\theta|x,y)$ by the average of beta distributions

$$\frac{\sum_{k=1}^{m}\pi(\theta|x,y_k)}{m}$$

In order to sample θ from this estimated distribution, we select a beta distribution at random from this sum, and sample θ from it.

An alternative to data augmentation in this example is to make use of the Learn Bayes algorithms as in the exercises in Albert [2009, p.111 and p.148]. For a modern application of data augmentation to experimental design and variable selection, see Ghosh and Clyde [2011].

13.2. Metropolis-Hastings Algorithm

MCMC methods have been discussed in several textbooks. We shall adapt the presentations appearing in Tanner [1993] as well as in articles [Andrieu et al., 2003; Geyer, 2005]. The previous methods of simulating data from a posterior distribution required the specification of a proper proposal density. This may be difficult to do especially for high multidimensional distributions. In this section, we describe the popular Metropolis-Hastings algorithm which relies on the theory provided by Markov chains.

Suppose we would like to draw a sample from a posterior density, denoted $p(\theta)$, known up to a normalizing constant. We may couch the problem within the context of a Markov chain with transition probability $q(\theta_{t-1}|\theta_t)$. The important feature of sampling in a Markov chain is that under certain conditions, there is ultimately a unique stationary distribution. Hence, the idea is to set up the chain so as to ensure that the stationary distribution is given by $p(\theta)$. This is done in two steps. In the first step, to ensure convergence to a unique stationary distribution, we require that we have a homogeneous chain, meaning that the chain is time independent or that it has stationary transition probabilities. As well, the chain should be irreducible, meaning that there is a positive probability of visiting any state from any other state, and finally, the chain should be aperiodic, meaning that it should not entangled in a cycle.

In the second step, we need to ensure that the stationary distribution is in fact given by the posterior distribution. In practice, it is difficult to verify that the invariant distribution of the Markov chain is $p(\theta)$. Instead, it is easier to verify a reversibility condition which implies it. In continuous state spaces, in order to ensure that the invariant distribution is given by $p(\theta)$, we require the reversibility condition given by

$$p(\theta_t)\, q(\theta_{t-1}|\theta_t) = p(\theta_{t-1})\, q(\theta_t|\theta_{t-1}) \tag{13.4}$$

Reversibility is not a necessary condition, but rather it is a simple way to show that an update mechanism has a specified stationary distribution [Geyer, 2005].

The sampling algorithm begins by choosing an initial value, say θ_0 along with a proposal density $q(\theta_t|\theta_{t-1})$ from which we can generate the next θ_t given θ_{t-1}. Sample a value u from a uniform distribution on $(0,1)$. Now given θ_{t-1}, sample a candidate θ_* value From $q(\theta|\theta_{t-1})$, and compute the acceptance probability

$$\alpha(\theta_*|\theta_{t-1}) = \min(1, R(\theta_{t-1}, \theta_*)) \tag{13.5}$$

where

$$R(\theta_{t-1}, \theta_*) = \frac{p(\theta_*)\, q(\theta_{t-1}|\theta_*)}{p(\theta_{t-1})\, q(\theta_*|\theta_{t-1})}$$

Then

$$\theta_t = \begin{cases} \theta_* & u < \alpha(\theta_*|\theta_{t-1}) \\ \theta_{t-1} & otherwise \end{cases}$$

This algorithm allows for the rejection of a selected value, and consequently, the chain is aperiodic. It is also irreducible since the support of q includes the support of p. We may show that the chain satisfies the reversibility condition (13.4) by first noting that the transition probability has two terms: the proposed probability and the probability of accepting it. This is given by $\alpha(\theta_*|\theta_{t-1})\, q(\theta_*|\theta_{t-1})$. Hence, we have

$$\begin{aligned}
[\alpha(\theta_*|\theta_{t-1})\, q(\theta_*|\theta_{t-1})]\, p(\theta_{t-1}) &= \min\left(\frac{p(\theta_*)}{p(\theta_{t-1})}\frac{q(\theta_{t-1}|\theta_*)}{q(\theta_*|\theta_{t-1})}, 1\right) q(\theta_*|\theta_{t-1})\, p(\theta_{t-1}) \\
&= \min\left(q(\theta_{t-1}|\theta_*)\, p(\theta_*), q(\theta_*|\theta_{t-1})\, p(\theta_{t-1})\right) \\
&= \min\left(1, \frac{q(\theta_*|\theta_{t-1})\, p(\theta_{t-1})}{q(\theta_{t-1}|\theta_*)\, p(\theta_*)}\right) q(\theta_{t-1}|\theta_*)\, p(\theta_*) \\
&= [\alpha(\theta_{t-1}|\theta_*)\, q(\theta_{t-1}|\theta_*)]\, p(\theta_*)
\end{aligned}$$

Consequently, the Markov chain is reversible, and the unique stationary distribution is given by $p(\theta)$.

It remains to specify the transition probabilities for which there are various examples. In a symmetric random walk, we suppose the transition density satisfies

$$q(\theta_*|\theta_{t-1}) = h(\theta_* - \theta_{t-1}) \tag{13.6}$$

where h is a symmetric density about the origin. Then

$$\alpha\left(\theta_*|\theta_{t-1}\right) = \min\left(1, \frac{p\left(\theta_*\right)}{p\left(\theta_{t-1}\right)}\right)$$

Another example arises when the density q satisfies the independence condition

$$q\left(\theta_*|\theta_{t-1}\right) = q\left(\theta_*\right) \tag{13.7}$$

For both cases, we may suppose that the density is given by a multivariate normal density. Albert [2009, p.111 and p.148] considers an interesting application to the genetic linkage.

13.3. Gibbs Sampling

Gibbs sampling is a popular method for setting up a Markov chain algorithm in order to sample from the joint posterior distribution $\pi\left(\theta|x\right)$ where $\theta = \left(\theta_1, \ldots, \theta_k\right)$ and $x = \left(x_1, \ldots, x_n\right)$ represents the data. We denote by θ_{-i} the vector θ with θ_i excluded

$$\theta_{-i} = \left(\theta_1, \ldots, \theta_i, \theta_{i+1}, \ldots, \theta_k\right)$$

For Gibbs sampling, we suppose that in a single cycle, we can sample successively from the following one-dimensional conditional distributions

$$\pi\left(\theta_1|\theta_{-1}, x\right)$$

$$\pi\left(\theta_2|\theta_{-2}, x\right)$$

$$\ldots$$

$$\pi\left(\theta_k|\theta_{-(k-1)}, x\right)$$

The stationary distribution remains the same and after the update

$$\pi\left(\theta_i|\theta_{-i}\right)\pi\left(\theta_{-i}\right) = \pi\left(\theta_i, \theta_{-i}\right) = \pi\left(\theta\right)$$

This procedure is repeated for a large number of cycles. Under certain conditions, this algorithm will converge to the desired posterior distribution [Geyer, 2005; Schervish and Carlin, 1992]. What is required is that the conditional densities used in the algorithm be strictly positive and that the kernel of the transition operator be square integrable with respect to a particular measure. Gibbs sampling may be viewed as a special case of the Metropolis-Hastings algorithm where the proposed value is always accepted.

In Gibbs sampling, we need to be able to sample from these one-dimensional conditional distributions. In many problems in practice, this is indeed the case. Programming the algorithm is easier, and there is no need to specify a proposal density. Albert [2009, Chapter 10] considers a variety of examples to illustrate the approach. See also Schervish and Carlin [1992].

13.4. EM Algorithm

The EM algorithm is one of the most popular tools in statistics for dealing with missing data problems. It appeared in various forms prior to its formal introduction in the literature by Dempster et al. [1977]. Suppose that we have a regular exponential family of densities

$$f(x|\phi) = h(x)\, C(\phi) \exp\left(\phi^T t(x)\right)$$

where ϕ, restricted to an r-dimensional convex set, and $t(x)$, a complete sufficient statistic, are $r \times 1$ vectors. Consider the situation where X is a complete data set observed indirectly through y so that there is a mapping $x \to y(x)$ from $\mathcal{X}$ to $\mathcal{Y}$. Hence x is known to lie in the subset of $\mathcal{X}$ determined by the equation $y = y(x)$. The complete data is related to the incomplete data by the function

$$g(y|\phi) = \int_{\mathcal{X}(y)} f(x|\phi)\, dx.$$

The implementation of the EM algorithm requires two steps: the E or estimation step and the M or maximization step. Simply put we have the E step whereby we estimate $t(x)$

$$t^{(p)} = E\left[t(x)\,|y, \phi^{(p)}\right]$$

and the M step whereby find $\phi^{(p+1)}$as the solution to

$$E[t(x)\,|\phi] = t^{(p)} \tag{13.8}$$

For the exponential family, the usual maximum likelihood estimation maximizes the log likelihood

$$\log f(x|\phi) = \log C(\phi) + \log h(x) + \phi^T t(x)$$

Since

$$E[t(x)\,|\phi] = -\frac{d}{d\phi} \log C(\phi)$$

the maximization is equivalent to (13.8).

We note that for the exponential family, the conditional density of x given y is

$$k(x|y, \phi) = h(x)\, C(\phi|y) \exp\left(\phi^T t(x)\right)$$

so that both k and f are members of the exponential family having the same natural parameter and sufficient statistic but different proportionality constants. Setting

$$\begin{aligned} L(\phi) &= \log \frac{f(x|\phi)}{k(x|y,\phi)} \\ &= \log C(\phi) - \log C(\phi|y) \end{aligned}$$

we see that

$$\frac{\partial}{\partial \phi} L(\phi) = E[t(x)|y,\phi] - E[t(x)|\phi] \tag{13.9}$$

Hence, in the EM algorithm, we see that if in the limit as $p \to \infty$, $\phi^{(p)} \to \phi^*$, then

$$E[t(x)|y,\phi^*] = E[t(x)|\phi^*],$$

and

$$\frac{\partial}{\partial \phi} L(\phi^*) = 0$$

When dealing with general families not necessarily exponential, we may define

$$Q(\phi'|\phi) = E[\log f(x|\phi')|y,\phi]$$

Then in the E step, compute $Q\left(\phi|\phi^{(p)}\right)$, and in the M step, choose $\phi^{(p+1)}$ as that value ϕ which maximizes $Q\left(\phi|\phi^{(p)}\right)$. Specializing the general case to the exponential family, we have

$$\begin{aligned} Q(\phi'|\phi) &= E[\log f(x|\phi')|y,\phi] \\ &= \log C(\phi) + E\left[h(x)|y,\phi^{(p)}\right] + \phi^T t^{(p)} \end{aligned}$$

so that in the M step, it means maximizing

$$\log C(\phi) + \phi^T t^{(p)}$$

In the next sections, we will consider two applications of the EM algorithm.

13.4.1. Application to Ranking

In this application, we develop an approach to exploit data originating in the form of incomplete rankings in order to construct a model for complete rankings. As an example, suppose we have a horse race involving t horses, and we would like to calculate the probabilities of various complete rankings from previous data involving outcomes of races for subsets of horses. In particular, we may want to calculate the probabilities of all the $t!$ complete rankings and choose the most likely one.

We begin by defining an exponential probability model for complete permutations. This model involves unknown parameters. Then, using the notion of compatibility for data stemming from incomplete rankings, we make use of the EM algorithm to obtain the maximum likelihood estimates of the parameters in the exponential model for complete rankings. In this way, the exponential model is reconstructed from knowledge of the incomplete rankings.

A ranking ν of t objects, expressed as a t-dimensional vector, represents the order of preference of the objects. In ranking t items, labeled 1,. . . ,t , a ranking ν is a mapping function from $1, \ldots, t$ to ranks $1, \ldots, t$, where $\nu(2) = 3$ means that item 2 is ranked third and the inverses $\nu^{-1}(3) = 2$ means that the item ranked third is item 2. It will be more convenient to standardize the rankings as:

$$x = \frac{\nu - \left(\frac{t+1}{2}\right)1}{\sqrt{\frac{t(t^2-1)}{12}}}, \tag{13.10}$$

where x is the $t \times 1$ vector with $\|x\| = 1$ and 1 is the $t \times 1$ vector all of whose components are equal to 1.

We consider the following exponential ranking model for $\theta = (\kappa, \mu)$:

$$p(x|\theta) = C(\kappa, \mu)\exp\left\{\kappa\mu^T x\right\} \tag{13.11}$$

where parameter μ is a $t \times 1$ vector with $\|\mu\| = 1$, scalar parameter $\kappa \geq 0$, and $C(\kappa, \mu)$ is the normalizing constant. In the null situation when each complete ranking has the same probability of being selected corresponding to $\kappa = 0$, the constant is exactly equal to

$$C(\kappa, \mu) = 1/t!$$

In the case of the distance-based Mallows model, the parameter μ can be seen as a modal ranking score vector. In other cases, μ provides a weighting of the ranks assigned to items. Since both $\|\mu\| = 1$ and $\|x\| = 1$, the term $\mu^T x$ can be seen as $\cos x$ where x is the angle between the modal ranking score vector μ and the observation x. The probability of observing a ranking is inversely proportional to its cosine of the angle from the modal vector. The parameter κ can be viewed as a concentration parameter. For small κ, the distribution of rankings will appear close to a uniform, whereas for larger values of κ, the distribution of rankings will be more concentrated around the modal ranking vector. Then

$$C^{-1}(\kappa, \mu) = \sum_{x \in \mathcal{P}} \exp\left\{\kappa\mu^T x\right\}. \tag{13.12}$$

where $\mathcal{P}$ is the $t!$ space of all permutations of the integers $1, \ldots, t$. When t is larger than 10, the exact calculation of the normalizing constant is prohibitive. Using the fact that the set of $t!$ permutations lie on a sphere in $(t-1)$-space, our model resembles the continuous von Mises-Fisher distribution which is defined on a $(p-1)$ unit sphere and arbitrary x not necessarily rankings:

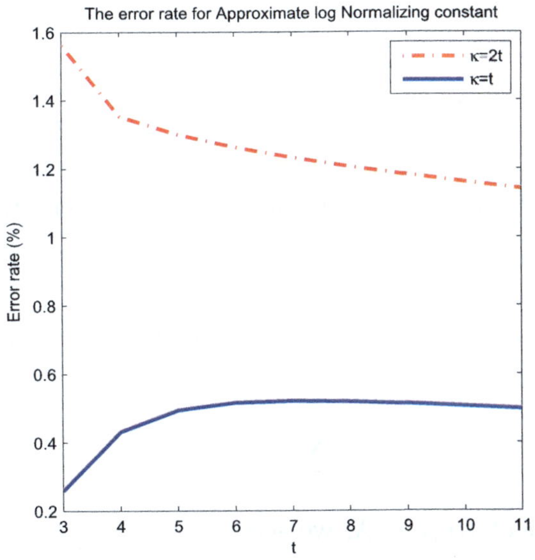

Figure 13.1.: The error rate for the approximate normalizing constant compared to the exact one computed by $t!$ summation. $\kappa = 0.5$ and t is computed from 4 to 9

$$p(x|\kappa, \mu) = \frac{\kappa^{\frac{p}{2}-1}}{(2\pi)^{\frac{p}{2}} I_{\frac{p}{2}-1}(\kappa)} \exp(\kappa \mu^T x) \tag{13.13}$$

where $I_{\frac{t-3}{2}}(\kappa)$ is the modified Bessel function of the first kind with order $\frac{t-3}{2}$. It was shown in Xu et al. [2018] approximately that we have for the constant in (13.12)

$$C(\kappa, \mu) \simeq C_t(\kappa) = \frac{\kappa^{\frac{t-3}{2}}}{2^{\frac{t-3}{2}} t! I_{\frac{t-3}{2}}(\kappa) \Gamma(\frac{t-1}{2})} \tag{13.14}$$

where $\Gamma(.)$ is the gamma function. Figure 13.1 displays the error rate for the approximate normalizing constant compared to the exact one computed by direct summation. Here, κ is chosen to be 0.5, and t ranges from 4 to 9. This approximation seems to be very accurate with an error rate of about 0.02% even when $t = 4$. The error decreases rapidly as t increases. The approximation allows us to approximate the first or second derivative of $C_t(\kappa)$ which will be needed in what follows.

We now proceed with the maximum likelihood estimation in the exponential model which may now be approximately expressed as

$$p(x|\theta) = C_t(\kappa) \exp\left\{\kappa \mu^T x\right\}. \tag{13.15}$$

Let $x = \{x_1, \ldots, x_n\}$ be a set of independent identically distributed standardized observed rankings from the approximate exponential model (13.15). The log-likelihood of

$\theta = (\mu, \kappa)$ is then given by

$$L(\theta) = n \log C_t(\kappa) + \sum_{i=1}^{n} \kappa \mu^T x_i \tag{13.16}$$

For the maximum likelihood estimators of μ and κ subject to $\|\mu\| = 1$ and $\kappa \geq 0$, we consider the objective function

$$L(\mu, \kappa, \lambda) = n \log C_t(\kappa) + \kappa \mu^T r + \lambda \left(1 - \mu^T \mu\right) \tag{13.17}$$

where

$$\sum_{i=1}^{n} x_i = r$$

Differentiating with respect to μ, κ, λ, we see

$$\begin{aligned} \hat{\mu} &= \frac{\hat{\kappa}}{2\hat{\lambda}} r \\ \hat{\mu}^T \hat{\mu} &= 1 \\ \frac{nC'(\hat{\kappa})}{C(\hat{\kappa})} &= -\ \hat{\mu}^T r \end{aligned} \tag{13.18}$$

From these equations, we get

$$\begin{aligned} \hat{\mu} &= \frac{r}{\|r\|} \\ -\frac{C'(\hat{\kappa})}{C(\hat{\kappa})} &= \frac{\|r\|}{n} = \tilde{r} \end{aligned}$$

Hence, the maximum likelihood estimators are given by

$$\hat{\mu}_{MLE} = \frac{\sum_{i=1}^{n} x_i}{\|\sum_{i=1}^{n} x_i\|}$$

and $\hat{\kappa}$ is the solution of

$$A_t(\kappa) \equiv \frac{-C_t'(\kappa)}{C_t(\kappa)} = \frac{I_{\frac{t-1}{2}}(\kappa)}{I_{\frac{t-3}{2}}(\kappa)} = \tilde{r} \tag{13.19}$$

In Banerjee et al. [2005], it was noted that a simple approximation to the solution of (13.19) is given by

$$\hat{\kappa}_{MLE} = \frac{\tilde{r}(t - 1 - \tilde{r}^2)}{1 - \tilde{r}^2}$$

which we shall take as the initial value κ_0 when applying the EM algorithm below. We note that since $\|x\| \leq 1$, it follows that $0 \leq \tilde{r} \leq 1$.

Suppose that we observe incomplete rankings $x_i^*, i = 1, \ldots, n$. Every incomplete ranking x_i^* gives rise to a compatibility class $C(x_i^*)$ of complete rankings. The members of the class $C(x_i^*)$ are not observed directly. In what follows, we shall make use of the model (13.15) and the EM algorithm to determine the complete ranking that gave rise to the incomplete ranking.

The implementation of the EM algorithm requires two steps: the E or estimation step and the M or maximization step. Simply put we have the E step: Begin with an initial value of θ, say θ_0. Calculate

$$
\begin{aligned}
Q(\theta;\theta_0) &= E \log \Pi_{i=1}^{n} p\left(x_i|\theta, x_i^*, \theta_0\right) \\
&= \sum_{i=1}^{n} \log C_t(\kappa) + \kappa\mu^T \left(\sum_{i=1}^{n} E\left(x_i|x_i^*, \theta_0\right)\right) \qquad (13.20) \\
&\cong \sum_{i=1}^{n} \log C_t(\kappa) + \kappa\mu^T \sum_{i=1}^{n} \left(\sum_{x_i \in C\left(x_i^*\right)} x_i p\left(x_i|\theta_0\right)\right) \qquad (13.21) \\
&= \sum_{i=1}^{n} \log C_t(\kappa) + \kappa\mu^T r_0 \qquad (13.22)
\end{aligned}
$$

where

$$
\begin{aligned}
r_0 &= \sum_{i=1}^{n} \left(\sum_{x_i \in C\left(x_i^*\right)} x_i p\left(x_i|\theta_0\right)\right) \qquad (13.23) \\
&= \sum_{i=1}^{n} \left(\left(\frac{k_i!}{t!}\right) \sum_{x_i \in C\left(x_i^*\right)} x_i\right) \qquad (13.24)
\end{aligned}
$$

since the summation over $\{x_i : x_i \in C(x_i^*)\}$ contains exactly $\frac{t!}{k_i}$ terms, namely the number of complete rankings compatible with x_i^*.

To obtain the maximum likelihood estimators of μ and κ subject to $\|\mu\| = 1$ and $\kappa \geq 0$, we consider as in the complete data situation, the objective function

$$L(\mu, \kappa, \lambda) = n \log C_t(\kappa) + \kappa\mu^T r_0 + \lambda\left(1 - \mu^T \mu\right) \qquad (13.25)$$

We obtain a completely analogous result:

$$\kappa^{(1)} = \frac{\tilde{r}_0(t - 1 - \tilde{r}_0^2)}{1 - \tilde{r}_0^2} \qquad (13.26)$$

with

$$\tilde{r}_0 \equiv \frac{\left\|\sum_{i=1}^{n}\left(\frac{k_i!}{t!}\right)\left(\sum_{x_i \in C\left(x_i^*\right)} x_i\right)\right\|}{n}. \qquad (13.27)$$

Also

$$\hat{\mu}^{(1)} = \frac{\sum_{i=1}^{n}\left(\frac{k_i!}{t!}\right)\left(\sum_{x_i \in C\left(x_i^*\right)} x_i\right)}{\left\|\sum_{i=1}^{n}\left(\frac{k_i!}{t!}\right)\left(\sum_{x_i \in C\left(x_i^*\right)} x_i\right)\right\|} \tag{13.28}$$

$$= \frac{r_0}{\|r_0\|} \tag{13.29}$$

Next, set

$$\sum_{i=1}^{n}\left(\sum_{x_i \in C\left(x_i^*\right)} x_i p\left(x_i | \kappa^{(p)}, \mu^{(p)}\right)\right) = r^{(p)}$$

and

$$\widetilde{r}^{(p)} = \frac{\left\|r^{(p)}\right\|}{\sum_{i=1}^{n}\left(\sum_{x_i \in C\left(x_i^*\right)} p\left(x_i | \kappa^{(p)}, \mu^{(p)}\right)\right)}$$

At stage $p+1$, we compute

$$\hat{\mu}^{(p+1)} = \frac{\sum_{i=1}^{n}\left(\sum_{x_i \in C\left(x_i^*\right)} x_i p\left(x_i | \kappa^{(p)}, \mu^{(p)}\right)\right)}{\left\|\sum_{i=1}^{n}\left(\sum_{x \in C\left(x_i^*\right)} x_i p\left(x_i | \kappa^{(p)}, \mu^{(p)}\right)\right)\right\|} \tag{13.30}$$

$$= \frac{r^{(p)}}{\left\|r^{(p)}\right\|} \tag{13.31}$$

and solve for κ from the equation

$$A(\kappa) = \frac{\left\|\sum_{i=1}^{n}\left(\sum_{x_i \in C\left(x_i^*\right)} x_i p\left(x_i | \kappa^{(p)}, \mu^{(p)}\right)\right)\right\|}{\sum_{i=1}^{n}\left(\sum_{x_i \in C\left(x_i^*\right)} p\left(x_i | \kappa^{(p)}, \mu^{(p)}\right)\right)} \tag{13.32}$$

where

$$A(\kappa) = \frac{I_{\frac{t-1}{2}}(\kappa)}{I_{\frac{t-3}{2}}(\kappa)} \tag{13.33}$$

A good approximation for $\kappa^{(p+1)}$would be for $\widetilde{r}^{(p)} \neq \pm 1$,

$$\kappa^{(p+1)} = \frac{\widetilde{r}^{(p)}\left(t - 1 - \left(\widetilde{r}^{(p)}\right)^2\right)}{1 - \left(\widetilde{r}^{(p)}\right)^2} \tag{13.34}$$

For computational purposes, we note that the normality constant may be dropped so that

$$\begin{aligned}
\widetilde{r}^{(p)} &= \frac{\left\| \sum_{i=1}^{n} \left(\sum_{x_i \in C(x_i^*)} x_i p\left(x_i | \kappa^{(p)}, \mu^{(p)}\right)\right)\right\|}{\sum_{i=1}^{n} \left(\sum_{x_i \in C(x_i^*)} p\left(x_i | \kappa^{(p)}, \mu^{(p)}\right)\right)} \\
&= \frac{\left\| \sum_{i=1}^{n} \left(\sum_{x_i \in C(x_i^*)} x_i \exp\left(\kappa^{(p)} \mu^{(p)T} x_i\right)\right)\right\|}{\sum_{i=1}^{n} \left(\sum_{x_i \in C(x_i^*)} \exp\left(\kappa^{(p)} \mu^{(p)T} x_i\right)\right)} \qquad (13.35)
\end{aligned}$$

and

$$\hat{\mu}^{(p+1)} = \frac{\sum_{i=1}^{n} \left(\sum_{x_i \in C(x_i^*)} x_i \exp\left(\kappa^{(p)} \mu^{(p)T} x_i\right)\right)}{\left\| \sum_{i=1}^{n} \left(\sum_{x \in C(x_i^*)} x_i \exp\left(\kappa^{(p)} \mu^{(p)T} x_i\right)\right)\right\|}. \qquad (13.36)$$

We illustrate this computation with an example.

Example Let $t = 3$ and *set* $\nu_1 = (1,2,3), \nu_2 = (1,3,2), \nu_3 = (2,1,3), \nu_4 = (2,3,1), \nu_5 = (3,1,2), \nu_6 = (3,2,1)$. Let $k = 2$ and suppose we observe the incomplete rankings $x_1^* = (1,2,-)$ and $x_2^* = (1,-,2)$. Then, the respective compatibility classes consist of

$$C(x_1^*) = \{(1,2,3), (1,3,2), (2,3,1)\} \qquad (13.37)$$

and

$$C(x_2^*) = \{(1,2,3), (1,3,2), (2,1,3)\}. \qquad (13.38)$$

For x_1, we have $\frac{k_1!}{t!} = \frac{2!}{3!} = \frac{1}{3}$

$$\begin{aligned}
\sum_{x_1 \in C(x_1^*)} x_1 \left(\frac{k_1!}{t!}\right) &= \frac{1}{3\sqrt{2}} [(-1,0,1) + (-1,1,0) + (0,1,-1)] \\
&= \frac{1}{3\sqrt{2}} (-2,2,0)
\end{aligned}$$

For x_2, we have $\frac{k_2!}{t!} = \frac{2!}{3!} = \frac{1}{3}$

$$\begin{aligned}
\sum_{x_2 \in C(x_2^*)} x_2 \left(\frac{k_2!}{t!}\right) &= \frac{1}{3\sqrt{2}} [(-1,0,1) + (-1,1,0) + (0,-1,1)] \\
&= \frac{1}{3\sqrt{2}} (-2,0,2)
\end{aligned}$$

Then,

$$\sum_{i=1}^{n}\left(\sum_{x_i\in C\left(x_i^*\right)} x_i\left(\frac{k_i!}{t!}\right)\right) = \frac{1}{3\sqrt{2}}\quad(-4,2,2)$$

$$= \quad r_0$$

Then,

$$\widetilde{r}_0 = \frac{\left((4)^2+(2)^2+(2)^2\right)^{1/2}}{6\sqrt{2}} = \frac{1}{\sqrt{3}}$$

$$\mu^{(1)} = \frac{(-4,2,2)}{\|16+4+4\|^{1/2}} = \frac{1}{\sqrt{6}}(-2,1,1)$$

In other calculations, we will use model values instead of $\frac{k_1!}{t!} = \frac{1}{3}$. Hence, $\kappa^{(1)} = \frac{2.5}{\sqrt{3}}$

13.4.2. Extension to Several Populations

Suppose that we now consider a mixture of several distributions of complete rankings given by

$$p\,(x|\Theta) = \sum_{k=1}^{K}\alpha_k p_k(x|\theta_k) \tag{13.39}$$

with densities

$$p_k(x|\theta_k) = C_t\,(\kappa_k)\exp\left\{\kappa_k\mu_k^T x\right\}, k = 1,2$$

and $\Theta = (\alpha_1,\ldots,\alpha_K\theta_1,\ldots,\theta_K)$. The weights $\{\alpha_h\}$ are non-negative and sum to 1. Suppose now that we only observe incomplete rankings $x_i^*, i = 1,\ldots,n$. Our goal to determine the mixture density parameters (13.39).

Define latent variables

$$Z_i = \begin{cases}1 & if\ population\ i\ is\ chosen\\ 0 & \text{otherwise}\end{cases}$$

Then,

$$\begin{aligned}P\,(Z_i = k|x_i) &= \frac{P\,(x_i|Z_i = k)\,P\,(Z_i = k)}{P\,(x_i)}\\ &= \frac{\alpha_k p_k(x_i|\theta_k)}{\sum_k \alpha_k p_k(x_i|\theta_k)} \equiv \gamma_{Z_i}\,(k)\end{aligned}$$

Now the likelihood function is given by

$$\prod_{i=1}^{n}\prod_{k=1}^{K}\alpha_k^{I(Z_i=k)}\left(p_k(x_i|\theta_k)\right)^{I(Z_i=k)}$$

Hence, forming the log-likelihood and the Lagrangian subject to $\mu_k^T\mu_k = 1$, $\kappa_k \geq 0$,

$$L = \sum_k\sum_{i=1}^{n} I\left(Z_i = k\right)\left(\log \alpha_k p_k(x_i|\theta_k)\right) + \sum_{k=1}^{K}\lambda_k\left(1-\mu_k^T\mu_k\right)$$

The maximization with respect to the $\{\alpha_k\}$ is unrelated to the other maximization with respect to the other components of Θ. Taking expected value we get

$$E\left(L|x^*\right) = \sum_k\sum_{i=1}^{n}\gamma_{Z_i}\left(k\right)\left(\log\alpha_k + E\left(\log p_k(x_i|\theta_k)\right)|x^*\right) + \sum_{k=1}^{k}\lambda_k\left(1-\mu_k^T\mu_k\right)$$

The EM algorithm now proceeds by first choosing initial values for $\{\alpha_k\}$ to evaluate γ_k. Then, with γ_k fixed, we proceed as before to maximize with respect to (κ, μ). We get since the $\{\alpha_k\}$ sum to 1

$$\alpha_k^{(p)} = \frac{\sum_{i=1}^{n}\gamma_{Z_i}\left(k\right)}{n}$$

The sum represents the effective number of points assigned to component k. Next, set

$$\sum_{i=1}^{n}\left(\sum_{x_i\in C\left(x_i^*\right)} x_i\alpha_k^{(p)}p_k\left(x_i|\kappa_k^{(p)},\mu_k^{(p)}\right)\right) = r_k^{(p)} \tag{13.40}$$

and

$$\tilde{r}_k^{(p)} = \frac{\left\|r_k^{(p)}\right\|}{\sum_{i=1}^{n}\left(\sum_{x_i\in C\left(x_i^*\right)}\alpha_k^{(p)}p_k\left(x_i|\kappa_k^{(p)},\mu_k^{(p)}\right)\right)} \tag{13.41}$$

At stage $p+1$,we compute

$$\hat{\mu}_k^{(p+1)} = \frac{\sum_{i=1}^{n}\left(\sum_{x_i\in C\left(x_i^*\right)} x_i\alpha_k^{(p)}p_k\left(x_i|\kappa_k^{(p)},\mu_k^{(p)}\right)\right)}{\left\|\sum_{i=1}^{n}\left(\sum_{x\in C\left(x_i^*\right)} x_i\alpha_k^{(p)}p_k\left(x_i|\kappa_k^{(p)},\mu_k^{(p)}\right)\right)\right\|} \tag{13.42}$$

$$= \frac{r_k^{(p)}}{\left\|r_k^{(p)}\right\|} \tag{13.43}$$

A good approximation for $\kappa_k^{(p+1)}$would be

$$\kappa_k^{(p+1)} = \frac{\widetilde{r}_k^{(p)}\left(t-1-\left(\widetilde{r}_k^{(p)}\right)^2\right)}{1-\left(\widetilde{r}_k^{(p)}\right)^2} \tag{13.44}$$

One thing to note is that in terms of calculations, the ratio $\frac{t!}{k!}$ could get very large even for modest values of k, t. For example for $t = 10, k = 4$, $\frac{t!}{k!} = 151200$. For this reason, we may instead of computing the entire compatibility class, we may choose a random sample generated as illustrated in the following example:

Suppose $x^* = (1 - -2 - -4 \text{-} 3 -)$

Step 1: Generate at random a complete permutation: $x = (2, 4, 8, 9, 7, 6, 10, 1, 5, 3)$

Step 2: Rearrange x to agree with x^* in those places where x^* ranks: Here, we look at entries $2, 9, 10, 5$.

These must be in the order $2, 5, 10, 9$.

Step 3: The new complete permutation compatible with x^* is $x = (2, 4, 8, 5, 7, 6, 10, 1, 9, 3)$

13.5. Variational Bayesian Methods

Variational Bayesian methods offer an alternative approach to approximate the posterior distribution of θ given x. This has been described in the literature by a number of authors, and here we will follow Blei et al. [2017].

Let $\pi(\theta)$ be the prior on θ, and let $f_\theta(x)$ be the density of x given θ. We would like to approximate the posterior density of θ given x denoted $\pi(\theta|x)$ by a simpler density $q(\theta)$ in such a way as to minimize the Kullback-Leibler divergence which is always nonnegative

$$D(q; \pi) = \int q(\theta) \log \frac{q(\theta)}{\pi(\theta|x)} d\theta \tag{13.45}$$

Let $f(x)$ be the marginal density of x. Then noting that f is independent of θ and $q(\theta)$ is a proper density, we see that

$$\begin{aligned} D(q; \pi) &= \int q(\theta) \left[\log \frac{q(\theta)}{f_\theta(x)} f(x)\right] d\theta \\ &= \int q(\theta) \log q(\theta)\, d\theta - \int q(\theta) \log f_\theta(x)\, d\theta + \log f(x) \end{aligned}$$

Hence, we have that

$$\log f(x) = D(q; \pi) + \mathcal{L}(q) \tag{13.46}$$

where $\mathcal{L}(q) = -E_q\left[\int q(\theta)(\log q(\theta) - \log f_\theta(x))\right]$ is called the evidence lower bound and $\log f(x)$ is called the log evidence with respect to q. The left side of (13.46) is fixed and hence maximizing the evidence lower bound with respect to q will minimize the Kullback-Leibler divergence.

To illustrate this approach, Xu et al. [2018] considered a Bayesian approach to modeling ranking data as

$$f_{\alpha,\theta}(y) = C(\alpha,\theta)\exp\left\{\alpha\theta^T y\right\}$$

where $y = y = \frac{R-\frac{t+1}{2}}{\sqrt{\frac{t(t^2-1)}{12}}}$ is a normalized vector of the ranks of t objects, θ is a $t\times 1$ vector parameter with $\|\theta\| = 1$, $\alpha \geq 0$ is parameter and $C(\alpha,\theta)$ is the normalizing constant. Using a Fisher-von Mises distribution, it was shown that approximately the normalizing constant is given by

$$C(\alpha,\theta) \simeq C_t(\alpha) = \frac{\alpha^{\frac{t-3}{2}}}{2^{\frac{t-3}{2}} t! I_{\frac{t-3}{2}}(\alpha)\Gamma(\frac{t-1}{2})}$$

Taking a Bayesian approach, we consider the following conjugate prior for (α,θ) as

$$\pi(\alpha,\theta) \propto [C_t(\alpha)]^{\nu_0}\exp\left\{\beta_0\alpha m_0^T\theta\right\} \tag{13.47}$$

where $\|m_0\| = 1$, $\beta_0 \geq 0$. Given y, the posterior density of (α,θ) can be expressed by

$$\pi(\alpha,\theta|y) \propto \exp\left\{\beta\alpha m^T\theta\right\} V_t(\beta\alpha)\frac{[C_t(\alpha)]^{n+\nu_0}}{V_t(\beta\alpha)}$$

where $\beta = \|\beta_0 m_0 + \sum_{i=1}^n y_i\|$, $m = (\beta_0 m_0 + \sum_{i=1}^n y_i)\,\beta^{-1}$ and $V_t(\lambda)$ is the normalizing constant for the von Mises-Fisher density with concentration parameter λ:

$$V_t(\lambda) = \frac{\lambda^{\frac{t-2}{2}}}{(2\pi)^{\frac{t}{2}} I_{\frac{t-2}{2}}(\lambda)} \tag{13.48}$$

The posterior density $\pi(\alpha,\theta|y)$ can be factored as

$$\pi(\alpha,\theta|y) = \pi(\theta|\alpha,y)\pi(\alpha|y) \tag{13.49}$$

where

$$\pi(\theta|\alpha,y) \sim vMF(\theta|m,\alpha\beta)$$

$$\pi(\alpha|y) \propto \frac{[C_t(\alpha)]^{n+\nu_0}}{V_t(\beta\alpha)} = \frac{\alpha^{\frac{t-3}{2}(v_0+n)} I_{\frac{t-2}{2}}(\beta\alpha)}{\left[I_{\frac{t-3}{2}}(\alpha)\right]^{\nu_0+n}(\beta\alpha)^{\frac{t-2}{2}}}$$

Here, $vMF(\theta|m,\alpha\beta)$ is the von Mises-Fisher density with mean direction m and concentration parameter $(\alpha\beta)$. The normalizing constant for $\pi(\alpha|y)$ is not available in closed form. Variational inference provides a deterministic approximation to the posterior distribution. We combined a vMF-Gamma distribution as the prior for α, θ:

$$\begin{aligned}\pi(\theta, \alpha) &= \pi(\theta|\alpha)\pi(\alpha) \\ &= vMF(\theta|m_0, \beta_0\alpha)G(a_0, b_0)\end{aligned}$$

where $G(a_0, b_0)$ is the Gamma $G(a_0, b_0)$ with shape parameter a_0 and rate parameter b_0 (mean equal to $\frac{a_0}{b_0}$), and $\pi(\theta|\alpha) = vMF(\theta|m_0, \beta_0\alpha)$. The choice of $G(a_0, b_0)$ for $\pi(\alpha)$ is motivated by the fact that for large values of α, $\pi(\alpha)$ tends to take the shape of a gamma density. In fact, for large values of α,

$$I_{\frac{t-3}{2}}(\alpha) \simeq \frac{e^{\alpha}}{\sqrt{2\pi\alpha}}.$$

which leads to the gamma density

$$\pi(\alpha) \propto \frac{[C_t(\alpha)]^{\nu_0}}{V_t(\alpha\beta)} \simeq \alpha^{(\nu_0-1)\frac{t-2}{2}} \exp(-(\nu-\beta)\alpha).$$

Under the usual variational Bayesian methods, all variables are assumed to be mutually independent. This is known as the mean-field approximation. However, inspired by the conjugate posterior distribution (13.49), we adopt a structural factorization of the posterior as

$$q(\theta, \alpha) = q(\theta|\alpha)q(\alpha)$$

which retains some measure of the correlation between θ and α.

13.5.1. Optimization of the Variational Distribution

In our model, the evidence lower bound with respect to q on the log marginal likelihood is given by the expectation

$$\begin{aligned}\mathcal{L}(q) &= E_{\theta,\alpha}\left[\ln \frac{f_{\alpha,\theta}(y)\pi(\theta|\alpha)\pi(\alpha)}{q(\theta|\alpha)q(\alpha)}\right] \\ &= E_{\theta,\alpha}[h(\theta,\alpha)] - E_{\theta,\alpha}[\ln q(\theta|\alpha)] - E_{\alpha}[\ln q(\alpha)] + constant\end{aligned} \tag{13.50}$$

where

$$\begin{aligned}h(\theta, \alpha) = &\sum_{i=1}^{n} \alpha\theta^T y_i + n\left(\frac{t-3}{2}\right)\ln\alpha - n\ln I_{\frac{t-3}{2}}(\alpha) + \alpha\beta_0 m_0^T\theta \\ &+ \left(\frac{t-2}{2}\right)\ln\alpha - \ln I_{\frac{t-2}{2}}(\alpha\beta_0) + (a_0-1)\ln\alpha - b_0\alpha\end{aligned}$$

For fixed α, the optimal posterior distribution $\ln q^*(\theta|\alpha)$ is

$$\ln q^*(\theta|\alpha) = \alpha\beta_0 m_0^T\theta + \sum_{i=1}^{n} \alpha\theta^T y_i + constant$$

where we recognize

$$q^*(\theta|\alpha) = vMF(\theta|m, \alpha\beta)$$

with $\beta = \|\beta_0 m_0 + \sum_{i=1}^n y_i\|$ and $m = (\beta_0 m_0 + \sum_{i=1}^n y_i)\,\beta^{-1}$.

Let $g(\alpha)$ denote the remaining terms in $f(\theta, \alpha)$ which only involve α:

$$\begin{aligned} g(\alpha) &= \left[n(\frac{t-3}{2}) + a_0 - 1\right]\ln\alpha - b_0\alpha - n\ln I_{\frac{t-3}{2}}(\alpha) \\ &- \ln I_{\frac{t-2}{2}}(\alpha\beta_0) + \ln I_{\frac{t-2}{2}}(\alpha\beta) \end{aligned}$$

It is still difficult to maximize $E_\alpha\left[g(\alpha)\right] - E_\alpha\left[\ln q(\alpha)\right]$ since it involves the calculation of modified Bessel function moments. By the method of Taghia et al. [2014], we need to find a tight lower bound $\underline{g(\alpha)}$ for $g(\alpha)$ so that

$$\mathcal{L}(q) \geq \underline{\mathcal{L}(q)} = constant + E_\alpha\left[\underline{g(\alpha)}\right] - E_\alpha\left[\ln q(\alpha)\right].$$

From the properties of the modified Bessel function of the first kind, we know that the function $\ln I_\nu(x)$ is strictly concave with respect to x and strictly convex relative to $\ln x$ for all $\nu > 0$. Then, we have the following two inequalities:

$$\ln I_\nu(x) \leq \ln I_\nu(\bar{x}) + \left(\frac{\partial}{\partial x}\ln I_\nu(\bar{x})\right)(x - \bar{x}), \tag{13.51}$$

$$\ln I_\nu(x) \geq \ln I_\nu(\bar{x}) + \left(\frac{\partial}{\partial x}\ln I_\nu(\bar{x})\right)\bar{x}(\ln x - \ln\bar{x}). \tag{13.52}$$

where $\frac{\partial}{\partial x}\ln I_\nu(x)|_{x=\bar{x}}$ is the first derivative of $\ln I_\nu(x)$ evaluated at $x = \bar{x}$.

Applying inequality (13.51) for $\ln I_{\frac{t-3}{2}}(\alpha)$ and inequality (13.52) for $\ln I_{\frac{t-2}{2}}(\alpha\beta_0)$, we have that

$$\begin{aligned} g(\alpha) \geq \underline{g(\alpha)} &= \left[n\left(\frac{t-3}{2}\right) + a_0 - 1\right]\ln\alpha - b_0\alpha \\ &+ \ln I_{\frac{t-2}{2}}(\beta\bar{\alpha}) + \frac{\partial}{\partial\beta\alpha}\ln I_{\frac{t-2}{2}}(\beta\bar{\alpha})\beta\bar{\alpha}\left(\ln\beta\alpha - \ln\beta\bar{\alpha}\right) \\ &- n\ln I_{\frac{t-3}{2}}(\bar{\alpha}) - n\frac{\partial}{\partial\alpha}\ln I_{\frac{t-3}{2}}(\bar{\alpha})\left(\alpha - \bar{\alpha}\right) \\ &- n\ln I_{\frac{t-2}{2}}(\beta_0\bar{\alpha}) - n\frac{\partial}{\partial\beta_0\alpha}\ln I_{\frac{t-2}{2}}(\beta_0\bar{\alpha})\beta_0\left(\alpha - \bar{\alpha}\right) \end{aligned}$$

Since equality holds when $\alpha = \bar{\alpha}$, we see that the lower bound of $\mathcal{L}(q)$ is tight. Rearranging the terms, we have the approximate optimal solution $\ln q^*(\alpha)$ as

$$\ln q^*(\alpha) = (a - 1)\ln\alpha - b\alpha + constant,$$

where

$$a = a_0 + n(\frac{t-3}{2}) + \beta\bar{\alpha}\left[\frac{\partial}{\partial\beta\alpha}\ln I_{\frac{t-2}{2}}(\beta\bar{\alpha})\right], \tag{13.53}$$

$$b = b_0 + n\frac{\partial}{\partial\alpha}I_{\frac{t-3}{2}}(\bar{\alpha}) + \beta_0\left[\frac{\partial}{\partial\beta_0\alpha}\ln I_{\frac{t-2}{2}}(\beta_0\bar{\alpha})\right]. \tag{13.54}$$

We recognize $q^*(\alpha)$ to be a gamma distribution as

$$q^*(\alpha) = G(a, b),$$

where a is the posterior shape parameter and b is the posterior rate parameter b_0. Note that we define $\bar{\alpha}$ to be the posterior mode obtained from the previous iteration as:

$$\bar{\alpha} = \begin{cases} \frac{a-1}{b} & if\, a > 1 \\ \frac{a}{b} & otherwise \end{cases}$$

Xu et al. [2018] applied these methods to a study of sushi data sets and to the APA data from Diaconis [1988].

13.6. Bayesian Nonparametric Methods

In nonparametric statistics, we were concerned with the subject of inference when data have been observed from a fixed distribution of interest. In classical Bayesian inference, prior distributions are placed on a finite set of parameters associated with fixed distributions from which data are observed. In Bayesian nonparametric inference, we go one step further and prior distributions are placed on the space of distributions. To motivate the need to consider nonparametric Bayesian methods, suppose that we observe data which stem from a mixture model. Such a model may involve many unknown clusters. How should one place a prior distribution on such a model? This raises a number of issues such as, for example, what are possible choices of priors and how can posterior distributions be computed? How can data be simulated from the resulting posterior distributions? The construction of a prior on the space of all distributions requires a deeper knowledge of probability than has been presented thus far. In the next sections, we sketch some of this necessary background.

13.6.1. Dirichlet Prior

We begin with the Dirichlet prior which was first used by Ferguson in this context.

Definition 13.1. A vector $(X_1, \ldots, X_k)$, $k \geq 2$, has a Dirichlet distribution with parameters $(\alpha_1, \ldots, \alpha_k)$ if it has joint density

$$f(x_1, \ldots, x_k) = \frac{\Gamma\left(\sum_{i=1}^k \alpha_i\right)}{\Pi_{i=1}^k \Gamma(\alpha_i)} \Pi_{i=1}^k x_i^{\alpha_i - 1}$$

where

$$\alpha_i > 0, x_i > 0, \sum_{i=1}^{k} x_i = 1$$

In the special case where $k = 2$ the vector (X_1, X_2) has a $Be\,(\alpha_1, \alpha_2)$ distribution. Some properties of the Dirichlet prior are

Lemma 13.1. *Suppose that* $p \sim$ *Dirichlet* $(\alpha_1, \ldots, \alpha_m)$

$$f\,(p|\alpha) = \frac{\Gamma\,(\sum \alpha)}{\prod \Gamma\,(\alpha_i)} \prod p_i^{\alpha_i - 1}, 0 < p_i < 1$$

with $\sum p_i = 1$.

Then,

$$\begin{aligned} E\,(p_i) &= \frac{\alpha_i}{\sum \alpha_i} \\ E\left(p_i^2\right) &= \frac{\alpha_i\,(\alpha_i + 1)}{(\sum \alpha_i)\,(\sum \alpha_i + 1)} \\ E\,(p_i p_j) &= \frac{\alpha_i \alpha_j}{(\sum \alpha_i)\,(\sum \alpha_i + 1)}, i \neq j \end{aligned}$$

Lemma 13.2. *Suppose that* $\Delta = (d_{ij})$ *is a matrix of pairwise distances. If* $p \sim$ *Dirichlet* $(\alpha_1, \ldots, \alpha_m)$ *then*

$$E\,(p'\Delta p) = \frac{\sum_i \alpha_i\,(\alpha_i + 1)\,d_{ii} + \sum_{i \neq j} \alpha_i \alpha_j d_{ij}}{(\sum \alpha_i)\,(\sum \alpha_i + 1)}$$

The following result can be demonstrated. Suppose that $Z_1, \ldots, Z_m$ are independent gamma $G\,(\alpha_i, 1)\,, i = 1, \ldots, m$ and $S = \sum_{i=1}^{m} Z_i$. Then, the distribution of

$$\left(\frac{Z_1}{S}, \ldots, \frac{Z_m}{S}\right)$$

is Dirichlet with parameters $(\alpha_1, \ldots, \alpha_m)\,, \alpha_i \geq 0, \sum_{i=1}^{m} \alpha_i > 0$.

Ferguson [1973] made use of the Dirichlet distribution to define the Dirichlet process.

Definition 13.2. Let $(\mathcal{X}, \mathcal{A})$ be a measurable space, and let H be a non-null finite measure (nonnegative and finitely additive) on $(\mathcal{X}, \mathcal{A})$. We say P is a Dirichlet process on $(\mathcal{X}, \mathcal{A})$ with parameter H if for every $k = 1, 2, \ldots,$ and measurable partition $\{A_1, \ldots, A_k\}$ of $\mathcal{X}$, the joint distribution of $(P\,(A_1)\,, \ldots, P\,(A_k))$ is a k-dimensional Dirichlet distribution with parameters $(H\,(A_1)\,, \ldots, H\,(A_k))$. By assumption, $H\,(A_i) = 0$ implies $P\,(A_i) = 0$.

We denote the Dirichlet process by

$$P \sim DP\,(H)$$

We shall denote expectation in this new setting by the symbol $\mathcal{E}$.

When $k = 2$, $P(A)$ has a $Be(H(A), (H(\mathcal{X}) - H(A)))$ distribution, and hence,

$$\mathcal{E}[P(A)] = \frac{H(A)}{H(A) + (H(\mathcal{X}) - H(A))} = \frac{H(A)}{H(\mathcal{X})}$$

and similarly,

$$\mathcal{V}ar[P(A)] = \frac{H(A)(H(\mathcal{X}) - H(A))}{H(\mathcal{X}) + 1}$$

In the Dirichlet process, the $\{P(A_i)\}$ behave as the random variables and the $\{H(A_i)\}$ as the corresponding parameters in the Dirichlet distribution. Proceeding to defining the concept of a posterior distribution, we follow Ferguson [1973] by introducing the notion of a random sample.

Definition 13.3. Let P be a random probability measure on $(\mathcal{X}, \mathcal{A})$. We say that $X_1, \ldots, X_n$ is a sample of size n from P if for any $m = 1, 2, \ldots$ and measurable sets $C_1, C_2, \ldots C_n$ and $A_1, A_2, \ldots, A_m$, in $\mathcal{A}$

$$P(X_1 \in C_1, X_1 \in C_2, \ldots X_1 \in C_n | P(A_1), \ldots, P(A_m), P(C_1), \ldots, P(C_n)) = \textstyle\prod_{i=1}^{n} P(C_i)$$

almost surely.

This definition implies that the events $\{X_i \in C_i, i = 1, \ldots, n\}$ are independent among themselves and from the rest of the process. The following theorem and subsequent examples are due to Ferguson [1973] who describes the posterior distribution under a Dirichlet prior.

Theorem 13.1. *Let P be a Dirichlet process on $(\mathcal{X}, \mathcal{A})$ with parameter H and let $X_1, \ldots, X_n$ is a sample of size n from $P \sim DP(H)$. Then, the conditional distribution of P given $X_1, \ldots, X_n$ is a Dirichlet process with parameter H^*, where*

$$H^* = H + \sum_{i=1}^{n} \delta_{X_i}$$

where $\delta_X(B) = 1$ if $X \in B$ and $= 0$ otherwise.

Example 13.1. Under the new paradigm, suppose that we wish to estimate a distribution function F with $\hat{F}$ using square error loss function

$$L\left(F, \hat{F}\right) = \int \left(F(t) - \hat{F}(t)\right)^2 dW(t)$$

where $W(t)$ is a given nonrandom weight function. Suppose F has a Dirichlet prior with parameter H. In distributional format, $F(t) = P(-\infty, t)$. When no observations are taken, $F(t)$ has a $Be(H(t), H(\mathcal{X}) - H(t))$ and hence,

$$\begin{aligned}\hat{F}(t) &= \mathcal{E}[F(t)] \\ &= \frac{H(t)}{H(\mathcal{X})} \equiv F_0(t)\end{aligned}$$

When a random sample of size n is taken, then the Bayes estimate is

$$\begin{aligned}\hat{F}_n(t) &= \mathcal{E}[F(t)|X_1, \ldots, X_n] \\ &= \frac{H(t) + \sum_{i=1}^n \delta_{[X_i,\infty)}(t)}{H(\mathcal{X}) + n} \\ &= p_n F_0(t) + (1 - p_n) F_n(t)\end{aligned}$$

where $p_n = \frac{H(\mathcal{X})}{H(\mathcal{X})+n}$ and F_n is the empirical distribution function given by

$$F_n(t) = \frac{\sum_{i=1}^n \delta_{[X_i,\infty)}(t)}{n}$$

Hence the Bayes estimate is a weighted average of the prior guess F_0 and the empirical distribution F_n.

Example 13.2. In a similar manner, we may estimate the mean of a distribution function with loss function

$$L(F, \hat{\mu}) = \left(\int t dF(t) - \hat{\mu}\right)^2$$

such that $\int t^2 dF(t) < \infty$. Assuming a Dirichlet prior with parameter H, we have for the posterior estimate

$$\begin{aligned}\hat{\mu}_n &= \mathcal{E}\left[\int t dF(t)|X_1, \ldots, X_n\right] \\ &= \int t d\mathcal{E}[F(t)|X_1, \ldots, X_n] \\ &= \int t d[p_n F_0(t) + (1 - p_n) F_n(t)] \\ &= p_n \int t dF_0(t) + (1 - p_n) \int t dF_n(t) \\ &= p_n \mu_0 + (1 - p_n) \hat{X}_n\end{aligned}$$

Consequently, the Bayes estimate is a weighted average of the prior mean and the sample mean.

13.6.2. The Poisson-Dirichlet Prior

Definition 13.4. For $0 \leq \theta < 1$, $a > -\theta$, let $(\beta_i, i \geq 1)$ be a sequence of independent *Beta* $(1-\theta, a+i\theta)$ distributed random variables. Define

$$\tilde{p}_1 = \beta_1, \hat{p}_i = \beta_i \prod_{k=1}^{i-1} (1-\beta_k), i \geq 2$$

Let $p_1 \geq p_2 \geq \ldots$ be ranked values of $(\tilde{p}_i, i \geq 1)$. Let $(\theta_i, i \geq 1)$ be a sequence of i.i.d. random variables with common distribution H, independent of $(\beta_i, i \geq 1)$. Then the random probability measure

$$P_{H,\theta,a}(.) = \sum_{i=1}^{\infty} p_i \delta_{\theta_i}(.)$$

is called a two-parameter Poisson-Dirichlet process on $(\mathcal{X}, \mathcal{A})$ with parameters θ, a and H. The special case of $\theta = 0$ leads back to the Dirichlet process. The two-parameter Poisson-Dirichlet distribution has many applications in different fields such as population genetics, ecology, statistical physics, and number theory.

13.6.3. Simulation of Bayesian Posterior Distributions

Simulation of the posterior distribution plays an important role in applications. In this connection, there have been considerable developments in the literature related to representations of the Dirichlet process which facilitate this task. We begin with the important representation by Sethuraman [1994].

Suppose that we have a stick of length one and that we "break" it into two parts of lengths β_1 and $1-\beta_1$, respectively. Set $\pi_1 = \beta_1$. Repeat the process on the length $1-\beta_1$ and $\pi_2 = \beta_2(1-\beta_1)$ so that we have

$$\pi_1 + \pi_2 + (1-\beta_1)(1-\beta_2) = 1$$

Repeated stick breaking and setting

$$\pi_k = \beta_k \prod_{i=1}^{k-1} (1-\beta_i)$$

leads to

$$\lim_{n\to\infty} \left[\sum_{i=1}^{n} \pi_i + \prod_{i=1}^{n} (1-\beta_i)\right] = \sum_{i=1}^{\infty} \pi_i = 1$$

The following theorem is by Sethuraman [1994].

Theorem 13.2. *Let $(\beta_i, i \geq 1)$ be a collection of independent identically distributed* *random variables have a Beta*$(1, a)$ *distribution*, $a > 0$. *Set*

$$\pi_k = \beta_k \prod_{i=1}^{k-1} (1 - \beta_i), i \geqq 2$$

Let $(\theta_i, i \geq 1)$*be a sequence of i.i.d. random variables with common distribution* H, *independent of* $(\beta_i, i \geq 1)$. *Then the random probability measure*

$$P_{H,\theta,a}(.) = \sum_{i=1}^{\infty} \pi_i \delta_{\theta_i}(.)$$

is a Dirichlet process with parameter aH.

In applications, only a finite number n of terms are used in which case it is necessary to set $\pi_n = 1$. One difficulty with the Sethuraman representation is that the weights are not monotone decreasing. A more efficient algorithm where the weights are monotone decreasing is by Zarepour and Al Labadi [2012] who first proved the following lemma. First, define

$$\begin{aligned} G_n(x) &= \int_x^{\infty} \frac{1}{\Gamma(a/n)} e^{-t} t^{\frac{a}{n}-1} dt \\ N(x) &= a \int_x^{\infty} e^{-t} t^{-1} dt, x > 0. \\ \Gamma_i &= \sum_{k=1}^{i} E_k, \end{aligned}$$

where $\{E_k\} \sim$ *i.i.d.* random variables having an exponential distribution with mean 1.

Lemma 13.3.
(i) $nG_n(x) \to N(x)$ *as* $n \to \infty$
(ii) $G_n^{-1}\left(\frac{x}{n}\right) \to N^{-1}(x)$ *as* $n \to \infty$
(iii) For fixed i, $G_n^{-1}\left(\frac{\Gamma_i}{\Gamma_{n+1}}\right) \underset{\longrightarrow}{a.s.} N^{-1}(\Gamma_i)$ *as* $n \to \infty$
where $G_n^{-1}(y) = \inf\{x : G_n(x) \geq y\}, 0 < y < 1$.

The lemma was then used to obtain

Theorem 13.3. *Let* $(\theta_i, i \geq 1)$*be a sequence of i.i.d. random variables with common distribution* H, *independent of* $(\Gamma_i, i \geq 1)$. *Then*

$$\sum_{i=1}^{n} \frac{G_n^{-1}\left(\frac{\Gamma_i}{\Gamma_{n+1}}\right)}{\sum_{i=1}^{n} G_n^{-1}\left(\frac{\Gamma_i}{\Gamma_{n+1}}\right)} \delta_{\theta_i} \underset{\longrightarrow}{a.s.} \sum_{i=1}^{\infty} \frac{N^{-1}(\Gamma_i)}{\sum_{i=1}^{\infty} N^{-1}(\Gamma_i)} \delta_{\theta_i} \tag{13.55}$$

The right-hand side of (13.55) is the Dirichlet prior representation by Sethuraman. In his doctoral thesis, Al Labadi [2012] proposed as well new algorithms for simulating various processes. Various processes such as the Beta process, the Beta-Stacy process, and the Polya tree process were described by Ghosh and Ramamoorthi [2003]. See also Habli [2016].

13.6.4. Other Applications

Hosseini [2019] in her doctoral dissertation considered the Goodness-of-fit problem of testing the null hypothesis that the distribution of the data is a given F_0. In analogy with the usual Pearson statistic, the sample space was partitioned into sets A_i, and the test statistic was defined to be proportional to

$$HO = \sum_{i=1}^{k} \frac{\left(P\left(A_i\right) - F_0\left(A_i\right)\right)^2}{F_0\left(A_i\right)}$$

The idea then is to use the observed sample to determine the Dirichlet posterior distribution from which to derive the distribution of $P\left(A_i\right)$. Exploiting the relationship between HO and the Kullback-Leibler distance between measures P and F_0, it is shown that asymptotically for large sample sizes $HO \xrightarrow{\mathcal{L}} \chi^2_{k-1}$.

Benavoli et al. [2014] proposed a Bayesian Wilcoxon signed-rank statistics. A variety of other topics including consistency, spherical distributions, and two-sample problems are discussed in the two textbooks by Ghosh and Ramamoorthi [2003] and Ghoshal and van der Vaart [2017]. Reviews of methods and applications are given in Muller and Quintana [2004] and Xuan et al. [2019]. See Fall [2019] for an interesting application to tomography. In his doctoral thesis, Yang [2014] describes nonparametric Bayes for *big data*.

Bibliography

C. Adcock, Z. Landsman, and T. Shushi. Stein's lemma for generalized skew-elliptical random vectors. *Communications in Statistics - Theory and Methods*, pages 1–16, 2019.

A. Agresti. *Categorical Data Analysis.* John Wiley and Sons, Inc., 2013.

J. Aitchison and S. D. Silvey. Maximum-likelihood estimation of parameters subject to restraints. *Annals of Mathematical Statistics*, 29:813–828, 1958.

Michael G. Akritas. The Rank Transform Method in Some Two-Factor Designs. *Journal of the American Statistical Association*, 85(409):73–78, 1990.

L. Al Labadi. *On new constructive tools in Bayesian nonparametric inference.* Doctoral thesis, University of Ottawa, Ottawa, Ontario, Canada, 2012.

J. Albert. *Bayesian Computation with R.* Springer, second edition, 2009.

M. Alvo. Nonparametric tests of hypotheses for umbrella alternatives. *Canadian Journal of Statistics*, 36:143–156, 2008.

M. Alvo. Bridging the gap: a likelihood: function approach for the analysis of ranking data. *Communications in Statistics - Theory and Methods*, 2015.

M. Alvo. Empirical bayes on a shoestring and other applications. *Communications in Statistics-Theory and Methods*, 2021. https://doi.org/10.1080/03610926.2021.1948061.

M. Alvo and M-P. Berthelot. Nonparametric tests of trend for proportions. *International Journal of Statistics and Probability*, 1: 92–104, 2012.

M. Alvo and P. Cabilio. Bayesian estimation of the difference between two proportions. *The Canadian Journal of Statistics*, 10 (2):139–145, 1982.

M. Alvo and P. Cabilio. On the balanced incomplete block design for rankings. *The Annals of Statistics*, 19:1597–1613, 1991.

M. Alvo and P. Cabilio. Rank correlations and the analysis of rank-based experimental design. In *Probability Models and Statistical Analyses for Ranked Data, eds. Fligner, M.A. and Verducci, S.J. Lecture Notes in Statistics, 80, 140–154, Springer-Verlag, New York.*, 1993.

M. Alvo, *Statistical Inference and Machine Learning for Big Data*, Springer Series in the Data Sciences, https://doi.org/10.1007/978-3-031-06784-6

M. Alvo and P. Cabilio. Rank test of trend when data are incomplete. *Environmetrics*, 5:21–27, 1994.

M. Alvo and P. Cabilio. Rank correlation methods for missing data. *Canadian Journal of Statistics*, 23:345–358, 1995a.

M. Alvo and P. Cabilio. Testing ordered alternatives in the presence of incomplete data. *Journal of the American Statistical Association*, 90:1015–1024, 1995b.

M. Alvo and P. Cabilio. Calculation of hypergeometric probabilities using Chebyshev polynomials. *The American Statistician*, 54:141–144, 2000.

M. Alvo and M. Charbonneau. The use of Spearman's footrule in testing for trend when the data are incomplete. *Communications in Statistics -Simulation and Computation*, 26:193–213, 1997.

M. Alvo and J. Pan. A general theory of hypothesis testing based on rankings. *Journal of Statistical Planning and Inference*, 61:219–248, 1997.

M. Alvo and E.C. Voldner. Bayesian interpolation of atmospheric deposition data in eastern North America. *The Journal of Official Statistics*, 15(1):87–101, 1999.

M. Alvo and P.L.H. Yu. *A Parametric Approach to Nonparametric Statistics*. Springer, 2018.

M. Alvo and P. Yu. *Statistical Methods for Ranking Data*. Springer, 2014.

M. Alvo, P. Cabilio, and P. Feigin. Asymptotic theory for measures of concordance with special reference to Kendall's tau. *The Annals of Statistics*, 10:1269–1276, 1982.

M. Alvo, P. L. H. Yu, and K.F. Lam. The use of rank correlations in contingency table analysis. In *International Conference on recent Advances in Survey Sampling.*, 2002.

M. Alvo, T.L. Lai, and P.L.H. Yu. Parametric embedding of nonparametric inference problems: theory and applications. *Journal of Statistical Theory and Practice, Special Issue in honor of Bimal and Bikas Sinha*, 12:151–164, 2018.

T.W. Anderson. *The Statistical Analysis of Time Series*. John Wiley and Sons, 1971.

C. Andrieu, N. De Freitas, and A. Doucet. An introduction to MCMC for machine learning. *Machine Learning*, 50:5–43, 2003.

R. G. Andrzejak, K. Lehnertz, C. Rieke, F. Mormann, P. David, and C. E. Elger. Indications of nonlinear deterministic and finite dimensional structures in time series of brain electrical activity: Dependence on recording region and brain state. *Phys. Rev. E*, 64(061907), 2001.

D. Anguita, A. Ghio, L. Oneto, X. Parra, and J. L. Reyes-Ortis. Energy efficient smartphone-based activity recognition using fixed-point arithmetic. *Journal of Universal Computer Science*, 19(9):1295–1314, 2013a.

D. Anguita, A. Ghio, L. Oneto, X. Parra, and J. L. Reyes-Ortis. A public domain dataset for human activity recognition using smartphones. *21th European Symposium on Artificial Neural Networks, Computational Intelligence and Machine Learning, ESANN 2013. Bruges, Belgium*, 2013b.

A.R. Ansari and R.A. Bradley. Rank-sum tests for dispersion. *Annals of Mathematical Statistics*, 31(4):1174–1189, 1960.

D. Bahdanau, K. Cho, and Y. Bengio. Neural machine translation by jointly learning to align and translate. *ICLR-2015, arXiv:1409.0473*, 2015.

A. Banerjee, I.S. Dhillon, J. Ghosh, and S. Sra. Clustering on the unit hypersphere using von Mises-Fisher distributions. *Journal of Machine Learning Research*, 6:1345–1382, 2005.

S. H. Barnes and M. Kaase. emphPolitical Action: Mass Participation in Five Western Countries. London: Sage, 1979.

S. Beggs, S. Cardell, and J. Hausman. Assessing the potential demand for electric cars. *Journal of Econometrics*, 16:1–19, 1981.

A. Benavoli, F. Mangili, G. Corani, M. Zaffalon, and F. Ruggeri. A Bayesian Wilcoxon signed-rank test based on the Dirichlet process. In *Proceedings of the 31st International Conference on Machine Learning*, volume 32, Beijing, China, 2014.

W. Benter. Computer-based horse race handicapping and wagering systems: a report. In W. T. Ziemba, V. S. Lo, and D. B. Haush, editors, *Efficiency of Racetrack Betting Markets*, pages 183–198. Academic Press, San Diego, 1994.

R. Bentoumi, M. Alvo, and M. Mesfoui. Dependence measure for length-biased survival data using copulas. *Dependence Modeling*, 7:348–364, 2019.

L. Billard and E. Diday. From the statistics of data to the statistics of knowkledge: Symbolic data analysis. *Journal of the American Statistical Association*, 98:470–487, 2003.

D.A. Binder and Z. Patak. Use of estimating equations for functions from complex surveys. *Journal of the American Statistical Association*, 89(427):1035–1043, 1994.

Y. M. Bishop, S. E. Fienberg, and P. W. Holland. *Discrete Multivariate Analysis: Theory and Practice*. Springer, 2007.

G.D. Birkhoff. Proof of the ergodic theorem. *Proc. Nat. Acad. Sci.* USA, 17:636, 1931.

D.M. Blei, A. Kucukelbir, and J.D. McAuliffe. Variational inference: a review for statisticians. *Journal of the American Statistical Association*, 112(518):859–877, 2017.

U. Bockenholt. Mixed-effects analysis of rank-ordered data. *Psychometrika*, 66(1):45–62, 2001.

D.D. Boos. On generalized score tests. *American Statist*, 46:327–333, 1992.

G.E.P. Box and G.C. Tiao. *Bayesian Inference in Statistical Analysis*. John Wiley and Sons, 1973.

G.E.P. Box, G.M. Jenkins, and G.C. Reinsel. *Time Series Analysis: Forecasting and Control*. Prentice Hall, third edition, 1994.

L. Breiman. *Probability*. SIAM, 1992.

L. Breiman. Random forests. *Machine Learning*, 45(1):5–32, 2001.

P.J. Brockwell and R.A. Davis. *Time Series: Theory and Methods*. Springer-Verlag, 1987.

J. Camacho, J.M. Garcia Gimerez, N.M. Fuentes Garcia, and G. Macia Fernandez. Multivariate big data analysis for intrusion detection: 5 steps from the haystack to the needle. *Computers & Security*, 87, 2018. https://doi.org/10.1016/j.cose.2019.101603.

G. Casella and R.L. Berger. *Statistical Inference*. Duxbury Press, second edition, 2002.

G. Casella, C. P. Robert, and M. T. Wells. Generalized accept–reject sampling schemes. *Institute of Mathematical Statistics Lecture Notes*, Monograph Series Vol. 45:342–347, 2004.

M. Chabert and J.-Y. Tourneret. Bivariate Pearson distributions for remote sensing images. In *IEEE International Geoscience and Remote Sensing Symposium*, pages 4038–4041, 2011.

R.G. Chapman and R. Staelin. Exploiting rank ordered choice set data within the stochastic utility model. *Journal of Marketing Research*, 19:288–301, 1982.

S.X. Chen, L. Peng, and Y-Li.. Qin. Effects of data dimension on empirical likelihood. *Biometrika*, 96(3):711–722, 2009.

T.L. Clark, R.L. Dennis, S. Seilkop, M. Alvo, E.C. Voldner, and M.P. Olson. The international sulfur deposition model evaluation: A summary. *Air Pollution Modeling and Its Application*, 6:305–320, 1988.

W. J. Conover. *Practical Nonparametric Statistics*. New York: John Wiley, 3rd edition, 1999.

W. J. Conover and Ronald L. Iman. Rank Transformations as a Bridge Between Parametric and Nonparametric Statistics. *The American Statistician*, 35(3):124–129, 1981.

D. Cox and D. Hinkley. *Theoretical Statistics.* Chapman Hall, New York, 1974.

D.R. Cox and D. Oakes. *Analysis of Survival Data.* Chapman & Hall, 1984.

B. M. Craig, J. J. V. Busschbach, and J. A. Salomon. Modeling ranking, time trade-off, and visual analog scale values for eq-5d health states: a review and comparison of methods. *Medical Care*, 47(6):634–641, 2009.

N.A.C. Cressie. *Statistics for Spatial Data.* John Wiley and Sons, second edition, 2015.

M. A. Croon. Latent class models for the analysis of rankings. In G. De Soete, H. Feger, and K. C. Klauer, editors, *New Developments in Psychological Choice Modeling*, pages 99–121, North-Holland, 1989. Elsevier Science.

G. Cybenko. Approximation by superpositions of a sigmoidal function. *Math. Control Signals Systems*, 2:303–314, 1989.

R.B. D'Agostino and M.A. Stephens. *Goodness-of fit Techniques.* Marcel Dekker Inc., New York, 1986.

W.W. Daniel. *Applied Nonparametric Statistics.* Duxbury Press., second edition, 1990.

H.E. Daniels. Saddlepoint approximations in statistics. *The Annals of Mathematical Statistics*, 25:631–650, 1954.

H.E. Daniels. Exact saddlepoint approximations. *Biometrika*, 67(1):59–63, 1980.

A. C. Davidson and D.V. Hinkley. Saddlepoint approximations in resampling methods. *Biometrika*, 75(3):417–431, 1988.

T. De Putter, M-F Loutre, and G. Wansard. Decadal periodicities of nile river historical discharge (a.d. 622–1470) and climatic implications. *Geophysical Research Letters*, 25(16), 1998.

L. T. Decarlo and S. S. Luthar. Analysis and class validation of a measure of parental values perceived by early adolescents: an application of a latent class models for rankings. *Educational and Psychological Measurement*, 60(4):578–591, 2000.

M.H. DeGroot. *Optimal Statistical Decisions.* McGraw-Hill, New York, 1970.

A. P. Dempster, N. M. Laird, and D. B. Rubin. Maximum likelihood from incomplete data via the EM algorithm. *Journal of Royal Statistical Society Series B*, 39(1):1–38, 1977.

A.F. Desmond. Optimal estimating functions, quasi-likelihood and statistical modelling. *Journal of Statistical Planning and Inference*, 60:77–121, 1997.

P. Diaconis. *Group Representations in Probability and Statistics.* Institute of Mathematical Statistics, Hayward, 1988.

P. Diaconis and R.L Graham. Spearman's footrule as a measure of disarray. *Journal of the Royal Statistical Society Series B*, 39:262–268, 1977.

T.L. DiCiccio, P. Hall, and J. Romano. Empirical likelihood is Bartlett-correctable. *Annals of Statistics*, 19:1053–1061, 1989.

E. Diday. An introduction to symbolic data analysis and its application to the SODAS project. *Revista De Mathematica: Theoria v.Applicationes*, 7:1–22, 2000.

E. Diday. *Symbolic Data Analysis and the SODAS software*, chapter The state of the art in symbolic data analysis: overview and future, pages 3–41. Jo, 2008.

E. Diday. Thinking by classes in data science: the symbolic data analysis paradigm. *WIREs Comput Stat*, 8:172–205, 2016.

R. Dittrich, W. Katzenbeisser, and H. Reisinger. The analysis of rank ordered preference data based on Bradley-Terry type models. *OR Spektrum*, 22:117–134, 2000.

R.J.M.M. Does. Higher order asymptotics for simple linear rank statistics. In Netherlands Organization for the Advancement of Pure Research, editor, *Mathematical Centre Tracts*, volume 151. 1982.

O. D. Duncan and C. Brody. Analyzing n rankings of three items. In R. M. Hauser, D. Mechanic, A. O. Haller, and T. S. Hauser, editors, *Social Structure and Behavior*, pages 269–310. Academic Press, New York, 1982.

J. Durbin. Incomplete blocks in ranking experiments. *British Journal of Psychology*, 4:85–90, 1951.

J. Durbin and S. J. Koopman. *Time Series Analysis by State Space Methods.* Oxford University Press, Oxford, 2012.

B. Efron. Logistic regression, survival analysis, and the Kaplan-Meier curve. *Journal of the American Statistical Association*, 83(402):414–425, 1988.

B. Efron. Tweedie's formula and selection bias. *Journal of the American Statistical Association*, 106(496):1602–1614, 2011.

M. D. Fall. Bayesian nonparametric and biostatistics: case of PET imaging. *The International Journal of Biostatistics*, 15(2):1–10, 2019.

T.S. Ferguson. A Bayesian analysis of some nonparametric problems. *The Annals of Satistics*, 1:209–230, 1973.

T.S. Ferguson. *A Course in Large Sample Theory.* John Wiley and Sons, 1996.

C. Field and E. Ronchetti. Small sample asymptotics. volume 13 of *Lecture Notes-Monograph Series*, pages 1–140. Institute of Mathematical Statistics, 1990.

S.E. Fienberg. *The Analysis of Cross-Classified Categorical Data*. The MIT Press, second edition, 1980.

R.A. Fisher. The use of multiple measurement in taxonomic problem. *Annals of Eugenics*, 7:179–188, 1936.

D.A.S. Fraser. On information in statistics. *The Annals of Mathematical Statistics*, 36(1):890–896, 1965.

E. Geelhoed. The sound of music and word of mouth: Hearing music and hearing about music. Technical Report HPL-2001-9TR1, Hewlett Packard, 2001.

C. Genest and J. Neslehova. A primer on copulas for count data. *Astin Buletin*, 37(2):475–515, 2007.

C. Genest and L-P. Rivest. Statistical inference procedures for bivariate Archimedian copulas. *Journal of the American Statistical Association*, 88(423):1034–1043, 1993.

C. Genest, J. Neslehova, and N. Ben Ghoral. Spearman's footrule and Gini's gamma: a review with complements. *Journal of Nonparametric Statistics*, 22(8):937–954, 2010.

R. Genuer, J. Poggi, C. Tuleau Malot, and N. Villa Vialaneix. Random forests for big data. *Big Data Research*, 9:28–46, 2017.

C. J. Geyer. Markov chain monte carlo. Lecture notes, University of Minnesota, 2005.

J. Ghosh and M.A. Clyde. Rao-blackwellization for Bayesian variable selction and model averaging in linear and binary regression: a novel data augmentation approach. *Journal of the American Statistical Association*, 106(495):1041–1052, 2011.

J.K. Ghosh and R.V. Ramamoorthi. *Bayesian Nonparametrics*. Springer-Verlag, New York, 2003.

S. Ghoshal and A. van der Vaart. *Fundamentals of Nonparametric Bayesian Inference*. Cambridge University Press, 2017.

D. Gingras, M. Alvo, and K. Adamowski. Regional flood relationships by nonparametric regression. *Nordic Hydrology*, 26:73–90, 1995.

V.P. Godambe. An optimum property of regular maximum likelihood estimation. *Annals of Mathematical Statistics*, 31:1208–1212, 1960.

A. I. Goldberg. The relevance of cosmopolitan / local orientations to professional values and behavior. *Sociology of Work and Occupation*, 3:331–356, 1975.

I.J. Good. Saddle-point methods for the multinomial distribution. *Annals of Mathematical Statistics*, 28(4):861–881, 1957.

I.J. Good. On the application of symmetric Dirichlet distributions and their mixtures to contingency tables. *The Annals of Statistics*, 4(6):1159–1189, 1976.

P. Goovaerts. Geostatistical analysis of disease data: estimation of cancer mortality risk from empirical frequencies using Poisson kriging. *International Journal of Health Geographics*, pages 1–33, 2005.

P. Goovaerts. *Geostatistical interpolation of rate data using Poisson kriging*, pages 767–774. Encyclopedia of GIS. Springer International Publishing, 2017.

I. C. Gormley and T. B. Murphy. Exploring voting blocs within the Irish electorate: A mixture modeling approach. *Journal of the American Statistical Association*, 103:1014–1027, 2008.

C. Goutis and G. Casella. Explaining the saddlepoint approximation. *The American Statistician*, 53(3):216–224, 1999.

J.C. Gower. Some distance properties of latent root and vector methods used in multivariate analysis. *Biometrika*, 53:325–338, 1966.

J.C. Gower. Multivariate analysis and multidimensional geometry. *The Statistician*, 17:13–25, 1967.

A. Graves, A. Mohamed, and G. Hinton. Speech recognition with deep recurrent neural networks. *ICASSP2013*, pages 6645–6649, 2013.

N. Habli. Nonparametric Bayesian modelling in machine learning. M.Sc. thesis, Department of Mathematics and Statistics Faculty of Science University of Ottawa, 2016.

J. Hajek. Asumptotic normality of simple linear rank statistics under alternatives. *The Annals of Mathematical Statistics*, 39:325–346, 1968.

J. Hajek and Z. Sidak. *Theory of Rank Tests.* Academic Press, New York, 1967.

M. Hallin and O. Tribel. The Efficiency of Some Nonparametric Rank-Based Competitors to Correlogram Methods. *Lecture Notes-Monograph Series*, 35:249–262, 2000.

R. J. Henery. Permutation probabilities as models for horse races. *Journal of the Royal Statistical Society Series B*, 43:86–91, 1981.

S. Hochreiter and J. Schmidhuber. Long short-term memory. *Neural Computation.*, 9(8):1735–1780, 1997.

S. Hochreiter, Y. Bengio, P. Frasconi, and J. Schmidhuber. Gradient flow in recurrent nets: the difficulty of learning long-term dependencies. *A Field Guide to Dynamical Recurrent Neural Networks. IEEE Press.*, 2001.

W. Hoeffding. A class of statistics with asymptotically normal distribution. *Annals of Mathematical Statistics*, 19:293–325, 1948.

W. Hoeffding. A combinatorial central limit theorem. *Annals of Mathematical Statistics*, 22:558–566, 1951.

P.G. Hoel, S.C. Port, and C.J. Stone. *Introduction to Stochastic Processes.* Houghton Mifflin Company, 1986.

R. Hosseini. *New development on Bayesian bootstrap for unrestricted and restricted distributions.* Doctoral thesis, University of Ottawa, Ottawa, Ontario, Canada, 2019.

T. K. Hsu and I. C. Yeh. Building real estate valuation models with comparative approach through case-based reasoning. *Applied Soft Computing*, 65:260–271, 2018.

R. Inglehart. *The Silent Revolution: Changing Values and Political Styles among Western Publics.* Princeton University Press, Princeton, 1977.

H. Inhaber and M. Alvo. World science as an input-output system. *Scientometrics*, 1(1):43–64, 1978.

G.M. Jenkins and D.G. Watts. *Spectral Analysis and its Applications.* Holden-Day, 1968.

J. Jesus and R.E. Chandler. Estimating functions and the generalized method of moments. *Interface Focus, The Royal Statistical Society*, 1(6):871–885, 2011. https://doi.org/10.1098/rsfs.2011.0057.

H. Joe. Multivariate extreme value distributions and coverage of ranking probabilities. *Journal of Mathematical Psychology*, 45:180–188, 2001.

J.A. John and E.R. Williams. *Cyclic Designs.* Chapman Hall, New York, 1995.

N.L. Johnson, S. Kotz, and N. Balakrishnan. *Continuous Univariate Distributions*, volume 1. John Wiley and Sons, 1994.

R.A. Johnson and D.W. Wichern. *Applied Multivariate Statistical Analysis.* Pearson Prentice Hall, sixth edition, 2008.

A.R. Jonckheere. A test of significance for the relation between m rankings and k ranked categories. *The British Journal of Statistical Psychology*, 7:93–100, 1954.

T. Kamishima and S. Akaho. Efficient clustering for orders. In *Proceedings of the 2nd International Workshop on Mining Complex Data*, pages 274–278, 2006.

A. Katharopoulos, A. Vyas, N. Pappas, and F. Fleuret. Transformers are RNNS: Fast autoregressive transformers with linear attention. *arXiv:2006.16236v2*, 2020.

M. Kendall. *Time Series.* Hafner Press, 1973.

M. Kendall, A. Stuart, and J.K. Ord. *The Advanced Theory of Statistics*, volume 3. Charles Griffin and Company Limited, fourth edition edition, 1983.

M.G. Kendall and A. Stuart. *The Advanced Theory of Statistics*, volume 2 of *Inference and Relationship*. Charles Griffin and Company Limited, fourth edition, 1979.

G. Koop and D. J. Poirier. Rank-ordered logit models: an empirical analysis of Ontario voter preferences. *Journal of Applied Econometrics*, 9(4):369–388, 1994.

S.J. Koopman and G. Mesters. Empirical Bayes methods for dynamic factor models. *The Review of Economics and Statistics*, 99(3):486–498, 2017.

N. V. Kovalevskaya, C. Whicher, T. D. Richardson, C. Smith, J. Grajciarova, J. Cardama, X.;and Moreira, A. A. Alexa, A..and McMurray, and F. G. G. Nielsen. Dnadigest and repositive: Connecting the world of genomic data. *PLOS Biology, Public Library of Science,*, 14:1–10, 2016.

P. F. M. Krabbe, J. A. Salomon, and C. J. L. Murray. Quantification of health states with rank-based nonmetric multidimensional scaling. *Medical Decision Making*, 27:395–405, 2007.

J.B. Kruskal. Multidimensional scaling by optimising goodness of fit to a non-standard hypothesis. *Psychometrika*, 29(1):1–27, 1964a.

J.B. Kruskal. Non-metric multidimensional scaling : a numerical method. *Psychometrika*, 29(1):115–129, 1964b.

S. Kullback. *Information Theory and Statistics.* John Wiley and Sons (Republished by Dover Publications in 1968; reprinted in 1978), 1968.

S. Kullback and R. A. Leibler. On information and sufficiency. *Annals of Mathematical Statistics*, 1:79–86, 1951.

T..L. Lai and J.Q. Wang. Edgeworth expansion for symmetric statistics with applications to bootstrap methods. *Statistica Sinica*, 3:517–542, 1993.

A.J. Lee. *U-Statistics.* Marcel Dekker Inc., New York, 1990.

D. Leger. A study of classification techniques. Master's thesis, University of Ottawa, 2009.

E.L. Lehmann. *Nonparametrics: Statistical Methods based on Ranks.* McGraw-Hill, New York, 1975.

E.L. Lehmann and G. Casella. *Theory of Point Estimation.* Springer, second edition edition, 1998.

E.L. Lehmann and J.P. Romano. *Testing Statistical Hypotheses.* Texts in Statistics. Springer, third edition edition, 2005.

K.Y. Liang and S.L. Zeger. Longitudinal data analysis using generalized linear models. *Biometrika*, 73(1):13–22, 1986.

K. Liu. Symbolic approach for big data analysis. University of Ottawa M.Sc. thesis, 2019. University of Ottawa M.Sc. thesis.

Y. Maesono. Asymptotic representations of skewness estimators of studentized U-statistics. *Bulletin of Informatics and Cybernetics*, 36:91–104, 2004.

G. Matheron. The theory of regional variables and its applications. No.5, Fontainebleau, France, 1971.

A. Maydeu-Olivares and U. Bockenholt. Structural equation modeling of paired-comparison and ranking data. *Psychological Methods*, 10(3):285–304, 2005.

C. McCabe, J. Brazier, P. Gilks, A. Tsuchiya, J. Roberts, A. O'Hagan, and K. Stevens. Use rank data to estimate health state utility models. *Journal of Health Economics*, 25:418–431, 2006.

P. McCullagh. Quasi-likelihood functions. *Ann. Statist*, 11:59–67, 1983.

T. Mikolov, I. Sutskever, K. Chen, G. Corrado, and J. Dean. Efficient estimation of word representations in vector space, 2013a.

T. Mikolov, I. Sutskever, K. Chen, G. Corrado, and J. Dean. Distributed representations of words and phrases and their compositionality. NIPS'13: Proceedings of the 26th International Conference on Neural Information Processing Systems, 2:3111–3119, 2013b.

R.G. Jr. Miller. Jackknifing variances. *The Annals of Mathematical Statistics*, 39(2):567–582, 1968.

G. Moors and J. Vermunt. Heterogeneity in post-materialists value priorities. evidence from a latent class discrete choice approach. *European Sociological Review*, 23:631–648, 2007.

D.F. Morrison. *Multivariate Statistical Methods*. McGraw-Hill, New York, third edition, 1990.

J. Mu. Covid-19 disease mapping based on Bayesian spatial-temporal models and poisson kriging models. Master's thesis, University of Ottawa, 2021.

P. Muller and F.A. Quintana. Nonparametric Bayesian data analysis. *Statistical Science*, 19(1):95–110, 2004.

T. B. Murphy and D. Martin. Mixtures of distance-based models for ranking data. *Computational Statistics and Data Analysis*, 41:645–655, 2003.

L.M.. Murray. Bayesian state-space modelling on high-performance hardware using LibBi. *Journal of Statistical Software*, 67(10):1–28, 2015.

Per A. Mykland. Bartlett identities and large deviations in likelihood theory. *The Annals of Satistics*, 27(3):1105–1117, 1999.

G. Napier and D. Lee. A Bayesian space-time model for clustering areal units based on their disease trends. *Biostatistics*, 20(4):681–697, 2019.

R.B. Nelsen. *An Introduction to Copulas*. Springer Series in Statistics, second edition, 2006.

J. Neyman. Smooth test for goodness of fit. *Skandinavisk Aktuarietidskrift*, 20:149–199, 1937.

S. W. Nombekela, M. R. Murphy, H. W. Gonyou, and J. I. Marden. Dietary preferences in early lactation cows as affected by primary tastes and some common feed flavors. *Journal of Diary Science*, 77:2393–2399, 1993.

A.B. Owen. *Empirical Likelihood.* Chapman & Hall, 2001.

E.B. Page. Ordered hypotheses for multiple treatments: a significance test for linear ranks. *Journal of the American Statistical Association*, 58:216–230, 1963.

A. Parikh, O. Täckström, D. Das, and Uszkoreit. J. A decomposable attention model. *Empirical Methods in Natural Language Processing*, 2016.

R. Pascanu, T. Mikolov, and Y. Bengio. On the difficulty of training recurrent neural networks. *ICML*, 2013.

K. Pearson. Contributions to the mathematical theory of evolution: skew variation in homogeneous material. *Transactions of the Royal Society*, 186:343–414, 1895.

K. Pearson. Mathematical contributions to the theory of evolution,x:supplement to a memoir on skew variation. *Philisophical Transactions of the Royal Society A.*, 197:287–299, 1901.

K. Pearson. Mathematical contributions to the theory of evolution, xix: second supplement to a memoir on skew variation. *Transactions of the Royal Society A*, 216:538–548, 1916.

A. A. O. Plumb, F. M. Grieve, and S. H. Khan. Survey of hospital clinicians' preferences regarding the format of radiology reports. *Clinical Radiology*, 64:386–394, 2009.

J. Qin and J.F. Lawless. Empirical likelihood and general estimating equations. *Annals of Statistics*, 22:300–325, 1994.

J. Qin and J.F. Lawless. Estimating equations, empirical likelihood and constraints on parameters. *The Canadian Journal of Statistics*, 23:145–159, 1995.

J. Ratcliffe, J. Brazaier, A. Tsuchiya, T. Symonds, and M. Brown. Estimation of a preference based single index from the sexual quality of life questionnaire (SQOL) using ordinal data. *Discussion Paper Series, Health Economics and Decision Science, The University of Sheffield*, 06:6, 2006.

J. Ratcliffe, J. Brazaier, A. Tsuchiya, T. Symonds, and M. Brown. Using DCE and ranking data to estimate cardinal values for health states for deriving a preference-based single index from the sexual quality of life questionnaire. *Health Economics*, 18:1261–1276, 2009.

J.C.W. Rayner, O. Thas, and D.J. Best. *Smooth Tests of Goodness of Fit Using R*. John Wiley and Sons, second edition, 2009.

M. Regenwetter, M. H. R. Ho, and I. Tsetlin. Sophisticated approval voting, ignorance priors, and plurality heuristics: A behavioral social choice analysis in a Thurstonian framework. *Psychological Review*, 114(4):994–1014, 2007.

N. Reid. Saddlepoint methods and statistical inference. *Statistical Science*, 3(2):213–238, 1988.

M. Riketta and D. Vonjahr. Multidimensional scaling of ranking data for different age groups. *Experimental Psychology*, 46 (4):305–311, 1999.

F. Rosenblatt. The perceptron: A probabilistic model for information storage and organization in the brain. *Psychological Review*, 65 (6):386–408, 1958.

N. Ross. Fundamentals of Stein's method. *Probabiliy Surveys*, 8:210–293, 2011.

D. Rothman. *Transformers for Natural Language Processing: Build innovative deep neural network architectures for NLP with Python, PyTorch, TensorFlow, BERT, RoBERTa, and more.* Packt Publishing, 2021.

J. A. Salomon. Reconsidering the use of rankings in the valuation of health states: a model for estimating cardinal values from ordinal data. *Population Health Metrics*, 1:1–12, 2003.

M.J. Schervish and B.P. Carlin. On the convergence of successive substitution sampling. *Journal of Computational and Graphical Statistics*, 1(2):111–127, 1992.

I. Schiopu-Kratina, H.Y. Liu, M. Alvo, and P-J. Bergeron. Estimation for recurrent events through conditional estimating equations, 2021. https://doi.org/arxiv.org/abs/2109.03932

M. Schuster and K. Paliwal. Bidirectional recurrent neural networks. *Transactions on Signal Processing*, 45(11): 2673–2681, 1997.

A. Schwab. *Beta-Binomial Kriging: A new approach to modeling spatially correlated proportions.* PhD thesis, University of Nebraska-Lincoln, 2015.

P.K. Sen.Asymptotically efficient tests by the method of n rankings. *Journal of the Royal Statistical Society, Series B*, 30:312–317, 1968.

P.K. Sen, I. Salama, and D. Quade. Spearman's footrule: asymptotics in applications. *Chilean Journal of Statistics*, 2 (1):3–20, 2011.

J. Sethuraman. A constructive definition of Dirichlet priors. *Statistica Sinica*, 4:639–650, 1994.

S.S. Shapiro and M.B. Wilk. An analysis of variance test for normality (complete samples). *Biometrika*, 52(3 and 4): 591–611, 1965.

R.H. Shumway and D.S. Stoffer. *Time Series Analysis and its Applications wit R Examples.* Springer, New York, fourth edition, 2016.

B.W. Silverman. *Density Estimation for Statistics and Data Analysis*, volume 26 of *Monograph on Statistics and Applied Probability.* Chapman Hall, New York, 1988.

A. Skrondal and S. Rabe-Hesketh. Multilevel logistic regression for polytomous data and rankings. *Psychometrika*, 68(2):267–287, 2003.

C. Spearman. The proof and measurement of association between two things. *American Journal of Psychology*, 15:72–101, 1904.

M. A. Stephens. *edf* statistics for goodness of fit and some comparisons. *Journal of the American Statistical Association*, 69:730–737, 1974.

H. Stern. Models for distributions on permutations. *Journal of the American Statistical Association*, 85:558–564, 1990.

H. Stern. Probability models on rankings and the electoral process. In M. A. Fligner and J. S. Verducci, editors, *Probability Models and Statistical Analyses for Ranking Data*, pages 173–195. Springer-Verlag, 1993.

I. Sutskever, O. Vinyals, and Q. V. Le. Sequence to sequence learning with neural networks. *NIPS-2014, arXiv:1409.3215*, 2014.

J. Taghia, M. Zhanyu, and A. Leijon. Bayesian estimation of the von-Mises-Fisher mixture model with variational inference. *IEEE transactions on pattern analysis and machine intelligence*, 36(9):1701–1715, 2014.

E. Takeshi and C. Yi-Hau. *Analysis of Survival Data with Dependent Censoring: Copula-Based Approaches.* Springer, 2018.

C.Y. Tang and C.L. Leng. Penalized high-dimensional empirical likelihood. *Biometrika*, 97(4):905–950, 2010.

M.A. Tanner. *Tools for Statistical Inference.* Springer-Verlag, second edition, 1993.

M.A. Tanner. *Tools for Statistical Inference: Methods for the Exploration of Posterior Distributions and Likelihood Functions.* Springer, 3rd edition, 1997.

G. L. Thompson. A note on the rank transform for interactions. *Biometrika*, 78(3):697–701, 1991.

Neil H. Timm. *Multivariate Analysis with Applications in Education and Psychology.* Wadsworth Publishing Company, Inc, 1975.

A. W Van der Vaart. *Asymptotic Statistics.* Cambridge Series in Statistical and Probabilistic Mathematics. Cambridge University Press, 2000.

C. Varin, N. Reid, and D. Firth. An overview of composite likelihood methods. *Statistica Sinica*, 21:5–42, 2011.

A. Vaswani, N. Shazeer, N. Parmar, J. Uszkoreit, L. Jones, A. Gomez, L. Kaiser, and I. Polosukhin. Attention is all you need. *arxiv.org/abs/1706.03762*, 2017.

J. K. Vermunt. Multilevel latent class models. *Sociological Methodology*, 33:213–239, 2004.

E. Vigneau, P. Courcoux, and M. Semenou. Analysis of ranked preference data using latent class models. *Food Quality and Preference*, 10:201–207, 1999.

E.C. Voldner and M. Alvo. On the estimation of sulfur and nitrogen wet deposition to the great lakes. *Environmental Science and Technology*, 23(10):1223–1232, 1989.

E..C. Voldner and M. Alvo. On the estimation of sulfur, nitrogen, cadmium and lead deposition to the great lakes. *Environmental Science and Technology*, 27:292–298, 1993.

A.S. Weigend and N.A. Gershenfeld. *Time Series Prediction: Forecasting the Future and Understanding the Past.* Addison-Wesley, 1994.

H. White. Maximum likelihood estimation of misspecified models. *Econometrica*, 1:1–25, 1982.

H. Xu, M. Alvo, and P.L.H. Yu. Angle-based models for ranking data. *Computational Statistics and Data Analysis*, 121:113–136, 2018.

J. Xuan, J. Lu, and G. Zhang. A survey on Bayesian nonparametric learning. *ACM Computing. Surveys*, 52(1):13:1–13:36, 2019.

A. Ya Khinchin. Zu Birkhoff's Losung des Ergodenproblems, *Mat. Ann.* 107:485, 1932.

Y. Yang. *Nonparametric Bayes for Big Data.* PhD thesis, Department of Statistical Science, Duke University, 2014.

F.W. Young and R.M. Hamer. *Multidimensional Scaling: History, Theory, and Applications.* Hillsdale, New Jersey: Lawrence Erlbaum Associates, 1987.

P. L. H. Yu and L. K. Y. Chan. Bayesian analysis of wandering vector models for displaying ranking data. *Statistica Sinica*, 11:445–461, 2001.

M. Zarepour and L. Al Labadi. On a rapid simulation of the Dirichlet process. *Statistics and Probability Letters*, 82:916–924, 2012.

Y. Zhao and H. Joe. Composite likelihood estimation in multivariate data analysis. *Canadian Journal of Statistics*, 33(3):335–356, 2005.

Index

M. Alvo, *Statistical Inference and Machine Learning for Big Data*, Springer Series in the Data Sciences, https://doi.org/10.1007/978-3-031-06784-6